The Role of Terrestrial and Aquatic Organisms in Decomposition Processes

The Role of Terrestrial and Aquatic Organisms in Decomposition Processes

The 17th Symposium of
The British Ecological Society
15–18 April 1975

edited by
J. M. Anderson
and
A. Macfadyen

Blackwell Scientific Publications
OXFORD LONDON EDINBURGH MELBOURNE

Osney Mead, Oxford,
8 John Street, London, WC 1
9 Forrest Road, Edinburgh,
P.O. Box 9, North Balwyn, Victoria, Australia.

First published 1976

British Library Cataloguing in Publication Data
The role of terrestrial and aquatic organisms
in decomposition processes—(British
Ecological Society. Symposia; 17th).
Bibl.—Index.
ISBN 0-632-00018-X
1. Anderson, J. M. 2. Macfadyen, Amyan
3. Series.
574·5 QP517.B5
Biodegration-Congresses.

Distributed in the United States of America by
Halsted Press, a Division of John Wiley & Sons, New York
and in Canada by J. B. Lippincott Company of Canada, Ltd., Toronto.

Printed and bound in Great Britain by
Butler & Tanner Ltd., Frome and London

Contents

Section 3
Decomposition processes as components of whole ecosystems

Section 4
Modelling of decomposition systems

Preface

A major purpose of the British Ecological Society Symposia has always been to promote discussion of topics which appear relevant to a number of different disciplines but do not fall in with current publications or conference programmes of one particular field of study. When this symposium was planned, biological decomposition appeared to be a classic example of an interdisciplinary topic. We believe that this is still true despite the interest focused on decomposition by recent publications under the International Biological Programme, comprising separate Freshwater, Marine and Terrestrial sections. Our aim, from the outset, was to bring together and to compare the approaches of these different areas. When we came to convert a theoretical scheme into a programme, it became evident that, in addition to the diverse approaches of the different 'biome' studies, those of the chemist, microbiologist, botanist and zoologist had also to be assimilated. Within this scheme our contributors then proceeded to reveal, both in their papers and in the ensuing discussion, a further range of ideas, many of which had not been fully anticipated. It is certainly for these more spontaneous themes that the symposium will be remembered by participants; it may be helpful to the reader to pinpoint some of these in a short introduction.

A theme which derives from ecologists' current interest in the stability of ecosystems is the relation between the rates of passage of chemical substances through ecosystems and the various environmental perturbations which are imposed on these systems. This applies particularly to rates of organic matter breakdown and mineral release. In practice, most fairly productive ecosystems retain a high proportion of carbon in the form of inactive microbial cells, nonliving, partially decomposed organic materials, and organic matter in solution. In soils, this material contributes to the water-retaining capacity of the soil, permeability to gases and is relevant to the performance of green plants in a number of other ways. Witkamp and Ausmus suggest that reserves both of carbon and of inorganic nutrient are manipulated to counteract the effects of seasonal cycles and major environmental change. In this way the stability and survival of the whole ecosystem is preserved. They demonstrate that the amounts of primary plant nutrients which leach out of complex forest ecosystems are surprisingly small, unless the systems are disrupted by man.

At a more analytical level, Saunders explores the implications for the stability of a complex system of sinusoidally varying photosynthetic inputs on daily and seasonal timescales and shows how slowly decomposing organic compounds of high molecular weight can serve as a further buffer system.

In a different context Nixon demonstrates the conservation of nitrate and ammonia sources in saltmarsh creeks where nitrogen is short, in contrast with an open estuary which, being enriched by fresh phytoplankton, is profligate with nitrogenous resources. A similar point is expressed for a terrestrial system by Coleman's examination of the effect of seasonal 'pulses' of photosynthate to the roots of a grassland.

One theme which reappeared repeatedly during the conference was the inadequacy of many generalizations which are current and widely accepted among ecologists in general but known to be inadequate by workers in particular fields. For example, there is the contrast between the trophic level classifications and practical difficulties encountered in separating the digestive functions of different groups. In most systems it appears that microbes and invertebrate animals interact in a number of ways which, on the whole, accelerate rates of decomposition. Material can be recycled between the two groups as occurs in 'microbe stripping' which is demonstrated by many detritus feeding animals. This situation is complicated by uncertainty as to whether microbial symbionts are at work in the animal's gut. (See papers by Hargrave, Berrie and Fenchel.) The leakage of metabolites from many organisms and the extracellular digestive activities of many microbes must also be considered. Here one becomes aware of the 'untidiness' which transgresses both the trophic level concept and generalizations about the ecological impact of systematic groupings.

Another very 'untidy' area concerns the relationship in aquatic systems between states of matter and chemical activities. Concepts of dissolved organic matter and particulate organic matter would seem clear enough at first sight and their separation from living microbes, a fairly simple matter. However, it proves that the first two are only separated in practice on the basis of ability to pass through certain mesh sizes; that large particles of material frequently flocculate from what appears to be a clear solution and that the attachment of microbes to apparently solid particles is highly dependent on physico-chemical factors which, in turn, decide levels of metabolic activity.

Evidently naïve concepts derived from our experience in the macroscopic world have severe limitations on the smaller scales appropriate, for instance, to the studies of soils and muds. In fact, the need to maintain a highly flexible approach to scale factors both in space and in time, was a further recurring theme of the meeting. Golterman, for example, describes the contrast in aquatic systems, between slow changes in physico-chemical factors through most of the vertical depth of the water body and the truly microscopic range within which such factors change both at the surface of bottom muds and at the water surface. Others, equally striking, occur in relation to particles and water films in soil as discussed by Swift. One is left with an overall concept that most biological systems contain large amounts of either very dilute or very refractory organic materials. At least most of the microbial decomposers exist in a condition of suspended animation for most of the time. It is only at relatively rare loci and for short periods of time that the full potentialities of these organisms are revealed because of combinations of circumstances resulting from physical factors or the activities of animals. Perhaps this diffuse and intractible characteristic of decomposition pathways in ecosystems helps to account for the surprising taxonomic diversity of their constituent organisms.

From the chemist's viewpoint, the precise agent which leads to the concentration of a chemical substrate may be unimportant. The application of particular enzymes in the presence of an appropriate micro-environment of oxygen, temperature and so on, is his sole concern. It could be argued that if, on the scale of a whole lake or a field of corn, such coincidences occur with finite probability, decomposition will take place and no more need be said. The ecosystem analyst may also be satisfied to know that the whole system operates in such a way as to conserve nutrients and to impede their loss from the system.

Most biologists, however, focus on the finer detail of these systems. We wish to learn precisely how these relatively unusual interactions take place, both because they constitute a real challenge to our ingenuity and imagination and also, we surmise, because we shall not be able to control decomposition processes on land or underwater, without the insight that such studies will provide.

J. M. Anderson
University of Exeter

A. Macfadyen
New University of Ulster

Section One
Physico-chemical aspects of decomposition

1 Zonation of mineralization in stratifying lakes

H. L. GOLTERMAN *Limnologisch Instituut, Nieuwersluis (U), The Netherlands*

Introduction

Deep lakes in temperate regions are often stratified; they have layers with distinct differences in physical factors such as irradiance, density and viscosity and of concentrations of many chemical constituents, during the warmer periods of the year. The upper layer, the epilimnion, receives radiant energy and thus becomes warmer and less dense than the lower layer, the hypolimnion. The thermocline or metalimnion, is a transitional layer which is always present between the epilimnion and hypolimnion except when the autumn and/or spring temperature inversions occur in seasonal climates (Figs. 1.1 and 1.2).

Thermal stratification prevents the mixing of surface waters and the

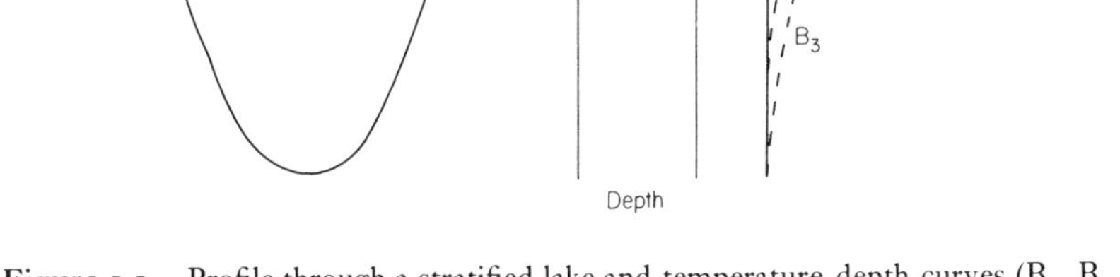

Figure 1.1. Profile through a stratified lake and temperature-depth curves (B_1, B_2, B_3). The broken lines (B_2, B_3) represent temperature-depth curves when the epilimnion is cooling in autumn. The boundaries of the thermocline (here only indicated for B_1) will gradually sink.

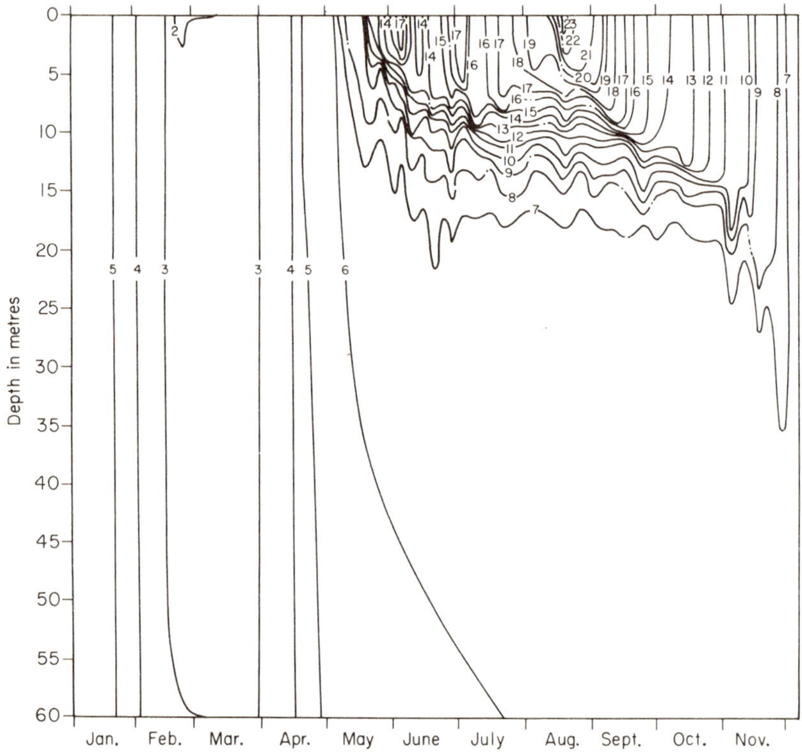

Figure 1.2. Temperature stratification in Windermere, North Basin (1947). Points of equal temperature are connected. (From Lund *et al.* 1963.)

mineral rich deep waters and also reduces oxygen diffusion to the benthos. Some tropical lakes are more or less permanently stratified and the hypolimnion is completely deoxygenated (Fig. 1.3).

Phytoplankton cells photosynthesize actively in the epilimnion where irradiance is high. Their growth, which is controlled by irradiance and nutrients, represents the main source of organic matter, both dissolved and particulate. Macrophytes only make a significant contribution to the production of organic matter in small lakes where the littoral zone represents a substantial area of the lake. In Lawrence Lake, for example, over half the surface area is covered by macrophytes and equal amounts of carbon are fixed by the phytoplankton and the higher plants (Wetzel *et al.* 1972).

This paper discusses the level in the water column at which the products of primary production are decomposed and the nutrients are released for further algal growth. Particular reference is made to data for a stratifying sandpit 'Vechten' in the Netherlands where these phenomena have been studied in detail (Figs. 1.4 and 1.5).

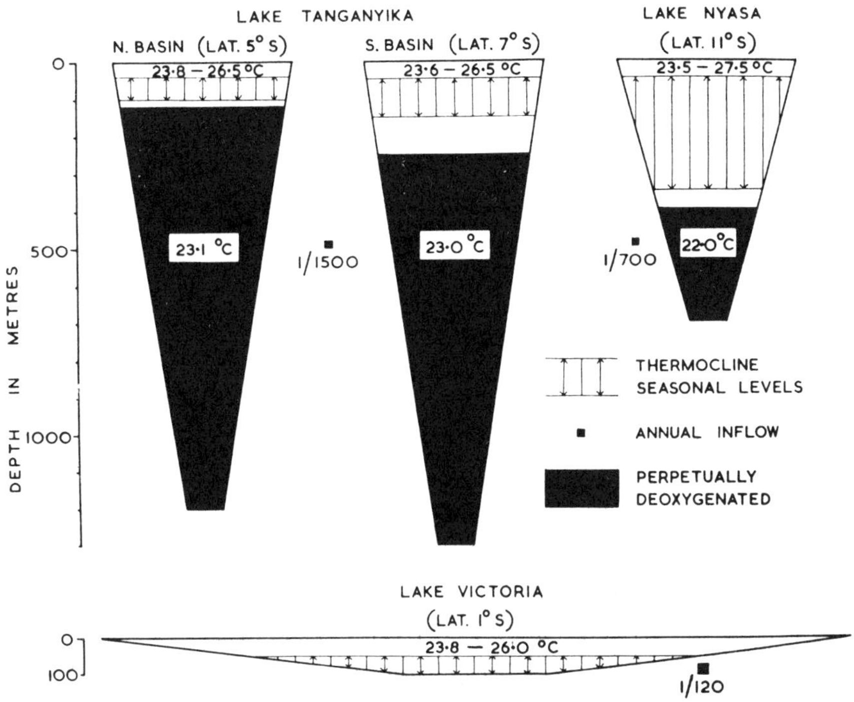

Figure 1.3. Profiles or vertical sections of some African lakes. Note the extension and seasonal variation of the thermocline and the small differences in temperature. (From Beauchamp 1964.)

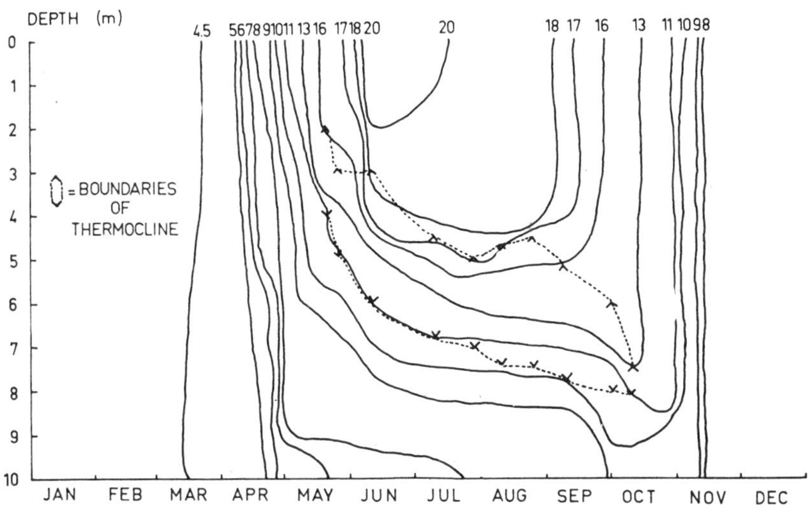

Figure 1.4. Isothermic curves of L. Vechten in 1970.

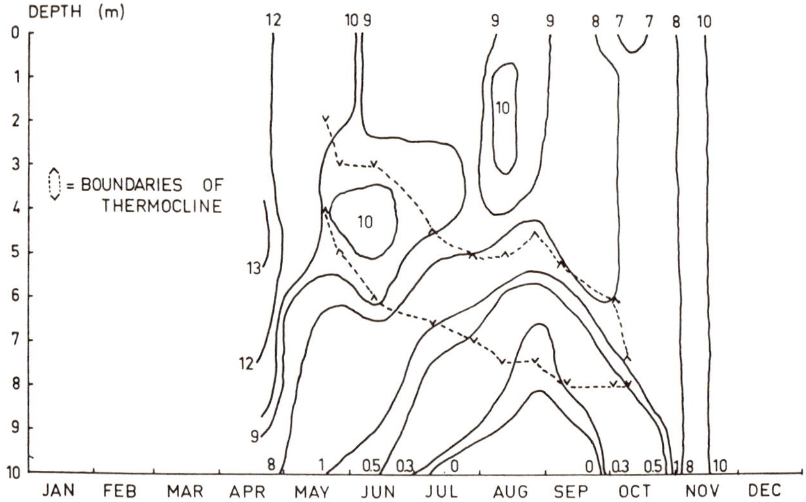

Figure 1.5. Oxygen isopleths of L. Vechten in 1970.

1. The regulation and measurement of algal production

Algal photosynthetic growth may be summarized by the equations:

$$5CO_2 + 2H_2O + NH_3 \longrightarrow C_5H_7NO_2 + 5O_2 \qquad (1.1a)$$

or

$$5CO_2 + 3H_2O + HNO_3 \longrightarrow C_5H_7NO_2 + 7O_2 \qquad (1.1b)$$

The formula for the overall products of photosynthesis, $C_5H_7NO_2$, allows for the fact that not only carbohydrates are formed, but also proteins. The formula does not mean that there are sufficient data to suggest that the mean 'global' algal composition can be calculated; it merely describes the empirical observation that the C/N ratio is often 5–6 and that the protein content is about 50% (Golterman 1975).

The major factors regulating the photosynthetic processes are frequently phosphate and nitrogen availability, temperature, light etc. The rate and yield of these processes in temperate waters are often controlled by the available phosphates.

Warming of the epilimnetic water and algal photosynthesis both depend on light. As eddy diffusion transports heat further down than the physical irradiance itself, the mixed waters of the epilimnion will extend deeper than the euphotic zone. Algal cells will be present, but photosynthetically inactive, in the dark zone of the epilimnion.

Algal populations are short lived; often blooms will last only some 6 to 8 weeks. The algae are eaten by the zooplankton or they die as a result of unfavourable physical and chemical conditions, e.g. the well-known example of *Asterionella* in Windermere (see Fig. 1.6; from Lund, 1949, 1950). When the algae are eaten not all organic matter is used by the zooplankton; con-

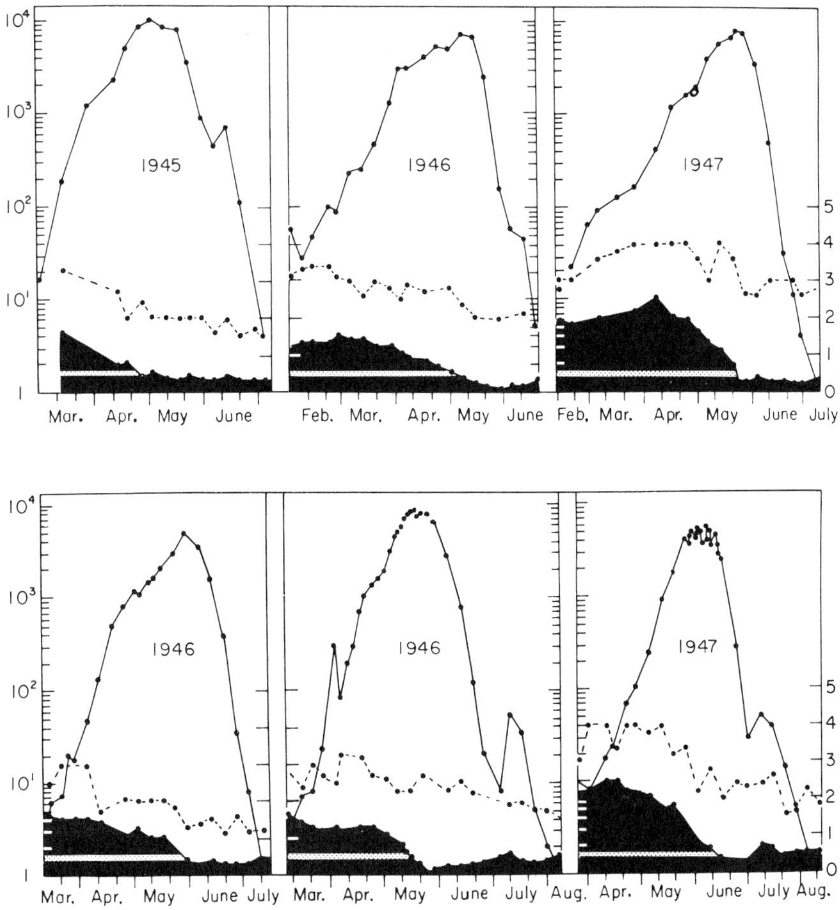

Figure 1.6. Live cells of *Asterionella* per ml (solid line) in Windermere, North Basin (upper figures) and South Basin (lower figures), in 1945–7. Nitrate nitrogen (× 10; interrupted line) and SiO_2 (mg l^{-1} solid black, with 0·5 mg l^{-1} cross-hatched).

siderable quantities remain as organic particles, either as dead cells or as faecal pellets. All dead organic matter is ultimately oxidized by bacteria. Algae killed by unfavourable physical and chemical conditions will firstly be subjected to autolysis, during which process they become permeable and lose soluble organic matter which will leach passively from the cells. Bacterial digestive enzymes may enter. During the autolysis compounds present in dissolved state in the cells and with a relatively low molecular weight probably go into solution. This leads to a loss of dry weight of only about 15% of the original algal dry weight (Fig. 1.7).

Apparently the bacterial digestive enzymes may enter the algal cells only after autolysis. Probably bacterial-algal complexes form and bacterial enzymes become active after adsorption onto or into the algal cells. If this

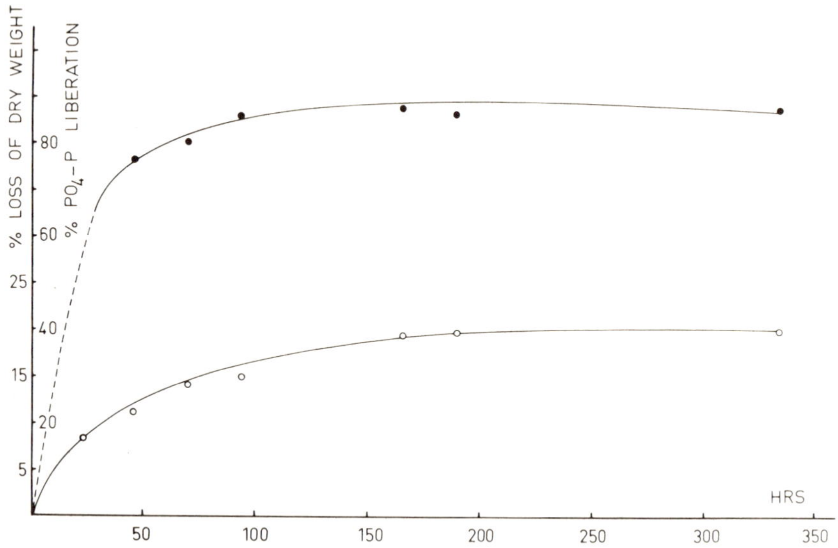

Figure 1.7. Loss of dry weight (○—○) and phosphate liberation ●—●) of *Scenedesmus* cells during autolysis with $CHCl_3$.

is so then there is no need for a considerable concentration of these enzymes to build up in the lake water itself. Thus the major part of the algal particulate organic matter and the dead organic particles produced by the zooplankton will be oxidized by bacteria. Their oxygen consumption is included in the oxygen uptake normally measured in a dark bottle. Another process included in the oxygen uptake is bacterial oxidation of dissolved organic matter, either from allochthonous or autochthonous sources. Thus two processes are actively working against the photosynthetic production of organic matter. They are *mineralization of dead material* by bacteria—the reverse of reaction I—and *algal respiration*. In both processes oxygen is consumed. Respiration is the process with which algae obtain their energy for cell maintenance even in darkness. As some of these processes have to continue in the same way in the light, the respiratory processes in the cells are separated in space from the photosynthetic processes. In this way respiration is ensured to continue both in dark and light, which fact must be accounted for during measurement of photosynthesis.

Rate of respiration is measured normally as oxygen uptake in time. Gross photosynthesis is often calculated by adding oxygen consumption in a dark bottle to that produced in a light bottle, a procedure which is theoretically correct if working in cultures. For natural algal populations this calculation is often unjustified due to secondary concomitant processes connected with the presence of other organisms and other organic matter such as detritus and dissolved organic matter.

Respiration is often thought to be about 5% of the photosynthetic capacity, i.e. maximally about 1 mg of O_2 per mg of chlorophyll *a* per hour. (Ganf 1974, Talling 1965). This uptake is derived from figures from a natural situation, where bacterial mineralization always occurs and the constituents of algal cells are constantly recycled. If respiration were the only cause of the oxygen consumption in darkness, which is all too often assumed to be the case, only carbon, hydrogen and oxygen would be lost from the particulate matter. If mineralization also takes place then nitrogen and phosphate will also recycle between dissolved inorganic and particulate organic matter. In earlier studies in shallow lakes Golterman (1971) showed that the oxygen uptake took place with a concomitant decrease of particulate carbon and nitrogen. As these two elements each decreased by about the same percentage of the total cellular carbon and nitrogen, it seems likely that mineralization by bacteria was the main cause of the oxygen uptake rather than algal respiration. This implies again that the algal cells have a short life and that their constituents turn over rapidly. During this breakdown nutrients are released, but are immediately readsorbed by living cells so that no increase in the dissolved concentrations can normally be demonstrated. This hypothesis is the only way to explain how nutrients are not completely depleted by photosynthetic processes in a few days. Naturally, if photosynthesis were sufficient only to compensate overnight respiration losses manifested by a decrease of cellular carbohydrates, or losses from cells below the euphotic zone, then the depletion would also not occur. In that case, however, it is not easy to understand why in moderately eutrophic lakes the inorganic nutrients are not completely used up by the algae, and it seems likely that the low concentrations that remain are the result of an equilibrium between uptake and release.

2. Mineralization in the epilimnion

From the above considerations, and assuming that mineralization in the epilimnion follows the same pattern as in shallow lakes, it seems likely that in most lakes a considerable amount of organic matter is oxidized by bacteria in the epilimnion. Oxygen uptake data to support this hypothesis are scarce because oxygen uptake is often too small to be measured. In earlier work indirect ways of estimating the degree of mineralization were therefore explored. Lund *et al.* (1963) estimated from the disappearance of oxygen in the hypolimnion that the major part of the primary production is mineralized in the hypolimnion, but this estimate was made before reliable data on primary productivity were available. Hargrave (1972) calculated that the O_2 uptake in the sediments of Lake Esrom equalled about one-third of the net phytoplankton production. Another approach was made by Ohle (1962, 1965), who collected sinking detritus in a sediment trap. He estimated the mineralization in the epilimnion to be up to 90% of the primary production. This high value may however be due to increased mineralization within the trap itself as a result of increased bacterial substrate concentration. Serruya & Serruya (1972) found a high oxygen uptake in the epilimnion of Lake Kinneret, but did not attempt to relate this quantity to the primary production.

Preliminary data on a stratifying sandpit, Lake Vechten, showed that oxygen uptake in the epilimnetic water was 80% ± 20% of the primary production (0·4–0·5 g m^{-2} d^{-1} of carbon). As the oxygen uptake rate in this lake is low, dinitrophenol was added for the two-day incubation period to avoid changes in bacterial populations. Golterman (1971) showed that this compound stabilizes the bacterial populations, but did not inhibit the proteolytic mineralization of organic matter by bacteria such as *Pseudomonas*. From the mean chlorophyll concentration in the epilimnion, and assuming an O_2 uptake of 1 mg per mg of chlorophyll per hour, a value of about 50% mineralization in the epilimnion can be calculated.

Although insufficient data were available, it seems likely that 50% to 80% of the phytoplankton were oxidized by bacteria in the epilimnion, and that easily decomposable compounds such as sugars, proteins and fats are oxidized in a few days by specialized bacteria. These bacterial populations will show a certain succession depending on the biochemical reactions they can carry out and on their substrates.

The exact value of the proportion of phytoplankton mineralized in the epi- or hypolimnion will probably depend on the ratio between the euphotic zone and the mixed depth of the epilimnion, and also on the concentration of the phytoplankton in the epilimnion as it seems likely that the rate of bacterial decomposition depends on the bacterial-substrate concentration. The importance of the ratio between 'mixed depth' and euphotic zone was demonstrated in an eutrophic, humus-rich lake, Vinkeveense Plassen, where the euphotic zone is only 2 m but the epilimnion about 15 m as the lake is exposed to strong winds. Although this lake has a productivity of 0·9 g m^{-2} d^{-1} of carbon, no oxygen depletion was found, suggesting that intensive mineralization occurred in the dark 2 m–15 m layer (unpublished data).

3. Bacterial succession in the epilimnion

Remarkably little is known about the succession of bacteria following the end of an algal bloom. More than 50% of the dry weight of dead algal cells is proteins, so one would expect proteolytic bacteria to dominate. Jones

(1971, 1972) found proteolytic bacteria to be 9% and 7% of the total number in Windermere and Esthwaite Water respectively. Dominant genera were *Acinetobacter*, *Xanthomonas*, *Corynebacterium*, *Flavobacterium* and *Pseudomonas*. Bacteria of the first two genera can also make amylase. Jones found no correlation between concentrations of viable bacteria and chlorophyll *a* concentrations, but enzyme activity occasionally increased with or soon after phytoplankton density. The correlation with primary production was higher than with algal biomass. Amylase-producing bacteria showed two peaks, one after the spring diatom maximum and one after the autumn maximum of *Ceratium* and blue green algae.

Jones (1973) found a relatively stable mixture of several species in Blelham Tarn. The total concentration of heterotrophic bacteria increased from 10^5 l^{-1} in April to 10^7—10^8 l^{-1} at the end of May with some later decrease to values around 10^7 l^{-1}. The numbers in two large experimental tubes, in which lake water was separated from the main lake (Lack & Lund 1974) were in the same order of magnitude but single species often dominated including genera tentatively identified as *Acinetobacter*, *Flavobacterium*, *Moraxella* and a member of the coryneform group.

We found concentrations of proteolytic bacteria in Lake Vechten between 10^5 l^{-1} and 10^7 l^{-1}, while somewhat higher values were found in the more productive Lake Tjeukemeer. Considering that the major part of the decomposable material is protein, the relative concentration of proteolytic bacteria seems low. This may be caused by the algal-bacterial congregations (see

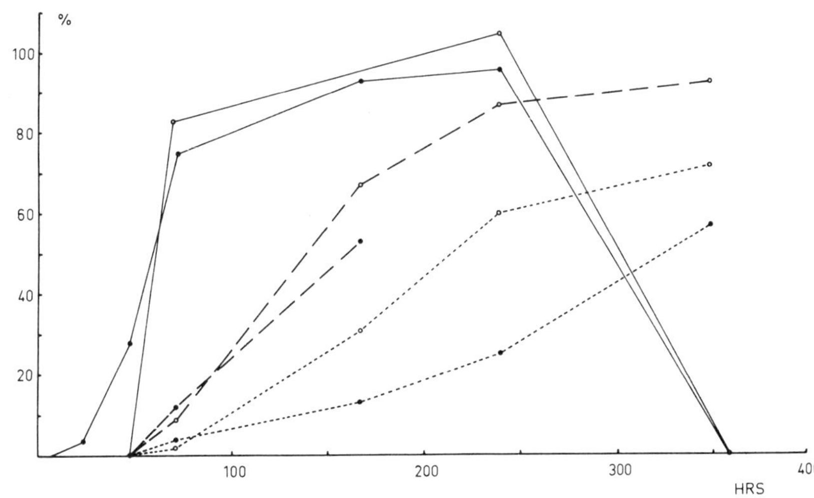

Figure 1.8. Ammonia produced from 'leached' algal cells after suspension in lake water at different temperatures. ○—○ : no treatment; ●—● : lake water heated to 65°C for 1 hr. ——: 30°C; – – –: 20°C; - - - - - -: 10°C.

above), so that plate counts for these bacteria are still more unreliable than for the other heterotrophic bacteria living on dissolved substrates.

Not many data are available on the kind and intensity of the biochemical reactions mediated by these bacteria. With an enrichment technique we isolated strains of *Pseudomonas* which could mineralize dead algal proteins com-

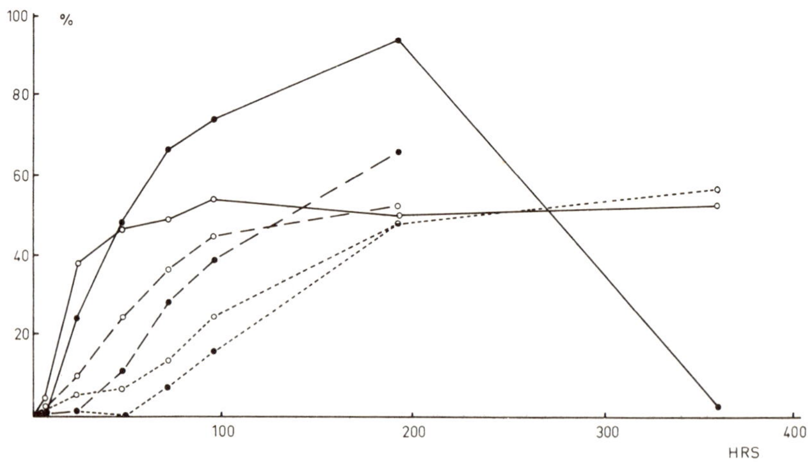

Figure 1.9. Release of ammonia from 'leached' *Scenedesmus* cells during mineralization in lake water (●—●) or in cultures of *Pseudomonas boreopolis* (○—○) at three temperatures. ——: 30°C; — — —: 20°C; - - - - - -: 10°C.

pletely to ammonia (Fig. 1.8, 1.9; Golterman, 1972). As these bacteria use the algal proteins both for their own proteins and for energy about 80% of the dry weight was lost and an equivalent amount of ammonia was released, which can be visualized in the following scheme:

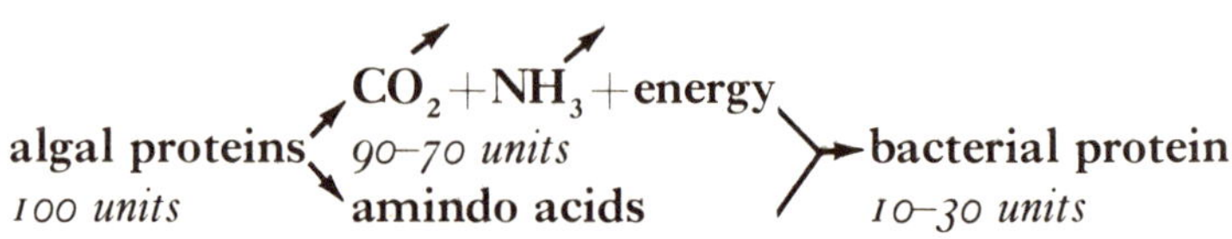

Some nitrogen will, of course, be converted into refractory matter. In the same way polysaccharide and lipid utilizing bacteria will only produce organic matter in the range of 10–30% of their substrate. One can easily understand why the larger part of the dying phytoplankton is mineralized in the epilimnion. Compounds such as cellulose and chitin which are not so easily broken down will sink slowly towards the metalimnion and ultimately into the hypolimnion.

4. Mineralization in the metalimnion

Ambühl (1969) found a high oxygen uptake just below or even in the metalimnion by calculating the differences between the March/April and the October oxygen concentrations. (Fig. 1.10). Usage of O_2 below the thermocline is even larger than this calculated amount due to diffusion of oxygen from above (Li, 1973). Serruya & Serruya (1972) also found a high oxygen consumption in the metalimnion of Lake Kinneret.

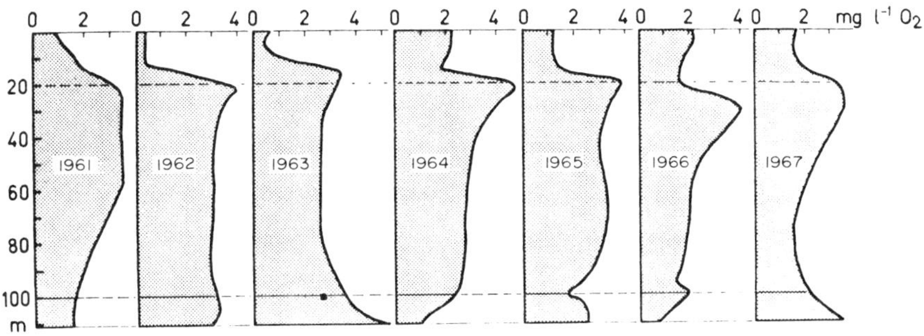

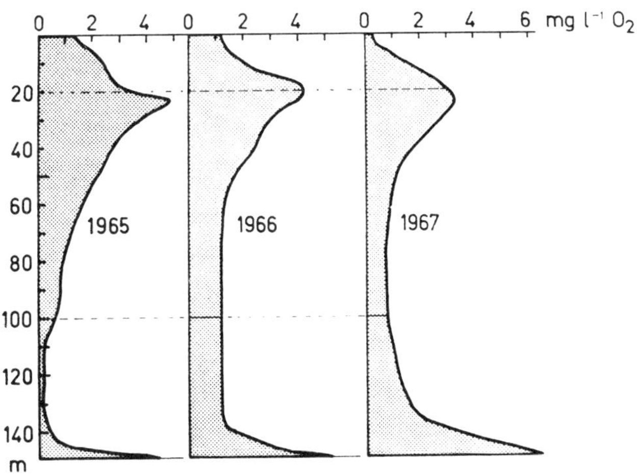

Figure 1.10. A. depletion of O_2 between the beginning and end of stratification; Lake of Lucerne, 1961–7, station I.
B. depletion of O_2 between the beginning and end of stratification; Lake of Lucerne, 1961–7, station II. (From Ambühl 1969.)

Anaerobic conditions developed even in the metalimnion causing denitrification and accumulation of Fe^{2+} and Mn^{2+}. Collins (1963, 1970) found the number of heterotrophic bacteria (plate counts) in the metalimnion to be greater than in the epilimnion. This is probably due to the fact that the rate at which particles sink decreases partly because the particles become smaller while sinking and also because the viscosity of the cooler water is greater in the hypolimnion. For a given change in temperature the change of viscosity affects the sinking rate more than does the difference in density. A decreased sinking rate in the metalimnion causes a higher concentration of particles there. It is an open question how far the progressive decaying of the particles in the metalimnion leads finally to a disintegration of the detritus particle with an apparent increase of bacterial concentrations. No information is yet available on the nature of the substrates decomposed in the metalimnion.

5. Mineralization in the hypolimnion

The organic particle finally arrives in the hypolimnion. It is not clear whether the oxygen consumption in the hypolimnion takes place only in the sediment, the overlying water becoming depleted by diffusion or whether oxygen is consumed in the water itself. The shape of the curves of oxygen consumption in Ambühl's (1969) figures does not suggest that diffusion to the mud is the sole cause of oxygen depletion. Hargrave (1972), however, found no biological O_2 demand in the hypolimnetic water in 6 hours, but it is possible that longer exposure times would have shown measurable oxygen uptake. The difference from Ambühl's results may also be explained by the great difference in depth. Ambühl measured O_2 uptake just under the thermocline at about 25 m, which is about the same depth as the bottom of Lake Esrom (maximum depth 22 m).

The degree of oxygen depletion in the hypolimnion depends on the total amount of oxygen present which varies with lake depth and with the amount of sinking dead organic material which is indirectly related to the primary production. The influence of depth is well demonstrated in Windermere where the deeper North Basin shows higher oxygen values than the shallower South Basin (Lund *et al.* 1963). The influence of primary production is generally less clear than that of depth. In Ambühl's profiles the curves at his four sampling stations show remarkable differences not related to the expected differences in primary production. This was probably caused by currents keeping the dead phytoplankton suspended for longer or shorter times in the epilimnion or by river currents influencing the distribution of oxygen by penetrating the hypolimnion.

As soon as the oxygen concentration falls below a value of about 4 mg l^{-1}, the bacteria will use not only oxygen in molecular form as electron acceptor, but also nitrate—the so-called denitrification reaction. As nitrate concentrations in lakes are mostly low compared with oxygen, this process does not contribute significantly to the oxygen-carbon balance. The importance of this process lies more in the loss of nitrogen that occurs, the reaction being:

$$NO_3^- + [H] \text{ organic matter} \rightarrow N_2 + CO_2 + H_2O \qquad (1.2)$$

Owing to this process nitrogen is less likely to accumulate than phosphate. As soon as oxygen and nitrate are depleted anaerobic conditions are established both in the completely deoxygenated water and in the sediments.

The quantity of reducing power that is needed to exhaust the oxygen is not negligible. If, before the hypolimnion develops, the water temperature is 7–8°C the oxygen concentration is 12 mg l^{-1} or 1·5 eq m^{-3}. In Lake Vechten, where water down to 3 m above the bottom is deoxygenated in about 100 days, 4·5 eq m^{-2} of reducing power is needed. Photosynthesis produces 0·45 g m^{-2} d^{-1} of carbon or 45 g m^{-2} in 100 days and, assuming 3 g of $C \simeq 1$ eq, this gives 15 equivalents of (photosynthetically fixed) reducing capacity. The disappearance of the oxygen is thus equivalent to 30% of the reducing power.

In Windermere (North Basin), where 3 g m^{-3} of O_2 is consumed between June and November in the water layer between 15 and 60 m (Lund *et al.* 1963) 135 g m^{-2} of O_2 or 17 eq m^{-2} of hydrogen-donor is needed. From Talling's (1965) paper a very rough estimate of gross photosynthesis for another year can be obtained which is approximately 0·9 g m^{-2} of C per day or 135 g m^{-2} in the 150 day period or 45 eq m^{-2} (confirmed by personal communication); i.e. roughly 35% of the primary production is used to deplete the hypolimnetic oxygen.

When comparing reactions such as oxygen consumption in the hypolimnion and oxygen production in the epilimnion the different surfaces of these two layers may play a role. Ohle (1962) introduced the idea of the 'funnel' effect. The importance of this effect is probably exaggerated by the practice of plotting lake profiles with exaggerated depth scales. A true scale profile of even a very deep lake resembles a pancake more than a funnel and material sinking in the littoral regions above the hypolimnion will probably be mineralized in the aerobic sediments.

6. Anaerobic processes

As soon as anaerobic conditions prevail bacterial respiration will continue with sulphate as substratum. The most important bacterium *Desulphovibrio desulphuricans* derives its energy by the following reaction:

$$2H^+ + SO_4^= + 8[H] - \text{donor} \rightarrow H_2S + 4H_2O \qquad (1.3)$$

As *Desulphovibrio* is a strict anaerobe bacterium the process is inhibited by traces of oxygen. No quantitatively reliable data are available to show which organic carbon compounds can be used. Compounds that are used in *cultures* are lactic acid, some alcohols, formate and others. Like other chemolithotrophic bacteria this species does not have a complete tricarboxylic acid (TCA) cycle and the cells can therefore not oxidize acetate. This not all reducing capacity of some organic compounds can be used, e.g.

$$\underset{\text{lactic acid}}{SO_4^{2-} + 2CH_3CHOHCOOH} \longrightarrow \underset{\text{acetic acid}}{2CH_3COOH} + 2CO_2 + H_2S + 2OH^- \quad (1.4)$$

In this case only one third of the reducing capacity is utilized. If formate is the substrate all the reducing capacity can be used:

$$SO_4^{2-} + 4HCOOH \rightarrow 4CO_2 + 2H_2O + H_2S + 2OH^-$$

In natural environments a mixture of several compounds will probably be used. Van Gemerden (1967) showed that dead cells of *Chromatium* and *Scenedesmus* could provide carbon and energy sources for *Desulphovibrio* growth; it is not known which part of natural mixtures of compounds containing amino acids, for example, is converted into acetate. There is a need for carbon balance studies of *Desulphovibrio* under ecological conditions to clarify these points.

Calculations of the fraction of primary production used for sulphate reduction must thus remain uncertain as long as the nature of the organic reductant is not known. Using lactate and formate as limits, however, some estimates can be made. In Lake Vechten 0·125 mole m^{-3} of sulphate is reduced in 100 days for which 3 eq m^{-3} of lactate but 1 eq m^{-3} of formate is needed. As the sulphate in a water layer of 2 m above the sediments, a reducing capacity of 2 or 6 eq m^{-2} is needed, or 13–39% of the primary production. Together with the reduction of the oxygen (30%) 43–70% of the organic matter is used in the hypolimnion, the lower figure therefore being the more likely one.

A third process, reduction of $Fe(OH)_3$, may use a small fraction of the reducing capacity. Originally it was suggested (see review by Golterman, 1967) that the $Fe(OH)_3$ would be reduced by the H_2S:

$$2Fe(OH)_3 + H_2S \rightarrow 2Fe^{2+} + S + 2H_2O + 4OH^-$$

In this case no extra reducing capacity is necessary as the flux of electrons continues, instead of stopping at the H_2S level. Golterman (1967, 1973) has argued that it is more likely that organic compounds are used directly. Tessenow (1972) holds the same opinion and Stumm & Morgan (1970) have discussed the thermodynamic feasibility of this reaction. Evidence that this process occurs in natural conditions has not yet been obtained. The quantities involved are not large. In Lake Vechten Fe^{3+} reduction may add about 0·5 eq m^{-2} to the demand for reducing capacity. It seems likely that an organic compound (rather than CO_2) is the oxidation product adding soluble organic matter to the hypolimnion.

Production of CH_4 will be the last process involving oxido-reduction of organic matter. The process can be described as:

$$CH_3COOH + 2H_2O \rightarrow 2CO_2 + 8[H] \quad (1.7a)$$

$$CO_2 + 8[H] \rightarrow CH_4 + 2H_2O \quad (1.7b)$$

Besides an internal conversion of acetic acid, other compounds through which CO_2 may be reduced are ethyl alcohol and hydrogen. In the first case the oxidation product is acetic acid, which enters the reaction (1.7a). Not many quantitative data on methane production are available. Birge & Juday (1911) found 38·5 ml l^{-1} of CH_4 in hypolimnetic water of Gravin Lake. Allgeier *et al.* (1932) measured a production of 28 ml of CH_4 per m^2 mud per day in Lake Mendota during 100 days (assuming an active depth of 1 dm). If one assumes a primary production of about 1 g m^{-2} d^{-1} of C in Lake Mendota it is clear that the methane production is at the uppermost a few percent of the primary production. Howard *et al.* (1971) estimated the rate of methane production to be 2,400 ml of CH_4 per m^2 of sediment in Lake Erie per day, a small part of which was oxidized at the 4 m depth.

This high figure for methane production in Lake Erie is not representative for the whole lake and can not be compared with primary production. It is only representative for areas where sediments were accumulating, while at another sediment covered area no methane production was found at all.

MacGregor & Keeney (1974) found a methane production of 4·5 to 250 ml per m^2 wet sediment in soft and hard water, eutrophic lakes respectively (assuming 5 cm as an active depth of methane production). They demonstrated that nitrate and sulphate inhibited methane synthesis.

Methane production is clearly competitive with the SO_4^{2-} reduction and the reaction may take place only after the SO_4^{2-} reaches a low level. Indeed Ohle (1968) found a lag phase between primary production and maximal methane production. Martens & Berner (1974) found no appreciable methane concentration in some marine sediments until about 90% of the seawater sulphate was removed by sulphate-reducing bacteria.

Methane will diffuse upwards and bubbles, if formed, will begin to rise into the hypolimnion. Methane oxidizing bacteria will oxidize the dissolved CH_4 if it comes sufficiently high so that it reaches the oxygenated layer. Kuznetsov (1959) and Anagnostidis & Overbeck (1966) suggested that the disappearance of O_2 from the hypolimnion is related to this process. Anagnostidis & Overbeck calculated that 70% of the oxygen consumption in the Plüss See may be caused by this methane oxidation during certain parts of the year. It should be remembered however that before methane oxidation can take place the lower layers of the hypolimnion must be anaerobic.

Deuser *et al.* (1973) found 50 km³ of methane in Lake Kivu and suggested that most of this methane was formed by bacteria from abiogenic CO_2 and H_2, both derived from volcanic action.

Hydrogen may normally be produced by fermentation of glucose, for example, derived from decomposition of cellulose, hemicellulose, starch and other polysaccharides. *Escerischia coli* can produce 44 mole of CO_2, 43 mole of H_2, 44 mole of acetic acid, 84 mole of lactic acid, 29 mole of succinic acid and 42 mole of ethyl alcohol from 100 moles of glucose fermented. The ratio of CO_2 and H_2 is 1 : 1, because they are produced from formic acid.

$$HCOOH \rightarrow H_2 + CO_2 \qquad (1.8)$$

Other carbohydrate fermenting bacteria are clostridia, which produce the same compounds. Some species do not produce mixtures, e.g. *Clostridium aceticum* and *C. thermoaceticum* produce acetic acid as the principal or only end product. Probably also important is another large group of clostridia which obtain their energy principally by the anaerobic breakdown of aminoacids. They may be the main producers of ammonia from the mud by a reaction such as

$$RCHNH_2COOH + 2RCHNH_2COOH + 2H_2O \longrightarrow RCOOH + CO_2 + 2RCH_2COOH + 3NH_3 \qquad (1.9)$$

The processes are summarized by Rheinheimer (1974) who found several of the above mentioned bacteria in mud. No quantitative data are yet available. Finally, part of the sinking organic matter will not be mineralized at all. Refractory compounds such as lignin, humic and fulvic compounds are formed in the sediments and in the dissolved state. A comparison with the primary production is not possible as organic refractory compounds washed into the lake will be included in this pool. The order of magnitude of these processes may be calculated from the following example.

From Mackereths's (1966) figure for decomposition of 0·4 mm a year and organic carbon content of 10%, a rough estimate can be made of a decomposition of about 10 g carbon $m^{-2}\,y^{-1}$. Approximately 50% of this carbon is adsorbed on the incoming allochthonous material so the carbon deposition may be about 5 g carbon $m^{-2}\,y^{-1}$ or only a few per cent of the primary production.

7. Calculation of increase of bicarbonate in the hypolimnion

Besides reducing the concentration of O_2 in the hypolimnion mineralization causes an increase in the CO_2 concentration.

$$C_5H_7NO_2 + 5O_2 \rightarrow 5CO_2 + NH_3 + 2H_2O \qquad (1.10)$$

The carbon dioxide will decrease the pH and may react with $CaCO_3$ in the sediments or react with $CaCO_3$ sedimenting from the epilimnion in calcium-rich lakes:

$$CaCO_3 + H_2O + CO_2 \rightarrow Ca(HCO_3)_2 \qquad (1.11)$$

After this reaction the pH will be lower than originally and some more CO_2 will remain in solution as free CO_2:

$$\ln[CO_2^-] \quad -pH + pk - \ln[HCO_3^-] \qquad (1.12)$$

Therefore, somewhat more than 50% of the increased amount of $[CO_2]+[HCO_3^-]$ is derived from mineralization. No great error is made in a relatively alkaline lake such as Lake Vechten if 50% is taken, but in soft waters, such as Esthwaite where HCO_3^- will be formed from CO_2+OH^- a much higher value (up to 100%) must be used.

The increase of $[CO_2]+[HCO_3^-]$ was measured in the hypolimnion of Lake Vechten at 0·5, 1·5, 2·5 and 3·5 m from the bottom during periods in 1968, 1969 and 1970. The differences in the three years are small and agree with isolated data from other years. The increasing amount of $[CO_2]_{tot}$ appeared to be equivalent to 75% of primary production during the first 6 weeks after the onset of stratification (17 Apr→28 May). During the second (28 May→9 Jul), third (9 Jul→20 Aug) and fourth (20 Aug→25 Sept) six-week periods a mean value of roughly one third was found. It is not unlikely that the higher value in the first period is partly due to mineralization of matter which had arrived at the bottom before the onset of stratification. The value of one third approaches the value derived from the oxygen depletion calculation. This agreement is seriously disturbed, however, by the fact that during the period (20 Aug→25 Sept) CO_2 increased at the same rate after the oxygen was already depleted. This is unexpected and not easily explained. There are two possibilities:

1. Diffusion of O_2, downwards.

   Li (1973) has shown that considerable amounts of O_2 arrive in the hypolimnion by molecular diffusion downwards. It is not possible to calculate the absolute amount, but it may be of some importance for the problem of the increasing amount of CO_2. The amount diffused would in that case be equal to the amount of CO_2 produced while the hypolimnion was anaerobic. Calculation would be easier in lakes deeper than Lake

Vechten, if the increasing amount of CO_2 could be measured. Often this would be much less than in a shallow stratifying lake and the necessary accuracy may be impossible.

2. Anaerobic production of CO_2:

$$\text{glucose} \rightarrow \text{ethanol} + CO_2$$

This would mean that the increase of dissolved organic matter would be twice that of the inorganic CO_2. Although the dissolved organic matter did increase it was a minute fraction of the inorganic CO_2. Furthermore it would mean that all primary production would be used, as the amount organic carbon must be twice the inorganic carbon.

The amount of dissolved organic matter in the hypolimnion is small, but the turnover time has not yet been measured. As these compounds are formed after the mineralization of easily decomposable matter came to an end it seems likely that they are refractory. But because the major fraction of primary production is already accounted for in either the epi- or hypolimnion it seems unlikely that refractory compounds, and end products like methane, may contribute significantly to the carbon budget. The same holds true for the organic matter in the sediments.

Conclusions

In stratifying lakes photosynthesis and mineralization are separated in time and place. The main part of the primary production takes place in the euphotic epilimnion and in the pelagial zone growth of phytoplankton accounts for 90–95% of the production of plant materials.

Life expectancy for the phytoplankton is short due to its restricted periodicity and grazing by zooplankton.

Mineralization is generally supposed to take place in the hypolimnion and benthic regions of the lake but it has been shown that mineralization also occurs in the euphotic zone, leading to a rapid turnover of nutrients. It is estimated that 50 to 80% of the phytoplankton were oxidized in the epilimnion, the true value probably depending upon the ratio of euphotic zone over mixed epilimnetic depth and phytoplankton concentration.

Part of the dead algal cells will sink. This material serves as a carbon and a hydrogen source inducing several different bacterial processes. Oxygen and sulphate will oxidize an important proportion of the sinking detritus; in Lake Vechten it may be estimated to fall in the range between 43% and 70% of the primary production. As long as the hypolimnion contains oxygen an agreement is found between CO_2 production and O_2 consumption. When anaerobic conditions occur, production of CO_2 continues, but it is not yet understood what happens to the equivalent amount of organic matter that apparently must be reduced. The quantity and nature of the anaerobic bacterial reactions are unknown. Methane production and the accumulation of organic matter in the sediments are quantitatively of no importance for the carbon cycle in most lakes.

References

ALLGEIER R.J., PETERSON W.H., JUDAY C. & BIRGE E.A. (1932) The anaerobic fermentation of lake deposits. *Int. Revue ges. Hydrobiol. Hydrogr.*, **26**, 444–61.

AMBÜHL H. (1969) Die neueste Entwicklung des Vierwaldstättersees (Lake of Lucerne) *Verh. int. Verein. theor. angew. Limnol.*, **17**, 219–30.

ANAGNOSTIDIS K. & OVERBECK J. (1966) Methanoxydierer und hypolimnische Schwefelbakterien. Studien zur ötologischen Biocönotik der Gewässermikro-organismen. *Ber. dt. Bot. Ges.*, **79**, (3), 163–74.

BEAUCHAMP R.S.A. (1964) The Rift Valley Lakes of Africa. *Verh. int. Verein. theor. angew. Limnol.*, **15**, 91–9.

BIRGE, E.A. & JUDAY C. (1911) The inland lakes of Wisconsin. The dissolved gases and their biological significance. *Bull. Wiscon. geol. nat. Hist. Surv.*, **22**, 259 pp.

COLLINS V. (1963) The distribution and ecology of bacteria in freshwater. *Proc. Soc. Wat. Treat. Exam.*, **12**, 40–73.

COLLINS V. (1970) Distribution and ecology of selected groups of bacteria in a stratified lake. *Abstract 10th int. Congr. Microbiol.*, Mexico, Aa–14.

DEUSER W.G., DEGENS E.T., HARVEY G.R. & RUBIN M. (1973) Methane in Lake Kivu: New data bearing on its origin. *Science*, **181**, 51–3.

GANF G.G. (1974) Community respiration of equatorial plankton. *Oecologia*, **15**, 17–32.

GEMERDEN H. VAN (1967) *On the bacterial sulphur cycle of inland waters.* Den Haag, J.H. Pasmans, 1967. Is: PhD. Thesis.

GOLTERMAN H.L. (1967) Influence of the mud on the chemistry of water in relation to productivity. *Proc. IBP-symp.* Amsterdam-Nieuwersluis, October 1966, pp. 297–313.

GOLTERMAN H.L. (1971) The determination of mineralization in correlation with the estimation of net primary production with the oxygen method and chemical inhibitors. *Freshwater Biol.*, 1(3), 249–56.

GOLTERMAN H.L. (1972) The role of phytoplankton in detritus formation. *Proc. IBP-UNESCO symp. Pallanza* (Italy), 1972. *Mem. Ist. Idrobiol., 29 Suppl.*, 89–103.

GOLTERMAN H.L. (1973) Vertical movement of phosphate in freshwater. In: *Environmental phosphorus handbook*, eds. E.J. Griffith, A. Beeton, J.M. Spencer & D.T. Mitchell, pp. 509–38. J. Wiley, New York, London.

GOLTERMAN H.L. (1975) *Physiological limnology*. Elsevier Scientific, Amsterdam.

HARGRAVE B.T. (1972) A comparison of sediment oxygen uptake, hypolimnetic oxygen deficit and primary production in Lake Esrom, Denmark. *Verh. int. Verein. theor. angew. Limnol.*, **18**, 134–9.

HOWARD D.L., FREA J.I. & PFISTER R.M. (1971) The potential for methane-carbon cycling in Lake Erie. Proc. 14th Conf. Great Lakes Res. 1971. *Internat. Assoc. Great Lakes Research*, pp. 236–40.

JONES J.G. (1971) Studies on freshwater bacteria: factors which influence the population and its activity. *J. Ecol.*, **59**, 593–613.

JONES J.G. (1972) Studies on freshwater bacteria: Association with algae and alkaline phosphatase activity. *J. Ecol.*, **60**, 59–77.

JONES J.G. (1973) Studies on freshwater bacteria: The effect of enclosure in large experimental tubes. *J. appl. Bact.*, **36**, 445–56.

KOUWE F.A. & GOLTERMAN H.L. (1976) Rol van bodemfosfaten in het eutrofiëringsproces. I. Onderzoek in een proefvijver in het Veluwerandmeer. H_2O, **9**(5), 84–6.

KUZNETSOV S.I. (1959) Mikrobiologische karakteristiek van de Volga-stuwmeren (translated from Russian) *Verh. Inst. Biol. Stuwmeren*, **1**, 69–81.

LACK T.J. & LUND J.W.G. (1974) Observations and experiments on the phytoplankton of Blelham Tarn, English Lake District. I. The experimental tubes. *Freshwater Biol.*, **4**(4), 399–415.

LI Y.H. (1973) Vertical eddy diffusion coefficient in Lake Zürich. Schweiz. Z. Hydrol., **35**(1), 1–7.

LUND J.W.G. (1949) Studies on *Asterionella*. I. The origin and nature of the cells producing seasonal maxima. *J. Ecol.*, **37**(2), 389–419.

LUND J.W.G. (1950) Studies on *Asterionella formosa* Hass. II. Nutrient depletion and the spring maxima. Part 1: Observation on Windermere, Esthwaite Water and Blelham Tarn. *J. Ecol.*, **38**(1), 1–14.

LUND J.W.G., MACKERETH F.J.H. & MORTIMER C.H. (1963) Changes in depth and time of certain chemical and physical conditions and the standing crop of *Asterionella formosa* Hass. in the North Basin of Windermere in 1947. *Phil. Trans. R. Soc. London, B*, **246**, 255–90.

MACGREGOR A.N. & KEENEY D.R. (1973) Methane formation by lake sediments during in vitro incubation. *Wat. Resour. Bull.*, **9**(6), 1153–8.

MACKERETH F.J.H. (1966) Some chemical observations on post-glacial lake sediments *Phil. Trans. R. Soc. London, B*, **250**, 165–213.

MARTENS C.S. & BERNER R.A. (1974) Methane production in the interstitial waters of sulfate-depleted marine sediments. *Science*, **185** (4157), 1167–9.

OHLE W. (1962) Der Stoffhaushalt der Seen als Grundlage einer allgemeinen Stoffwechseldynamik der Gewässer. *Kieler Meeresforsch.*, **18**(3), 107–20.

OHLE W. (1965) Primärproduktion des Phytoplanktons und Bioaktivität holsteinischer Seen, Methoden und Ergebnisse. *Limnologisymposium*, Finland 1964, 24–43.

OHLE W. (1968) Chemische und mikrobiologische Aspekte des biogenen Stoffhaushaltes der Binnengewässer. *Mitt. int. Verein. theor. angew. Limnol.*, **14**, 123–33.

RHEINHEIMER G. (1974) *Aquatic microbiology*. J. Wiley, London, New York, 1974.

SERRUYA C. (1972) Metalimnic layer in Lake Kinneret, Israel. *Hydrobiologia*, **40**(3), 355–9.

SERRUYA C. & SERRUYA S. (1972) Oxygen content in Lake Kinneret: Physical and biological influences. *Verh. int. Verein. theor. angew. Limnol.*, **18**, 580–7.

STUMM W. & MORGAN J.J. (1970) *Aquatic chemistry; an introduction emphasizing chemical equilibria in natural waters*. J. Wiley, New York, London.

TALLING J.F. (1965) Comparative problems of phytoplankton production and photosynthetic productivity in a tropical and a temperate lake. *Mem. Ist. Ital. Idrobiol.*, **18** Suppl., 399–424.

TESSENOW U. (1972) Lösungs-, Diffusions- und Sorptionsprozesse in der Oberschicht von Seesedimenten. I. Ein Langzeitexperiment unter aeroben und anaeroben Bedingungen im Fliessgleichgewicht. *Arch. Hydrobiol./Suppl.*, **38**(4), 353–98.

WETZEL R.G., RICH P.H., MILLER M.C. & ALLEN H.L. (1972) Metabolism of dissolved and particulate detrital carbon. *Mem. Ist. Ital. Idrobiol.*, **29** Suppl., 185–243.

2 Microsites and time changes in soil microbe ecology

R. HISSETT *Department of Microbiology, West of Scotland Agricultural College, Auchincruive, Ayr, KA6 5HW, U.K.*
T. R. G. GRAY *Hartley Botanical Laboratories, University of Liverpool, P.O. Box 147, Liverpool, L69 3BX, U.K.*

Abstract

The distribution of soil bacteria appears to be discontinuous, with some species showing marked preferences for certain types of particle as sites for growth. Generally, heterotrophic bacteria are more common on organic than on mineral surfaces. In some instances, these preferences may be associated with localized physico-chemical irregularities, such as pH changes. The existence of these microsites may be the result of activity by other micro-organisms. The irregular distribution of micro-organisms in soil may be matched by equal irregularity in their activity. It is clear that if all the organic matter reaching a soil is shared out equally between all the decomposer micro-organisms in a soil, this can only result in a very slow growth rate; with a doubling time in excess of two or three weeks in most cases. Rapid growth seems to be a rare phenomenon, associated with the initial colonization of fresh substrates.

Introduction

Progress in soil microbiology has always been restricted by a hiatus between theories of how the micro-organisms are functioning, and the experimental means available for testing these theories. In particular, it seems especially desirable, at present, to gain information about the behaviour of individual microbial cells, or microcolonies. However, most of the experimental techniques used by soil microbiologists only give information on the behaviour of large sections of the soil population. While it is possible to draw conclusions about cell or microcolony events from such data, it is important that the validity of such data is not overstretched in a rush of theoretical zeal.

In the present paper, an attempt will be made to see what conclusions about some aspects of the behaviour of soil micro-organisms can be drawn from currently available data, with particular reference to studies at two sites with which the authors have been connected. The first of these sites is the sand dune soil at Freshfield in south Lancashire. This site consists of an area of coastal sand dunes which were planted with *Pinus nigra* ssp. *laricio* about 1924. Dense growth by the trees has almost completely eliminated the ground flora, and the soil appears to be developing a podzol profile (Goodfellow *et al.* 1968). The second site, Meathop Wood, is in the extreme north of Lancashire, and consists of a mixed deciduous woodland on an outcrop of Carboniferous limestone, between 30 and 55 metres above sea level. The soil is predominantly a brown earth, varying greatly in depth, and carries a rich and varied ground flora. The dominant trees are *Quercus petraea*, *Q. petraea* × *robur*, *Fraxinus excelsior*, *Betula pubescens* × *verrucosa* and *Acer pseudoplatanus* with a well developed understorey of *Corylus avellana*. Further details of the site are given in Hissett & Gray (1973).

Spatial distribution of micro-organisms

Data on the distribution of micro-organisms within a soil may be markedly affected by the scale on which the original investigations were carried out—i.e. by the quadrat size. This effect, which is well known in relation to studies on higher organisms (Kershaw 1964), is perhaps even more important in microbiology, because of the extreme smallness of the organisms studied in relation to the normally observable environment. Most studies of soil micro-organisms have involved sampling above the milligram or millimetre level, and can therefore be referred to as macroscopic. Such studies include transects across shore lines, or comparisons between points within a study area. Even investigations of variation in microbial numbers or types down a soil profile are still macroscopic, in the present context. Only studies which are capable of revealing the position and nature of individual cells or micro-colonies, can begin to fully describe the relationship of soil micro-organisms to their environment. While the latter type of work may be more interesting for the microbial ecologist, data acquired by macroscopic investigations may adequately supply data required for integrating micro-organisms into whole ecosystem studies.

MACROSCOPIC DISTRIBUTION

The ability of macroscopic sampling to provide data on whole soil microbial populations, while completely obscuring local differences, was demonstrated by an experiment in which plate counts were performed to measure the bacterial population of nine sampling sites within the type hectare at Meathop Wood. All nine samples were collected at the same depth, in the A_1 horizon, on the same day. Despite the great heterogeneity of the site, the standard deviation of the nine counts, 28% of the mean, was no greater than normal sampling errors for replicate counts from a single sample (Gray *et al.* 1974). It cannot be inferred from this, however, that the bacteria were evenly distributed throughout the A_1 horizon soil.

Since, in a natural soil, most of the substrate for heterotrophic bacteria and fungi, is added to the surface layers, in the form of litter, counts generally decrease with depth. This general trend can be modified by adverse physical or chemical conditions. The presence of tannins or other inhibitory compounds may depress the microbial activity in some types of litter, at least temporarily (see below). Table 2.1 shows the numbers of bacteria, counted on peptone/yeast extract agar, at various depths in Meathop Wood soil. These figures show a marked decrease through the organic and mineral A horizons, but little or no change with depth within the B horizon. The figures also show a greater variability in the humus layer, suggesting a greater variety of habitats for microbial growth. This conclusion was supported by the fact that bacterial isolates from the O_h and A_h horizons were taxonomically more diverse than isolates from the A_1 horizon or from *Quercus* or *Fraxinus* litter (Hissett & Gray 1973).

Table 2.1. Numbers of viable bacteria in three soil profiles in Meathop Wood. Counts expressed as millions per g dry weight of soil. (Gray *et al.*, 1974.)

Depth (cm)	Horizon	Profile 1	Profile 2	Profile 3	Mean and Standard deviation
0	O_h/A_h	15·3	55·1	14·4	28·3±23·3
3	A	9·85	6·75	4·63	7·08±2·63
6	A/B	4·00	3·48	3·22	3·57±0·80
10	B	4·90	3·27	1·75	3·31±1·58
20	B	4·60	5·02	3·65	4·42±0·71
30	B	—*	1·45	1·59	—
40	B	—	3·70	—*	—
50	B	—	2·24	—	—
60	B	—	3·76	—	—
70	B	—	4·18	—	—
80	B	—	5·43	—	—

*Profile 1 reached bed-rock at 30 cm and profile 3 reached bed-rock at 40 cm.

MICROSCOPIC DISTRIBUTION

Investigations of the distribution of micro-organisms at the microsite level have only rarely been carried out. This is due, not only to the laborious nature of such work, but also to the difficulty of observing micro-organisms in most soil types. Soils with a high content of clay minerals or finely divided solid organic matter can easily mask an appreciable part of the bacterial population, when prepared for microscopic examination. The advantages of choosing a soil which is easy to work with, and the sort of information which can be gained, are shown by studies carried out over a period of several years on a sand dune soil at Freshfield. Microscopic studies of this soil were facilitated by the fact that sand grains accounted for more than 99% of the mineral particles in both A and C horizons.

Numbers of bacteria counted by incident-light fluorescence microscopy, on mineral and organic particles in the A_1 horizon are shown in Table 2.2a. These show that even though organic particles account for less than 15% of all the particles in the soil they are the site of well over half of the attached bacterial population. Even so, only a small fraction of the available surface is occupied—approximately 0·17% of the organic and 0·02% of the sand grain surfaces (Table 2·2b). It is interesting to conjecture why more than one third of the bacteria in the soil should be attached to sand grains, when

Table 2.2a. Distribution of bacteria on mineral and organic particles in Freshfield A_1 horizon soil. Bacteria counted by fluorescence microscopy after staining with acridine orange. (Gray *et al.* 1968.)

Soil sample	Organic matter content	No. of organic particles as a percentages of all particles	Numbers of bacteria g^{-1}			
			Mineral particles		Organic particles	
			No.	%	No.	%
1	4·3	14·9	$3·9 \times 10^5$	36	$7·0 \times 10^5$	64
2	2·9	13·4	$4·0 \times 10^5$	43	$5·3 \times 10^5$	57

Table 2.2b. Occupation of surfaces of sand grains by bacteria in Freshfield soil, based on an assumed average bacterial cell size of $1·0 \times 0·8\ \mu m$. (Gray *et al.* 1968.)

	A_1 horizon	C horizon
No. of soil particles per g of sand	$5·6 \times 10^4$	$5·6 \times 10^4$
Surface area of sand grains, $cm^2\ g^{-1}$	72·24	72·45
No. of bacteria per g of sand	$1·85 \times 10^6$	$2·39 \times 10^6$
No. of bacteria per cm^2	$2·57 \times 10^4$	$3·30 \times 10^4$
% of area colonized	0·02	0·026

there is still so much available space on the preferred organic particles. Possible reasons for this could be: (i) that the two populations are different, and have different physiological requirements; (ii) that many of the organic surfaces are hostile due to the presence of inhibitory substances or lack of nutrients; or (iii) that some of the mineral particles have thin organic coatings which are not readily visible under the microscope. A further possibility is that even organic particles may only become hospitable after modification by other localized effects (see discussion of localized pH effects later in this section).

In a recent paper, Siala *et al.* (1974) have given a more detailed picture of the distribution of one of the dominant bacteria in the Freshfield soil, *Bacillus subtilis*. They used a specific fluorescent antiserum to check a very large number of mineral and organic particles for the presence of vegetative cells of *B. subtilis*. Initial plate counts of pasteurized and unpasteurized soil suspensions, showed that 67·5% and 19% of *B. subtilis* propagules in the A_I and C horizons respectively, were in the vegetative form, despite the fact that the gross pH value of A_I horizon soil, at 4·5, was lower than the minimum for growth on laboratory media.

Data shown in Table 2.3, demonstrate that *B. subtilis* cells showed an

Table 2.3. Mean density of *Bacillus subtilis* cells on organic and mineral particles in three profiles through Freshfield soil. (Modified from Siala *et al.* 1974.)

	No. of cells per mm² of particle surface		
Horizon	Organic	Mineral	Total
Upper A_I	1·5	0·01	0·16
Lower A_I	0·60	0	0·06
Upper C	0·65	0·004	0·06
C (8–10 cm)	0·18	<0·001	0·03

even more marked preference for organic particles, than the bacterial population as a whole. At 4 levels investigated within the mineral horizons of the soil, the density of colonization was more than 100 × greater on organic particles than on mineral particles. Petrographic analysis of the mineral particles showed that these included cryptocrystalline quartz with iron oxide inclusions, cordierite, calcite, calcite with iron oxide inclusions, quartz and aluminosilicate. However, *B. subtilis* cells were almost completely confined to particles of cryptocrystalline quartz with iron oxide inclusions, even though these provided only 5–6% of the total mineral surface. Only at a depth of 8–10 cm in the C horizon were appreciable numbers of cells detected on

Table 2.4. Density of *Bacillus subtilis* cells on different types of organic particle in Freshfield soil. (Siala *et al.* 1974.)

	No. of cells per mm² particle surface	
	A_1 horizon	C horizon
Pine branches—Cortex	12·5	0·5
—Stele	13·1	0·5
Pine roots—Living	0·02	0·007
—Dead—Cortex	4·1	2·4
—Stele	6·3	1·6
Mycorrhizal rootlets—Living	0·06	0·08
—Dead	0·9	0·6
—Detached and dead	0·3	0·3
Charcoal	0·003	0·07
Organic matter/hyphal aggregates	2·7	3·0
Fungal hyphae	0	0
Sclerotia	0	0
Insect exoskeletons	0	0
Mollusc shells	0	0
Resin aggregates—HCl soluble, crumbly	0·1	0·1
—HCl insoluble, hard	0·02	0·1
Unidentifiable	3·6	0·6

another mineral surface, with 11% on quartz. A similarly selective distribution was also found for cells on organic particles (Table 2.4). The greatest cell densities were generally found on fragments of dead roots and branches of *Pinus*. Living roots and mycorrhizal rootlets were only slightly colonized. Cells were also found on aggregates of organic matter and fungal hyphae. No evidence was found for colonization of purely fungal structures or animal remains. Throughout the soil, *B. subtilis* cells occurred generally in very small colonies, averaging 4–5 cells on organic particles and only 2 cells per colony on mineral particles. Single cells were common on mineral particles. It was also noted that cells on organic particles were larger than those on sand grains (Hill & Gray 1967). Investigations of this type are probably not feasible, at present, for soils with a more usual clay and silt content, where the micro-organisms would be harder to observe, and the soil particles harder to identify. However, it does show that individual bacterial populations in soil may be more highly structured than has often been realized in the past. Clearly, if a representative collection of isolates is to be obtained from a soil, it is very important that samples should include all the microhabitats in appropriate quantities. Otherwise, some important components of the microflora could either be excluded altogether, or have their importance overlooked.

Localized pH effects

Further evidence for the importance of considering the nature of a soil at the microsite level, when discussing its suitability for microbial growth, comes from studies of the pH requirements of soil isolates. The bulk pH value of a soil, measured by electrode in a suspension or paste of the soil, may not accurately reflect the conditions experienced by individual cells of the microorganisms present. Colloidal clay particles can attract hydrogen ions, resulting in regions of increased acidity. These would be of particular importance where an organism, or its extracellular enzymes, are also adsorbed on to the clay particles (McLaren 1960). A more common situation in soil bacteriology, is the occurrence of apparently non-acid tolerant strains or species in acid soils. In these situations, the bulk pH level of the soil may be 1–2 units below the minimum for growth in laboratory media.

Williams *et al.* (1971) have investigated this situation, as it affects *Streptomyces* spp. in the Freshfield sand dune soil and in a mature podzol under *Pinus sylvestris* at Delamere, in Cheshire. Table 2.5 shows the bulk pH values of all horizons in these two soils, with plate count figures for neutrophilic and acidophilic *Streptomyces*. These were counted on starch-casein medium (Kuster & Williams 1964), used at its original pH value of 7·0, and in a

Table 2.5. Numbers of acidophilic and neutrophilic *Streptomyces*, in developing and mature podzol soils at Freshfield and Delamere respectively. (Modified from Williams *et al.* 1971.)

		No. of propagules per g dry wt. of soil (× 10^3)			
Horizon	Mean pH	Acid medium	Neutral medium	Total	Percentage of neutrophiles in population
Freshfield					
L	3·7	30	4	34	12
F_1	3·7	270	3	273	1
F_2-H	3·5	2,100	12	2,112	0·6
A_1	4·2	6	80	86	93
Upper C	7·5	0	500	500	100
Lower C	8·1	0	250	250	100
Delamere					
L	4·2	16	24	40	60
F_1	3·8	12	19	31	61
F_2-H	3·6	11	20	31	65
A_1	3·7	19	1	20	5
A_2	4·0	21	5	26	19
B_1	3·6	25	1	26	4
B_2	4·4	25	0·5	25·5	2

modified form at pH 4·5. Tests on isolates from the two media, showed that few of the organisms isolated on the acid medium could grow above pH 6·5 and conversely, few of those isolated on the neutral medium would grow below pH 5·5. They could thus be regarded as largely separate populations, so that the two counts could be combined to give a total count and an approximate estimate of the proportion of neutrophiles in the combined population. The authors also noted that all the Freshfield isolates from the acid medium appeared to belong to a single species, whereas the Delamere soil yielded 16 distinct 'species'. Williams *et al.* (1971) suggest that this may be related to the ages of the two soils. Whereas the Delamere soil is a mature podzol and has been in existence for 2,000 years (Kendrick 1958); acidification of the upper horizons at the Freshfield site only began about 60 years ago, with the planting of the alkaline sand dunes with pines. This suggestion would clearly bear investigation at analogous sites, particularly in view of the common assumption by microbial ecologists, that microbial populations in open ecosystems are capable of rapid change and adaptation. The time scale for such changes may sometimes be longer than is commonly supposed. The Freshfield and Delamere results also shed some light on the mobility of soil bacteria between soils. The two sites are 45 km apart, in a north–south direction. Assuming that, at least in the two decomposing pine litters, there must be some similar acidic microsites, this distance would appear to be an insuperable barrier to spread.

An explanation of the survival of the dominant neutrophilic *Streptomyces* sp. (F1) in the acid horizons at Freshfield, was provided by experiments reported by Williams & Mayfield (1971). They showed that F1 occurred most frequently on fragments of dead pine roots, particularly mycorrhizal roots. When various nitrogenous materials were added to nonsterile A_1 hori-

Table 2.6. Effects of amendment of nonsterile A_1 horizon Freshfield soil with nitrogen-containing substances. (Williams & Mayfield 1971.)

	pH value of		
Substance added	Mineral grains	Organic particles	Bulk soil
No amendment	4·2	4·2	4·2
Ammonium sulphate	4·1	4·1	4·1
Ammonium phosphate	4·1	4·1	4·1
Glucosamine hydrochloride	4·2	4·2	4·2
L-lysine hydrochloride	4·1	4·2	4·1
L-lysine (free base)	4·1	5·6	4·5
Casamino acids	4·5	6·6	4·8
Casein	4·3	5·6	4·6
Chitin	4·1	5·6	4·3
Dead fungus mycelium	4·6	6·8	4·7

zon soil in the laboratory, only minor changes in the bulk pH value were recorded, but some substances caused a much greater increase in the pH level around organic particles (Table 2.6). Counts of F1 in the amended soil over a 7-week period, showed that only those substances which raised the pH level of organic particles to around 6·0 (casamino acids, chitin and fungus mycelium) stimulated an increase in the population. Where such stimulation did occur, increases of three orders of magnitude were common. Samples of heat-sterilized A_I horizon soil were amended with either chitin or dead fungal mycelium (*Trichoderma viride*) and inoculated with pure cultures of bacteria and fungi, or with a suspension of A_I soil. Increases in pH value around organic particles correlated with the appearance of free ammonia in the soil, (Table 2.7). Of the organisms tested, only the four

Table 2.7. Ability of selected micro-organisms to liberate ammonia, and create microsites with high pH value, in sterile A_I horizon Freshfield soil amended with either chitin or dead *Trichoderma viride* mycelium. (Williams & Mayfield 1971.)

Organism	pH around organic particles		Nessler's reagent
	Soil+ chitin	Soil+dead mycelium	(both systems)
Trichoderma viride	6·8	6·8	+
Penicillium decumbens	6·8	6·8	+
Mortierella sp.	6·6	6·8	+
Mucor sp.	6·6	6·6	+
Bacillus sp.	4·8	4·8	−
Gram negative rod	5·2	5·4	±
Streptomyces F1	4·2	4·1	−
Streptomyces F6	4·1	4·1	−
Streptomyces C1	4·1	4·1	−
A_I soil suspension	6·8	6·8	+
Uninoculated soil	4·2	4·3	−

fungi, and the A_I soil suspension, liberated sufficient ammonia to create microsites at pH 6·0 or greater. The three *Streptomyces* spp. (all neutrophiles) did not alter the pH value of the organic particles and had probably died without any growth occurring, despite the fact that all three were known to be capable of decomposing chitin and fungus mycelium in suitable conditions. During sterilization in the autoclave, localized sites of near-neutral or alkaline reaction appear to be destroyed, so that only the acid-tolerant fungi were capable of initiating growth. When sterilized soils were first inoculated with a fungus, such as *Trichoderma viride*, to create microsites at pH 6·0 or above, the neutrophilic *Streptomyces* could also grow. A similar dependence on initial fungal growth to create favourable microsites in sterile A_I horizon soil has also been shown for *Staphylococcus*, *Micrococcus* and *Arthrobacter* isolates (Lowe & Gray 1973): and probably also applies to *Bacillus subtilis* (Siala & Gray 1974).

It is not clear whether this requirement for acid-tolerant decomposers, such as fungi, to liberate ammonia, and thus lead to the creation of high pH regions around organic particles, before growth of non-acid tolerant species, is a laboratory artefact. If the neutrophilic species are themselves capable of liberating ammonia from nitrogenous compounds, they might be able to maintain the high pH microsites once these are formed. However, ammonia would presumably be continually lost from such sites by leaching, nitrification and possibly by gaseous diffusion. Thus, a constant supply of suitable nitrogenous materials would be required. It is, perhaps, unlikely that such a constant supply would be maintained continuously in natural soils. In fact, Lowe & Gray (1973) showed that at least some of the neutrophilic bacteria in nonsterile A_I horizon soil were pH-limited, and showed a burst of growth if the soil were adjusted to neutrality (bulk pH) by the addition of KOH or $CaCO_3$. It is likely, therefore, that microsites suitable for neutrophilic bacteria within an acid soil cannot always be maintained, but are created by the soil fungi in response to a new input of a suitable substrate.

The time factor in soil microbiology

Growth and death of soil micro-organisms can be viewed from two standpoints—that of the individual cell, and that of the population as a whole. Although the two are obviously connected, it may be deceptively difficult to use information relating to individual cells to infer what is happening to whole populations, and vice versa. At any one time, at the microsite level, microbial cells will either be growing, lying dormant (but, with the possible exception of bacterial endospores, still respiring and using energy), or dying or being eaten by the soil microfauna. The changes in population observable at the macroscopic level will be the resultant of these three processes, which may all be occurring simultaneously in different microsites. Furthermore, because of difficulties in rapid identification, particularly in bacteriology, most counts will cover a wide range of species. So population changes recorded are normally only the means of a large number of individual species fluctuations.

Since the rate of addition of fresh substrates to soil usually shows marked seasonal fluctuations, with a peak at the time of leaf-fall in the autumn, it might be expected that populations of decomposer micro-organisms would show parallel fluctuations. While such fluctuations usually do occur in the litter horizons, they are not always detectable in the mineral soil horizons (Gray *et al.* 1974). Seasonal changes have been noted in soils subject to regular periods of drought (Meiklejohn 1957), and leaf litter decomposition can also be delayed if the litter remains dry on the ground (Minderman & Daniels 1967). It would seem that large seasonal population variations are more likely to be caused by climatic factors than by substrate availability (Gray & Williams 1971a). Lack of measurable population change, of course, does not necessarily imply a constant level of activity on the part of the soil microflora. Changes in cell turnover rate could occur without altering the standing crop of cells.

COLONIZATION OF FRESH SUBSTRATES IN SOILS

One of the best pointers to the possible growth rates of soil micro-organisms during periods when they are not substrate limited, is the rate of colonization of new substrates, particularly plant litter. Calculations of maximal growth rates from the initial increase in populations on such substrates rest on two main assumptions: (i) that the population observed at each sampling time consists predominantly of actively growing cells; and (ii) that predation is negligible.

That the proportion of viable, and therefore probably growing, cells may be high during the initial colonization of leaf litter was supported by a count of the total number of cells observable on *Fraxinus* leaflets, after about one month on the ground at Meathop Wood. A direct count, using fluorescein isothiocyanate staining (Babiuk & Paul 1970), gave a count only 5–10 times greater than dilution plate counts (Hissett 1970). This is much lower than comparable viable to total cell ratios for *Quercus* litter and O_h+A_h and A_I horizon soils (Hissett 1970) which were generally in the 200–1,000 range; and, in view of the probable inefficiency of the plate count technique when applied to bacteria on leaf litter, suggests that most of the cells were probably growing. Analogy with chemostat studies would also suggest that viability levels should be high in a rapidly growing culture (Tempest *et al.* 1967).

With regard to the rate of predation experienced by bacteria during the initial colonization of fresh leaf litter, there appears to be no information on the rate of increase of populations of protozoa or other microbial feeders on leaf litter. However, studies on model aquatic systems suggest that the build-up of predatory protozoa populations may lag behind those of their prey bacteria by several days. (Curds 1971; Bick & Muller 1973). *A priori*, there seems no reason why a similar delay should not occur on leaf litter. With special regard to the bacterial population on *Fraxinus* leaf litter at

Meathop Wood, it has been shown that 45% of isolates were *Arthrobacter* spp. (Hissett & Gray 1973). This genus has been shown to be unsuitable for supporting growth of a common soil ciliate (Coler & Gunner 1969), and may, therefore, be less subject to heavy predation than some other soil bacteria.

Assuming the correctness of the above arguments, the generation time for bacteria colonizing air-dried *Fraxinus* leaflets placed on the soil surface at Meathop Wood, can be calculated from plate count data (Table 2.8). In November, 1967, the count increased almost tenfold in the first three days, giving a mean doubling (i.e. generation) time of 21·6 hours. Since some time will have been lost before the air-dried litter became moist enough to permit bacterial growth, the actual growth rate may have been higher than this estimate. In November, 1968, the count of bacteria on the litter only increased by a factor of 2·5 during the first three days, giving a mean generation time of 53·4 hours. However, since there was no recordable rainfall during the first week after the litter was placed on the ground in 1968 (compared with 111 mm in 1967), the poor growth rate probably reflects the dry state of the litter.

Some other estimates of growth rates for soil bacteria under highly favourable conditions, drawn from published data, are given in Table 2.9.

Table 2.8. Dilution plate counts of bacteria on *Fraxinus* leaflets placed on the ground in Meathop Wood in early November. Counts in millions per g of oven dry leaf litter. (Gray *et al.* 1974.)

	Number of bacteria	
Age of litter	1967/8	1968/9
Initial count	59·1	51·1
3 days	561	130
1 week	1,625	222
2 weeks	2,040	108
4 weeks	800	—
8 weeks	1,200	—

These are not intended to be typical of mean growth rates in natural soils, but show what the peak growth rates may be. In all the examples given, the amounts of available substrate have been artificially increased, either by drying, sterilizing and remoistening, or by actual amendment. In addition, all but the last experiment (Shields *et al.* 1973) were carried out at a temperature well above the average for a temperate region soil. The doubling times given for the *Streptomyces* sp. are probably a slight overestimate, since the plate counting technique used would not have adequately measured

Table 2.9. Growth rates for soil bacteria under favourable conditions, calculated from published data.

Organism	Site	Experimental details	Temp °C	Generation (or doubling) time (h)	Ref*
Arthrobacter sp.	Laboratory	Sterile A_1 horizon sand dune soil.			
		pH adjusted to 7·5 with $CaCO_3$.	25	6·6	1
		Ditto pH adjusted with KOH	25	6·5	
		Sterile C horizon sand dune soil pH 8·1	25	12	
Micrococcus sp.	Laboratory	Sterile A_1 horizon sand dune soil.			
		pH adjusted to 7·5 with $CaCO_3$.	25	8·8	1
		Sterile C horizon sand dune soil pH 8·1	25	13·2	
Streptomyces sp.	Laboratory	Non-sterile A_1 horizon sand dune soil (period of maximum growth)			
		i. Amended with dead fungus mycelium	25	13	2
		ii. Amended with chitin	25	29	
		iii. Amended with casamino acids	25	44	
Mixed	Field	Brown chernozemic soil in open cylinder embedded in field soil.			
		Amended with glucose	?†	19	3

*References—1. Lowe & Gray 1973
2. Williams & Mayfield 1971
3. Shields *et al.* 1973.

†Experiment carried out in early May in Saskatchewan, Canada.

mycelial production. However, since the population changes used for the calculations were large, the proportional errors, from this source, in the doubling time estimate, are unlikely to be very great.

MEAN GROWTH RATES FOR SOIL BACTERIA

Siala & Gray (1974) measured the rate of growth of microcolonies of *Bacillus subtilis* on glass microscope slides buried in nonsterile unamended Freshfield A_I horizon soil. After a decrease in the number of cells during the first week after they were placed in the soil, growth was recorded during the second and third weeks. The mean generation time during this second period was 230 h. Despite the inevitable disturbance involved in bringing soil samples into the laboratory, and the slightly unnatural glass surface on which the bacteria were growing, this estimate is probably much closer to actual mean growth rates for soil bacteria than any of the figures given in Table 2.9.

Attempts have been made to estimate the mean doubling times for soil microflora, from the fluctuations observed in regular counts (Parinkina 1973, 1974). This method is based on the assumption that if significantly large fluctuations are observed, these must reflect real changes in the population, and that microbial production can be calculated by summing all the increases over an experimental period. The mean doubling time can be calculated from the ratio of production to mean biomass. However, before meaningful figures can be produced, it is necessary to accurately determine the experimental errors involved in sampling and counting. The apparent 'production' caused by the sampling error fluctuations in population estimates can then be subtracted from the calculated production to give an estimate of real production. If such corrections are not made, the mean growth rate will be overestimated, and will be determined by the frequency of counting, as much as by the actual growth of soil bacteria. If fully corrected for counting errors, this approach should underestimate growth, since it assumes that growth and death do not occur simultaneously. It is doubtful whether the method has any general validity, apart from measurements of the initial response to a change in the environment, such as the addition of fresh substrates to a soil.

Since there appear to be no very satisfactory methods of directly measuring normal rates of bacterial production in undisturbed soil, a few attempts have been made to calculate possible growth rates, based on the mean standing crop or biomass and the rate of addition of fresh substrates to the soil. Such calculations have used the familiar equation of Marr *et al.* (1963):

$$\frac{dx}{dt}+ax \quad Y\frac{ds}{dt}$$

where x = biomass of cells, a = specific maintenance constant, Y = yield coefficient (weight of biomass produced per unit weight of substrate utilised), and s = weight of substrate available.

Accurate figures for the two constants, a and Y, are not available for soil bacteria growing *in situ*, but some recent results enable earlier assumption, presumably as microbial biomass or stable metabolites. Chemostat exused a value of 0·001 h^{-1} for the specific maintenance constant, a. This has recently been supported by Shields *et al.* (1973) who calculated figures of 0·001–0·002 from studies on a Canadian field soil.

Babiuk & Paul (1970) and Gray & Williams (1971b) used a yield coefficient (Y) of 0·35. However, Payne (1970) suggested that coefficients were commonly as high as 0·60. Shields *et al.* (1973) found that 60% of the carbon from glucose added to a field soil was retained after the first flush of respiration, presumably as microbial biomass or stable metabolities. Chemostat experiments on *Arthrobacter globiformis* have also suggested a yield coefficient of about 0·5 (Luscombe & Gray, 1974). In view of these findings, the earlier figure Y = 0·35 may be slightly too low.

Brock (1971) has briefly discussed the relative merits of using total or viable counts as the basis for estimating microbial growth rates by different methods. Babiuk & Paul (1970) used total counts as a basis for their calculations, and found that more than half of the substrate reaching the soil would be required for maintenance of the soil bacteria alone. This would leave little or no substrate for growth, once the soil fungi and other decomposers are considered. They concluded that plate counting probably gives a better estimate of metabolizing cells.

If the bacterial biomass and substrate input rates for Meathop Wood, given by Gray *et al.* (1974), are used with a yield coefficient Y = 0·50 and a maintenance constant a = 0·001 h^{-1}; the bacteria (excluding actinomycetes) appear to require 646 kg of substrate per hectare year for maintenance. Anderson & Domsch (1973) found that bacteria only accounted for 22% of the respiration in an agricultural soil, the other 78% being fungal in origin. In an acidic forest soil, the bacterial share of the total substrate input is unlikely to be much higher than this. Assuming (i) that the bacteria utilize 25% of the organic matter reaching the soil in Meathop Wood, and (ii) that all the bacterial biomass produced becomes available, at some time, as substrate for further growth; the mean generation time for bacteria on the site would be 12 days.

General conclusions

Although there are large errors and uncertainties involved in all estimates of growth rates for soil bacteria, it seems that soil bacteria seldom, if ever, grow at rates commonly observed in the laboratory. Even on fresh plant litter which does not contain inhibitory substances, it is unlikely that doubling times often fall much below 15 hours. The doubling time for the whole viable population of natural soils is probably in excess of 8–10 days in most instances. In view of these very low growth rate estimates, the question whether most soil micro-organisms show autochthonous or zymogenous growth patterns, would seem to be as big a challenge to soil microbiologists as it has ever been.

References

ANDERSON J.P.E. & DOMSCH K.H. (1973) Quantification of bacterial and fungal contributions to soil respiration. *Arch. Mikrobiol.* **93**, 113–27.

BABIUK L.A. & PAUL E.A. (1970) The use of fluorescein isothiocyanate in the determination of the bacterial biomass of grassland soil. *Can. J. Microbiol.* **16**, 57–62.

BICK H. & MULLER H.P. (1973) Population dynamics of bacteria and protozoa associated with the decay of organic matter. In *Modern methods in the study of microbial ecology*, ed. T. Rosswall. *Bull. Ecol. Res. Comm.* (Stockholm) **17**, 379–86.

BROCK T.D. (1971) Microbial growth rates in nature. *Bact. Rev.* **35**, 39–58.

COLER R.A. & GUNNER H.B. (1969) Microbial populations as determinants in protozoan succession. *Wat. Res.* **3**, 149–56.

CURDS C.R. (1971) Computer simulations of microbial population dynamics in the activated sludge process. *Wat. Res.* **5**, 1049–66.

GOODFELLOW M., HILL I.R. & GRAY T.R.G. (1968) Bacteria in a pine forest soil. In: *The ecology of soil bacteria*, eds. T.R.G. Gray & D. Parkinson, pp. 500–15. Liverpool University Press.

GRAY T.R.G., BAXBY P., HILL I.R. & GOODFELLOW M. (1968) Direct observation of bacteria in soil. In: *The ecology of soil bacteria*, eds. T.R.G. Gray & D. Parkinson, pp. 171–92. Liverpool University Press.

GRAY T.R.G., HISSETT R. & DUXBURY T. (1974) Bacterial populations of litter and soil in a deciduous woodland II.—Numbers, biomass and growth rates. *Rev. d'Écol. Biol. Sol*, **10**, 15–26.

GRAY T.R.G. & WILLIAMS S.T. (1971a) *Soil Micro-organisms*. Oliver & Boyd, Edinburgh.

GRAY T.R.G. & WILLIAMS S.T. (1971b) Microbial productivity in soil. In 'Microbes and biological productivity', *Symp. Soc. gen. Microbiol.* **21**, 255–86.

HILL I.R. & GRAY T.R.G. (1967) Application of the fluorescent antibody technique to an ecological study of bacteria in soil. *J. Bact.* **93**, 1888–96.

HISSETT R. (1970) *The bacterial population of leaf litter and soil in a deciduous woodland.* Unpublished PhD thesis. University of Liverpool.

HISSETT R. & GRAY T.R.G. (1973) Bacterial populations of litter and soil in a deciduous woodland—I Qualitative studies. *Rev. d'Écol. Biol. Sol*, **10**, 495–508.

KENDRICK W.B. (1958) *Microfungi in pine litter*. Unpublished Ph.D. thesis. University of Liverpool.

KERSHAW K.A. (1964) *Quantitative and dynamic ecology*. Edward Arnold, London.

KÜSTER E. & WILLIAMS S.T. (1964) Selection of media for isolation of streptomycetes. *Nature, Lond.* **202**, 928–9.

LOWE W.E. & GRAY T.R.G. (1973) Ecological studies on coccoid bacteria in a pine forest soil—II. Growth of bacteria introduced into soil. *Soil Biol. Biochem.* **5**, 449–62.

LUSCOMBE B.M. & GRAY T.R.G. (1974) Characteristics of *Arthrobacter* grown in continuous culture. *J. gen. Microbiol.* **82**, 213–22.

MCLAREN A.D. (1960) Enzymic activity in structurally restricted systems. *Enzymologia* **21**, 356–64.

MARR A.G., NILSON E.H. & CLARK D.J. (1963) The maintenance requirement of *Escherichia coli. Ann. N.Y. Acad. Sci.* **102**, 536–48.

MEIKLEJOHN J. (1957) Numbers of bacteria and actinomycetes in a Kenya soil. *J. Soil Sci.* **8**, 240–7.

MINDERMAN G. & DANIËLS L. (1967) Colonisation of newly fallen leaves by micro-organisms. In: *Progress in soil biology*, eds O. Graff & J.E. Satchell, pp. 3–9. North Holland Publishing Co.

PARINKINA O.M. (1973) Determination of bacterial growth rates in tundra soils. In: *Modern methods in the study of microbial ecology*, ed. T. Rosswall. *Bull. Ecol. Res. Comm.* (Stockholm) **17**, 303–309.

PARINKINA O.M. (1974) Bacterial production in tundra soils. In: *Soil organisms and decomposition in tundra*, eds. A.J. Holding, O.W. Heal, S.F. MacLean Jr. & P.W. Flanagan, pp. 65–77. Tundra Biome Steering Committee (Stockholm).

PAYNE W.J. (1970) Energy yields and growth of heterotrophs. *Ann. Rev. Microbiol.* **24**, 17–52.

SHIELDS J.A., PAUL E.A., LOWE W.E. & PARKINSON D. (1973) Turnover of microbial tissue in soil under field conditions. *Soil Biol. Biochem.* **5**, 753–64.

SIALA A. & GRAY T.R.G. (1974) Growth of *Bacillus subtilis* and spore germination in soil observed by a fluorescent-antibody technique. *J. gen. Microbiol.* **81**, 191–8.

SIALA A., HILL I.R. & GRAY T.R.G. (1974) Populations of spore-forming bacteria in an acid forest soil, with special reference to *Bacillus subtilis*. *J. gen. Microbiol.* **81**, 183–190.

TEMPEST D.W., HERBERT D. & PHIPPS P.J. (1967) Studies on the growth of *Aerobacter aerogenes* at low dilution rates in a chemostat. In: *Microbial physiology and continuous culture*, eds. E.O. Powell, C.G.T. Evans, R.E. Strange & D.W. Tempest, pp. 240–53. H.M.S.O., London.

WILLIAMS S.T., DAVIES F.L., MAYFIELD C.I. & KHAN M.R. (1971) Studies on the ecology of actinomycetes in soil—II. The pH requirements of streptomycetes from two acid soils. *Soil Biol. Biochem.* **3**, 187–95.

WILLIAMS S.T. & MAYFIELD C.I. (1971) Studies on the ecology of actinomycetes in soil—III. The behaviour of neutrophilic streptomycetes in acid soil. *Soil Biol. Biochem.* **3**, 197–208.

3 Carbon and nitrogen levels as factors influencing fungal decomposers

D. PARK *Botany Department, The Queen's University, Belfast, U.K.*

Abstract

One of the variables influencing the rate and extent of decomposition of biodegradable materials is the supply of mineral nutrients, especially of nitrogenous nutrients, since micro-organisms have a relatively high requirement for nitrogen, and nitrogenous compounds are often limiting in decomposer habitats.

C/N (or C/mineral) ratio is much used as an indicator of potential decomposability. It is argued here that insufficient consideration has been given to the validity of this coefficient, particularly in relation to microbial growth on insoluble carbonaceous substrata. Concentration of nitrogen (or mineral) supply is suggested as being, in some situations, more relevant to microbial activity. Data presented support the connection of nitrogen source concentration with the time-course of decomposition and with the sequence of activity of different micro-organisms during decomposition.

Introduction

The importance of mineral supply for micro-organisms during decomposition was recognized early in the present century. Hutchinson & Richards (1921) introduced the term 'nitrogen factor' to quantify the relationship between available N and amount of decomposition. The C/N ratio of the decomposable material was later used (e.g. Waksman & Tenney 1928; Tenney & Waksman 1929; Jensen 1931a), and the subsequent development of this coefficient as an index of decomposability and the degree of immobilization as against mineralization may have led to some neglect of the significance of the simpler parameter of concentration of mineral source in the microbial environment. Some data obtained by the author (Park 1976) were best explained on this more direct basis, and led to the present examination.

Media of different N levels in the isolation of fungi

Examinations of decomposing materials for the presence of decomposers are usually made using an isolation medium containing N at about 100 mg l^{-1}

Table 3.1. Cellulose-decomposing fungi isolated from several sources using medium of two different N levels.

	Medium	
	Standard	Low N
N mg l^{-1}	142	38
C/N ratio	31	117
Source		
Viable seed	*Alternaria* sp.	Nil
Decomposing *Crataegus* log	*Chrysosporium pannorum* *Trichoderma polysporum* *T. viride* *Trichosporiella cerebriformis*	*Monocillium* sp. Sterile mycelium (S2)
Sisal door mat	*Botrytis cinerea* *Cladosporium cladosporioides* *Chaetomium* sp. *Gliocladium roseum* *Humicola grisea* *Mammaria echinobotryoides* *Pyrenochaeta acicola* Sterile mycelium (S1) *Varicosporium elodeae*	*Humicola grisea* *Fusarium oxysporum*
Dead *Prunus* leaf	*Cladosporium cladosporioides* *Epicoccum purpurascens* *Phialophora* sp. *Varicosporium elodeae*	*Cladosporium cladosporioides* *Phialophora* sp. *Phoma* spp. (3 types) *Volucrispora aurantiaca*
Live *Polypodium* pinna	Sterile mycelia (S3 & S4)	*Alternaria* sp. *Cephalosporium* sp. Sterile mycelium (S5)
Senecio rhizosphere	*Cladosporium cladosporioides* *Pellicularia* sp. *Penicillium terrestre* *Pyrenochaeta acicola* *Phoma* sp. *Scytalidium lignicola* Sterile mycelia (S6 & S8)	Sterile mycelium (S7)

or more, and with a C/N ratio of around 40. Several habitats were here investigated by making otherwise comparable platings on two media, one being standard cellulose agar (Park 1973)*, the other having one fiftieth the amount of ammonium salt and yeast extract. Calculated N levels are 142 mg l^{-1} and 38 mg l^{-1} respectively, and C/N ratios 31 and 117 respectively. The range of N content had been intended to be wider, but allowance had not been made in the design for the relatively high N content of the agar-agar used. Even within this narrow range of difference between the media the results of the isolations showed interesting differences. Table 3.1 is a qualitative summary of the results and shows that from most sources fewer species were obtained on the low N medium, and that these were mostly different from those obtained on the standard medium. Out of 40 isolate-types only three appeared on both series of plates. Estimates of biomass and activity of micro-organisms during decomposition depend in part on a knowledge of the species present, yet this present result indicates that our knowledge of this aspect may be limited by the conventional use of methods that are not always appropriate.

Elective culture microcosms at different N levels

Baiting methods, by exposing decomposable substrata at a natural site, and subsequent isolation or identification, are a common way of obtaining information about active microbial decomposers. Control of the N level of the baits is, however, not easy. An elective culture method was here used to detect how fungi might differ in their ability to become active in cellulosic material at different N levels. Sterile flasks (250 ml) with 0·2 g commercial cellulose and 20 ml mineral salts medium at three N levels (106, 10·6 and 1·06 mg l^{-1}; C/N ratios 42, 420 and 4,200) were inoculated with 0·02 ml of a blended suspension of a natural substratum, and incubated at 13° for 60–65 days. Dilutions ($2{\cdot}5 \times 10^{-4}$) of blended microcosm contents from each set were plated, 1 ml over each of 10 standard cellulose agar plates and subsequent fungal growth identified and counts made. Fungi represented by more than one occurrence in any set are included in the summary in Table 3.2. Out of the 28 isolate-types, 12 had made successful activity at each N level, and six had been active at both the lower N levels. Among these two groups there were several types not usually found in published lists of cellulose-decomposing fungi. In contrast *Penicillium spinulosum* and *Trichoderma koningii* were found only at the highest level, and *T. harzianum* and *T. viride* only at the intermediate N level. These four species belong to types common

*[$(NH_4)_2SO_4$, 0·5 g; KH_2PO_4, 1·0 g; KCl, 0·5 g; $MgSO_4.7H_2O$, 0·2 g; $CaCl_2.2H_2O$, 0·1 g; yeast extract, 0·5 g; ball-milled cellulose, 10·0 g; agar, 10·0 g; water to 1 l].

Table 3.2. Development of cellulolytic fungi in microcosms at different N levels.

Inoculum site No.	1			2			3			4			5			6		
N level	H	M	L	H	M	L	H	M	L	H	M	L	H	M	L	H	M	L
Candida humicola		1				1	3	2	1		1						1	3
Cephalosporium spp.					2	2								1				
Chrysosporium spp.							3				3			1			3	1
Chrysosporium pannorum	3	3	1															
Cladosporium cladosporioides		1			1	1						1			1			
Cylindrocarpon destructans	2	1	1					2							1			
Cylindrocarpon magnusianum				1					2									
Fusarium oxysporum				1	2	1			1									
Fusidium griseum										1	1		1					
Gliocladium roseum	2							2										
Humicola sp.							1	3					2					
Humicola grisea		1	2											2			2	1
Mammaria echinobotryoides	1	1	1	1				1										
Penicillium spinulosum							3											
Phialophora spp.				1	3	1		1		1			3	2				1
Scytalidium lignicola		1	1					2	1									1
Sphaeropsidales			1		1	2		1		1	2	2			1			
Sporotrichum	2		1		3				2									
Trichoderma harzianum								1										
Trichoderma koningii	2															3		
Trichoderma viride					1									1				
Trichosporiella cerebriformis		1	1	2		1		2	1	2	1							1
Trichosporon cutaneum																	2	2
Verticillium sp.			1			1			1			1	1	2	2			
Verticillium lateritium										3	1	1						
Verticillium nigrescens			1						1									
Volucrispora aurantiaca	2									3	2							
Volutella ciliata								3	1					2			2	

N levels (mg l^{-1}): H, 106; M, 10·6; L, 1·06.
Sites: 1, kitchen garden soil; 2, rotten branch; 3, acid, stony soil; 4, angiosperm leaf litter; 5, pine litter; 6, shrub border soil.
Species entries: no entry—0 or 1 colony/10 plates
1—2–10 colonies/10 plates
2—11–100 colonies/10 plates
3—over 100 colonies/10 plates

in records of cellulose-decomposers, and much used in experimental studies of cellulolysis. In addition to these a further four species failed to develop at the lowest N level although showing success where higher concentrations of N were present. One species, *Verticillium nigrescens*, was recorded only from the lowest N level, and that in two of the site sets.

The results indicate that the fungal community that develops in a substratum from a mixed inoculum such as occurs in soil or litter may be influenced by the N level of the exposed substratum. However, the single observation after a rather lengthy time interval obscures the interpretation. An organism could be absent from a set at a particular time because the given conditions were not suitable for its activity, because it was repressed by activity of other organisms or because it had declined after a previous period of activity. Further, an organism, unable by itself to be active under the given conditions, could have its activity promoted by previous activity of other organisms. The data in Table 3.2, while demonstrating that N level affects the species present and their relative abundance, also reveal differences between the inoculum series in the response of individual species to the different N levels. Some of these differences seem likely to relate to the interactions between species initially selected by the original conditions. To analyse the situation in more detail a single site was chosen for the following experiments.

Time-course of colonization at different N levels

SITE EXAMINATION

An accessible site at the Botany Department experimental grounds, Derryvolgie, Belfast was chosen, and the soil examined on several occasions by soil plating (Warcup 1950) and by dilution plating (8×10^{-3}) using cellulose agar at three nitrogen levels (142, 46 and 38 mg l^{-1}). The differences (Table 3.3) between the results from the media were less marked than some of those described above (Tables 3.1 and 3.2), though a number of isolate types failed to appear at the lowest N level (e.g. *Aspergillus fumigatus*, *Botrytis cinerea*, *Phialophora* sp, Sphaeropsidales). More marked were the differences between the method of plating, *Humicola grisea*, *Monodictys levis*, *Trichocladium asperum* and *Ulocladium botrytis* failing to appear on the dilution plates, and *Aspergillus fumigatus*, *Cephalosporium* spp., *Trichosporiella cerebriformis* and Sphaeropsidales not appearing on the soil plates.

ELECTIVE CULTURE MICROCOSMS

Flasks (100 ml) with 1·0 g commercial cellulose and 20 ml mineral salts medium were used as the units. Four levels of N were used: 400, 100, 20 and 4 mg l^{-1} (C/N ratios 10, 40, 200 and 1,000 respectively). After inoculation with 1 ml of a 10^{-2} dilution of soil the flasks were incubated at 11°–

Table 3.3. Fungi isolated from Derryvolgie soil.
S=soil plates (+; present)
D=dilution plates (numbers $10^{-3} g^{-1}$ soil).

Plating medium: mg N l^{-1}	142		46		38	
Isolation method	S	D	S	D	S	D
Aspergillus fumigatus		11		4		
Botrytis cinerea		4	+	1	+	
Cephalosporium spp.		2		6		7
Chrysosporium pannorum	+	1		1		
Cladosporium cladosporioides	+	39	+	16	+	16
Cylindrocarpon destructans	+	15	+	7	+	11
Epicoccum purpurascens		1				
Fusarium culmorum	+	27	+	10	+	1
Fusarium oxysporum		1				
Gliocladium roseum				1		
Humicola spp.	+		+		+	
Mammaria echinobotryoides		1	+	3	+	4
Metarrhizium anisopliae		3				
Monodictys levis	+		+		+	
Mortierella vinacea	+		+		+	8
Penicillium spp.		8	+	7	+	4
Phialophora spp.		3				
Phoma spp.		18		10		
Pyrenochaeta acicola		17		17		
Scopulariopsis brevicaulis	+	3		4		
Scytalidium lignicola	+	6	+	4	+	6
Sporothrix sp.	+	2				
Trichocladium asperum	+					
Trichoderma harzianum	+	4				
Trichosporiella cerebriformis	+	1		1		
Ulocladium botrytis	+		+		+	
Verticillium sp. 1	+	61	+	34	+	17
Verticillium lateritium	+	1			+	1
Volucrispora aurantiaca		1				
Volutella ciliata	+	43	+	52	+	31

12° and observations made at intervals from 14 to 70 days by blending whole microcosm contents and plating convenient dilutions, 1 ml to each of ten standard cellulose agar plates, from which identification and counts were made after incubation. The results (Fig. 3.1) showed successional patterns at each N level. At 400 mg l^{-1} N early activity by *Monodictys*, *Humicola*, *Cylindrocarpon*, *Chrysosporium* and *Penicillium* was followed by a phase of increasing dominance by *Penicillium*, with *Chrysosporium* following closely for a time then dropping out and giving way to *Trichoderma*. At 100 mg l^{-1} N initial activity was similar but *Volutella*, *Volucrispora* and *Trichocladium*

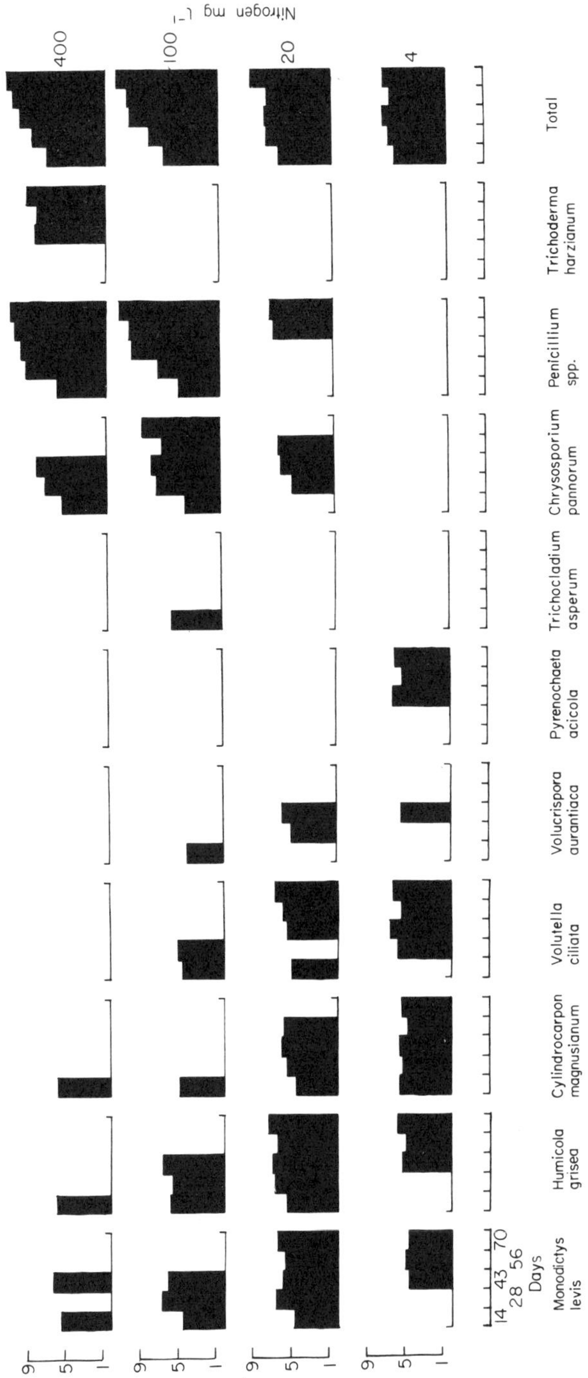

Figure 3.1. Elective culture microcosms at different N levels; patterns of colonization (11°–12°C).

were also present. *Penicillium* and *Chrysosporium* continued to increase, the other species dropping out. *Trichoderma* did not appear at this or at any lower N level. At 20 mg l^{-1} N *Monodictys* and *Humicola* again appeared as dominants initially, and here persisted throughout. *Cylindrocarpon* maintained its contribution for longer than at higher N levels, but was not represented at the final observation. *Penicillium* and *Chrysosporium* were later in appearing and were lower in numbers, *Chrysosporium*, indeed, being absent from the 70-day population. At 4 mg l^{-1} N *Cylindrocarpon* was dominant early and persisted, unlike its pattern at the higher N levels. *Monodictys*, *Humicola* and *Volutella* appeared later in this set, but contributed largely to the less diverse population through the remainder of the observations. *Pyrenochaeta acicola* appeared at this N level only. A two-dimensional pattern is apparent, in time and in N level.

The same system was used in a repetition with closer time intervals from 4 days to 28 days. Unfortunately some lack of control of temperature was experienced in this series and the range (13°–15°C) was higher. Possibly related to this is a condensation of the time-course of events, and the results (Fig. 3.2) differ in some other details, but a similar pattern is evident. *Monodictys*, *Humicola*, *Cylindrocarpon*, *Chrysosporium* and *Penicillium* present sequences of the same form as previously. *Trichoderma* showed a similar temporal pattern to that given earlier, but was present at lower N levels. *Scytalidium* and *Cladosporium*, neither of which appeared in the earlier series, were active here as temporary representatives at the upper two N levels, and *Cladosporium* also entered at the last observation at 20 mg l^{-1} N. *Trichocladium* was again a transient contributor restricted by N level. *Mortierella vinacea* was recorded here. The species had also appeared in the platings in the earlier series but records had not been made of its occurrence, because of doubts about its cellulolytic nature (Gams 1960). In this experiment its consistency demanded that consideration be given to its contribution to the population. Its pattern follows closely that of *Penicillium*.

The method used here, blending and dilution plating, has limitations, particularly since some of the contributory species were not revealed at the site by this method, but the technique does permit relative numerical comparison within a species. A major disadvantage is that as total 'numbers' increase the increasing dilution necessary reduces the sensitivity for organisms with a lower 'numerical' level than those making the major contribution to 'numbers'. Because of this a disappearance with time of a species from the records may mean a decline of its activity or may mean no more than a persistence at about the previous level. Despite the difficulties of interpretation and despite some discrepancies between the two runs, possibly five categories of organism can be identified:

I Early colonizers, soon declining at high N levels, but persisting longer

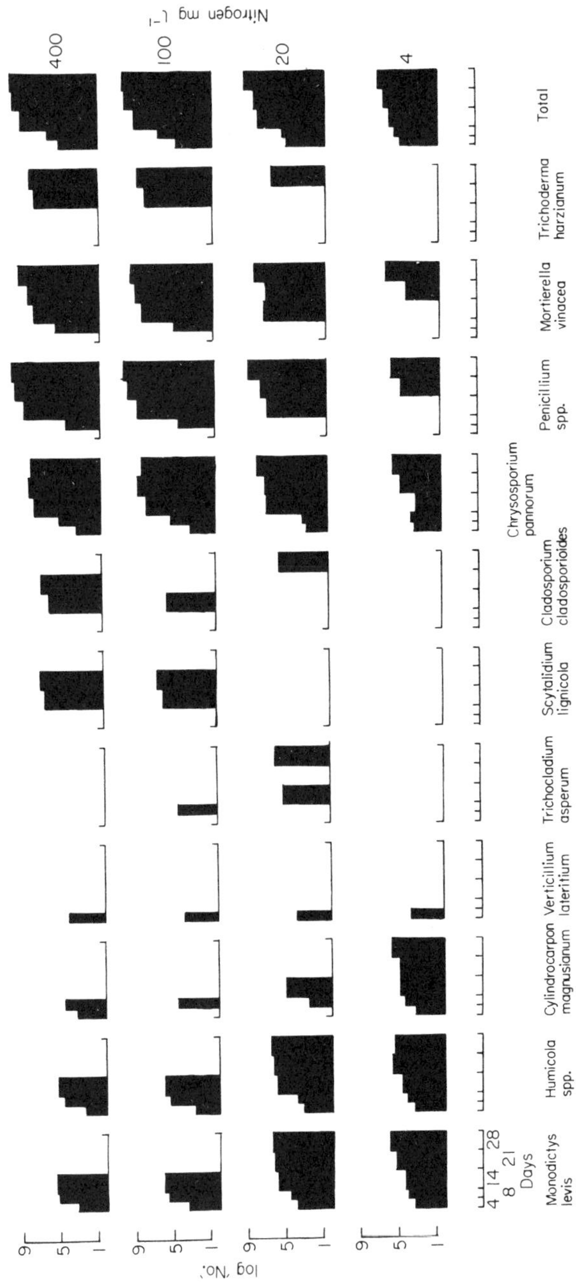

Figure 3.2. Elective culture microcosms at different N levels; patterns of colonization (13°–15°C).

at low N levels, and dominating there (*Monodictys levis*, *Humicola* spp., *Cylindrocarpon magnusianum*)

II Organisms starting activity early at high N levels, but later at low N levels, and either increasing throughout to become numerically dominant (*Penicillium* spp, *Mortierella vinacea*) or declining after a period of high representation (*Chrysosporium pannorum*)

III Transient representatives whose incidence, period and duration relates, at least in some, to N level (*Trichocladium asperum*, *Volucrispora aurantiaca*)

IV Organisms active late and only at the higher N levels (*Scytalidium lignicola*, *Trichoderma harzianum*), and

V Organisms active late and only at low N levels (*Pyrenochaeta acicola*).

Categories IV and V may be particular forms of category III.

A relevant point of significance is the absence of many of the accepted characteristic cellulose decomposers that were recorded as present in the research site (Table 3.3).

CELLULOSE DECOMPOSITION

An examination of cellulose weight loss under the conditions of this experiment was made (Fig. 3.3). Only at 4 mg l^{-1} N were the losses significantly different from those at the other levels, although indications of more decomposition at higher N levels were beginning to be indicated at 42 days. Here, as found by Park (1976) with a pure culture of a cellulolytic fungus at low N levels the rate of decomposition was higher per unit N available. However, decomposition at the higher N levels appears not to be complete, and also there are no data on the proportion of N assimilated.

CELLULOLYSIS AT DIFFERENT N LEVELS

Clearly a number of species from the site examined are able to be active in growth and reproduction at very low N levels. However, ability to decompose and utilize cellulose at different N levels may not directly be deduced from the results described. Because of species interactions an organism may fail to be represented in counts in a given situation although having a potential for cellulose decomposition under those conditions; furthermore it would be possible for an organism to be active at a later stage in a regime that initially did not meet its requirements. An attempt was made to detect effects of N level on the cellulolytic capacity of different species, using as a criterion the clearing of cellulose agar containing different levels of N. The mineral content of the media was based on that of standard cellulose agar

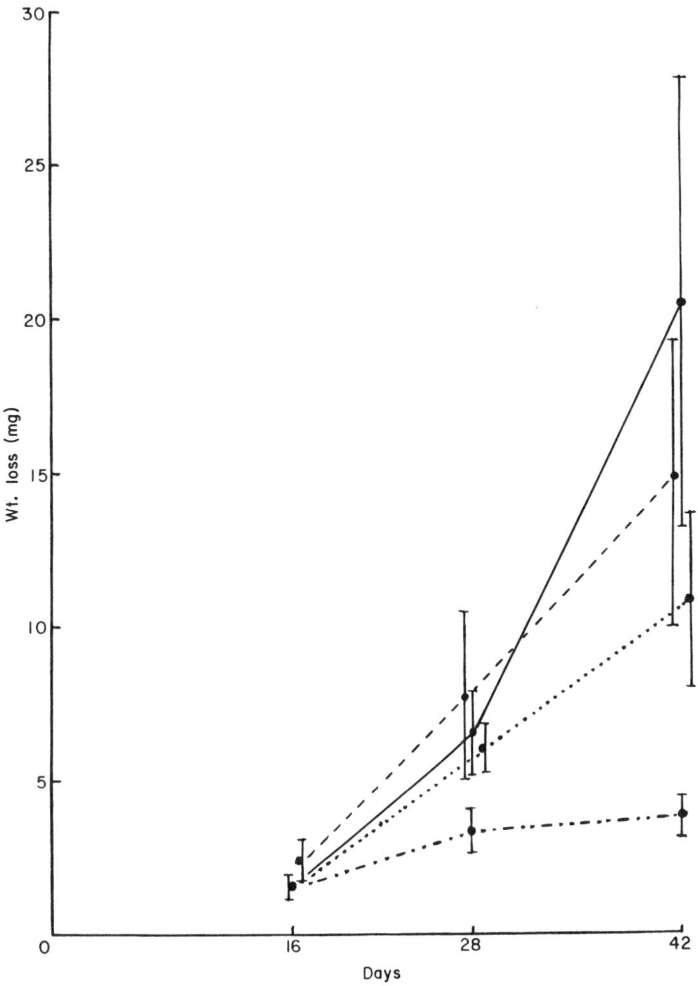

Figure 3.3. Cellulose decomposition at different N levels (11°–12°C); –··– 4 mg N l^{-1}, ···· 20 mg N l^{-1}, --- 100 mg N l^{-1}, —— 400 mg N l^{-1}.

with amounts of N source adjusted to provide 106, 10 and 2 mg l^{-1} N. Not until after the results were recorded was it recognized that agar-agar contains an amount of N appreciable in relation to these concentrations. The agar used (Oxoid No. 3) is recorded as containing 0·3% N, which in the 1·2% concentration used yields 36 mg l^{-1} to the medium. This fact helps to relate the good cellulolytic ability of one investigated fungus in agar medium without added N source with the inability of the same fungus to decompose cellulose in liquid medium of the same composition (Park 1975, 1976). The three N levels provided in this experiment are calculated, then, as 142, 46 and 38 mg l^{-1}, which are higher than intended, and have the two lower levels relatively closer than had been hoped for.

Table 3.4. Dependence of fungal cellulolysis on N level (see text).

Minimum N requirement for cellulolysis (mg l^{-1})	Number of tested isolates
142	4
46	21
38	66

Table 3.4 shows that of 91 isolates tested four showed cellulolytic activity only at the highest level of N, while a further 21 showed it at the two upper levels but not at the lowest. The remaining 66 of the 71 isolates were able to make some clearing of the cellulose even at the lowest N level obtained. Differences in intensity of cellulolytic activity were apparent but were not put on a quantitative basis. In repetitions of this experiment some isolates fell into the second category in some tests and into the third category in others. Where this happened the isolate is included with category three at the lowest N level. This observation, together with the small interval in N concentration between the two media, suggestes that the threshold N concentration sufficient for production of cellulases by different species may be somewhat sharply defined, and that the lower limit might be around 40 mg l^{-1} for a number of species. A number of the fungi failing to exhibit cellulolysis at the lowest N level provided here are regarded as characteristic and widespread cellulose decomposers, whereas some of the species in the third category are less regarded as having any great significance in decomposition of cellulose. A better separation of quantitative mineral requirements for cellulolysis among species is desirable.

Discussion

Organisms are commonly given an overall classification as having or not having ability to decompose particular substrates. Thus we have the designation 'cellulose-decomposing fungi'. But although a wide variety of cellulose-decomposing fungi may be known to be present in a site yet when a particular form of cellulosic material is placed at that site it becomes colonized by only a few of those fungi (e.g. Tribe 1960), and the species activity may be determined by the local conditions (Griffiths & Jones, 1963). The properties and requirements of potential decomposers are mostly not understood enough for us to account for this phenomenon. In addition some species may be popularly regarded as better decomposers than others, thus we find some decomposers described as having 'a high degree of cellulolytic ability' (Garrett 1963). Such statements are conventionally related to good laboratory conditions of temperature, nutrition, pH, etc., but a ranking arrived at under such conditions may not apply where conditions deviate from the optimum. Some attention has been given to the importance of immediate conditions, thus Warcup (1965) states that, 'because a large part of the plant material added to soil is cellulosic, the decomposition of this carbohydrate has attracted considerable attention.... The heterogeneity of the cellulose-decomposing microflora permits attack over a wide range of environmental conditions, with or without oxygen, in acid or alkaline soils, at various moisture levels, and at temperatures ranging from near freezing point to the top of the thermophilic range'. The omission of mineral nutrients from this list might be indicative of its neglect in some areas yet cellulose differs from some other insoluble substrata such as keratin, chitin and lignin in its lack of elements other than carbon, hydrogen and oxygen, but the importance of mineral supply for decomposition has been frequently stressed. Nitrogen is recognized as being expecially relevant, and as pointed out in the introduction to this paper considerations have been dominated by the concept of C/N ratio of the decomposable material. While the implications of this ratio for overall decomposition have been explored in considerable detail its relevance for growth and metabolism of individual organisms and of mixtures of species has not been fully followed. A number of investigators have pointed the way, and the series of papers, Merrill & Cowling (1966), Levi, Merrill & Cowling (1968) and Levi & Cowling (1969), is especially useful in this direction. A number of aspects, however, deserve closer attention. One of these is the way in which the solubility of the carbon source relates to any C/N ratio effects. Where the carbon source is soluble and readily accessible in solution to microbial surfaces then a higher relative amount of that source may act to increase the inter- and intra-cell competition for the nitrogen source, and may result in a nitrogen deficiency at the molecular reaction level which could reduce the total rate of decomposition. But where the carbon source is insoluble the same mechanism may not apply, particularly where a decomposer can concentrate nitrogen in certain areas so permitting production of hydrolases largely in those regions. In such a situation the areas of activity may provide a nutrient C/N ratio different from that of the overall material, with much of the insoluble carbon source acting as an inert matrix. The overall *rate* of substrate decomposition may here be little affected by the excess carbon source. If this holds then C/N ratio lacks a measure of relevant information for decomposition, namely the concentration of the components available to the microbial cell surface. A high C/N ratio may describe a habitat where the N concentration is so low as to limit microbial activity, or one where an insoluble carbon source is present in excess but where the N concentration is favourable for activity. Thus, varying the C/N ratio by altering the amount of insoluble carbon source may have a different effect from doing it by altering the amount of nitrogen source (Park 1976). Another relevant feature is that, as shown by Levi & Cowling (1969) at low concentrations of N a more conservative utilization of this resource may lead to greater decomposition per unit N, even with C/N ratio held constant. Present results indicate that a low N level may not only give a high decomposition ratio, but may also control in the selection of initial decomposer species. Some fungi appear better able to convert cellulose at low concentrations of N, while others for best development appear to require either an initially high N concentration (40–100 mg l^{-1}) or some previous activity by fungi without this high requirement. The previous activity by early colonizers is likely to include some concentration of the initially diffuse N source, and may reach a point where more demanding organisms can succeed. The comment of Tribe (1960) to the effect that the presence and activity of specific fungi may have an important bearing on the extent of decomposition of some residues seems especially relevant in this connection. In the microcosms examined here the course of species succession was longer at lower N levels, and the point along the sequence at which an ecological succession begins in a material may be determined by its concentration of mineral nutrients rather than by its C/N ratio.

An examination of N levels in different materials and habitats is appropriate. Griffiths & Jones (1963) have pointed out that although many isolates of *Trichoderma* are recorded as being able to utilize cellulose in culture yet Siu (1951) states that *Trichoderma* species are rarely recorded as growing on cellulosic material in soil. While occurrences are recorded (e.g. Jensen 1931b), this observation is likely to relate to the requirements of *Trichoderma* species, which are regarded as characteristic colonizers of woody tissues. *Scytalidium lignicola* appears from records to have a similar habitat preference. Wood is stated to be a material deficient in N, yet these fungi in the present study were found to be more characteristic of high N level habitats. Attention to N concentration may he helpful here. Table 3.5 shows the N content of timbers calculated from data given by Levi & Cowling, assuming the wood to have a 50% moisture content which is a reasonable

Table 3.5. Concentration of N ($\mu g\, g^{-1}$ or ppm) in wood: calculated from Levi & Cowling (1969, Table 1) on the basis of 50% moisture content as suitable for decay.

	sapwood	heartwood
Quercus falcata	3,636	1,435
Fraxinus americana	2,000	1,600
Pinus strobus	1,634	909
Picea sitchensis	1,081	641

Table 3.6. N levels ($\mu g\, g^{-1}$ or ppm) in soils and soil solutions.

[1]Soil solutions		26·7 to 940	
[2]Arable soils		8 to 240	
[3]Forest soils		Mull	Mor
Beech	Old leaf litter	1,284	272
	Wormcasts	272	—
	Upper humus	52	338
	Lower humus	10	95
Spruce	Upper humus	106 (Upper and Middle humus)	115
	Middle humus		32
	Upper topsoil	7	—

Data calculated from:

1. Chapman H.D. (1965).
2. Russell E.W. (1973) p. 335.
3. Bornebusch C.H. (1930). Cited by Russell (1973).

Table 3.7. Fresh-water N levels.

Fresh-water source	N $\mu g\, g^{-1}$ or ppm
[1]Storm rainwater	0·05–0·07
[2]Volga—summer	0
—winter	>1
[3]Blelham Tarn Feeders	
Ford Wood Beck	1·40
Tock How Beck	2·81
Wray Beck	0·75
Tarn water (0–5 m)	0·31
[4]Windermere	0·3–1·2
[5]Eutrophic lakes	
Lough Neagh	0·9
Kinnego Bay (L. Neagh)	1·76
Loch Leven	1·8
Lake Washington	0·57
Lake Erie	0·40
Lyngby So	0·80
Esthwaite Water	0·59

Data from:

1. Voigt G.K. (1960).
2. Shadin V.I. (1956). Cited by Hynes H.B.N. (1970).
3. Willoughby & Redhead (1973).
4. Macan T.T. & Worthington E.B. (1972).
5. Wood R.B. & Gibson C.E. (1973).

environmental value, and one suitable for decay. The N levels are all high, indeed all are greater than the upper level used in the artificial microcosms in the present study. Soils and soil solutions (Table 3.6) have on the whole lower concentrations of N, in these values ranging from 7 to 1,284 ppm N. Aquatic environments have much lower N concentrations (Table 3.7) and in still waters levels of around 1 mg l^{-1} may indicate a polluted or eutrophic condition (Table 3.8). Thus microbial environments vary widely in N concentration.

Table 3.8. Limnological classification on N levels.

Nutrient Class	N $\mu g g^{-1}$ or ppm
ultra oligotrophic	0·2
oligo- to mesotrophic	0·2–0·3
meso- to eutrophic	0·3–0·5
eu- to polytrophic	0·5–1·5
polytrophic	1·5

In these terms the amount of N in wood is hardly small, as is claimed by Merrill & Cowling (1966). And in their study of the role of nitrogen in the decomposition of wood Levi & Cowling (1969) used media that cover a much wider range of N content than that of the wood being considered (Table 3.9). At the upper end the N level of their media is comparable with

Table 3.9. Data calculated from Levi & Cowling (1969).

Levi & Cowling (1969) media	C/N ratio	N mg l^{-1}	Carbon source
Table 2	16:1 to 4:1	1,250	soluble
	32:1 to 8:1	125	soluble
	160:1 to 40:1	12·5	soluble
	1,600:1 to 400:1	1·25	soluble
Table 3	75:1	118	soluble
Table 4	1,600:1	5·56	soluble
	7,500:1	1·18	soluble
Table 6	20:1	50	cellulose
	200:1	5·0	cellulose
	2,000:1	0·5	cellulose

wood and with some soils at 1,250 mg l^{-1}, while the lowest value, 0·5 mg l^{-1} is lower than the lowest value used in this present study. It is also noteworthy that in all but one of their experiments soluble carbon sources were used, so that the results may not apply directly to cellulose decomposition.

Nevertheless many of their data show a correlation with N concentration at least as good as that with C/N ratio.

It has been established that there may be an absolute increase in the amount of N in a substratum during its decomposition. This can occur in terrestrial (Gosz, Likens & Bormann 1973) and in aquatic (Willoughby & Redhead 1973) environments. Such a process, operating against an activity gradient, may depend in its rate and extent upon the steepness of the gradient between the N concentration of the substratum and that of the environment so that nutrient *concentrations* may be appropriate for this particular consideration also. The increase in amount of minerals in a substratum is considered to be in part a function of microbial immobilization, and since microbial growth and synthesis will lower the free concentration of minerals in the substratum a steeper activity gradient may result. Another consequence of microbial assimilation of minerals is the creation of a differential in their concentrations within the substratum itself that may have important consequences for overall decomposition and substratum conversion. Values for N in culture-grown fungi are around 2·6% dry weight; and at 80% moisture content this converts to an N level of 5,000 ppm. Even at the very low mycelial concentrations of N obtained by Levi & Cowling (1969) in wood-destroying fungi grown at a C/N ratio of 7,500 (1·18 mg N l^{-1} medium) the conversion gives a mycellial N level of 200–100 ppm, indicating a more than 100-fold concentration from the medium. The point to be recognized is that the distribution of mineral elements within a substratum may be as important for decomposition as the overall concentration.

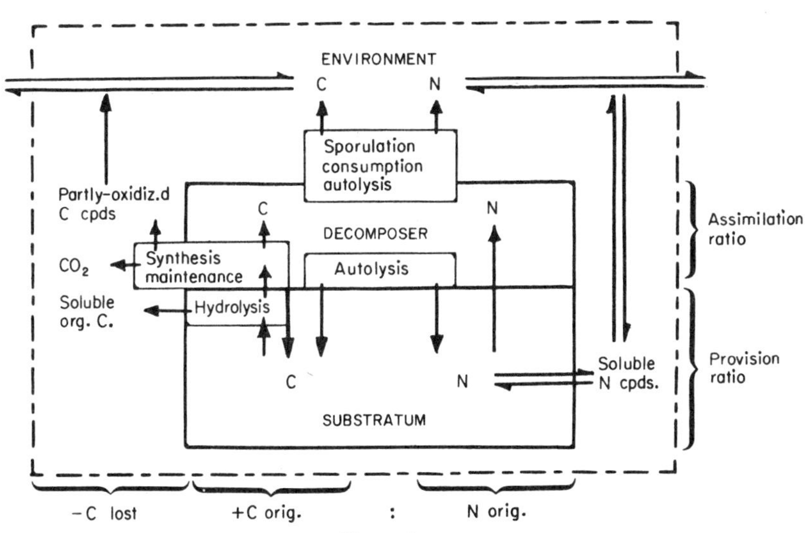

Figure 3.4. Schema for the three different C/N ratios relevant to decomposition.

Returning to the starting point, C/N ratio, and, since these arguments apply to other mineral elements, C/mineral ratio, these are usually expressed in terms of the elemental proportions in the original material, and may conveniently be termed *provision ratios* (Fig. 3.4). The amount of carbon in the decomposer biomass per unit of nutrient element assimilated, gives what can be called *assimilation ratios*, that especially with limiting nutrient, will give information of a different significance. Thirdly *decomposition ratios*, the amount of carbon decomposed per unit mineral element is yet another index that has been used (Jensen 1931b) in studies of this subject. Values and ratios obtained in any decomposition study, should be related to the aim of the study and to the sort of information being sought. In addition, in model systems using representative organisms the species should be chosen with care. Veldkamp & Kuenen (1973) have drawn attention to the fact that most micro-organisms isolated in pure culture have been obtained under conditions that select for a high maximum specific growth rate (μ max) whereas in nature in nutritionally poor environments organisms may be more successful with a relatively low maximum specific growth rate but a high substrate affinity (Ks). Steensland (1973) has made a similar point.

References

Chapman H.D. (1965) Chemical factors of the soil as they affect micro organisms. In: *The Ecology of Soil-Borne Pathogens*, eds. K.F. Baker & W.C. Snyder, pp. 120–39. University of California Press.

Gams W. (1960) Studium zellulolytischer Bodenpilze mit Hilfe der Zellophanstreifen-Methode und mit Carboxymethyl-Zellulose. *Sydowia, Ann. Mycol. Ser.* II, **14**, 295–307.

Garrett S.D. (1963) A comparison of cellulose-decomposing ability in five fungi causing cereal foot rots. *Trans. Brit. Mycol. Soc.* **46**, 572–6.

Gosz J.R., Likens G.E. & Bormann F.H. (1973) Nutrient release from decomposing leaf and branch litter in The Hubbard Brook Forest, New Hampshire. *Ecol. Monographs* **43**, 173–91.

Griffiths E. & Jones D. (1963) Colonization of cellulose by soil micro-organisms. *Trans. Brit. Mycol. Soc.* **46**, 285–94.

Hutchinson H.B. & Richards E.H. (1921) Artificial farmyard manure. *J. Min. Agr.* (G. Br.) **28**, 398–411.

Hynes H.B.N. (1970) *The Ecology of Running Waters.* Liverpool University Press.

Jensen H.L. (1931a) The microbiology of farmyard manure decomposition in soil—I. Changes in the microflora and their relation to nitrification. *J. Agric. Sci.* **21**, 38–80.

Jensen H.L. (1931b) The microbiology of farmyard manure decomposition in soil—II. Decomposition of cellulose. *J. Agric. Sci.* **21**, 81–100.

Levi M.P. & Cowling E.B. (1969) Role of nitrogen in wood deterioration—VII. Physiological adaptation of wood-destroying and other fungi to substrates deficient in nitrogen. *Phytopathology* **59**, 460–8.

Levi M.P., Merrill W. & Cowling E.B. (1968) Role of nitrogen in wood deterioration—

VI. Mycelial fractions and model nitrogen compounds as substrates for growth of *Polyporus versicolor* and other wood-destroying and wood-inhabiting fungi. *Phytopathology* **58**, 626–34.

MACAN T.T. & WORTHINGTON E.B. (1972) *Life in Lakes and Rivers.* Collins-Fontana New Naturalist, London.

MERRILL W. & COWLING E.B. (1966) Role of nitrogen in wood deterioration: Amount and distribution of nitrogen in fungi. *Phytopathology* **56**, 1083–90.

PARK D. (1975) A cellulolytic Pythiaceous fungus. *Trans. Brit. Mycol. Soc.* **65**, 249–57.

PARK D. (1976) Cellulose decomposition by a Pythiaceous fungus. *Trans. Brit. Mycol. Soc.* **66** in press).

RUSSELL E.W. (1973) *Soil Conditions and Plant Growth* (10th ed.) Longmans, London.

SIU R.G.H. (1951) *Microbial decomposition of cellulose.* Reinhold, New York.

RUSSELL E.W. (1973) *Soil Conditions and Plant Growth* (10th ed.) Longmans, London.

SIU R.G.H. (1951) *Microbial decomposition of cellulose.* New York, Reinhold.

STEENSLAND H. (1973) Continuous culture of a sewage fungus, *Fusarium aquaeductum. Arch. Mikrobiol.* **93**, 287–94.

TENNEY F.G. & WAKSMAN S.A. (1929) Composition of natural organic materials and their decomposition in the soil—IV. The nature and rapidity of decomposition of the various organic complexes in different plant materials under aerobic conditions. *Soil. Sci.* **28**, 55–84.

TRIBE H.T. (1960) Aspects of cellulose decomposition in Canadian soils—I. Observations with the microscope. *Canada. J. Microbiol.* **6**, 309–16.

VELDKAMP H. & KUENEN J.G. (1973) The chemostat as a model system for ecological studies. In: *Bull. Ecol. Res. Comm. (Stockholm)*, ed. T. Rosswall, **17**, pp. 347–55.

VOIGT G.K. (1960) Alteration of the composition of rainwater by trees. *Amer. Midl. Nat.* **63**, 321–6.

WAKSMAN S.A. & TENNEY F.G. (1928) Composition of natural organic materials and their decomposition in the soil—III. The influence of nature of plant upon the rapidity of its decomposition. *Soil Sci.* **26**, 155–71.

WARCUP J.H. (1950) The soil plate method for isolation of fungi from soil. *Nature, Lond.* **166**, 117–18.

WARCUP J.H. (1965) Growth and reproduction of soil micro-organisms in relation to substrate. In: *Ecology of Soil-Borne Plant Pathogens*, eds. K.F. Baker & W.C. Snyder, pp. 52–67. University of California Press.

WILLOUGHBY L.G. & REDHEAD K. (1973) Observations on the utilization of soluble nitrogen by aquatic fungi in nature. *Trans. Brit. Mycol. Soc.* **60**, 598–601.

WOOD R.B. & GIBSON C.E. (1973) Eutrophication and Lough Neagh. *Water Res.* **7**, 173–87.

4 Aggregation responses of soil arthropods in relation to the soil environment*

M. B. USHER *Department of Biology, University of York, York YO1 5DD, U.K.*

Summary

The majority of field studies of Collembola and Acarina have shown that aggregations are a common feature of the distribution of these arthropods. In attempting to find the cause of these aggregations, the use of the correlation coefficient between the number of arthropods and measures of the physical environment is problematical since cause and effect cannot be established from the significance or otherwise of this coefficient. In controlled laboratory environments the location of a food source tends to be of importance in determining the site if not the size of an aggregation.

Data are presented for a Scots pine (*Pinus sylvestris*) forest soil in Perthshire, Scotland. Areas of high nutrient status have different size dimensions. Aggregated distributions of arthropods are frequent, and these are divided somewhat subjectively into large (gradients) and small (distinct) aggregations. In general, it will be shown that predatory species are less strongly aggregated than detritus and fungus feeders. The distinct aggregations can be uniformly or randomly distributed or aggregated themselves. It is shown that in some species more aggregations are formed at higher population densities, whilst in other species the number of arthropods per aggregation increases with population density. The whole community is analysed by

*In order to avoid confusion over the use of the word 'Factor' in several different contexts, the following convention has been used in this paper:

'Factor' has been retained for any one of the mathematical abstractions derived from a 'factor analysis' of the data.

'Feature' has been used in the context usually covered by 'environmental factor', i.e. environmental features are individually measured components of the environment.

'Influence' has been used in the context usually covered by 'the factor affecting sp. A was . . .', where in this paper such a sentence would be written 'the influence affecting sp. A was . . .'.

both principal component and factor analyses. The results of these two methods are not particularly different, though rotation of the axes in a factor analysis might assist in the biological interpretation. The general conclusion that can be drawn from these analyses is that the most important effect on the arthropod community is structural, a combination of depth in the soil, soil water content and pore space, but that these influences are so closely correlated that their individual effects, if any, are inseparable.

There is a discussion of aggregation in soil arthropods and pattern in plant communities. The causes of both are complex and may relate to the establishment of a larger population density than in an unaggregated population. The comparative stability of the soil environment may contribute to its species richness.

Introduction

The majority of studies on the arthropods which live in litter or soil have shown a nonrandom distribution, with the arthropods occurring together in small localized pockets of high density. Such a feature of animal distributions has usually been termed 'aggregation': the word 'aggregate' is defined by the Oxford English Dictionary as 'collect together' and seems to imply some active process on the part of the individuals. The word 'pattern' has been used to describe nonrandomness in plant distributions: this word tends to imply a passive, or certainly less active, process generating the nonrandomness. The two words do, however, describe the same phenomenon: the occurrence together in a few localized areas of more organisms than would be expected if the organisms were distributed at random.

Field studies in relation to aggregation behaviour of soil arthropods tend to have had one of three aims. First, some studies seem content to demonstrate aggregation. That aggregations exist has now been demonstrated for so many species that there is no need to list the species. Secondly, aggregation imposes statistical problems on sampling—how many samples are needed to give an accurate estimate of the mean population density, or what is the underlying probability distribution so that an efficient transformation can be carried out to normalize the data before applying tests of significance? It has been shown that the negative binomial distribution often gives a good fit to data consisting of counts of mites and Collembola (Nef 1962; Berthet & Gérard 1965), though Hartenstein (1961) obtained a better fit with the Neyman A distribution for some species. Thirdly, some studies have endeavoured to determine the causes of aggregations.

Perhaps the most important environmental influence acting upon the soil arthropods is the soil water. Hale (1967) considers that the Collembola, whilst not in general indicating soil conditions, might reflect the water content of soils. Several studies have shown positive correlations between soil water content and the numbers of arthropods in forest soils. Poole (1961, 1962) has demonstrated such a relation for the Collembola *Friesea mirabilis*, *Tullbergia krausbaueri*, *Megalothorax minimus* and *Isotoma notabilis* (some of these relations are based on correlation coefficients, others on partial correlation coefficients after elimination of such environmental features as the dry weight or depth of the organic matter). *Hypogastrura viatica* actively aggregates in places of high humidity (van der Kraan 1971). Lebrun (1971) has shown a positive correlation between soil water content and the logarithm of the numbers of some of the Cryptostigmata that he studied. Lebrun (1969) also suggested that the population density of *Nothrus palustris* was positively related to soil water content. In an old field ecosystem, Knight & Read (1969) have suggested that an optimum soil water content, 10–20% of dry weight, exists for maximum density of six species of *Tomocerus*. Some contrary results have been found in salt marshes, where Luxton (1967a) has shown that the total Cryptostigmata and total Mesostigmata are negatively correlated with water content (due both to tide and precipitation), though in a Puccinellietum soil he found no correlation between any group of mites and water content (Luxton 1967b). Joosse (1966) observed that the maritime Collembola *Anurida maritima* was stimulated to aggregate in cavities by the incoming tide or by rainfall during low water. Water, in excess, can also cause aggregations: van der Kraan (1971) has shown that *Hypogastrura viatica* is drifted into large aggregations with densities reaching 1,000 cm^{-2} when it is floating on flood water.

Three theoretical features of the relation between soil water and arthropod distribution are of interest. First, Vannier (1970) has been concerned with pF thresholds acting as a stimulus for arthropods to move under desiccating conditions. He has shown (Vannier 1967) that at about pF 5 the Isotomidae respond more quickly than the Cryptostigmata. Secondly, Gifford (1968) has been concerned with the precise nature of the relation: whether soil water is in the form of vapour or liquid, and the water relations of the arthropod species (i.e. their cuticle structure, respiration and their ability to live in a water film). Thirdly, Joosse (1970, 1971) and Joosse & Verhoef (1974) have developed a theory that Collembola come together by undirected movements into small areas of optimum humidity conditions. They suggest that the formation of aggregations is significant since moulting and egg laying occur in these localized areas, and the greater population density increases the probability that the opposite sexes will be near each other for fertilization. It is suggested that the species of Collembola which aggregate most strongly are those with no courtship behaviour.

Local variation in temperature can also affect the distribution of soil arthropods. Edwards & Lofty (1971) experimented by artificially heating soil, but since the heated plot was drier, there is likely to be an interaction between temperature and water content in these experiments. Lebrun & van Ruymbeke (1971) suggest that, since development time in Cryptostigmata is related to temperature, the maximum population density of *Damaeus* sp. occurs in sites with an optimal temperature. The data for cold temperatures are somewhat conflicting. Whilst Healey (1967) observed activity at −6°C and feeding and defaecation at −4°C in *Onychiurus procamptus* in environments in which the temperature had dropped slowly, he also showed that sudden drops in temperature to 5°C could render the animals inactive. Tillbrook (1967) considered that the pattern of distribution of antarctic arthropods could result from winter deaths due to cold, whilst Remmert & Wisniewski (1970) have shown that Collembola are resistant for several hours to subzero temperatures during the arctic summer.

Temperature might be the influence causing aggregations of *Isotoma viridis* and various species of mites under white as opposed to black boards (Jensen & Corbin 1966), though it is suggested that time of day, past weather and humidity are also responsible for these aggregations. Jensen (1968) later showed that white boards do attract more species of Collembola than black boards. Schalk (1968), discussing the physical structure of the environment, has shown that there are larger populations of arthropods in sandy than in loamy areas, and Haarløv (1960) has illustrated relations between pore structure and arthropod size. There is some evidence, however, that Collembola can modify the structure of their environment by building cells of faecal material (Poinsot 1966; Massoud *et al.* 1968). Poole (1961, 1962) has shown that many species of Collembola are positively correlated with the depth of the organic layer in coniferous forest soils. Luxton (1967b) shows negative correlations between all groups of mites and pH in saltmarsh soils, whereas Wakerley (1963) found an optimum pH for Collembola density in grassland plots at pH 6·5. He also showed that the density of Collembola increased linearly with phosphate content of the soil, and thus that maximum population density occurred under conditions of high phosphate at pH 6·5, the conditions required for a good grass sward. Edwards & Lofty (1969), however, found that an increase in pH increased the population density of some species, and decreased the density of other species. They also found that whereas inorganic fertilizers slightly increased the population densities of both mites and Collembola, organic fertilizers more than doubled the populations of both groups. This result is in accord with Gisin (1955) who found that organic manure increased the populations of all Collembola in vineyard soils.

These studies, based on field data, would indicate that the physical and chemical environment of the soil have a particularly strong effect on the location of aggregations. However, laboratory studies with *Pseudosinella impediens* (Christiansen 1970; Barra & Christiansen 1975), *Hypogastrura bengtssoni*, *Proisotoma minuta*, *Folsomia candida* and *Sinella coeca* (von Törne 1974) and *F. candida* (Usher & Hider 1975) have all shown that the location of a food source is the most important influence determining the site of an aggregation. The studies on *P. impediens* show that food as an influence is more important to the young stages than to the adults. The studies on *F. candida* have shown that water is also an important influence.

Although food can be measured in laboratory studies, it is a feature that is difficult to measure in the field. Results or reviews on the feeding habits of Mesostigmata are given by Karg (1961), Cryptostigmata by Wallwork (1958) and Hartenstein (1962), mites generally by Wallwork (1967), and Collembola by Poole (1959), Bödvarsson (1970) and Petersen (1971). In general these studies show that soil arthropods are not very specific feeders. However, laboratory studies have shown that there are preferred foods, particularly with fungus-feeding species. Thus, *Nothrus biciliatus* preferred one out of 17 fungi offered to it, and in general it found young fungal cultures more attractive than old ones (Saichuae *et al.* 1972). Mills & Sinha (1971) found that *Hypogastrura tullbergi* showed a preference for four out of 43 fungi and actinomycetes. These four species were characterized by forming a mycelial mat which permitted free movement of the Collembola. The generality of feeding, shown by field studies, can be partly explained by lack of availability of the preferred food: thus in the laboratory Knight & Angel (1967) found *Tomocerus* predominantly fed on fungal spores, eating rather little litter, humus or fungal hyphae, whilst in the field where spores were scarce the species ate more litter and fungal hyphae. Despite the observed generality of feeding, it is thus possible that aggregations are located where there are particularly favoured fungi, etc.

Preference for certain species of microflora might also be responsible for the succession of arthropods on decaying vegetable material. Winston (1956) has described the succession of arthropods on decomposing acorns, whilst Hayes (1963) has shown that some Phthiracarid mites occur at an earlier stage of the decomposition of coniferous needles than other species. The spatial arrangement of acorns or needles would thus be capable of determining the aggregation patterns of species. Pande & Berthet (1973) show that whilst *Phthiracarus* spp. are more usually associated with twigs in the litter, *Microtritia minima* is more usually associated with needles and *Rhysotritia duplicata* with bark, though the three species are not confined to these preferred components of the litter. A positive correlation between the roots of *Artemisia campestris* and *Rhysotritia* (*Pseudotritia*) *ardua* has been found in the field by Führer (1961), who also demonstrated this relation in laboratory experiments with water extracts of dead and decaying *Artemisia* roots.

Curry (1973) has shown that species of *Isotoma* can select litter of various plant species under field conditions: *I. notabilis* is most frequent in *Dactylis glomerata* litter and *I. viridis* in *Rumex crispus* litter. Bagnall (1932) recorded *Sminthurinus cingulatus* in Norway and Britain only around *Pedicularis palustris* plants. Blackith (1974) suggests that aggregations are formed in the rhizosphere of the favoured higher plant.

Besides the food and environmental influences acting to cause aggregations, there would also appear to be some inbuilt sociability in the arthropods themselves. Some of this might result from egg clusters since juveniles are frequently aggregated (e.g. *Folsomia quadrioculata* juveniles, Macfadyen 1952). Joosse (1970, 1971) and Joosse & Verhoef (1974) have suggested that the formation of aggregations is important in the moulting process, as well as being a replacement for courtship behaviour prior to egg laying. Wallace (1967) has shown that at high density the newly hatched nymphs of *Sminthurus viridis* eat dead and dying adults, in which there are large quantities of uric acid, and die. He suggests that this is density-dependent mortality. It might be appropriate to speculate that some means of population control is achieved through the formation of aggregations. There is some evidence to suggest that aggregations are multispecific in content. Poole (1961) has suggested this in the Collembola of a coniferous forest soil, and Gridelet & Lebrun (1973) have demonstrated a positive correlation between *Dermatophagoides pteronyssinus* and *Cheyletus* sp. However, the mechanism for the existence of several species in the same aggregation is difficult to explain. Laboratory experiments by Christiansen (1967) and Longstaff (1974) have demonstrated the difficulty of maintaining cultures of two or more species of Collembola.

Aggregations would therefore appear to be caused by a variety of influences. Perhaps the most important of these is the location of favoured food sources, though soil moisture is also likely to be of importance. The correlation between food and water under field conditions is unknown, and hence field data suggesting water and laboratory data suggesting food may be manifestations of the same influence, i.e. that the preferred food grows best at some particular water content. The causes of aggregations under field conditions are analysed in the remainder of this paper. The sampling scheme was designed to provide information on a number of these possible influences acting to cause aggregations, in relation to both the chemical and physical environment in which the arthropods were living. It was, however, impossible to investigate the local distribution of food sources since the preferences of each species were unknown and since the small soil sample was destroyed during the extraction of the arthropods and the analysis of the physical and chemical environment.

THE PHYSICAL AND CHEMICAL ENVIRONMENT

Site factors

The data relate to a sampling site, approximately 20 m square, on Cnoc Eòghainn, a southerly outlyer of the Black Wood of Rannoch in Perthshire, Scotland (56° 40′N, 4° 22′W; National Grid Reference NN551546). The cnoc is at an elevation of approximately 380 m, and is covered with a stand of pine trees, *Pinus sylvestris* L. ssp. *scotica* (Schott) E.F. Warburg, which were mostly 80 to 90 years old. The ground is not wholly covered with herbaceous vegetation due to winter grazing by sheep and deer (a small fenced enclosure on the cnoc quickly became fully vegetated). The most frequent herbaceous species are wavy hair-grass, *Deschampsia flexuosa* (L.) Trin., heath bedstraw, *Galium hercynicum* Weigel and wood-sorrel, *Oxalis acetosella* L. The sparseness and uniformity of this ground flora considerably eased the problems of sampling the soil fauna. The climate is cold and moist, with an annual precipitation of approximately 118 cm and a mean annual temperature of 6°C (the mean temperatures for January and February are close to 0°C). The soil is a humus iron podsol, which has a litter horizon about 5 mm deep (where no drifting of litter has occurred), a fermentation horizon of approxi-

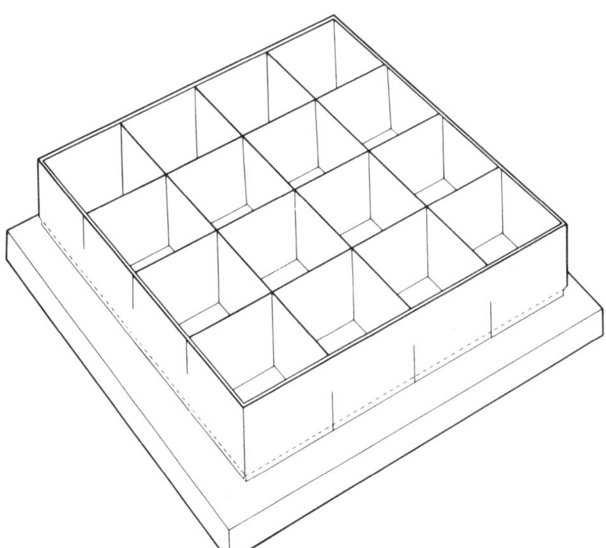

Figure 4.1. The sampling tool used to cut sixteen contiguous, square soil cores. The tool was constructed of stainless steel blades which were hollow ground to facilitate sharpening. Each core was divided into three horizontal layers, each 1 cm thick, before extraction of the arthropods and measurement of the environment features.

mately 10 cm depth, and a humus horizon of approximately 28 cm depth. Stones are absent in the organic horizons.

Blocks of 16 contiguous square soil cores, arranged in a square four cores by four cores, were cut with a sampling tool (see Fig. 4.1) to a depth of 3 cm. Thus, this study concentrates on the arthropods and environment of the litter and upper fermentation horizons (extraction from deeper cores indicated that approximately 95% of the arthropods occurred in the surface 3 cm). Each core was 4 cm square, and was divided into three layers, each of 1 cm thickness. The sampling tool thus yield 48 soil samples; the location of this block of samples being randomly selected within the sampling area.

Horizontal distribution in the soil

Two physical features (soil water content and soil pore space) and four chemical features (nitrogen, phosphorous, potassium and calcium concentrations) were measured in all 1,152 samples that were collected. The techniques of analysis of these samples have been described by Usher (1970a). Although a seasonal distribution or longer term trends were observed for these environmental features, they also showed characteristic horizontal distributions.

The water content of two blocks of samples, one taken in December at the time of greatest mean water content and the other taken in July at the time of least water content, are shown in Fig. 4.2. In the December data,

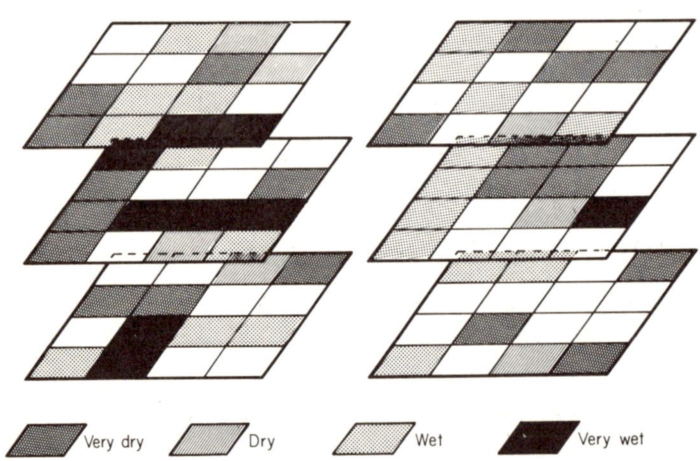

Figure 4.2. The distribution of wet and dry samples in two blocks of samples collected in December (left—the time of the greatest water content) and in July (right—the time of the least water content).

the mean water content of the surface, expressed as weight of water to oven dry weight of soil, is 257% (95% confidence limits are 238–276%). Samples with below 238% water content have been designated 'dry', and samples with above 276% as 'wet'. Similarly, dry and wet samples have been designated with respect to the mean for each other layer. It can be seen in Fig. 4.2 that there is considerable variability in the surface and centre layers, with the wet and dry samples being more or less randomly arranged, though there are strong vertical similarities between the two layers. In the lowest layer the wet and dry samples are not randomly arranged, but show moisture gradients, probably due to the greater environmental stability of this layer. The pore space showed no definite horizontal distribution. Vertically, pore space showed a distinct distribution, being greatest in the surface layer.

All four nutrients showed some degree of heterogeneity in their distributions. The nitrogen distribution in all seasons of the year typically shows samples with high or low nitrogen concentration randomly distributed. Since high nitrogen occurs in single samples or pairs of samples, it can be seen that the 'concentration aggregation' of nitrogen occupies a volume of 16–32 cm^3. These 'concentration aggregations' of nitrogen are the smallest observed at Rannoch. The three other nutrients measured also formed 'concentration aggregations', but of a larger volume. Such 'concentration aggregations' for potassium occupy a horizontal area of 32–64 cm^2, and vertically they are reflected to a depth of at least 3 cm. Thus the volume of these 'concentration aggregations' is of the order of 100–200 cm^3. Phosphorus occurred in the blocks of samples as a gradient running from areas of high to low phosphorus concentration. Such 'concentration aggregations' are approaching the volume of the whole block of samples, and thus they are of the order of 750 cm^3 in volume. There is no season of the year when the distribution of calcium can be considered homogeneous. Calcium distribution can be in the form of small patches of high concentration, these patches being the same size or a little larger than the patches of high potassium concentration. Also, gradients of calcium concentration can be seen across some of the blocks of samples resembling the gradients of phosphorus concentration. Neither of these types of 'concentration aggregation' are found at any particular season of the year, and thus it would appear that the 'concentration aggregations' of calcium are more variable in volume than those of nitrogen, potassium or phosphorus.

Soil arthropods are thus faced not with a uniform soil environment, but with an extremely heterogeneous environment. A vertical distribution of environmental features could be predicted since the horizons change from litter to fermentation, and the deeper in the fermentation layer the longer the time lag between the litter falling and that stage of decomposition. However, there is an extremely heterogeneous horizontal distribution of environmental features, these environmental features only being weakly correlated one with the other.

THE ARTHROPODS (COLLEMBOLA AND ACARINA)

When considering the distribution of individuals within the soil environment, three separate distributions can be considered, namely seasonal, vertical and horizontal. In relation to the formation of aggregations, the horizontal distribution is of greatest importance, though the vertical distribution of the arthropods can affect the distances between the centres of aggregations. Seasonal distribution is probably more closely related to the microclimate, though it must be remembered that as the population density changes so might the aggregation responses of the species. The seasonal and vertical distribution of the Collembola (Usher 1970b), of the Mesostigmata (Usher 1971b) and of the Cryptostigmata (Usher 1975a) have already been described.

If a series of independent samples of an insect population is available, Taylor (1961) showed that there was a relation between the sample variances and the sample means of the form

$$s^2 = a\,m^b \tag{4.1}$$

where s^2 is the sample variance, m the sample mean, and a and b are constants. Two facts emerge from equation 4.1. First, simple logarithmic transformation of both sides implies that the coefficients a and b can be estimated by regression analysis. Secondly, the coefficient b can be used as a measure of the aggregating tendency of the species (if a species is distributed according to the Poisson distribution then $b=1$). Since a total of 24 blocks of samples were collected for each of which a sample mean and a sample variance was calculated, the coefficients of equation 4.1 have been estimated. These have previously been described for the Mesostigmata (Usher 1971a) and the Cryptostigmata (Usher 1975b) and are listed for the Collembola in Table 4.1. In this table the values of the correlation coefficient are all large and significant, indicating that the relation in equation 4.1 generally fits the Collembola data well.

In order to compare the various groups of arthropods, data for the most frequent species are given in Table 4.2. The values of b are usually large (in excess of about 1·4) for the species of Collembola and Cryptostigmata. An exception to this is *F. quadrioculata*, but Usher (1969) has found that this is the only species of Collembola ever to show a uniform (regular) distribution in the soil. The values of b for the Mesostigmata are usually closer to unity and are probably related to the feeding habits of the Mesostigmata.

Table 4.1. The parameters of equation 4.1, where a and b have been estimated by regression analysis. The value of the correlation coefficient, r, between the logarithm of s^2 and the logarithm of m is given with the sample size in brackets.

Species	a	b	r	
Anurida granaria (Nicolet)	2·345	1·258	0·983	(9)
Anurophorus binoculatus (Kseneman)	1·432	1·171	0·984	(24)
Dicyrtoma ornata (Nicolet)	1·593	1·132	0·982	(18)
Folsomia quadrioculata (Tullberg)				
Juveniles	2·341	1·396	0·921	(24)
Adults	1·330	1·233	0·972	(24)
Friesea mirabilis (Tullberg)				
Juveniles	1·647	1·398	0·972	(24)
Adults	1·646	1·258	0·968	(24)
Isotoma sensibilis (Tullberg)				
Juveniles	3·592	1·732	0·955	(24)
Pale adult form	2·079	1·228	0·930	(20)
Dark adult form	1·706	1·313	0·952	(20)
Total adults	2·015	1·448	0·911	(24)
Isotomiella minor (Schäffer)	1·668	1·259	0·988	(24)
Isotomurus palustris (Müller)	5·128	1·500	0·958	(20)
Lepidocyrtus curvicollis Bourlet	1·881	1·201	0·984	(14)
Neanura muscorum (Templeton)	1·645	1·146	0·984	(14)
Neelus minimus Willem	1·403	1·119	0·993	(19)
Onychiurus absoloni (Börner)	1·737	1·452	0·961	(24)
Onychiurus latus Gisin	1·293	1·085	0·991	(20)
Sminthurides pumilis (Krausbauer)	2·107	1·166	0·950	(7)
Tullbergia callipygos Börner	1·716	1·450	0·978	(24)
Willemia anophthalma Börner	1·255	1·079	0·974	(20)

The species with the largest b value, *O. minimus*, is not predatory, whilst both *P. lapponicus* and *V. nemorensis* are predatory species and each of these have a value of b very close to unity.

Thus, if the value of b can be used as an index of the aggregation tendency, two general statements can be made about the Rannoch fauna. First, predatory species, although showing a heterogeneous distribution into aggregations, are less strongly aggregated than humus-, detritus- or fungus-feeding species. Secondly, the values of b are very similar for many species of Collembola and Cryptostigmata, and hence it would appear that the non-predatory species are showing about the same degree of tendency to aggregate.

By taking three-dimensionally contiguous soil samples, it was possible to investigate the size and distance relations of aggregations. By studying the statistic

$$\chi^2 = s^2(n-1)/m \qquad (4.2)$$

Table 4.2. A comparison of the most frequently occurring species of Collembola, Cryptostigmata and Mesostigmata at Rannoch. In all cases adults and juveniles have been pooled to give total counts for the species.

Group and species	Total number extracted from all samples	Value of *b*
Collembola		
Onychiurus absoloni	7,146	1·452
Isotoma sensibilis	4,309	1·944
Friesea mirabilis	4,305	1·371
Folsomia quadrioculata	3,363	1·290
Tullbergia callipygos	1,995	1·716
Cryptostigmata		
Tectocepheus velatus	12,165	1·571
Platynothrus peltifer	3,285	1·495
Steganacarus magnus	2,784	1·556
Ceratoppia bipilis	1,406	1·703
Adoristes poppei	1,358	1·623
Mesostigmata		
Olodiscus minimus	2,230	1·254
Veigaia nemorensis	917	1·034
Parazercon radiatus	872	1·152
Trachytes pyriformis	726	1·181
Pergamasus lapponicus	497	1·019

where s^2 and m are defined above and n is the number of samples, Usher (1969, 1971a, 1975b) has recognized five types of spatial distribution, namely uniform (χ^2 significant with $p \geqslant 0{\cdot}95$), random (χ^2 not significant), spatially large aggregations or 'gradients' (χ^2 significant with $p \leqslant 0{\cdot}05$, and with an analysis of variance of the data for a block of samples showing significant differences between rows and/or columns), distinct aggregations (χ^2 significant with $p \leqslant 0{\cdot}05$, but with no significant differences between rows and/or columns in the analysis of variance. Distinct aggregation can be seen by inspection of the data, as in Fig. 4.3), and other aggregations (significance as above, but distinct aggregations were not discernable in the data).

A summary of such analyses of the arthropods at Rannoch is given in Table 4.3. Comparisons of the various groups shows that a random distribution is more common in the Mesostigmata than in the Cryptostigmata and Collembola, and that a distribution into spatially large aggregations, the gradients, is less common in the Collembola than in the mites. A uniform distribution is particularly rare. As far as individual species are concerned, many of the Collembola (Usher 1969) and Cryptostigmata (Usher 1975b) occurred only in aggregations. The juveniles of both groups showed less

Table 4.3. A summary of the form of the distribution of the arthropods in the soil. The five types of distribution are discussed in the text. The data are presented for all species, and for the groups of five most frequent species listed in Table 4.2.

	Number of species included	Number of blocks of samples analysed	Distributions				
			Uniform	Random	Gradients	Distinct aggregations	Other aggregations
Collembola	12	234	2	64	9	121	38
	5 most frequent	198	2	49	9	103	35
Cryptostigmata	12	212	0	17	48	99	48
	5 most frequent	133	0	7	32	54	40
Mesostigmata	11	103	0	43	18	33	9
	5 most frequent	88	0	34	16	29	9

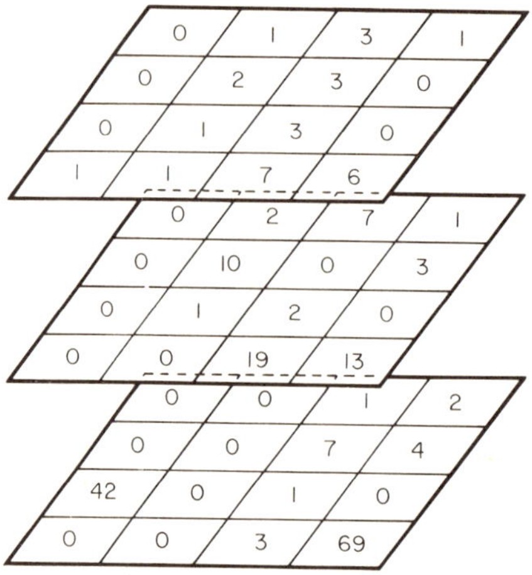

Figure 4.3. The distribution of juvenile *Isotoma sensibilis* in a block of samples collected in February. Large, distinct aggregations can be seen, particularly in the lowest layer where there are large isolated counts of 42 and 69.

tendency to occur at random than the adults. In the predatory Mesostigmata, although the tendency was towards randomness many species still exhibited a considerable degree of aggregation (for example *V. nemorensis* was aggregated in 13 of the 20 blocks analysed).

The aggregated distributions were further examined to determine the nature of the aggregation response. Amongst the Collembola the only species to occur more than once with a gradient type of aggregation was *O. absoloni*. It is perhaps a feature of this genus that they form territorially large aggregations (Glasgow 1939). Amongst the mites many species occurred in a gradient situation, and in *T. velatus* no 'distinct' aggregations were found by inspection of the data. Since these territorially large aggregations were demonstrated by a gradient across a block of samples, it can be seen that the size of the aggregation is at least the same as that of the block of samples, i.e. that such aggregations have a horizontal dimension of at least 250 cm^2.

The data of distinct aggregations can be subjected to further analysis. A block of samples with juvenile *I. sensibilis* is shown in Fig. 4.3, from which it can be seen that there are four aggregations. The analysis of the distinct aggregations was performed to determine the number of aggregations in a block of samples, the distance between aggregations (assuming that the arthropods were at the centre of the samples in which they occur), the distribution of the aggregations themselves, and the relation between population density and the aggregation response.

It is a feature of all three groups that the most frequently occurring

Table 4.4. The numbers of distinct aggregations formed in a block of samples. The data are given for the whole of a group and for the five* most frequently occurring species in the group as listed in Table 4.2.

	Number of aggregations in a block					
	1	2	3	4	≥5	Mean
Collembola, all species	14	31	45	19	13	2·94
Collembola, 5 species	7	25	32	17	12	3·10
Cryptostigmata, all species	33	34	18	11	3	2·16
Cryptostigmata, 4 species	10	21	13	7	3	2·48
Mesostigmata, all species	18	13	6	2	2	1·95
Mesostigmata, 5 species	12	10	3	2	2	2·04

* Only 4 species of Cryptostigmata since *T. velatus* did not occur as distinct aggregations.

species tend to have more aggregations per block of samples than the less frequently occurring species. The means in Table 4.4 indicate that the horizontal area occupied by an aggregation is about 80 cm^2 in the Collembola, about 100 cm^2 in the Cryptostigmata and of the order of 120 cm^2 in the Mesostigmata. However, these areas measure both the area of high density and the surrounding area of low density: visual inspection of the data indicated that whereas the area of high density was frequently in two or more adjacent samples in the Collembola and Cryptostigmata it was frequently confined to a single sample in the Mesostigmata. Thus, the horizontal size of the aggregation is often less than 20 cm^2 in the predatory species and frequently of the order of 30 to 50 cm^2 in the non-predatory species.

The distances between aggregations can be used to investigate the function of the aggregation response in these arthropods. First, if the distances between the centres of aggregation are less variable than would be expected at random and relatively large in length, then this might suggest that there is some form of aggregation territorality (cf. Wynne-Edwards, 1962 for a general discussion of territorality) whereby the group of arthropods and not the individual establishes a territory. Such a relation cannot, however, really be demonstrated in the soil arthropods. Usher (1969) indicated that uniformity of distance between aggregations occurred in the dark adult form of *I. sensibilis*, and the data are indicative of an approach towards uniformity in the pale adult form. Also *O. minimus* showed uniform spacing between aggregations (Usher 1971a) and there is an approach towards uniformity with the aggregations of adult *C. bipilis*.

Secondly, the distances between the centres might be rather irregular and bimodal, with a number of short distances and a number of long distances. Such a distribution is exhibited by the adults of *P. peltifer* and is indicative of aggregations being clustered together so that the aggregations are themselves aggregated. This form of distribution in the soil could be an intermediate form of distribution between the territorially small distinct aggregation and the territorially large gradient, and might reflect local pockets of particularly high density in an area of generally high density. This doubly aggregated distribution is not particularly common, being shown only by *P. lapponicus* in the Mesostigmata and by the juveniles of the Collembola *I. sensibilis*, *F. mirabilis* and *F. quadrioculata*. The latter observation, relating to juveniles, might suggest that Collembola search out more favourable environments for the deposition of their eggs.

Thirdly, the distances between aggregations might be more or less random. This is an assumption of the negative binomial distribution, and hence it has been assumed by many soil biologists (e.g. Hartenstein 1961; Berthet & Gérard 1965; Gérard & Berthet 1966). However, the fact that the aggregations of most species are randomly scattered throughout the soil is likely to make the detection of causes of aggregations more difficult.

Finally, from the analysis of the distinct aggregations, the relation between aggregation tendency and population density can be investigated. Usher (1969) proposed three types of relation namely:

Type I, as the population density increases so does the number of aggregations per unit of surface area or volume. If r_n is the correlation coefficient between the population density and the number of aggregations, and r_s the correlation coefficient between the population density and the mean number of individuals in an aggregation, then for a Type I relation r_n would be positive and significant and r_s would be non-significant (or possibly negative);

Type II, as the population density increases so does the mean number of individuals per aggregation. Type II relations would be demonstrated if r_n was nonsignificant and r_s positive and significant;

Type III, is a composite of the previous two types, probably occurring if niches became unsuitable at high population densities. Type III relations would be demonstrated if both r_n and r_s were positive and significant (or possibly due to data shortage if neither were significant).

The results of analysis of the more common species are shown in Table 4.5. The Collembola and Mesostigmata, whenever the relation could be identified, showed a Type I response. Since this implies that more aggregations are being formed with increasing population density, it could be argued that the species in these groups have wide niches. Many species of Cryptostigmata had a Type II relation. Since this implies that new sites for aggregations are not being formed as the population density increases, these species might in general have relatively narrow niches. Type III relations, being intermediate between Types I and II, might thus occur with species with intermediate sized niches: the only species at Rannoch to show a Type III relation are the adults and juveniles of *P. peltifer*.

Table 4.5. The correlation coefficients between population density and the mean number of arthropods within distinct aggregations (r_s) and the number of aggregations (r_n), and the type of relation between population density and aggregation response. Only the most frequently occurring species are included. Levels of significance are indicated by n.s.—not significant, *—0·01 < p ⩽ 0·05, **—0·001 < p ⩽ 0·01 and ***—0·001 ⩾ p.

Species	Sample size	r_n	r_s	Type
Collembola				
O. absoloni	8	0·63 n.s.	−0·66 n.s.	I?
I. sensibilis				
Juvenile	23	0·63 ***	0·18 n.s.	I
Pale adult	17	0·65 **	0·40 n.s.	I
Dark adult	8	0·31 n.s.	−0·35 n.s.	?
F. mirabilis				
Juvenile	10	0·84 **	0·29 n.s.	I
Adult	8	0·53 n.s.	0·27 n.s.	?
F. quadrioculata				
Juvenile	14	0·49 n.s.	−0·22 n.s.	?
Adult	5	0·93 *	0·71 n.s.	I
T. callipygos	8	0·33 n.s.	−0·13 n.s.	?
Cryptostigmata				
P. peltifer				
Juvenile	10	0·78 **	0·53 n.s.	III?
Adult	8	0·84 **	0·85 **	III
S. magnus	5	−0·34 n.s.	0·73 n.s.	?
C. bipilis				
Juvenile	6	0·37 n.s.	−0·16 n.s.	?
Adult	14	0·18 n.s.	0·71 **	II
A. poppei	11	−0·17 n.s.	0·78 **	II
Mesostigmata				
O. minimus	6	0·34 n.s.	−0·29 n.s.	?
V. nemorensis	7	0·58 n.s.	0·16 n.s.	?
P. radiatus	5	0·89 *	0·78 n.s.	I
T. pyriformis	6	0·21 n.s.	−0·24 n.s.	?
P. lapponicus	5	−0·21 n.s.	−0·72 n.s.	?

The analyses of the individual species thus show that aggregations occur very frequently and that they are very heterogeneous in their size. The aggregation responses are partially determined on a seasonal basis, but more importantly there is a relation between the form of the aggregation and the population density. These analyses thus quantify the aggregation process, but they cannot determine the causes of aggregations.

INTERRELATIONS BETWEEN ARTHROPODS AND THE ENVIRONMENT

Correlation

For each of the 48 samples within a block there are counts of the numbers of each of the species of Collembola, Cryptostigmata and Mesostigmata and estimates or direct measurements of seven features of the environment. Since on each sampling date two blocks of samples, each located at random, were taken, a correlation coefficient can be estimated for the 96 samples. However, treating each sample as being randomly located might introduce an error, which has been eliminated by the following technique. For each block independently, an analysis of variance or analysis of covariance has been performed on each variate and on each pair of variates, removing the variance or covariance due to position (row or column) within the block. The correlation coefficient was calculated for the residual variances and covariance by the formula

$$r_{ij} = \frac{\text{residual covariance } (i, j)}{\sqrt{(\text{residual variance } (i) \times \text{residual variance } (j))}} \qquad (4{\cdot}3)$$

where i and j relate to any pair of variates ($i \neq j$), and the residual variances and covariances have 41 d.f. The r_{ij}s for both blocks of samples were transformed to z_{ij} by

$$z \quad 0{\cdot}5\, ln\left(\frac{1+r}{1-r}\right) \qquad (4{\cdot}4)$$

the two values for z_{ij} averaged, and a pooled correlation coefficient calculated from

$$r = \frac{e^{2z}-1}{e^{2z}+1} \qquad (4{\cdot}5)$$

The effective number of degrees of freedom of such a pooled correlation coefficient is 78.

In order to illustrate the methods of analysis, the data for the Collembola and environmental features in samples collected on 16th January 1966 will be discussed (Table 4.6). Inspection of the correlation coefficients shows some of the relations between the Collembola and the environment: adult *F. mirabilis* is positively related to nitrogen content and *A. binoculatus* is negatively related to potassium content. These two relations might be real since the species show just the one relation and are not correlated to a complex of environmental features. However, the correlations of juvenile *F. quadrioculata* are positive for depth, water content and nitrogen, and negative

Table 4.6. The pooled correlation matrix, calculated from equations 4.3, 4.4 and 4.5, for the relation between the Collembola and the environmental features (data from samples collected 16 January 1966). For statistical significance, correlation coefficients would have to exceed in absolute value the following: 0·22 ($p = 0·05$), 0·29 ($p = 0.01$) and 0·36 ($p = 0·001$). Values that are statistically significant are set in bold type.

Variates	Variates 1	2	3	4	5	6	7	8	9	10	11	12	13	14	15	16	17	18
1. *F. mirabilis* (juv.)																		
2. *F. mirabilis* (ad.)	**·32**																	
3. *O. absoloni*	·16	·19																
4. *O. latus*	·07	—·02	·19															
5. *T. callipygos*	**·28**	·08	—·05	·18														
6. *A. binoculatus*	·01	—·14	·13	·14	·05													
7. *F. quadrioculata* (juv.)	·10	·18	·02	·01	·01	·11												
8. *F. quadrioculata* (ad.)	—·01	·07	·07	**·26**	—·10	**·43**	·11											
9. *I. minor*	**·31**	**·27**	**·36**	·12	0	**·22**	·18	**·29**										
10. *I. sensibilis* (juv.)	—·07	·13	**·46**	**·32**	·01	—·02	—·17	·05	·02									
11. *I. sensibilis* (pale ad.)	—·07	—·19	**·35**	·13	·04	·02	—**·42**	—·09	—·13	**·31**								
12. *I. sensibilis* (dark ad.)	—·11	—·03	**·24**	—·18	0	·04	—·20	—**·22**	—·10	·19	**·60**							
13. Depth	—·02	·16	—**·41**	—·21	—·02	·09	**·44**	—·06	·08	—**·47**	—**·55**	—**·23**						
14. Water content	—·04	·09	—**·39**	—**·25**	—·11	·11	**·40**	—·07	·18	—**·48**	—**·57**	—**·25**	**·88**					
15. Pore space	—·08	—·10	**·44**	·21	—·05	·04	—**·34**	·07	—·01	**·43**	**·54**	**·29**	—**·89**	—**·67**				
16. Nitrogen	·12	**·32**	—·19	—·11	·16	—·19	**·22**	—·17	—·09	—**·23**	—·14	—·09	**·44**	**·29**	—**·47**			
17. Phosphorus	—·01	·10	**·48**	**·27**	·05	—·04	—**·28**	·13	—·02	**·49**	**·54**	**·22**	—**·74**	—**·73**	**·68**	—·14		
18. Potassium	·09	—·13	·20	·04	—·12	—**·30**	—**·36**	—·08	—·13	**·24**	**·48**	·16	—**·76**	—**·73**	**·55**	—**·28**	**·56**	
19. Calcium	·14	·12	**·29**	·20	—·03	—·21	—**·27**	—·08	—·12	**·28**	**·39**	·03	—**·53**	—**·61**	**·48**	—·03	**·61**	**·42**

for pore space, phosphorus, potassium and calcium. The positive and negative correlations are reversed for juvenile *I. sensibilis*. Inspection of the correlations between the environmental features shows that strong correlations exist, particularly between depth, water content and pore space. The main problem with interpreting data such as in Table 4.6 is the concept of 'cause and effect'. If, for example, any species correlates positively with depth, then it can be predicted (i) that there would be a positive correlation with water content (r between water and depth is 0·88) and (ii) that there would be a negative correlation with pore space (r between depth and pore space is −0·89). Such a pattern of correlations can be seen in the data, and it is thus impossible to say whether the number of individuals can be attributed to either depth, water content or pore space. There are, therefore, in the environmental data complexes of highly correlated influences acting either singly or together on the arthropods. Calculation of partial correlation coefficients, as undertaken by Poole (1961, 1962), is generally not very helpful when the environmental influences are so strongly correlated: for example, if the partial correlation coefficient between juvenile *I. sensibilis* and each of the three environmental features, depth, water content and pore space are calculated (Table 4.7), only one of the coefficients is significant

Table 4.7. Correlation and partial correlation coefficients for the relation between the juveniles of *I. sensibilis* and the three environmental features, depth, water content and pore space. In the table the following notation has been used: *I*, *Isotoma*; *W*, water content; *D*, depth; and *P*, pore space; and a dot in a list of subscripts implies that variates after the dot have been eliminated.

Correlation coefficients	Partial correlation coefficients	
$r_{IW}=-·48$	$r_{IW.D}=-·16$	$r_{IW.DP}=-·20$
	$r_{IW.P}=-·29$	
$r_{ID}=-·47$	$r_{ID.W}=-·11$	$r_{ID.PW}=·05$
	$r_{ID.P}=-·21$	
$r_{IP}=·43$	$r_{IP.W}=·17$	$r_{IP.DW}=·13$
	$r_{IP.D}=·03$	
$r_{WD}=·88$	$r_{WD.P}=·84$	
$r_{WP}=-·67$	$r_{WP.D}=·52$	
$r_{DP}=-·89$	$r_{DP.W}=-·85$	

after elimination of one feature ($r_{IW.P}$ in Table 4.7), and none of the correlations are significant after elimination of two features.

Such results with matrices of correlation coefficients or partial correlation coefficients indicate the need for analysis by some method that can demonstrate the combined effects of several variables.

Factor and principal component analysis

In a multivariate situation one is concerned with a set of *m* variates, which might include numbers of species and environmental features, each of which has been recorded in *n* samples. If a variate is denoted by X_i ($i=1, 2, \ldots m$), then the basic model of factor analysis is

$$X_i = \sum_{r=1}^{k} l_{ir} f_r + e_i \qquad (4.6)$$

where there are k ($k<m$) common factors f_r, l_{ir} are the factor loadings, and e_i represent a residual source of variation, the unique variances.

Ecologically, the factor analysis model thus states that the individual biological attributes of such a multivariate system can be expressed in terms of a small number, k, of common factors and a residual unique source of variation. The analysis estimates values of l_{ir} and e_i, and aims to interpret the meaning of these common factors. Although the model is not based on any hypothesis about the distribution of the variables x_{ij}, in the data matrix, the use of the product moment correlation coefficient, the techniques of maximum likelihood estimation (Lawley & Maxwell 1971; Jöreskog 1967; Jöreskog & Lawley 1968) and the testing of significance do require that the variables x_{ij} are drawn from a multivariate normal distribution. Since this is not the case with counts of soil arthropods, logarithmic transformation of the data has been used. Also, in biological situations the assumption of linearity of the common factors is often doubtful, but it suffices as a first approximation. Correlation matrices have been used in preference to covariance matrices since they eliminated the problems of scaling of the original variates (which would have been impossible in data where some of the variates are counts of animals and other variates are measures of the environment). The computer programme referred to by Jöreskog (1967) has been used for the estimation of all factor loadings and unique variances quoted in this paper.

The factor analysis model can be compared with principal component analysis, which transforms the *m* variates X_i into a set of *m* uncorrelated variates Y_i according to the following rule. Y_1 has a maximum possible variance, Y_2 has a maximum variance subject to being uncorrelated with Y_1 (i.e. orthogonal axes), Y_3 has a maximum variance subject to being uncorrelated with Y_1 and Y_2, and so on. Principal component analysis can thus be represented by

$$X_i = \sum_{r=1}^{m} l_{ir} Y_r \qquad (4.7)$$

where l_{ir} are the loadings. The ease of the principal component model lies in the fact that the loadings are given by the latent vectors of the correlation matrix **R**, and the variance accounted for by each new variate, Y_i, is given by the ratio of the appropriate latent root of **R** to the the trace of **R**.

A principal component analysis of the data in Table 4.6, extended to include the mites, is shown in Figs. 4.4 and 4.5. The first component accounts for over 30% of the variance, and is clearly much larger than the

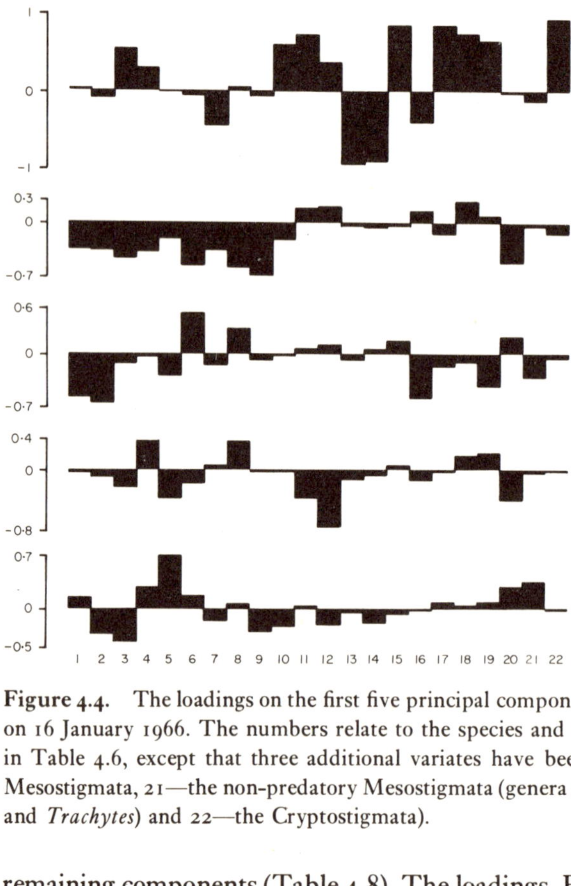

Figure 4.4. The loadings on the first five principal components for the samples collected on 16 January 1966. The numbers relate to the species and environmental features listed in Table 4.6, except that three additional variates have been added (20—the predatory Mesostigmata, 21—the non-predatory Mesostigmata (genera *Olodiscus*, *Zercon*, *Parazercon* and *Trachytes*) and 22—the Cryptostigmata).

remaining components (Table 4.8). The loadings, Fig. 4.4, indicate that the group of three environmental features, depth, water content and pore space, are important in interpreting this component since the first two are large

Table 4.8. A principal component analysis of the data in Table 4.6, extended to include the mites.

Component	1	2	3	4	5
Latent root	6·64	2·47	1·95	1·46	1·32
Percentage of variance	30·2	11·2	8·8	6·6	6·0

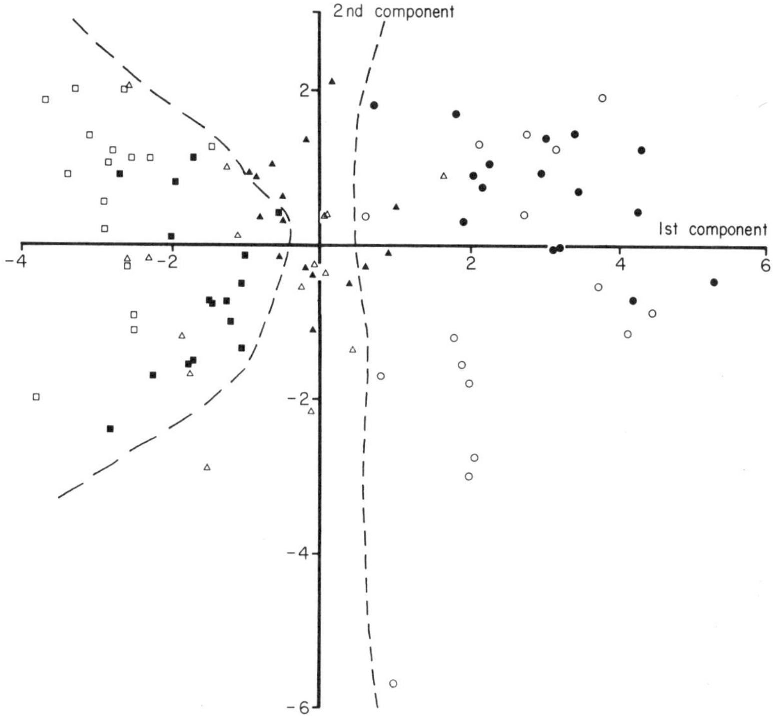

Figure 4.5. The first two principal components, derived from the data in Table 4.6 and Fig. 4.4. Circles represent samples from the surface layer, triangles from the centre layer and squares from the lowest layer. Open symbols relate to one block of samples, filled symbols to the other block of samples. The two dashed lines delimit the rightward extent of square symbols and the leftward extent of the circle symbols.

and negative and the third is positive. The second component is of interest, since all the environmental variates have near-zero loadings, whilst the majority of species of Collembola and the predatory Mesostigmata have large (negative) loadings. Although this component accounts for only little more than 11% of the variance, it would appear to be related to some general sociability of the soil arthropods. The remaining components in Fig. 4.4 all account for rather small percentages of the variance, and from a single analysis their interpretation is uncertain, except in so far as the loading for nitrogen in the third component is reasonably large. Plotting the first two components, whose interpretation is straightforward, produces a scatter of points without any clearly defined clusters (Fig. 4.5). The increase in depth from right to left in Fig. 4.5 makes it clear that, ecologically, the surface and lowest layers of the soil are distinct, since there is no overlapping of samples from these two layers. It would therefore appear that the arthropod community of the surface litter horizon is different from the community of the fermentation horizon.

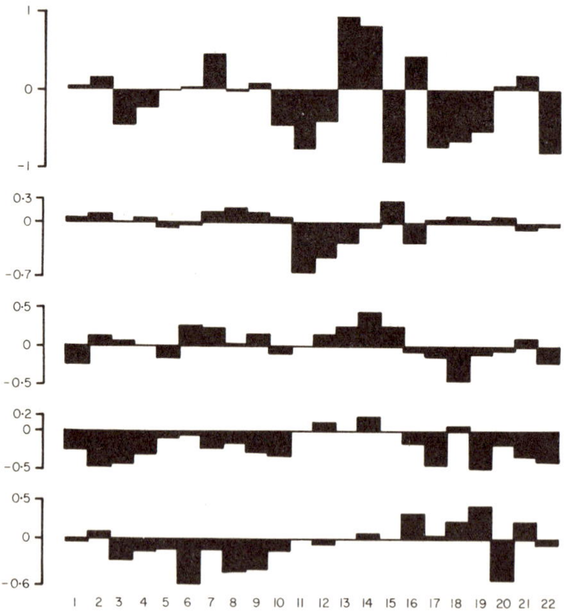

Figure 4.6. The loadings of five common factors derived from the correlation matrix in Table 4.6. The numbers relate to species and environment features and are listed in Tables 4.6 and 4.9, extended as in Fig. 4.4.

Maximum likelihood factor analysis indicated that five common factors were sufficient to account for the covariance in the data. The loadings are shown in Fig. 4.6 and the unique variances in Table 4.9. The unique variances for depth and pore space are zero since all of their variation has been

Table 4.9. The unique variances derived from a factor analysis of the data in Table 4.6, extended to include the mites.

Species	Unique variance	Environmental feature	Unique variance
1. *F. mirabilis* (juv.)	·78	13. Depth	0
2. *F. mirabilis* (ad.)	·33	14. Water content	·11
3. *O. absoloni*	·50	15. Pore space	0
4. *O. latus*	·72	16. Nitrogen	·58
5. *T. callipygos*	·95	17. Phosphorus	·25
6. *A. binoculatus*	·54	18. Potassium	·30
7. *F. quadrioculata* (juv.)	·70	19. Calcium	·25
8. *F. quadrioculata* (ad.)	·69	20. Predatory	
9. *I. minor*	·62	Mesostigmata	·71
10. *I. sensibilis* (juv.)	·64	21. Non-predatory	
11. *I. sensibilis* (pale ad.)	·26	Mesostigmata	·81
12. *I. sensibilis* (dark ad.)	·22	22. Cryptostigmata	·13

accounted for by the five common factors. The unique variances of the other environmental features are generally small, indicating that their distribution is accounted for reasonably well by the five common factors. The Collembola are variable: whereas the adults of *F. mirabilis* and *I. sensibilis* have small unique variances, other species, particularly *T. callipygos*, have large unique variances. Unlike principal component analysis, where the components could be ordered, the factors form a set of five that cannot be ordered into the largest, etc. However, rotation of the factors is possible, and one such

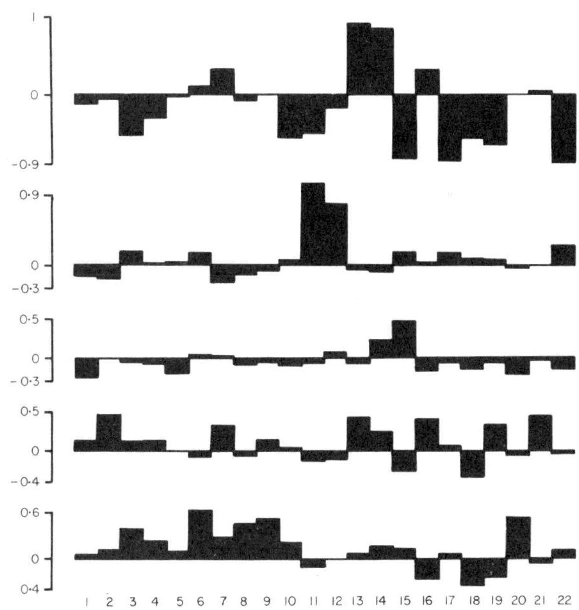

Figure 4.7. The same loadings as in Fig. 4.6 after rotation of the common factors by the Varimax method. The numbers relate to the same variates as in Fig. 4.6.

rotation by the widely-used Varimax method is shown in Fig. 4.7. This method of rotation has the general characteristic of increasing the size of a few of the loadings, whilst making many more loadings near zero, and thus aiding the interpretation of the factors.

Interpretation

The number of principal component and factor loadings shown in Figs. 4.4 and 4.6 is limited for ease of presentation, but there is no limit to the size of matrix analysed except that imposed by the computer store available. The unique variances for species in the two groups of mites, Mesostigmata and Cryptostigmata, are listed in Table 4.10. Comparing each group of vari-

Table 4.10. The unique variances for factor analyses of the Mesostigmata and Cryptostigmata data collected on 16 January 1966.

Mesostigmata species	Unique variance	Cryptostigmata species	Unique variance
O. minimus	·78	*T. velatus*	·02
V. nemorensis	·91	*P. peltifer*	
P. radiatus	·87	Juvenile	·59
T. pyriformis	·92	Adult	·44
Juvenile *Zercon*, etc.	·78	*S. magnus*	·52
Eugamasus sp.	·69	*C. bipilis*	
		Juvenile	·30
		Adult	·14
		A. poppei	·23
		Chamobates schützi	·68
		Eupelops sp.	·45
		Camisia spinifer	·30
		Porobelba spinosa	·80

ates, the mean unique variance of the Mesostigmata is 0·82, of the Collembola 0·58, of the Cryptostigmata 0·41, and of the seven environmental features 0·21. These means indicate that the factor analysis model is giving a reasonably good fit to the Cryptostigmata data since the unique variances are generally reasonably small and the common factors account for a large amount of the correlation. In the Mesostigmata data the common factors are accounting for very little of the observed correlation, whilst the Collembola are intermediate between the Cryptostigmata and Mesostigmata. The environmental features have generally small loadings indicating that the common factors can be interpreted in terms of environmental influences.

Since blocks of samples were collected on twelve occasions during the year, some general statements can be made about the interpretation of the principal component and factor loadings. In all instances the principal component accounting for the largest amount of the variance, as well as one of the factors, is structural, showing large loadings on depth and water content, and a large contrary loading on pore space. In all instances this factor or component reflects the vertical distribution of the species, separating the species of the soil surface (often large species such as adult *I. sensibilis*, *Tomocerus minor*, *A. poppei* and *C. bipilis*) from the species occurring deeper in the soil (such as *T. calipygos* or the juveniles of *F. quadrioculata*). Since this first factor is concerned with the vertical distribution of the microarthropods, it contains relatively little information concerning the causes of aggregation except in so far as specifying the depth at which most aggregations are likely to occur.

In the majority of analyses three other factors or components could be interpreted. First, one of these is associated with the general 'sociability' of the species, since large loadings of the same sign were found with the majority of species of Collembola and mites, and the loadings on the environmental features were small. This factor indicates that when arthropods are frequent, many species contribute to this abundance, or conversely that the majority of species tend to avoid some area. Since the environmental features have near zero loadings on this factor, the cause of this general sociability cannot be investigated.

Secondly, a factor relating to nitrogen was frequently seen in the analysis. In this factor the loading on the nitrogen content was large whilst the loadings on other environmental features were small (except sometimes phosphorus and calcium which also showed large loadings). Amongst the arthropods, *C. spinifer*, *F. mirabilis* and *O. absoloni* (summer months only), showed large loadings, though at some seasons of the year many other species had large loadings on this factor. The interpretation of this factor ecologically is more difficult: nitrogen has been shown to occur in rather small 'concentration aggregations'. However, a likely explanation is that high nitrogen content is related to faecal content. If this is correct, this factor might suggest aggregations where the arthropods are living and defaecating (a trivial result) or it might suggest aggregation at sites of former biological activity. The association with this factor of *F. mirabilis*, or species with sucking mouthparts and no visible gut contents, suggests possibly the latter since arthropod faeces are known to be good substrates for microbial activity.

Thirdly, a factor relating to potassium content was frequently seen in the analysis. In this factor the loading on the potassium content was large whilst the loadings on other environmental features, except calcium on occasions, were small. Rather few species of arthropods had large loadings, though *O. absoloni* (autumn and winter only) and *P. peltifer* had loadings of the same sign as potassium and adult *I. sensibilis* tended to have loadings of the contrary sign. Ecologically, the potassium content might be a reflection of the active roots or mycorrhiza in the soil. The 'concentration aggregations' of potassium were larger than those of nitrogen, and would approach the volume occupied by a surface feeding root of a pine tree. Tentatively, then, this factor can be interpreted in relation to the rhizosphere.

The interpretation of the factor analyses thus indicates four influences acting upon the community of soil arthropods. These relate to the vertical distribution of the individual species, to the general sociability of the species, and tentatively to the sites of former biological activity and the rhizosphere. When Collembola were examined, a fifth factor could sometimes be seen: this related to the predatory species of the Mesostigmata, which tended to show negative loadings with species of the family Onychiuridae and positive loadings with species of Isotomidae. Harris (1975) has demonstrated that

Pergamasus longicornis will eat Isotomids in preference to Onychiurids. This factor thus does not relate to the causes of aggregation, but rather to the fact that predatory species tend to be distributed in relation to their preferred prey species.

Discussion

Studies on aggregation or pattern of species are interesting for two reasons. First, is there any biological significance in the clumping together of individuals, or is it merely an accident of the biology of the species concerned (such as, for example, an aggregation of juveniles at the site of an egg cluster)? Secondly, does the formation and causation of aggregations throw any light on the co-existence within a limited area of a number of species that appear to be ecologically similar?

The nature of the clumping together of individuals may indicate the biological significance of the aggregation process. Five types of pattern have been described by Hutchinson (1953), who named them:

Vectorial—the distribution is determined by external influences, such as temperature, humidity, density gradients, winds, etc.

Reproductive—the distribution may be determined by the offspring remaining near the parents.

Social—the distribution is determined by signalling of various kinds, leading either to spacing or clumping.

Coactive—the distribution may be determined by interaction between species in competition.

Stochastic—the distribution is determined by random forces.

There are a considerable number of studies of pattern in plant populations, and these tend to show that small-scale pattern is reproductive, due to vegetative spread of species and limited seed dispersal, and larger-scale pattern is vectorial being imposed on the vegetation by the pattern of the environment. In studies of soil arthropods, any one of these five categories of pattern may occur, though aggregations are likely to result only from the first four listed. Hale (1966) suggested that aggregations could be explained by a slow dispersion from the site of an egg cluster (reproductive pattern) and by a coming together at a food source (vectorial pattern). Laboratory observations on the behaviour of juvenile *Isotoma sensibilis*, remaining at the egg cluster site for three instars before dispersing (Usher 1969), suggest that reproductive pattern is frequent amongst juveniles, though such a pattern could not exist in the predatory Mesostigmata which lay single eggs apparently at random (Harris 1975). A change from reproductive to other forms of pattern over the life span of a soil arthropod seems likely (Takeda 1973; Barra & Christiansen 1975).

The majority of laboratory and field studies indicate that aggregation in adults is vectorial pattern. This is particularly clearly shown by Joosse's (1966) study of *Anurida maritima*, which is stimulated to aggregate in air-filled crevices either by the incoming tide or by rain at low tide. Whatever the vectorial influence, it would appear that an ability to respond to this influence is important for the survival of the individual since it will be forced into a micro-environment in which its chances of survival are greater than in any random micro-environment. Thus, aggregation allows a species to achieve a larger population density than if it were not aggregated. Studies at Rannoch have shown that the more frequent species of Collembola and Cryptostigmata tend to be more strongly aggregated than the rarer species: the values of b in Table 4.1 are correlated with the overall abundance of the Collembola, ($r=0{\cdot}57$, 17 d.f., $p=0{\cdot}01$). A similar correlation with the Cryptostigmata fails to be significant ($r=0{\cdot}44$, 14 d.f., $p>0{\cdot}05$). The data from Rannoch therefore tend to support the hypothesis that an aggregated species can maintain a greater population density than a non-aggregated or weakly aggregated species. Improved methods in estimating the tendency to aggregate (Patil & Stiteler 1974) might be useful in proving or disproving this hypothesis.

Social and coactive patterns are much more difficult to demonstrate under field or laboratory conditions due to correlation between any features of the system that are measured. Observations on moulting groups of Collembola (Strebel 1932) suggest that in this instance social aggregations are transitory, and have some protective function for the moulting Collembola. Strebel demonstrated this form of aggregation for *Hypogastrura purpurascens*. Harris (1975) has shown that predatory mites find this genus distasteful, and B.C. Longstaff claims (personal communication) to be able to detect an unpleasant smell in high density cultures of *H. denticulata*. Joosse (1970, 1971) has suggested that aggregations of Collembola that have no courtship behaviour are essentially vectorial, but that they have an important social role since they increase the chance of females making contact with spermatophores. Thus the location of the aggregation is vectorial although the stimulus is social. In the Rannoch data the species with courtship behaviour are infrequent, and hence the data cannot be used to test Joosse's hypothesis. There are no field or laboratory studies in which coactive pattern in soil arthropods has been demonstrated. Such a pattern is very difficult to prove under field conditions where two species might appear to have separate and clearly distinct niches although this might be due to interspecific competition forcing one of the species out of its optimal niche. Coactive patterns in soil arthropods can really only be proved in controlled laboratory experiments.

The different niches occupied by soil arthropods at first sight seem hard to enumerate. There is perhaps a conflict in the methods of analysis used. If a multivariate approach is selected (Bonnet *et al.* 1970; Hill 1971) then the fundamental idea is to explain the distribution of several species in terms of a few environmental factors or components. If, however, one assumes that there must be the same number or more niches or resources than species (Levins 1968), then a multivariate approach can only be inappropriate since one wants to increase and not decrease the influences to be considered. Blackith (1974) undertook such an analysis for the Collembola of Irish blanket bogs, but he had to make many assumptions which might be questioned.

The concept of the niche width of individual species may explain the observed relations between aggregation and population density listed in Table 4.5, where relations in the sequence I, III, II indicate a decreasing width of the species niche. The majority of species of Collembola and Mesostigmata had a type I relation (as population density increases so does the number of aggregations), whilst many of the Cryptostigmata showed a type II relation (as population density increases so does the number of individuals within an aggregation). *Folsomia candida*, which occurs in Britain in rather specialized habitats like greenhouses, has a type III relation (Usher & Hider 1975). These observations suggest that the niche widths of soil arthropods are variable. Investigation of multivariate data by factor analysis provides a series of k common factors and m unique variances, and is thus more realistic in the assessment of niches since there are $m+k$ possible niches determined by the analysis. The Collembola data, Fig. 4.6 and Table 4.9, indicate that although the 5 common factors have some effect on the community, each species also has a relatively large unique variance. An interpretation of factor analysis could be that the species own niche is represented by its unique variance, and that the overlap of niches is represented by the common factors. All the species of Collembola (Table 4.9) and mites (Table 4.10) thus show that there is some overlap of niches. These results are in accord with the conceptual framework on niches described by May (1974). The soil environment, particularly under a forest canopy as at Rannoch or in the Irish blanket bogs (Blackith 1974) is extremely steady, so there is an effective limit to niche overlap. Blackith invoked the environmental stability to explain the large proportion of pale coloured species, indicating that the observed proportion placed the Irish bogs in the tropics (by applying Gloger's rule). The soil environment is thus, for temperate climates, extremely steady. To quote May 'In a perfectly stable deterministic environment, arbitrarily close packing and rich speciation is possible, and to a certain limited extent the greater the environmental steadiness, the closer the packing, and the richer the consequent assembly of species.' It would therefore appear that environmental stability and aggregation behaviour promote a large population density of an apparently large number of species in the superficially homogeneous soil environment.

Acknowledgements

Although all analyses reported in this paper have been performed in the University of York, the data were collected whilst I was a post-graduate student in the Department of Forestry and Natural Resources, University of Edinburgh. I should therefore like to thank Dr. D.R. Gifford and Dr. D.N. Lawley who supervised this work, the Nuffield Foundation and Edinburgh University for financial support, and the Forestry Commission for access to the Black Wood of Rannoch.

References

BAGNALL R.S. (1932) A Scottish record of a rare springtail (*Sminthurinus cingulatus*, Bagn.). *Scott. Nat.* **193**, 20.

BARA J.A. & CHRISTIANSEN K. (1975) Experimental study of aggregation during the development of *Pseudosinella impediens* Gisin and Gama (Collembola Entomobryidae). *Pedobiologia*, **15**, 343–7.

BERTHET P. & GÉRARD G. (1965) A statistical study of microdistribution of Oribatei (Acari). Part I. The distribution pattern. *Oikos* **16**, 214–27.

BLACKITH R.E. (1974) The ecology of Collembola in Irish blanket bogs. *Proc. R. Ir. Acad.*, sect. B **74**, 203–26.

BÖDVARSSON H. (1970) Alimentary studies of seven common soil-inhabiting Collembola of Southern Sweden. *Ent. Scand.* **1**, 74–80.

BONNET L., CASSAGNAU P. & Izarra D.C. de (1970) Étude écologique des Collemboles muscicoles du Sidobre (Tarn). II. Modèle mathématique de la distribution des espèces sur un rocher. *Bull. Soc. hist. nat. Toulouse* **106**, 127–45.

CHRISTIANSEN K. (1967) Competition between Collembolan species in culture jars. *Rev. Ecol. Biol. Sol* **4**, 439–62.

CHRISTIANSEN K. (1970) Experimental studies on the aggregation and dispersion of Collembola. *Pedobiologia* **10**, 180–98.

CURRY J.P. (1973) The arthropods associated with the decomposition of some common grass and weed species in the soil. *Soil Biol. Biochem.* **5**, 645–57.

EDWARDS C.A. & LOFTY J.R. (1969) The influence of agricultural practice on soil microarthropod populations. In: *The Soil Ecosystem*, ed. J.G. Sheals, pp. 237–47. Systematics Association Publication No. 8.

EDWARDS C.A. & LOFTY J.R. (1971) The influence of temperature on numbers of invertebrates in soil, especially those affecting primary production. *Proc. 4th. Coll. Soil. Zool.*, 545–55. I.N.R.A. Paris.

FÜHRER E. (1961) Der Einfluss von Pflanzenwurzeln auf die Verteilung der Kleinarthropoden im Boden, untersucht an *Pseudotritia ardua* (Oribatei). *Pedobiologia* **1**, 99–112.

GÉRARD G. & BERTHET P. (1966) A statistical study of microdistribution of Oribatei (Acari). Part II. The transformation of the data. *Oikos* **17**, 142–9.

GIFFORD D.R. (1968) The significance of soil water to soil invertebrates. In *The Measurement of Environmental Factors in Terrestrial Ecology*, ed. R.M. Wadsworth, pp. 171–80. Blackwell Scientific Publications, Oxford & Edinburgh.

GISIN H. (1955) Recherches sur la relation entre la faune endogée de Collemboles et les qualités agrologiques de sols viticoles. *Revue suisse Zool.* **62**, 601–48.

GLASGOW J.P. (1939) A population study of subterranean soil Collembola. *J. anim. Ecol.* **8**, 323–53.

GRIDELET D. & LEBRUN P. (1973) Contribution a l'étude écologique des acariens des poussières de maisons. *Acarologia* **15**, 461–76.

HAARLØV N. (1960) Microarthropods from Danish soils: Ecology, phenology. *Oikos*, suppl. 3, 1–176.

HALE W.G. (1966) A population study of moorland Collembola. *Pedobiologia* **6**, 65–99.

HALE W.G. (1967) Collembola. In: *Soil Biology*, ed A. Burgess & F. Raw, pp. 397–411. Academic Press, London & New York.

HARRIS J.R.W. (1975) *Aspects of the kinetics of predation with reference to a soil mite* Pergamasus longicornis *Berlese*. Unpublished D. Phil. Thesis: University of York.

HARTENSTEIN R. (1961) On the distribution of forest soil microarthropods and their fit to 'contagious' distribution functions. *Ecology* **42**, 190–4.

HARTENSTEIN R. (1962) Soil Oribatei—I. Feeding specificity among forest soil Oribatei (Acarina). *Ann. ent. Soc. Am.* **55**, 202–6.

HAYES A.J. (1963) Studies on the feeding preferences of some Phthiracarid mites (Acari: Oribatidae). *Entomologia exp. appl.* **6**, 241–56.

HEALEY I.N. (1967) An ecological study of temperatures in a Welsh moorland soil, 1962–1963. *J. anim. Ecol.* **36**, 425–34.

HILL J.R. (1971) An ordination of Oribatid mite fauna under differing forest covers. *Proc. 3rd. Int. Congr. Acarology*, 85–90.

HUTCHINSON G.E. (1953) The concept of pattern in ecology. *Proc. Acad. nat. Sci. Philad.* **105**, 1–12.

JENSEN P. (1968) Changes in cryptozoan numbers due to systematic variation of covering boards. *Ecology* **49**, 409–18.

JENSEN P. & CORBIN K.W. (1966) Some factors affecting aggregation of *Isotoma viridis* Bourlet and *Arion fasciatus* Nilsson. *Ecology* **47**, 332–4.

JOOSSE E.N.G. (1966) Some observations on the biology of *Anurida maritima* (Guérin), (Collembola). *Z. Morph. Ökol. Tiere* **57**, 320–8.

JOOSSE E.N.G. (1970) The formation and biological significance of aggregations in the distribution of Collembola. *Neth. J. Zool.* **20**, 299–314.

JOOSSE E.N.G. (1971) Ecological aspects of aggregation in Collembola. *Rev. Ecol. Biol. Sol* **8**, 91–7.

JOOSSE E.N.G. & VERHOEF H.A. (1974) On the aggregational habits of surface dwelling Collembola. *Pedobiologia* **14**, 245–9.

JÖRESKOG K.G. (1967) Some contributions to maximum likelihood factor analysis. *Psychometrika* **32**, 443–82.

JÖRESKOG K.G. & LAWLEY D.N. (1968) New methods in maximum likelihood factor analysis. *Br. J. Math. Stat. Psychol.* **21**, 85–96.

KARG W. (1961) Ökologische Untersuchungen von edaphischen Gamasiden (Acarina, Parasitiformes), 1 & 2. *Pedobiologia* **1**, 53–74, 77–98.

KNIGHT C.B. & ANGEL R.A. (1967) A preliminary study of the dietary requirements of *Tomocerus* (Collembola). *Am. Midl. Nat.* **77**, 510–17.

KNIGHT C.B. & REED V. (1969) Microstratification of *Tomocerus* (Collembola) in a pine open-field continuum. *Rev. Ecol. Biol. Sol.* **6**, 221–34.

KRAAN C. VAN DER (1971) Some aspects of field dependent distribution in a population of *Hypogastrura viatica* Tullberg 1872. *Rev. Ecol. Biol. Sol.* **8**, 99–102.

LAWLEY D.N. & MAXWELL A.E. (1971) *Factor Analysis as a Statistical Method.* 2nd ed. Butterworths, London.

LEBRUN P. (1969) Écologie et biologie de *Nothrus palustris* (C.L. Koch) (Acari Oribatei). Densité et structure de la population. *Oikos* **20**, 34–40.

LEBRUN P. (1971) Écologie et biocénotique de quelques peuplements d'arthropodes édaphiques. *Institut Royal de Sciences Naturalles de Belgique*, Memoire **165**, 1–203.

LEBRUN P. & RUYMBEKE M. VAN (1971) Intérêt écologique de la relation entre la température et la durée de développement des Oribates. *Acarologia* **13**, 176–85.

LEVINS R. (1968) *Evolution in changing environments.* Princeton University Press, Princeton.

LONGSTAFF B.C. (1974) *Experimental and computer studies of competition in soil Collembola.* Unpublished D. Phil. Thesis: University of York.

LUXTON M. (1967a) The ecology of saltmarsh acarina. *J. anim. Ecol.* **36**, 257–77.

LUXTON M. (1967b) The zonation of saltmarsh acarina. *Pedobiologia* **7**, 55–66.

MACFADYEN A. (1952) The small arthropods of a *Molinia* fen at Cothill. *J. anim. Ecol.* **21**, 87–117.

MASSOUD Z., POINSOT N. & POIVRE C. (1968) Contribution à l'étude du comportement constructeur chez les Collemboles. *Rev. Ecol. Biol. Sol.* **5**, 283–6.

MAY, R.M. (1974) On the theory of niche overlap. *Theor. Popul. Biol.* **5**, 297–332.

MILLS J.T. & SINHA R.N. (1971) Interactions between a springtail, *Hypogastrura tullbergi*, and soil-borne fungi. *J. econ. Ent.* **64**, 398–401.

NEF L. (1962) The distribution of Acarina in the soil. In: *Progress in Soil Zoology*, ed. P.W. Murphy, pp. 56–8. Butterworths, London.

PANDE Y.D. & BERTHET P. (1973) Studies on the food and feeding habits of soil Oribatei in a black pine plantation. *Oecologia* **12**, 413–26.

PATIL G.P. & STITELER W.M. (1974) Concepts of aggregation and their quantification: a critical review with some new results and applications. *Res. Popul. Ecol.* **15**, 238–54.

PETERSEN H. (1971) Collembolernes ernaeringsbiologi og dennes økologiske betydning. *Ent. Meddel.* **39**, 97–118.

POINSOT N. (1966). Existence d'un comportement constructeur chez un Collembole Isotomidae *Subisotoma variabilis* (Gisin 1949). *Rev. Ecol. Biol. Sol.* **3**, 173–8.

POOLE T.B. (1959) Studies on the food of Collembola in a Douglas fir plantation. *Proc. zool. Soc. Lond.* **132**, 71–82.

POOLE T.B. (1961) An ecological study of the Collembola in a coniferous forest soil. *Pedobiologia* **1**, 113–37.

POOLE T.B. (1962) The effect of some environmental factors on the pattern of distribution of soil Collembola in a coniferous woodland. *Pedobiologia* **2**, 169–82.

REMMERT H. & WISNIEWSKI, W. (1970) Low resistance to cold of polar animals in summer. *Oecologia* **4**, 111–12.

SAICHUAE P., GERSON U. & HENIS Y. (1972) Observations on the feeding and life history of the mite *Nothrus biciliatus* (Koch). *Soil Biol. Biochem.* **4**, 155–64.

SCHALK V. (1968) Zur Bodenfauna von Wiesen und Luzernebeständen unterschiedlicher Standorte unter besonderer Berücksichtigung der Oribatiden. *Pedobiologia* **8**, 424–506.

STREBEL, O. (1932) Beiträge zur Biologie, Ökologie und Physiologie einheimischer Collembolen. *Z. Morph. Ökol. Tiere* **25**, 31–153.

TAYLOR L.R. (1961) Aggregation, variance and the mean. *Nature, Lond.* **189**, 732–5.

TAKEDA H. (1973) A preliminary study of Collembolan populations in a pine forest. *Res. Popul. Ecol.* **15**, 76–89.

Tillbrook P.J. (1967) Arthropod ecology in the maritime Antarctic. *Antarctic Res. Ser.* **10**, 331–56.

Törne E. von (1974) Experimentelle Motivation systemökologischer Deutungen von zootischen Aggregationen im Boden. *Pedobiologia* **14**, 324–38.

Usher M.B. (1969) Some properties of the aggregations of soil arthropods: Collembola. *J. anim. Ecol.* 38, 607–22.

Usher M.B. (1970a) Pattern and seasonal variability in the environment of a Scots pine forest soil. *J. Ecol.* 58, 669–79.

Usher M.B. (1970b) Seasonal and vertical distribution of a population of soil arthropods: Collembola. *Pedobiologia* **10**, 224–36.

Usher M.B. (1971a) Properties of the aggregations of soil arthropods, particularly Mesostigmata (Acarina). *Oikos* **22**, 43–9.

Usher M.B. (1971b) Seasonal and vertical distribution of a population of soil arthropods: Mesostigmata. *Pedobiologia* 11, 27–39.

Usher M.B. (1975a) Seasonal and vertical distribution of a population of soil arthropods: Cryptostigmata. *Pedobiologia* **15**, 364–74.

Usher M.B. (1975b) Some properties of the aggregations of soil arthropods: Cryptostigmata. *Pedobiologia* **15**, 355–63.

Usher M.B. & Hider M. (1975) Studies of populations of *Folsomia candida* (Insecta: Collembola): the causes of aggregations. *Pedobiologia* **15**, 276–83.

Vannier G. (1967) Étude *in situ* des rélations de la microfauna au dessèchment progressif d'un type de sol donné. *C. r. Acad. Sci., Paris,* ser. D. **265**, 2090–2.

Vannier G. (1970) Réactions des microarthropods aux variations de l'état hydrique de sol. Techniques relatives à l'extraction des arthropods du sol. *Rev. Ecol. Biol. Sol.* **7**, 289–309.

Wakerley S.B. (1963) Effect of lime and phosphate on soil Collembola. *Ann. appl. Biol.* **51**, 171–2.

Wallace M.M.H. (1967) The ecology of *Sminthurus viridis* (L.) (Collembola)—I. Processes influencing numbers in pastures in Western Australia. *Aust. J. Zool.* **15**, 1173–206.

Wallwork J.A. (1967) Acari. In: *Soil Biology*, eds. A. Burgess & F. Raw, pp. 363–95. Academic Press, London & New York.

Wallwork J.A. (1967) Acari. In *Soil Biology*, ed. A. Burgess & F. Raw, pp. 363–95. Academic Press, London & New York.

Winston P.W. (1956) The acorn microsphere, with special reference to arthropods. *Ecology* **37**, 120–32.

Wynne-Edwards V.C. (1962) *Animal Dispersion in Relation to Social Behaviour*. Oliver & Boyd, Edinburgh & London.

Section Two
The action and interaction of organisms in decomposition processes

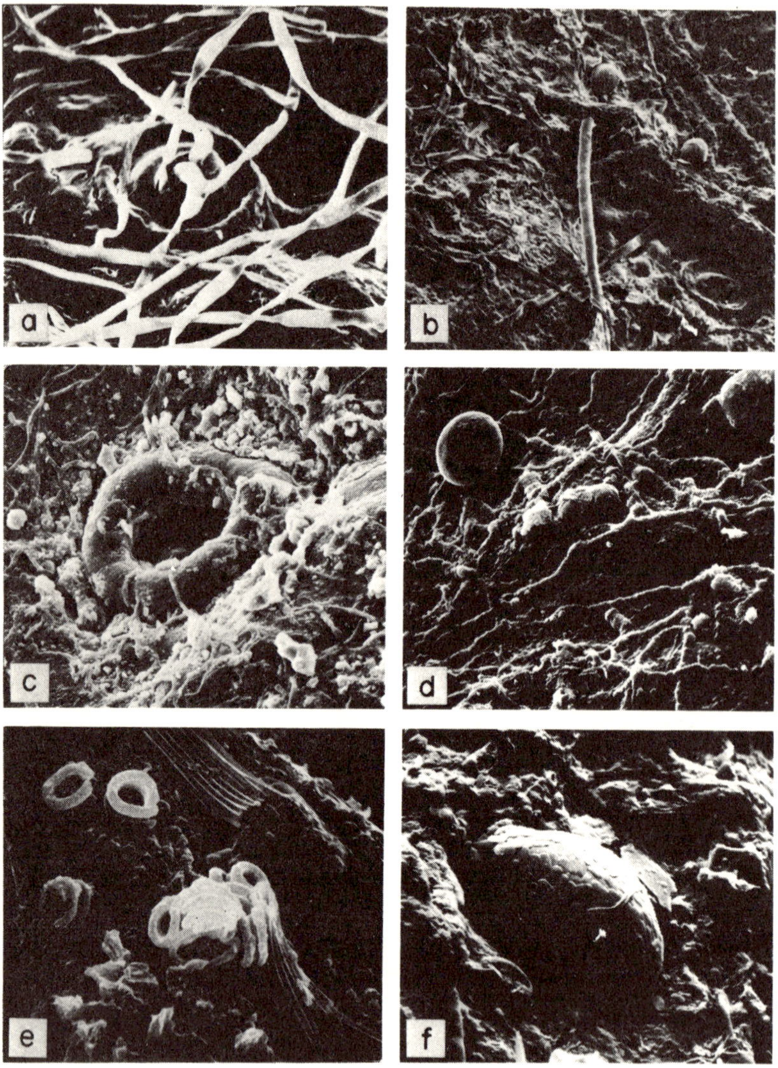

Plate 5.1.

a. Beech (*Nothofagus truncata*): leaf on tree; showing mycelial growth from stoma and smooth wax on leaf surface.

b. Beech (*Nothofagus truncata*): leaf in litter; showing accumulation of surface debris and fungal growth tending to cover the leaf surface.

c. Kauri (*Agathis australis*): leaf on tree; showing fungal growth and fragments of wax on leaf surface.

d. Kauri (*Agathis australis*): leaf in litter; showing accumulation of microbial debris and loss of wax.

e. Snow tussock grass (*Chionochloa macra*): tussock leaf; showing epiphytic growth of bacteria.

f. Snow tussock grass (*Chionochloa macra*); rotting leaf base; showing protozoan colonization (*Nebela collaris*) and accumulation of debris.

(Stereoscan electronmicrographs)

5 Decomposition processes in New Zealand soils with particular respect to rates and pathways of plant degradation

J. D. STOUT, K. R. TATE and L. F. MOLLOY
Soil Bureau, Department of Scientific and Industrial Research, Lower Hutt, New Zealand

Introduction

Long before a leaf falls to the ground, it begins a cycle of decay which in many ways is as complex and as fascinating as the cycle of growth that brought it into being (Fig. 5.1). The resorption of nutrients into the permanent parts of the plants and the autolysis of leaf tissue that marks the period of senescence is associated with the primary invasion of saprophytes, initiating a process of decomposition which may take thousands of years to complete (Plate 5.1). The greater part of this process of decomposition may take place above ground, on the plant itself or in the litter covering the soil surface. Because of the protracted period of decomposition in temperate soils however, the greater part of the decomposing organic matter in temperate soils is found in the mineral soil horizons, in a form remote from that of its plant origin and in intimate association with the organisms comprising the decomposer network, and the mineral soil matrix. This aggregation of living and dead organic matter, which defines the scope of decomposition, also defines the difficulties of its study, for any decomposition process tends not only to transform the substrate but also to synthesize new animal and microbial tissue which, with time, constitutes an increasing part of the decomposing matter. The aim of present researches in New Zealand is to delineate these pathways of degradation, to identify the synthesis of fresh compounds, and to measure the rate of these processes.

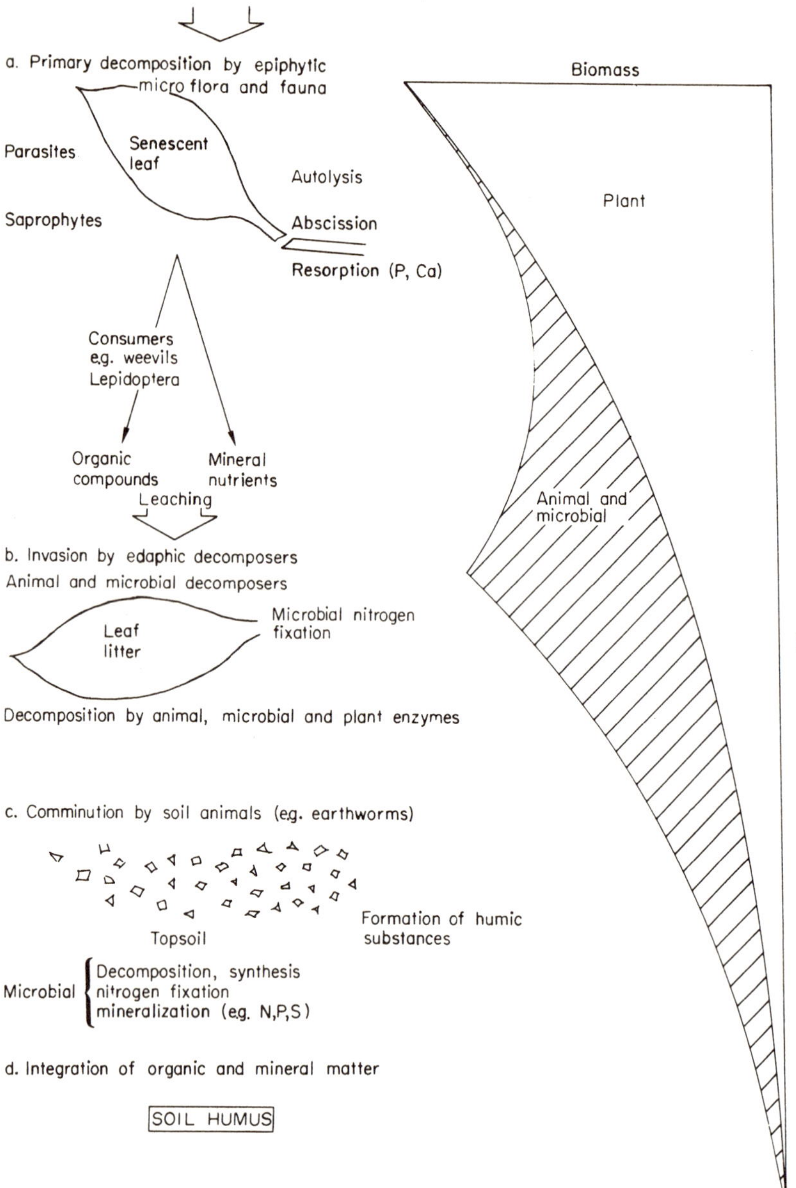

Figure 5.1. Diagrammatic representation of terrestrial decomposition cycle, depicting:
a. initial stage of senescence, resorption and loss of nutrients; loss to primary consumers; colonization by epiphytic saprophytes and parasites.
b. leaf litter stage, metabolism of readily available nutrients by residual plant enzymes and the zymogenous microflora; colonization by edaphic populations.
c. topsoil stage, loss of obvious physical identity of leaf material but persistance of humic substances; comminution by soil animals; breakdown of more resistant materials (cellulose) by autochthonous population; cycling of nutrients in the rhizosphere.
d. subsoil stage, apparent uniformity of soil humus, intimate association of mineral and organic material; diminished biomass; slow rate of decomposition of the most resistant plant, animal and microbial residues.

Soil organic matter

ORIGIN OF SOIL ORGANIC MATTER

Under natural conditions, soil organic matter is derived from the plant and animal residues of the terrestrial ecosystem, normally cycled in situ. Further accretions of organic matter may be derived from the atmosphere (Wilson 1959), particularly in areas close to the sea, as is much of New Zealand, and from wind- or waterborne debris translocated from their point of origin in loess, or alluvial or colluvial deposits.

By far the greater part of soil organic matter is normally derived from the residues of higher plants characterized by a high carbon, hydrogen and oxygen content, and relatively low nitrogen, phosphorus and sulphur content. Much smaller amounts are derived from animal faeces, of which the greater part in herbivorous animals consists of relatively unchanged plant material, and urine. Animal tissues have a much higher proportion of protein than plant tissue and generally a much higher proportion of nitrogen, phosphorus and sulphur, while urine is high in nitrogen and potassium. Thus, there is considerable diversity in the sources of soil organic matter, and perhaps the most remarkable feature of well-humified soil organic matter is its relatively stable chemical composition.

COMPOSITION OF PLANT RESIDUES CONTRIBUTING TO SOIL ORGANIC MATTER

Although the relative amounts of the major plant constituents can vary widely between species, most plants usually consist of the following: polysaccharides (nonstructural carbohydrates, such as sugars, starches and fructosans, and structural carbohydrates, mainly cellulose and hemicellulose), protein, lignin, and lesser amounts of tannins, terpenes, resins, waxes, and inorganics.

Following senescence and death of the plant tissue, the constituents more readily metabolized by animals and microorganisms are rapidly consumed, while the more resistant substances (e.g. lignin) undergo only minor changes (Fig. 5.2). Indeed some plant constituents, including waxes and resins can apparently survive for very long periods without significant alteration (Bergman 1963).

In terms of decomposition kinetics, these more resistant substances will tend to limit the rate at which the turnover of total organic carbon in the plant tissue can take place (Minderman 1968) (Fig. 5.2).

The longevity of some plant fractions in soils has been amply demon-

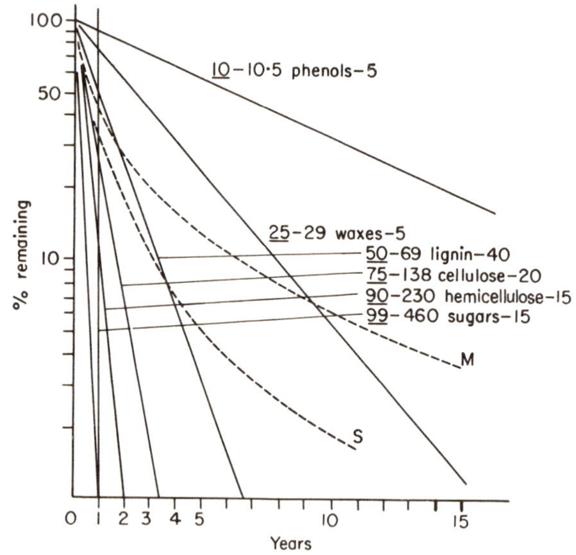

Figure 5.2. The decomposition curves of the various groups of constituents, if their decomposition could be represented by a logarithmic function (the straight lines from the point 100%). The first number in front of the name of the constituent indicates the loss after 1 year. The second number represents the value of p (average annual rate of decomposition × 100), in this case varying between 10·5 and 460. The number after the constituent represents its percentage in weight of the original litter (these values are rough averages and they do not represent a specific analyses). The line S shows the summation curve obtained by annual summation of the residual values of the separate components. The line M gives an approximation, based on some analyses, of the probable course of the decomposition in the mor-type at Hackfort (after Minderman 1968).

strated (Cheshire *et al.* 1971, Goh 1972). Some phenols released by chemical degradation from soil organic matter have indicated the survival (>100 years) of hardwood lignin, or lignin derivatives, in a highly productive New Zealand pasture soil, where turnover of organic carbon is relatively rapid (Tate 1973b).

In an attempt to understand the contribution of decomposing plant tissues to soil organic matter, a so-called 'light fraction' has been separated by Ford *et al.* (1969). The 'light fraction' is considered to be partly humified plant and microbial remains, so that the yield of this fraction should be highest where decomposition is least, and vice versa. Table 5.1 summarizes data which show a highly significant positive correlation between 'light fraction' yield and percent carbon. Although the 'light fraction' represented only a small fraction by weight (<4%) of most soils, it was nevertheless a rich source of carbon and nitrogen, accounting for 11–26% of the total soil carbon, and 5–17% of the total nitrogen. The bulk of the total soil nitrogen

Table 5.1. Carbon and nitrogen data for climosequence soils in New Zealand Tussock Grasslands. The light fraction of some soils under tussock grassland and its contribution to total soil carbon and nitrogen. L.F. Molloy, unpublished data.

Classification (Taylor & Pohlen 1968)	Soil Name	Whole Soil C(%)	Yield of Light Fraction %	Contribution of Light Fraction to Total Soil Carbon and Nitrogen C(%)	N(%)
Brown-grey earth	Conroy	2·7	2·0	26	14
Yellow-grey earth	Cluden	2·7	0·9	11	5
	Tima	3·6	3·6	21	12
Yellow-brown earth	Tawhiti	5·1	2·1	16	6
	Lammerlaw	6·6	3·0	17	10
	Maungatua	22·1	43·4	68	63
	Carrick	4·9	1·9	12	6
	McKerrow	7·3	3·5	14	8
	Obelisk	5·3	4·2	25	17

appeared to be in intimate association with the mineral soil, because the 'light fraction' contributed about twice the amount of total soil carbon as it did nitrogen.

Plant residues may also be recognized from the chemical degradation of the classical soil organic matter fractions (Kononova 1966), namely humic acid (alkali soluble, acid insoluble), fulvic acid (alkali soluble, acid soluble) and humin (alkali and acid insoluble). The humin fraction appears to contain the bulk of the 'light fraction' (Drs K.R. Tate & L.F. Molloy, unpublished results, Goh *et al.* 1975). Although the humin and fulvic acid fractions tend to contain the youngest plant material, differences do occur with different vegetation on the same soil, and between different soils (K.M. Goh & J.D. Stout, unpublished results). The rapid equilibration of age of these fractions (Goh *et al.* 1975) suggests that the differences are as much a reflection of the extraction technique as of the decomposition pathway. Nevertheless, promising results have been obtained with the identification of certain phenolic compounds released by acid hydrolysis from soil humic acids (Tate 1973b). This technique has not only enabled the plant origin of some of the phenolics to be identified, but also provided means by which broad vegetational type may be distinguished.

Investigations comparing the organic matter in the topsoils and subsoils under a hard beech (*Nothofagus truncata*) forest, and under pasture at an adjacent site, showed that enrichment of the soil organic matter of the latter by bomb radiocarbon was essentially confined to the top 10 cm of the soil (Tate 1973b). The soil at the pasture site, which now produces 20,000 kg ha^{-1} herbage dry matter annually (Saunders & Metson 1971), formerly supported a hard beech forest. The ratio of vanillic acid to 4-hydroxybenzoic acid in the acid hydrolysates of the topsoil and subsoil humic acids decreased down the profile under hard beech, but markedly increased down the profile under the pasture. These phenolic acids are probably derived from lignin residues, and, in the light of radiocarbon data, the ratios suggested that the imprint of the former forest vegetation was still retained in the pasture subsoil after more than 100 years. Subsequent investigations of soil humic acids (K.R. Tate & H.A. Anderson, unpublished results) have shown that the phenolic substances are released from the highest molecular weight fractions, which are predominantly plant-like in character. However, at least one major compound found in some humic acid hydrolysates, notably from a podzol Bh horizon, seems to be of possible fungal origin. Soil and fungal humic materials can produce the same or similar compounds upon treatment with sodium amalgam (Martin & Haider 1971; Schnitzer & Ortiz de Serra 1973).

An insight into the dynamics of litter breakdown and humus formation can be gained by quantitative analysis of the phenolics released by acid hydrolysis from litter and soil humic acids. Piper & Posner (1972) have shown that humic acids from organic residues in different states of decay within a soil yield slightly more 'lignin' phenolic acids, and appreciably more phenolics altogether, when reductively degraded with sodium amalgam. When the humic acid hydrolysis patterns from a New Zealand soil profile under hard beech (*Nothofagus truncata*) and from a podzol profile under Scots pine (*Pinus sylvestris*) in Aberdeenshire, Scotland were compared, the A horizon of the former, and the litter H horizon of the latter were very similar. Some aspects of the chemistry of the podzol profile have been described (Anderson 1970). Although differences in vegetation and parent material are important, the climatic differences between central New Zealand and north-east Scotland are probably the major factors causing the observed differences between the soils. The lack of a significant macrofaunal population in the litter under the Scots pine compared to the larger numbers of active decomposer organisms found in the litter and soil under hard beech has probably been the key factor contributing to the morphological and chemical differences between the two soils.

As well as releasing phenolic substances, acid hydrolysis of soil humic acids can be used in studies of the organic nitrogen status of soils. In general, the yield and distribution of amino acids in soil hydrolysates appear to vary with the climatic conditions and soil types from which they are extracted (Bremner 1955, Carles & Decau 1960). The amounts and nature of the individual amino acids in humic acid hydrolysates do not appear to vary greatly. Nevertheless, an insight into the microbiological activity in specific soil horizons can be gained from their analysis. For example, in comparing the topsoil and subsoil humic acids under the hard beech forest and the pasture, the following observations were made:

1 Concentrations of individual amino acids in the hydrolysates were higher from the topsoil than from the subsoil humic acids.
2 Levels of individual amino acids were generally higher in the pasture topsoil than in the forest topsoil humic acid hydrolysates.
3 Levels of individual amino acids in the pasture and forest subsoil humic acid hydrolysates were identical within experimental error.
4 The levels of glucosamine and galactosamine, both of microbiological origin, were higher by a factor of 4–5 in the topsoil humic acid from the pasture than from the forest. In the subsoil humic acid hydrolysates, the levels of these two amino sugars were only about twice as high in the soil under pasture.

These results clearly reflect the higher level of microbial activity, and therefore turnover (decomposition) rate, in the soil under pasture, particularly in the topsoil. They tend to support the other data available for these soils.

ACCUMULATION OF ORGANIC MATTER

The accumulation of soil organic matter is a function of three inter-related factors:
1 the accretion of organic residues;
2 the decomposition of organic residues;
3 the bonding or 'fixation' of organic residues to the inorganic fraction of the soil, particularly the clay fraction. In a sense, this bonding or 'fixation' constitutes a 'sink' since the bound organic material is substantially removed from the normal organic cycling process.

Under relatively stable conditions a soil tends to reach an equilibrium with regard to organic matter accumulation. This equilibrium tends to be determined by two main factors:
1 the soil—vegetation complex which determines potential plant productivity, in so far as soil fertility and plant growth are concerned;
2 climate, which affects the two main parameters, namely rate of growth and rate of decomposition, but affects them differently, since lack of moisture may limit plant growth before it limits microbial decomposition (Fig. 5.3). Similarly, decomposition rates may increase exponentially at temperatures when plant growth tends to be exceeded by plant respiration.

In a sense, organic accumulation is a function of the soil energy status. The energy status of the soil is related partly to radiation and partly to the hydrological cycle. A major loss of radiation energy is effected by evapotranspiration and since evapotranspiration rates are largely a function of the vegetation there are significant differences in the energy status of the same

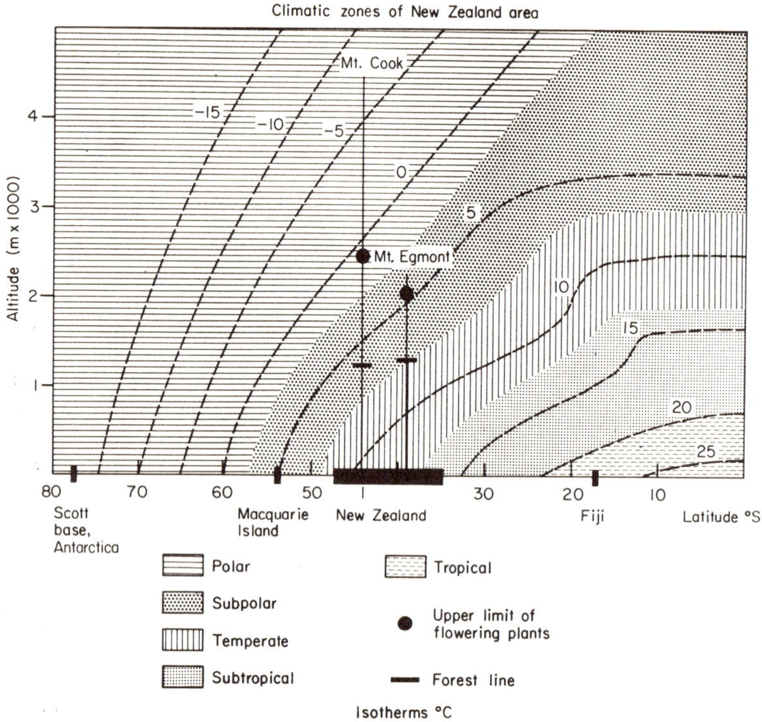

Figure 5.3. Energy status of the zonal soils of the New Zealand area determined by climatic parameters. Note the contrast of high altitude in the temperate region with high insolation and the low insolation of high latitudes. (After N.Z. Soil Bureau 1968.)

soil under different vegetations. For example, comparison of areas under pasture and forest at Lower Hutt, New Zealand, suggested that although the annual net radiation was greater for forest than for grassland, about 80% of the forest radiation was used in evaporation, and further, that the heat storage in the forest soil was less than for the grassland soil. The total evaporation of the forest was greater than the potential evaporation from the grassland but the grassland showed a much wider seasonal variation (R.J. Jackson, personal communication).

Generally the greater part of the soil organic matter is in the topsoil where plant debris normally first accumulates (Miller & Blakemore 1968). In very wet soils, however, such as peats and gley podzols, there is a higher proportion in the subsoil, because although organic matter has accumulated throughout the soil profile, it decays slowly and the greater bulk density of the subsoil results in a greater mass per unit volume of organic matter. Where there is massive accumulation of litter on top of the mineral soil, as in a New Zealand kauri podzol, a substantial part of the organic debris may remain in the litter layers due to the slow rate of decomposition and the absence of soil animals incorporating the litter into the mineral topsoil. Because accumulation is largely a function of plant growth and plant decay, there are differences between soils under forest vegetation, where the deep rooting of trees extends into the subsoil, and pasture with shallow rooting grasses where root growth is predominantly in the top 5–10 cm of the soil.

The principal overall trend is for organic matter to increase with increasing moisture status of the soil, but when increased rainfall is associated with increased leaching, the amount of organic matter diminishes. Similarly, soils with low fertility in most cases show limited accumulation of organic matter.

Most New Zealand soils have about 15,000 g m^{-2} of organic carbon. Lower quantities are found in drier soils, soils of low fertility, or soils of low bulk density (e.g. peats or pumice soils). Higher quantities are found in podzolized soils or gley podzols, the highest figure of about 100,000 g m^{-2} being recorded for gley podzols developed under podocarp forest in the mild hydrous climate of the west coast of South Island.

CYCLING OF CARBON, SULPHUR AND PHOSPHORUS

Carbon

Carbon is central to the whole organic cycle. With hydrogen and oxygen, it comprises the major bulk of living material. The photosynthetic fixation of carbon dioxide also provides the principal source of energy for plant growth. Fortunately, therefore, carbon occurs in nature as three isotopes, two stable (^{13}C and ^{12}C) and one radioactive (^{14}C, radiocarbon). Their existence enables us to use isotopic fractionation as a measure of change, and radiocarbon decay as a measure of time, and both are firmly built into the soil organic cycle.

The two stable isotopes of carbon (^{13}C and ^{12}C) occur in the proportion of about 1 : 99 and the carbon dioxide of the atmosphere has a $\delta^{13}C$ value of −6·3‰ (Stout *et al.* 1975). However, fractionation occurs during photosynthesis, when the lighter isotope is preferentially selected. This fractionation varies with the plant species, and with environmental conditions. The greatest difference is found between plants with the Hatch-Slack photosynthetic pathway and those with the Calvin pathway, the latter showing the greater fractionation. Thus, the photosynthetic pathway from which the soil organic residues are derived imprints an initial $\delta^{13}C$ value which may change during the process of decomposition. In general, there appears to be little change in the litter horizons (Fig. 5.10) or even in topsoil values compared with the original plant material. However, in most mineral subsoils, the organic material appears to be a little enriched in ^{13}C relative to the original

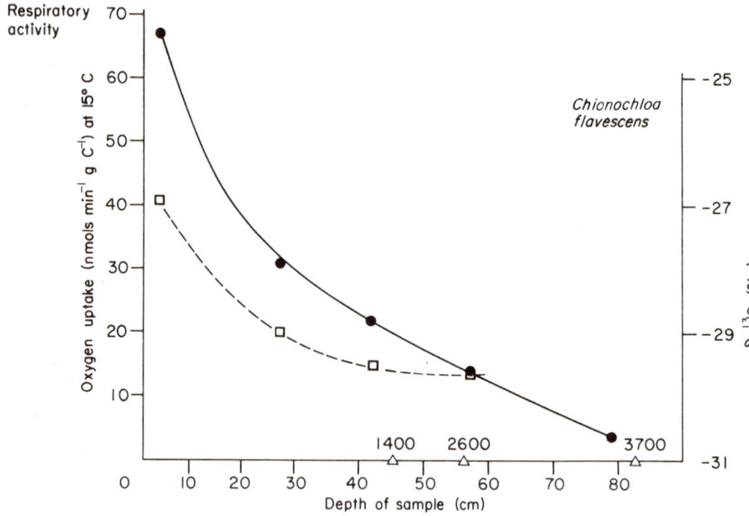

Figure 5.4. Respiratory rate (measured electrometrically) ●, relative radiocarbon enrichment of the stable carbon isotope ^{13}C [δ ^{13}C] □, and age (years B.P. △) of a peat profile under tussock (*Chionochloa flavescens*) on the Rimutaka Range, New Zealand. Note the greater age of this organic horizon compared with Fig. 5.5. (Stout *et al.* 1975.)

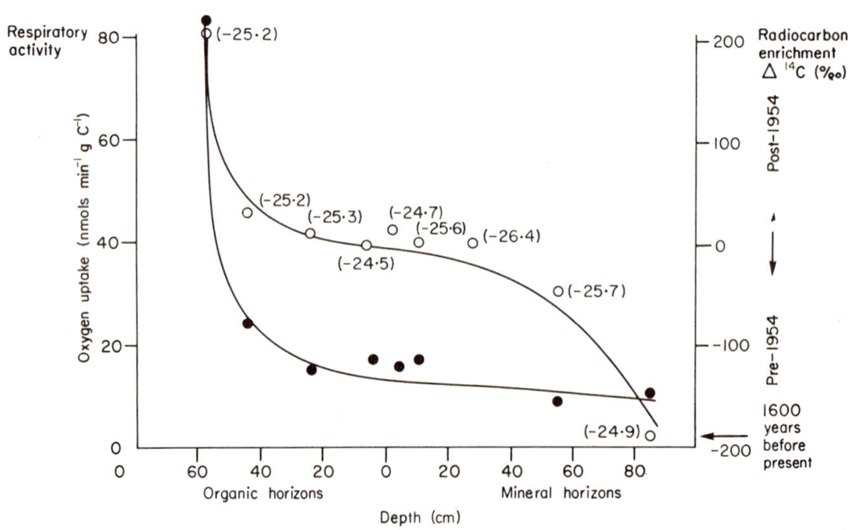

Figure 5.5. Respiratory rate (measured electrometrically) ●, relative enrichments of the stable carbon isotope ^{13}C [δ ^{13}C] (values in brackets) and the radioactive carbon isotope, ^{14}C (o), of a soil profile under kauri (*Agathis australis*) in North Auckland, New Zealand. Note the deep organic horizons and comparative youth of the organic matter indicating a relatively high rate of turnover. (Stout & O'Brien 1973.)

plant residue, i.e. the decomposer organisms also appear to prefer the lighter isotope, leaving the heavier isotope to accumulate in the undecomposed organic material (Stout *et al.* 1975). In organic soils however (e.g. peats) the organic material is lighter at greater depths, corresponding to greater age (Fig. 5.4). The rate of decomposition for a peat is also shown in Fig. 5.4 which indicates that it varies with its age, being very low in the deeper horizons. In mineral soils where there is marked litter accumulation, as in the kauri podzol, there is a pattern (Fig. 5.5) which would be compatible with a decomposition process intermediate between the normal soil organic cycle and the cumulative build-up of peat.

Radiocarbon provides a time scale for the last 40,000 years, covering the most important period of soil formation in New Zealand. Where a discrete event has occurred, such as a volcanic ash shower, a soil marker bed may be provided and where this event can be dated (e.g. by a piece of charred wood) we have a means of measuring the rate of pedogenetic processes. Alternatively, it is possible to 'date' the soil carbon itself, i.e. not a discrete event but a series of temporally associated events. Measurements of the different genetic horizons of the soil profile give an indication of the rate of soil processes, and particularly of accumulation and decomposition rates. There are difficulties in interpreting the past vegetation and climatic history, but the data available on New Zealand soils suggest that there is a relationship between radiocarbon ages and the known rates of soil formation and soil processes which are indicated by increasing age with increasing altitude or latitude (Stout & O'Brien 1973).

Because atmospheric levels of radiocarbon were dramatically increased from 1954 following the detonation of thermonuclear bombs, a global enrichment experiment was initiated in which, first the plants, then the consumers and decomposers, and finally the soil organic matter showed an increase in radiocarbon. This made two experiments possible: first, measurement of the rate of infiltration of organic matter into the soil profile, which was found to be only a few millimetres a year, and second, measurement of the rate of turnover of organic matter, where the total accretion of organic matter to the soil was known. Jenkinson's (1971) classical experiment is used as a model for the global enrichment by 'bomb' radiocarbon. Data for a pasture soil of moderate productivity indicated the close relationship between atmospheric levels and plant levels of radiocarbon, the greatly enriched level of the earthworms feeding predominantly on young plant debris, and the slow enrichment of the topsoil, followed by the subsoil, as the post-bomb organic residues penetrate the soil horizon (Fig. 5.6) (Stout 1971).

Carbon cycling may be measured directly, as well as isotopically, as the uptake of oxygen or the evolution of carbon dioxide in respiration or fer-

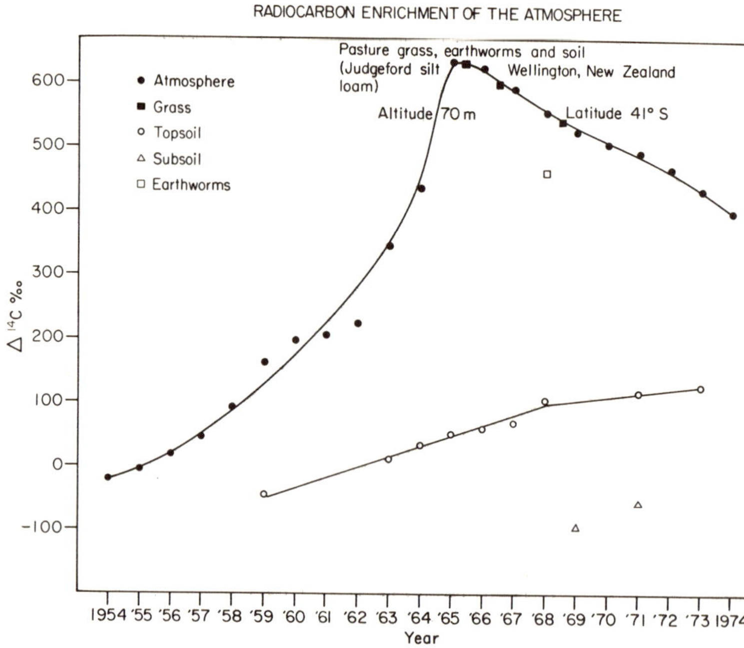

Figure 5.6. Enrichment of atmosphere, pasture grass and soil earthworms with 'bomb' radiocarbon. Note the concomitant enrichment of the atmosphere and the growing pasture; the slower enrichment of the earthworms and topsoil due to the dilution of the enriched carbon by old carbon; and the greatly delayed enrichment of the subsoil due to the slow rate of diffusion of the organic matter down the soil profile. (Stout 1971.)

mentation. No entirely satisfactory measurement technique is possible for all soils because of different rates of respiration, different factors controlling respiration, or simply different measurement objectives. There can be considerable differences in the respiration of the same soil using different techniques and also under varying conditions and at different times of the year. Nevertheless, it is possible to recognize some factors of major importance in most soils. First, the loss of carbon dioxide is generally relative to the amount of soil organic carbon, so that values comparing different soils are most usefully normalized by expressing them on a unit weight of carbon basis. This value, for unamended soil samples, may be termed the basal respiratory rate. Secondly, the loss of carbon dioxide is relative to the size and activity of the biomass: the vegetation, the animals and the micro-organisms. The vegetation influences the size and rate of primary production, which is the basis for all further biological activity, and there is generally a greater release of carbon dioxide from a soil under high-producing vegetation than from the same soil under low-producing vegetation. However, the magnitude of the difference is not always revealed by the basal respiratory rate, since it may be largely a function of the cyclic response of the soil populations to fresh substrate. This is known as the zymogenic response, and it varies with the size and nature of the biomass, for the death of much of the biomass—the partial sterilization effect—can release fresh, readily accessible substrate to support further growth and to stimulate activity of the residual population (Jenkinson 1966). Thus, differences in respiratory activity of plots of the same soil under grassland of greatly differing productivity is reflected more by the zymogenic response than by the basal respiratory rate (Stout & Dutch 1968). Thirdly, loss of carbon dioxide is a function of temperature, the rate normally doubling for every increase of 10°C.

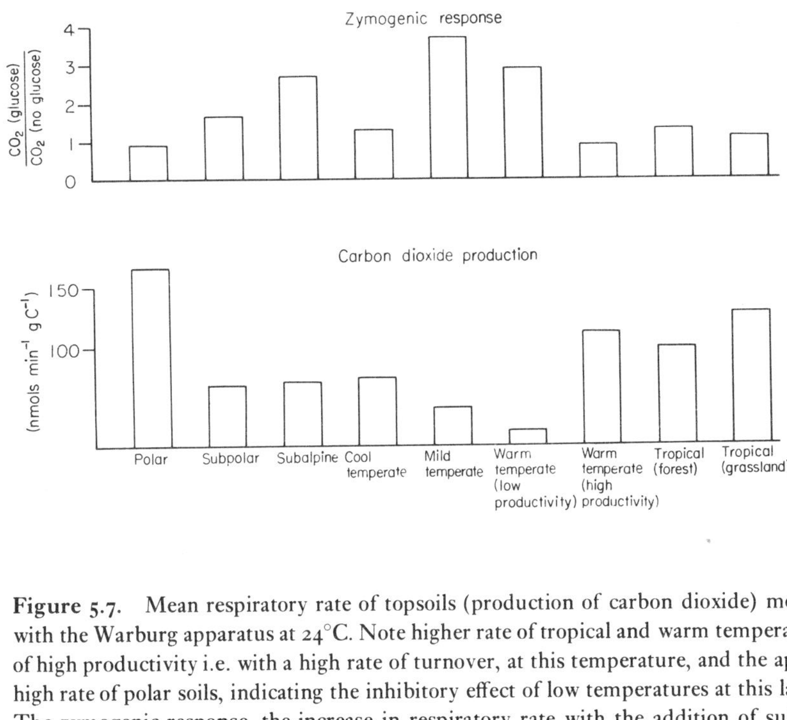

Figure 5.7. Mean respiratory rate of topsoils (production of carbon dioxide) measured with the Warburg apparatus at 24°C. Note higher rate of tropical and warm temperate soils of high productivity i.e. with a high rate of turnover, at this temperature, and the apparent high rate of polar soils, indicating the inhibitory effect of low temperatures at this latitude. The zymogenic response, the increase in respiratory rate with the addition of substrate, expressed as the ratio of carbon dioxide production with added glucose to carbon dioxide production without added glucose, also shows zonal variation. (Stout & Dutch 1968.)

When values obtained from a wide latitudinal range of soils using a uniform technique are compared (Fig. 5.7) they show a broad contrast between tropical soils, which tend to have a high respiratory rate, and soils at higher latitudes, except for the polar soils. This technique (Warburg) measures respiration at a temperature which approximates to that of tropical soils but is much greater than the mean temperature prevailing in polar soils. Thus the results for polar soils indicate a *potentially* high respiratory rate, which is normally inhibited by low temperatures and arid conditions.

Similarly, because of their lower mean field temperatures, rates for temperate soils would be only about half the experimental values shown in Fig. 5.7. As Fig. 5.7 also shows, there tends to be a difference in the zymogenic response of the tropical and temperate soils. This reflects differences in the size of the soil organism biomass. Responses are greater in temperate soils with a large biomass than in tropical soils with a smaller animal and microbial biomass. On the other hand, because of the higher rate of turnover, organic material in tropical soils tends to be younger than that in temperate soils. This explains, why, even at the same temperature, tropical soils tend to have higher respiratory rates. However, when a temperate soil under a highly productive vegetation and a high rate of turnover, such as a New Zealand high-producing pasture, is compared with a tropical soil under a relatively low-producing forest, the temperate soil may have a higher respiratory rate than the tropical soil (Stout & O'Brien 1973).

Just as Jenkinson's experiment (1971) showed the rapid decrease in the rate of decomposition of plant residues in soil over a short period of time, so respiratory activity of different horizons also show a decrease with depth, i.e. with time. This is shown with a New Zealand soil profile under kauri forest where the values fall steeply in the litter horizons and then level out at greater soil depths (Fig. 5.5). Respiration was measured here as rate of oxygen uptake in an electrolytic respirometer in which low rates of respiration can be measured over a long period of time (Birch & Melville 1969). The radiocarbon values and the $\delta^{13}C$ values for this profile are also shown. The decline in respiratory rate with age and depth for a peat profile under snow tussock (*Chionochloa flavescens*) together with radiocarbon values and the $\delta^{13}C$ values are also given in Fig. 5.4. Respiratory activities for two soils under tussock grassland are shown in Fig. 5.8. Although the relative rate for each soil is correlated with the age of the organic material, the absolute rate varies, since this is not simply a function of time, but of the actual decomposition processes that have taken place in that time. As Fig. 5.8 shows, respiratory activity, particularly of litter horizons, is also greatly influenced by the moisture regime.

One distinction in respiratory measurements indicative of differences in patterns of decomposition is given by the Respiratory Quotient (R.Q.), the ratio of carbon dioxide produced to oxygen taken up. In most soils this tends to be less than one under field conditions, though it may increase when the soil is moistened and therefore becomes anaerobic. Where there is an active anaerobic flora and adequate substrate the R.Q. is greater than one (Stout & Dutch 1968).

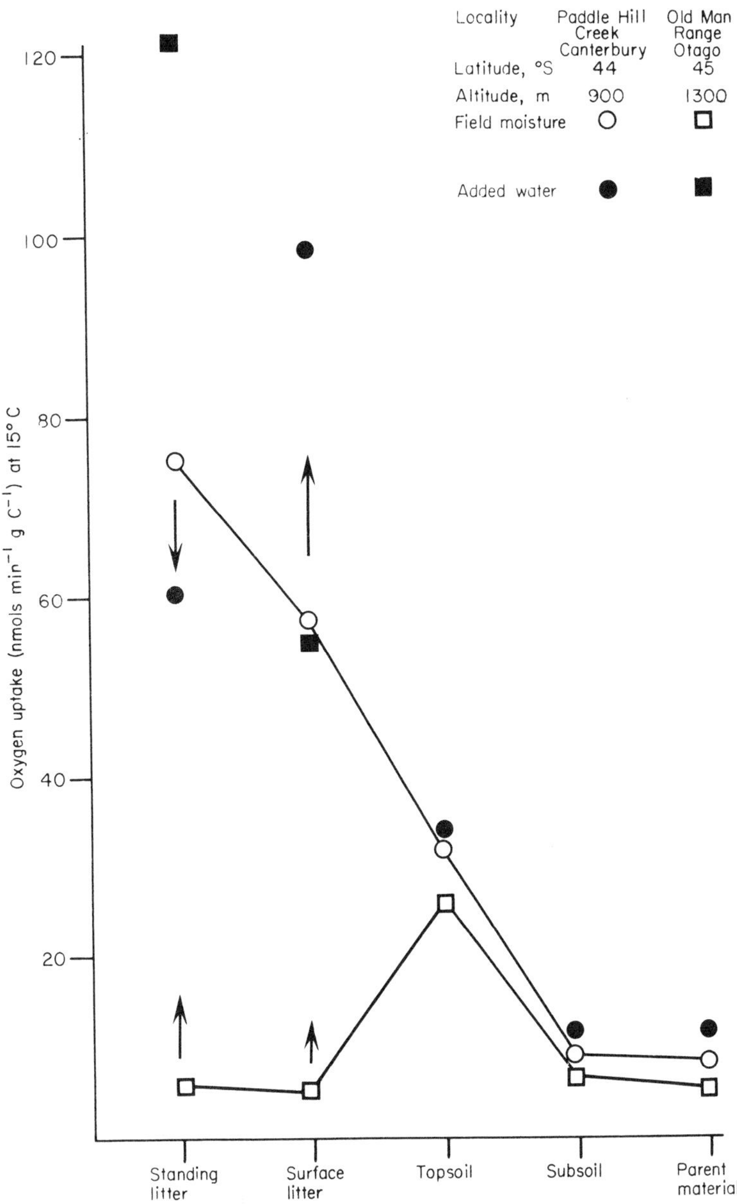

Figure 5.8. Respiratory activity (measured electrometrically) of two snow tussock (*Chionochloa rigida*) grassland profiles in Canterbury and Otago, South Island, New Zealand, with and without added water.

Sulphur

Sulphur is required for the metabolism of all organisms. Nevertheless, knowledge of the nature, amount, fate and interrelationships of sulphur compounds in soils, plants and micro-organisms is far from complete (Freney 1967). Most sulphur in surface soils accumulates in soil organic matter, and must generally be mineralized before it is available to plants. Soil sulphur deficiency in New Zealand soils was considered for many years to be confined to areas of low rainfall, since wind-spray and precipitation are the principal source of sulphur for soils, and to the unimproved tussock grasslands of South Island (Widdowson & Blakemore 1974). In these latter areas, decomposition of organic matter may be restricted by cold or acidity. In New Zealand, accession of sulphur from industrial sources is limited. It is now recognized that there is a widespread need for sulphur to be added for optimum pasture growth, and that sulphur deficiency in the intensively and semi-intensively farmed areas of both main islands has been largely prevented through the use of superphosphate fertilizers.

Measurements so far made of the isotopic ratios ($\delta^{34}S$) of adsorbed and soluble sulphate sulphur in a chronosequence of soils formed on accumulating sand dunes on the west coast of North Island, New Zealand, have indicated that these are similar to contemporary sea water or marine evaporates, and since there is very little sulphur in the parent material of the soils, it must therefore be derived almost wholly from marine sulphate, either directly at shore level, as sea spray, or by precipitation. These ratios for both plant material and soil organic sulphur differ significantly from sea water sulphate values (M. Kusakabe unpublished data).

The turnover of plant sulphur is normally slower than that of microbial tissue, and since there appears to be little sulphur fractionation by most plants (Mekhiteya & Pankina 1968; Kusakabe *loc. cit.*), the microbial fractionation of sulphur is probably the major form of biological fractionation in soils (Stout *et al.* 1975).

There is generally a close correlation between nitrogen and sulphur in soils, because these elements are both components of proteins, and N/S ratios in most New Zealand soils are about seven (Blakemore *et al.* 1968). In contrast, C/S and C/N ratios may vary widely among soils, presumably because carbon is a more ubiquitous element, occurring in many compounds where nitrogen and sulphur are absent (e.g. polysaccharides, lignin).

Mineralization of organic sulphur is mainly accomplished by soil micro-organisms, who use part of the sulphur for cell synthesis; only the sulphur in excess of their requirements is released for plant growth. The factors affecting the balance between mineralization and immobilization, which must occur simultaneously, are not clearly defined (Freney 1967).

It is encouraging to note, however, that measurements in a climosequence of soils in tussock grasslands, of soil sulphatase activity, which apparently originates solely from decomposer micro-organisms, may provide a good indication of the rates of organic S mineralization (T.W. Speir, personal communication).

Phosphorus

Both sulphur and phosphorus are involved in the chemical reaction system and biological systems of soils, in contrast to nitrogen, which is almost exclusively confined to the biological system. Processes within the chemical system, including precipitation and solubilization, are determined by the law of mass action, and therefore by the chemical composition of the soil solution; depletion of nutrients such as phosphorus and sulphur in the soil solution by plant uptake therefore initiates release of further nutrients from the soil solid phase (Jansson 1971).

There is, however, a marked influence of micro-organisms on the uptake and utilization of phosphate by plants (Barber 1968; Bowen & Rovira 1969), and microbial phosphate also plays a significant role in the phosphorus nutrition of plants.

Most work with New Zealand soils has been restricted to determining total, inorganic, and organic phosphorus (Saunders 1968). Whereas the N/S ratio of soil remains remarkably constant, C/N ratios and particularly C/organic P ratios increase with increasing soil development. In a study of the Franz Joseph chronosequence of soils in South Island, New Zealand, which range in age from 0 to 22,000 years, the total phosphorus in the soil profiles (including stones and boulders) decreased from 14,000 kg ha^{-1} for a 75 cm deep profile at 0 years, to about 1,000 kg ha^{-1} for a 22,000-year-old soil, when the only material $>$72 mm particle size was quartz (Stevens & Walker 1970). Peak values of 7,500 kg N and 500 kg organic phosphorus were reached after about 12,000 years.

In studies of a climosequence of soils in tussock grasslands, in South Island, New Zealand, forms of phosphorus have been used with other chemical parameters to rank the soils in a development sequence (Molloy & Blakemore 1974).

The patterns of accumulation of organic carbon, nitrogen, sulphur, and phosphorus in a range of soils under ryegrass (*Lolium* spp.)—white clover (*Trifolium repens* L.) pasture was studied by Jackman (1964a, b) to determine where accumulation and mineralization took place in the soil profile, when a steady state was reached, and what differences occurred among different soils. He compared two groups of soils, those with substantial quantities of amorphous clay, derived from volcanic materials, and those lacking substantial quantities of amorphous clay (Fig. 5.9). It has been suggested that the accumulation of organic matter in the first group was due to its immobilization by bonding with the clay or by the adsorption of enzymes by the clay (Ross & McNeilly 1972a). In Tokomaru silt loam, without amorphous

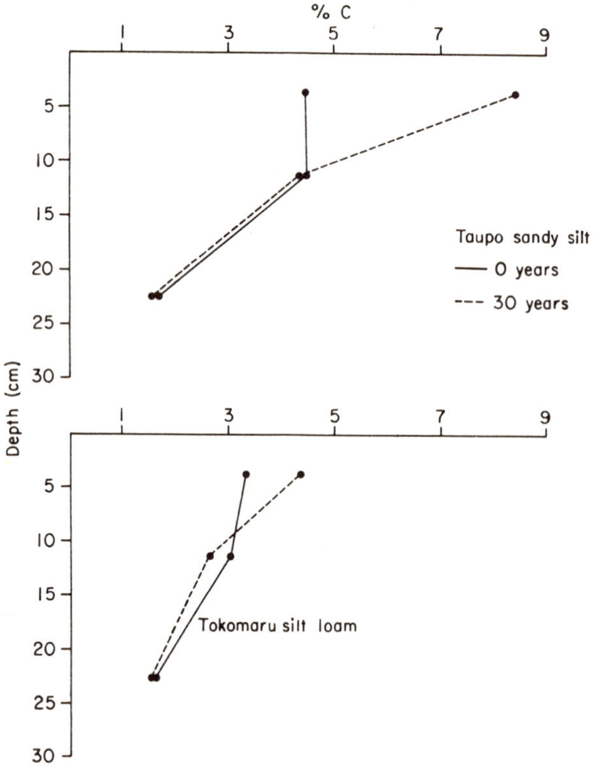

Figure 5.9. Patterns of change in levels of organic carbon content of two soils, one, Taupo sandy silt, with amorphous clay and one, Tokomaru silt loam, without, 0 and 30 years after ploughing. (After Jackman 1964a.)

clay, accumulation exceeded decomposition in the top 7·5 cm of soil; there was either little or no change or accumulation was slightly less than decomposition in the next 7·5 cm; and no change occurred below 15 cm (Table 5.2). The turnover time, or half-life, of the major elements (C, N, S, P) varied with the differing soils. Soils containing amorphous clay had higher turnover rates. An indication of the magnitude of decomposition rates in these soils is given by the calculated annual rates of change of organic nitrogen, sulphur, and phosphorus averaged over the first 10 and 30 years under pasture (Jackman 1964a, b), (Table 5.2).

Similar patterns of organic accumulation have been followed in soil

Table 5.2. Steady state and rate of turnover of organic matter (C, P, N and S) in Tokomaru silt loam, North Island, New Zealand (Jackman, 1964a, b).

Depth (cm)	Carbon	Nitrogen	Sulphur	Phosphorus
		Steady State (g m^{-2}) *(50-year-old pasture)*		
0–7·5	2,910	318	39	40
0–30	9,990	845	102	119
		Rate of Turnover *(half life ($t\frac{1}{2}$) years)*		
0–7·5	14 (10–22)	6 (3–18)	2 (1·5–4)	6 (4–8)
7·5–15	12 (6–00)	6 (4–10)	3 (1·6–8)	n.d.

n.d.—not determined

chronosequences, developed under native or exotic vegetation (Stevens & Walker 1970), and soil climosequences. Unlike the decomposition of a piece of a single plant species, as in Jenkinson's (1971) experiment, where there was an exponential rate of decay, the chronosequences exhibit an exponential rate of accumulation, signifying that plant growth and organic accumulation initially exceed the capacity of the decomposer network, and that accumulation and decomposition reach a steady state only after a period of time. In part this is caused by limitation of plant nutrients, particularly phosphorus. In the youngest member of a chronosequence plant nutrients are in a mineral form, but with time, they tend to accumulate increasingly in organic forms which are not generally available for plant growth, although they may be available for microbial growth. The relative diminution both of rates of accumulation and rates of decomposition is also a characteristic of chronosequence development (Fig. 5.10). This is related to the increasing age of the soil organic matter; the respiratory rate falls as the proportion of younger material with more accessible nutrients diminishes (Goh & Stout 1973). Similarly the respiratory rate is lower for increasing depths down a soil profile. In the Franz Josef Chronosequence, formed in the wake of a retreating glacier, mild superhumid conditions lead to the oldest member of the chronosequence having a vast accumulation of organic matter of increasing age down the profile (Stevens & Walker 1970); in the Manawatu sand chronosequence, formed on accumulating sand dunes in a mild humid climate, there is less accumulation of organic matter, though the oldest organic matter is still thousands of years old (Goh *et al.* 1975).

In a climosequence, accumulation initially tends to increase with increasing rainfall, but the rate of accumulation may diminish when leaching reduces the availability of plant nutrients or when low temperatures severely restrict plant growth (Molloy & Blakemore 1974). But either temperature or moisture may tend to limit decomposition before there is a significant

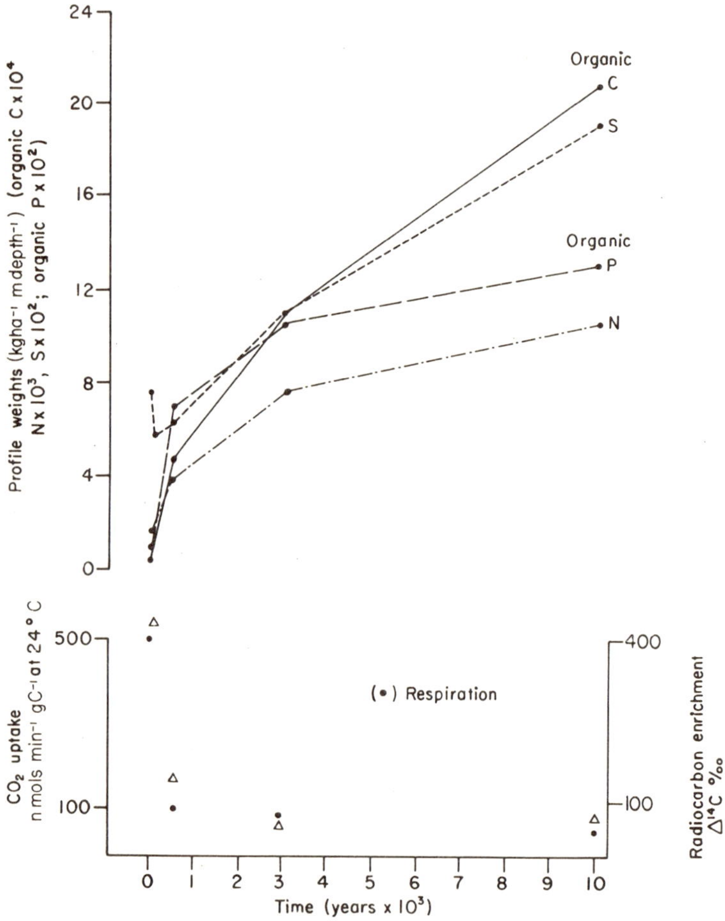

Figure 5.10. Accumulation of organic C, organic P, N and S in a chronosequence of soils developed on aeolian sand (Syers *et al.* 1970). Respiratory rate (Warburg) (●) and 'bomb' radio-carbon enrichment (△) of the topsoils under grass indicate a higher rate of turnover in the early stages of the chronosequence, despite a high rate of accumulation of organic matter. (Goh & Stout 1973.)

diminution of plant growth. The point of steady state of soil organic matter reached in climosequences, therefore, tends to be the modal value of the zonal soil of the particular climate.

Decomposition studies in particular ecosystems

The dominant New Zealand ecosystems are native forests; exotic forests; tussock grasslands (increasingly invaded by introduced grasses and herbs);

exotic pasturelands; and the urban ecosystem of man with its prodigious use and recycling of natural products. More information is available on productivity of the exotic ecosystems than on native ecosystems (Egunjobi 1969).

NATIVE FORESTS

Most of New Zealand enjoys a sufficiently humid climate to favour forest growth, although the treeline is at lower altitudes than in North America or Europe (Poole 1968). In some situations severe frosts and high winds throughout the year limit tree growth below the regional timberline. Originally, before extensive human settlement, forests covered over two-thirds of the land area. Native forests now cover less than a quarter of the land surface, largely confined to the steep mountain chains, where they afford a valuable protective function against erosion and excessive run-off.

There is a close association between the dominant trees of a forest and the soil. Broadly, the conifers, including kauri (*Agathis australis*) and podocarps tend to form mors, and the broadleaved trees, such as tawa (*Beilschmiedia tawa*) and puriri (*Vitex lucens*), form mulls. The beeches (*Nothofagus* spp.) tend to be predominantly mor-forming trees growing on soils of low natural fertility, with particularly low levels of phosphorus. Because of this, beech forests have so far remained least affected by human settlement. Most of the broadleaf-podocarp forests which dominated the fertile lowlands have been cleared, over the last century, and the land developed for pasture.

The native forest litters show a wide diversity in their chemical composition (Miller & Blakemore 1968). Physically, these litters range from soft broad leaves, similar to much tropical vegetation, through the highly siliceous small hard leaves of beech, to the very small leaves of podocarps such as rimu (*Dacrydium cupressinum*). There is a similar range in their mineral content; from leaves and litter of broadleaved trees relatively high in bases, to those low in bases, such as beech, podocarps and kauri. In the natural succession, the tendency for the soil to lose fertility with age is reflected in the vegetation sequence, with broadleaved trees giving way to podocarps or kauri (Wardle 1973).

Much of the most common zonal soils of New Zealand—the yellow-brown earth, similar to the acid brown forest soil of Europe—has developed under beech forest which has been a characteristic forest vegetation since the Cretaceous. Although forming a mor-type litter with generally limited soil animal biomass, the lack of great seasonal fluctuation in the microclimate, and the generally high mean annual temperatures, result in a relatively high rate of turnover, as with tropical soils. Despite the relatively unpalatable character of the leaves, indicated by their relatively long period in the litter, and the constant secretion of polyphenols and other substances which may be slow to decompose, there is a reasonable balance between accumulation and decomposition. Under more humid conditions however, accumulation tends to exceed decomposition, as on the superhumid west coast of South Island. Here, populations of many soil animals, including earthworms, are restricted in distribution, and the high water table tends to inhibit microbial decay.

Slow rates of decomposition are also found in most situations under kauri, and some of the podocarp trees, notably rimu (*Dacrydium cupressinum*), where vast mounds of litter accumulate and contain much of the soil organic matter. Here, it seems probable that the terpenes secreted by the tree may be toxic to many soil animals, so that this factor, combined with the unpalatability of the leaves, prevents the decomposition and incorporation of the litter into the mineral soil. (K.E. Duncan, personal communication.)

Radiocarbon enrichment by 'bomb' carbon can provide a measure of the rate of incorporation and turnover of carbon in the soil profile. However, the profiles studied in New Zealand have often been on steep slopes, where infiltration of the soil through cracks, and the lateral movement of water and organic material down the slope are important. Differences that affect the rate of diffusion of carbon down the profile have however, been observed, with a higher rate of diffusion occurring in a silty soil than in a clayey soil. Generally, the radiocarbon figures on hill soils suggest an excessive rate of turnover. An annual increment of carbon of about 1,000 g m^{-2}, and a total soil content of 50,000 g m^{-2} or more suggest that the overall turnover rate should be about 50–100 years. On a level lowland site, without these drainage complications, the mean subsoil organic matter age is about 500 years, indicating a reasonably fast rate of turnover (Tate 1973b).

Some quantitative studies of forest ecosystems have been made, but largely near urban areas (Daniel 1975). Much of the data has been gathered at Silverstream, Wellington, where some low altitude beech forests (largely 'hard' beech *N. truncata*) remains under a relatively mild climate (Miller 1968). Such an ecosystem contains a plant biomass of about 25–30 kg m^{-2}, but an animal biomass of less than 10 g m^{-2}. The soil and litter organic matter amount to about 4·5 kg m^{-2}; the annual litter fall is about 500 g m^{-2}. The roots are approximately 13% of the total plant biomass and presumably root production is of a comparable magnitude (Miller 1963). Permanent annual tree growth amounts to about a third of the litter fall.

In the Wellington area, mean litter temperatures are about 10°C and vary little throughout the year, largely because the evergreen canopy and high wind movement prevent extreme conditions from arising in the forest floor. Consequently, the moisture regime is more significant than the temperature regime, for, although rainfall is fairly evenly spread throughout the year, evapotranspiration leads to much drier conditions during the summer than winter.

Decomposer network

The mature leaf of *Nothofagus* is small and hard, and not particularly palatable to invertebrate animals. Generally, less than 10% is consumed on the tree. Coccids are an important endemic parasite (Hoy 1962) and a variety of fungi attack both the timber and the leaves (Lancaster 1955).

In the beech forest floor, there is a wide diversity of invertebrates (Dugdale 1974; Dutch & Stout 1968; McColl 1975). Important decomposers include amphipods, isopods, millipedes, adult and larval Coleoptera, lepidopteran larvae and earthworms. Microarthropods are amongst the most numerous groups and other animal groups, working over partly decomposed or fragmented material, include enchytraeid and other small oligochaete worms and many insects, including dipteran larvae. There is a large aquatic microfauna, or terrestrial plankton (Stout 1963), including both harpacticid and cyclopoid copepods, Cladocera, ostracods, microdrili, and the usual assemblage of nematodes, tardigrades, gastrotrichs and Protozoa (Stout 1973b, c, 1974).

The microbiology of a New Zealand beech forest, unlike that of other native forests, shows a close resemblance to that of European beech forests. This is shown by the initial saprophytic fungi colonizing the leaves and by the litter and soil mycoflora (Ruscoe 1971a, b), and also by the bacteria on the leaves and fresh litter, which show the same predominance of pseudomonads as recorded in Europe. Spore-forming and pleomorphic bacteria predominate in the mineral horizons (Dutch & Stout 1968; Stout 1971).

Rates of decomposition

Rates of decomposition in hard beech forests in the Wellington area have been measured by: the use of nylon net bags of litter, various respiratory and enzyme measurements, and by radiocarbon measurements.

Experiments with nylon net bags show that up to 20% weight loss may take place within six months, but that only about a third of the litter has been decomposed (Fig. 5.11) after four years. However, the total litter content of about 1,400 g m^{-2} and the annual litter fall of 500 g m^{-2} imply a litter turnover of about three years. Part of this discrepancy may be attributed

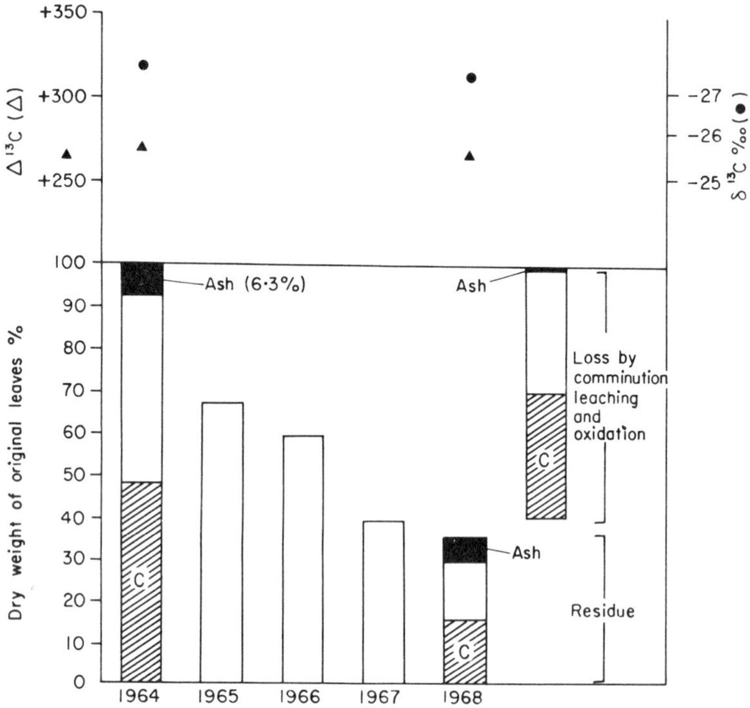

Figure 5.11. The loss of dry weight of 'hard' beech leaves (*Nothofagus truncata*) in nylon net bags (16 holes per cm^2) in the litter horizon of a beech forest soil, Wellington, North Island, New Zealand. Note the relatively small loss of ash indicating little comminution by animals. There was no fractionation of the carbon isotopes over this period. (Dutch & Stout 1968; Rafter & Stout 1970.) C=carbon.

to the exclusion of the larger decomposers from the bags. The initial rapid loss of weight may have been largely due to the loss of readily soluble or readily metabolized plant residues (Dutch & Stout 1968).

Warburg respiratory measurements (Dutch & Stout 1968; Ross & McNeilly 1972b; Ross 1973) have shown that the rates of oxygen uptake and carbon dioxide production declined with increasing age of the decomposing plant material. However, respiratory rates were appreciably affected by moisture so that the upper litter layers, which are more prone to desiccation, showed greater fluctuation in their rates. The lower litter layers showed a greater response to added substrate because the populations of zymogenic bacteria tended to be higher, and also because their more sustained respiratory rate could exhaust the natural supply of nutrients more readily than with drier litter. Measurements of relatively undisturbed soil cores with an electrolytic respirometer have confirmed a higher rate of oxygen uptake per unit weight of organic matter in the topsoil than in the subsoil (Stout & O'Brien 1973).

Studies have been made of the activities of dehydrogenase, invertase, amylase and polyphenol-oxidizing enzymes in 'hard' beech leaves, litter and soil (Ross 1973; Ross & Roberts 1973; Ross & McNeilly 1973). The carbohydrate and organic acid content of the leaves and litter have also been determined, as well as the amounts of polyphenols occurring in the litter, and in the canopy leachate. Glucose, fructose, sucrose and raffinose were detected in appreciable quantities in fresh leaves. Significant quantities of xylose and quinic acid were also detected. These disappeared, or were greatly reduced in concentration, in the litter, particularly on incubation. The polyphenol content of the litter fluctuated throughout the year, being greatest in the summer (K.R. Tate, unpublished results). Invertase activity decreased down the soil profiles; although this activity was significantly correlated with bacterial numbers it may also have originated from leaf enzymes. Amylase activity was significantly correlated with the distribution of fungal hyphae; it was not detected in green leaves. Polyphenol-oxidizing enzymes occurred throughout the soil profile and appeared to originate principally from micro-organisms; their activity was lowest in summer and highest in winter, the inverse of the concentrations of orthodihydric phenols.

EXOTIC FORESTS

Although a wide range of exotic trees have become established in New Zealand over the last century, the most economically important have been the conifers, and particularly Monterey pine (*Pinus radiata*). Extensive plantings of this species cover almost a quarter of a million hectares of the central plateau of North Island, and the success of these plantations has encouraged further planting in almost all parts of New Zealand. The Monterey pine reaches maturity in about thirty years in most situations. Under favourable conditions, it may attain a peak production 15–20 years after planting, of between 3,000 and 4,000 g m^{-2} yr^{-1} of dry matter. The greater part of this (60%) is wood and only about 9% is roots (Will 1966). Although this represents a large nutrient demand in forest that is systematically cropped, most of the nutrients taken up from the soil serve to build up the canopy and only a small proportion is removed in logs for timber. Nutrient demand is high in the early years of the forest but once the trees have a closed canopy the quantity of foliage stabilizes and the full litter decomposition cycle is established within a few years. At this stage, nutrients cycled in rain water (and canopy drip) and slash litter from prunings and thinning dramatically reduce the nutrient demand on the soil (Knight & Will 1970). Thus, decomposition in this commercially exploited ecosystem is of special importance. Recently, Florence & Lamb (1970) have concluded that the failure of second crop *P. radiata* grown on sand dune soils in South Australia may be due to the slow rate of decomposition and nutrient release from plant remains.

A ten-year-old *P. radiata* crop, after thinning to waste, has about 200 tonnes ha^{-1} of dry matter, with the greater part (about two thirds) in undecomposed litter and slash, about 50 tonnes in standing trees, and about 14 tonnes in roots. A 35-year-old crop, on clear felling, yields about 300 tonnes ha^{-1} of timber and about 150 tonnes of undecomposed litter, slash, thinnings and stumps and roots. Thus, there is substantial recycling during the last quarter century of the forest's life (Knight & Will 1970).

Litter fall averages about 550 g m^{-2} yr^{-1}, to which must be added the slashings and prunings of a commercial stand, which are about 1,000 g m^{-2} yr^{-1}. The litter consists mostly of needles (about 400 g m^{-2}), and the remainder twigs, branches and cones. About 1,200–1,500 g m^{-2} of litter occurs on the forest floor, and, on the pumice soils in the centre of North Island, it takes about 3–4 years to decompose. This rate of decomposition is similar to that of native beech forest in comparable situations, and that of conifer litter in other parts of the world (Will 1966).

Decomposer network

Little work has been done on the microbiology of decomposing litter in the exotic forests of New Zealand, although there is a great deal of incidental information on organisms of obvious economic importance, such as forest pests and mycorrhizae. One interesting experiment related the presence and activity of the roots, and consequently mycorrhizae, of the pine forest to the decomposition of the litter (Gadgil & Gadgil 1971). This compared the loss of weight of litter in plots subjected to various treatments to reduce root and mycorrhizal activity, with a control plot in an unthinned *Pinus radiata* stand. The results (Table 5.3) showed a dramatic *increase* in the decomposition of the litter where mycorrhizal growth had been restricted. The basidiomycete mycelial wefts, normally very conspicuous in the litter,

Table 5.3. Decomposition of Monterey Pine (*Pinus radiata*) Litter, North Island, New Zealand (after Gadgil & Gadgil, 1971).

Treatment	Dry Weight of litter after 12 months (g)
A. Plot cut and roots removed	474·8±211·5
B. Plot cut and soil dug to 30 cm	351·6±133·7
C. Plot cut	659·0±328·4
D. Control	1,681·7±213·9

were absent from the experimental plots, and it was thought that their absence permitted the more active proliferation of the saprophytic micro-organisms responsible for litter decay.

The fauna of the exotic forests appear to resemble that reported from northern conifer forests. There is the same succession of microarthropods associated with the ageing of the needles; the same predominance of arthropods in the litter horizons; and the same reliance on the larger fauna, whether tipulid larvae or earthworms, for the final comminution and incorporation into the mineral soil. Recently, comparative studies have been made of the indigenous and exotic forests of the West Coast of South Island. These have shown that the greatest faunal diversity is in the podocarp-beech forest, and that a less diverse range of fauna is found in the beech forests and cut-over forests. The fauna is again less diverse in the exotic forests (*Pinus* spp.), although relics of the former indigenous forest fauna may still be found where pockets of litter from the previous vegetation remain (Dugdale 1974; McColl 1974).

Rates of decomposition

The only detailed study that has so far been completed on the decomposition of pine litter followed the loss of dry weight, and loss of N, P, K, Mg and Ca from needles litter held in nylon net bags (1·0 mm mesh) on the forest floor of a 33-year-old *Pinus radiata* stand over a six-year period. The changes of microfauna and mesofauna extracted from the net bags were followed over a four-year period, by which time the fauna had diminished greatly. Further comminution of the litter in the bags was clearly being prevented by the inability of the larger fauna to penetrate the fine nylon mesh because the surrounding and underlying litter had been reduced to fine colloidal insect frass, and was being incorporated into the mineral horizons (Will 1967; Styles 1967).

The results of this decomposition study are summarized in Fig. 5.12. These show that the dry weight decreased steadily for the first three years but thereafter remained relatively constant. The total amount of nitrogen remained relatively constant over the first 15 months, but then rose abruptly and after a further three months the total nitrogen was 167% of the original amount. The amount of phosphorus initially dropped within the first three months, probably due to loss of remaining mobile phosphorus by leaching, but it then remained fairly constant throughout the experiment. A slight transient increase in phosphorus coincided with the increase in nitrogen, and was possibly associated with an increase of the decomposer biomass. Only 20% of the initial amount of potassium remained after the first three

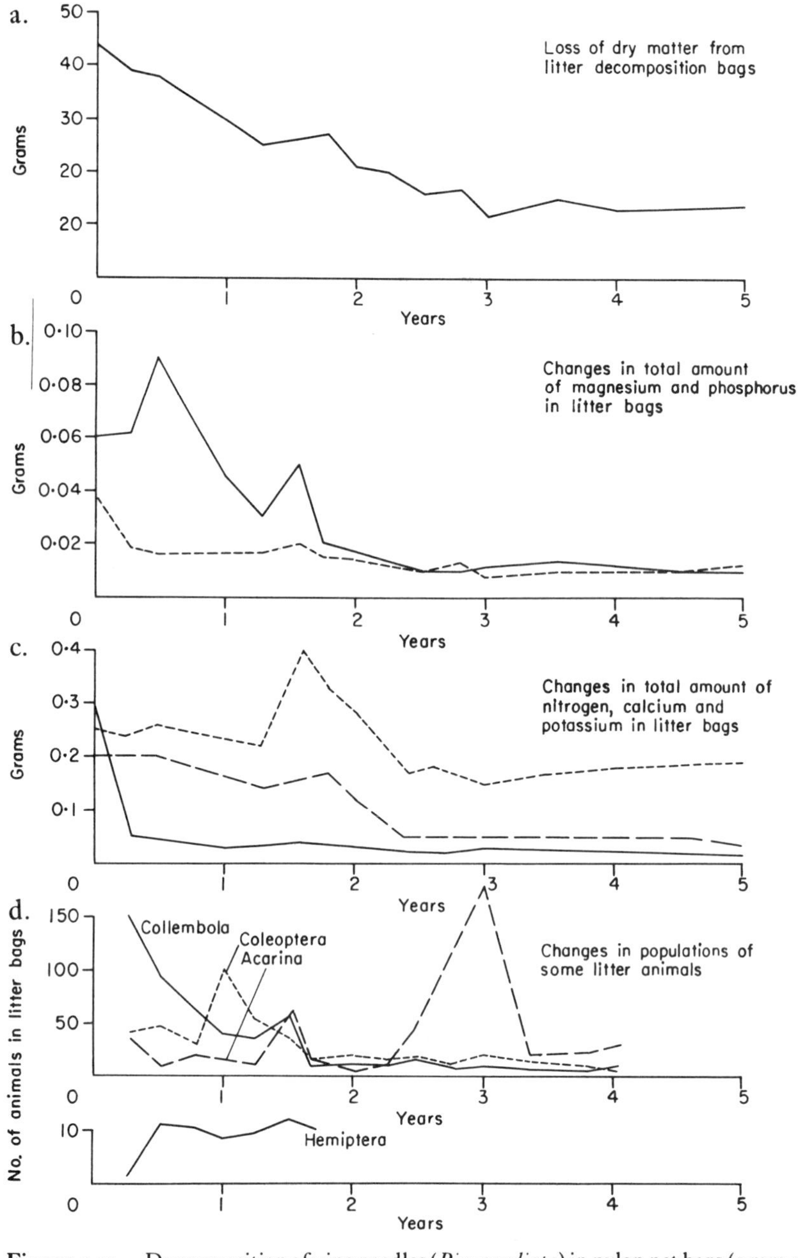

Figure 5.12. Decomposition of pine needles (*Pinus radiata*) in nylon net bags (1 mm mesh) in the litter horizon of a pine forest, North Island, New Zealand. Note the peak of mineral concentration coinciding with the peak biomass after about 18 months. (Will 1967; Styles 1967.)

—— dry matter (a), magnesium (b), potassium (c).

- - - - phosphorus (b), nitrogen (c).

– – – calcium (c).

months. This initial rapid loss of potassium was followed by a slow decline. Magnesium levels fluctuated more than the other elements, with a 50% increase in the first nine months; the amount tended to fall subsequently, but a second increase coincided with the nitrogen and phosphorus peaks. Changes in calcium levels showed a similar pattern to changes in the total dry matter content, most of the loss occurring in the first two years. Calcium levels showed a small transient peak after 18 months coinciding with the similar peaks in the nitrogen, phosphorus and magnesium data.

The populations of the microarthropods and mesofauna showed a typical faunal succession. In the first year there were insects feeding on the soft epidermis and juices of the undecomposed needles; these included some Hemiptera, and a large number of Collembola. As decomposition proceeded with the proliferation of fungal growth, the number of entomobryid Collembola decreased considerably. In the second year, the litter consisted of partly broken-down needles with patches of fungal mycelium and matted needles penetrated by plant rootlets. Hypogasturid Collembola, though fewer than in the first year, were present in number, as were small trichopterygid beetles. The number of omnivorous and predacious mites increased slightly; Lepidoptera larvae and juvenile millipedes and symphylids were present. By the third year, the Hemiptera had disappeared; numbers of Collembola and beetles had decreased markedly, but the number of omnivorous mites, particularly Cryptostigmata (oribatids) had slightly increased. Only a few Diptera and Lepidoptera larvae were collected. By the fourth year, the fauna had diminished appreciably, and all groups were much more poorly represented; by this time the nylon bags were covered with freshly-fallen needles, decomposing needles and accumulated insect frass.

NATIVE TUSSOCK GRASSLANDS

The tussock grasslands of New Zealand have had a varied history. At the time of European settlement, low tussocks grew on the semi-arid and some of the subhumid soils, and tall tussocks on some of the soils of the humid area of South Island. Tall tussock also flourished above the forest line in both islands, but otherwise was limited in North Island to recent soils derived from volcanic ash, and the pumice soils of the central plateau. The vegetation had been modified by fire prior to European settlement, which had induced the spread of tussock grassland, as in the prairies of North America, and this change may have been consolidated by coincidental climatic change. The policy of burning became widespread with the advent of European colonization, and was accompanied by grazing and overgrazing by introduced mammals; the vertebrate grazers had previously been flightless birds such as the moa (*Dinornis* spp.), now extinct, and takahe (*Notornis mantelli*). While these changes accelerated the retreat of the forest, they also led to the deterioration of the tussock grasslands. This deterioration has been partially remedied by the introduction of exotic grasses and the extensive use of fertilizers. However, it is now difficult to find a tussock grassland either materially unmodified by human land usage or at a steady state in terms of ecosystem development.

The tussock grasslands show a comparable physiological diversity to that of the native forests. The pattern of mineral cycling and mineral accumulation varies with the tussock species and its distribution, and its physiological character. Unlike most native forests, however, the natural tussock grasslands tend to maintain their fertility.

Productivity appears relatively low in the short tussock grassland (O'Connor 1959) but higher productivity has been observed in tall tussock grassland, particularly narrow-leaved snowgrass (*Chionochloa rigida*), with values similar to other natural temperate grasslands, i.e. *c.* $2\,g\,d\,wt\,m^{-2}\,day^{-1}$ (P.A. Williams, personal communication). However, at higher altitudes, snowgrass such as *C. macra* has only about half the productivity of *C. rigida* growing in the forest zones. Annual root production may almost equal annual tiller production. These species also differ in their biomass, with the above-ground biomass of *C. rigida* being greater than the below-ground biomass, in contrast to *C. macra* (P.A. Williams, personal communication). Tussocks grow from tillering, and have a growing season of from five to eight months (Mark 1965). They have very low nett assimilation rates but show tolerance to sub-freezing temperatures (Mark 1975). Normally, no massive litter accumulation occurs in well-drained sites. A relatively high rate of turnover, at least of above-ground material, is therefore suggested, perhaps five years for *C. rigida* and seven years for *C. macra*, the latter resembling alpine herbfield (Bliss & Mark 1974). An interesting contrast is provided by the tussock grasslands of Macquarie Island, which lies south of New Zealand in the subpolar zone, latitude 54° (Fig. 5.3). The tussock grasslands there consist of two dominants, one of which is *Poa foliosa*, a tussock grass which develops a high pedestal of peat under its crown, and appears to go through a cyclic development of pioneer, building, mature, degenerate and gap stages. It is associated with other small herbs, grasses and bryophytes, or the associated dominant, the large herbaceous *Stilbocarpa* (Ashton 1965). At an altitude of 45 m, the biomass of *P. foliosa* is $920\,g\,m^{-2}$ above ground, and $1{,}700\,g\,m^{-2}$ below ground; the annual replacement above ground is 91% (Jenkin & Ashton 1970). This relatively high rate of productivity, which is apparently greater than that of the New Zealand snow tussocks, is associated with a high leaf area, high chlorophyll content, a low temperature optimum, and a long growing season. The consequence is an accumulation of plant debris forming the pedestal, which may attain a height of about 50 cm or more. Radiocarbon measurements suggest that such a pedestal may accumulate over about 50 years, and observations indicate that the peat may accumulate at the rate of about 2·5 cm yr^{-1} (Ashton 1965).

Decomposer network

Prior to European colonization, consumers were restricted to birds and invertebrates. The takahe favours the base of the leaf and its grazing consequently added a good deal of uneaten leaf material to the ground litter. Similarly, the grasshopper (Acrididae), probably the most important insect consumer, damages far more of the tussock than it eats, since, like the takahe, it feeds on the base of the leaf (White 1975). The distribution, and consequently importance of most other consumers, whether domestic or feral mammals, or other insects, such as Lepidoptera, Diptera and Coleoptera, is patchy and local. Consequently, the decomposer network is important, as in native forests. Decomposer populations in the tussock grasslands are closely related to climate, which is generally limiting for both animal and plant growth, and which affects both the activity and composition of the microbial populations. In arid or semi-arid zones, microbial populations are limited both in the phyllosphere and rhizosphere, and soil animals may be restricted to microarthropods, especially mites and Collembola (Lee 1968). In the subhumid zone, microbial populations are more active and more diverse, and the animal fauna includes earthworms, and frequently larger insects (Stout 1973b). In humid situations where temperatures are not severe, earthworms are present, and the litter may harbour an extensive fauna; microbial invaders of the leaf surface can include not only bacteria, fungi and protozoa but also algae. With more severe temperatures, the fauna diminishes in diversity; earthworms and enchytraeids may be present. Nematodes also exhibit a definite relationship to climatic parameters in soils in tussock grasslands (Yeates 1974). Sequences of soils taken through a range of climate also show consistent fluctuations in microbial populations (such as nitrogen-fixers, both symbiotic and free-living, and nitrifiers) and in the overall importance of psychrophils (Stout 1958).

Such differences in animal and microbial populations are correlated with development of the soil profile and the amount of soil organic matter.

Rates of decomposition

One sequence of soils currently under study lies on a southeast–northwest transect across southern South Island, from the mouth of the Taieri river to Haast (Fig. 5.13). It cuts across the main divide, and consequently covers a wide range of climatic conditions, from the wet West Coast to the dry

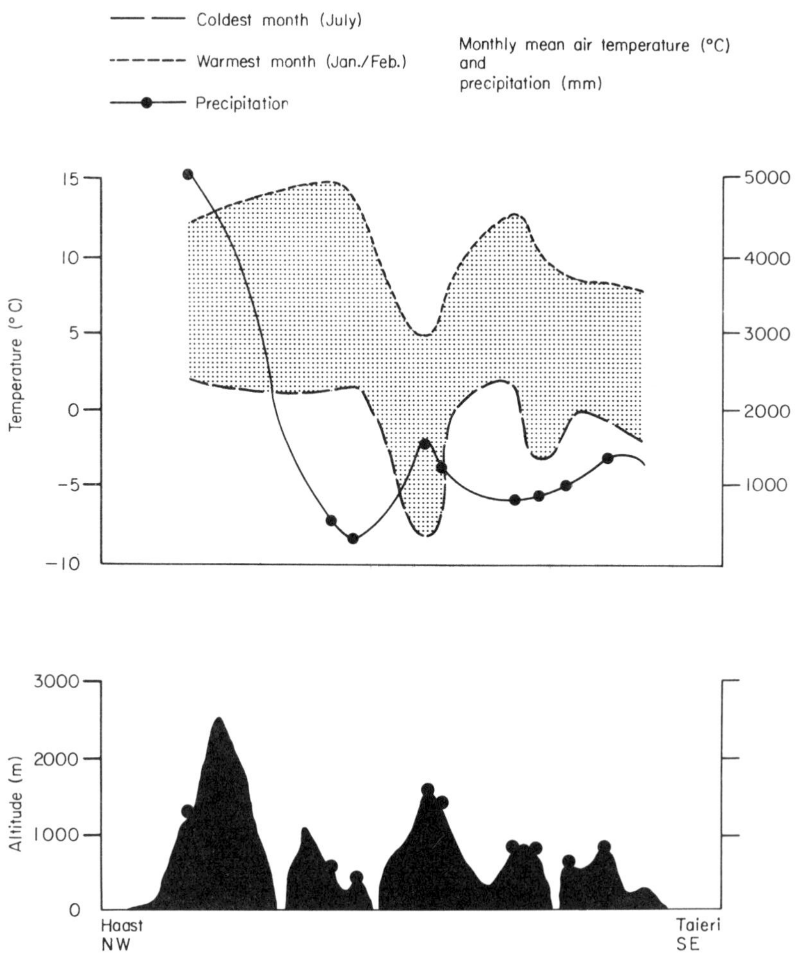

Figure 5.13. Transect of the southern half of the South Island showing topography, precipitation, temperature and location of the tussock grassland sites. Hatching shows extreme monthly temperatures. (Molloy & Blakemore 1974.)

interior basin; from mild temperatures at low altitudes to the more rigorous alpine climate of the axial ranges, with severe frost and persistent strong winds. The soils studied include representatives of typical zonal soils developed from schist. The development reflects the climate and vegetation history (Molloy & Blakemore 1974).

Three techniques are being used for decomposition studies in these soils. They are:

1 Weight loss and chemical changes of a standard snowgrass (*Chionochloa rigida*) litter held in nylon net bags of about 3 mm mesh;

2 Changes in the amounts and chemical composition of three substrates (cellulose, hemicellulose, and lignin) isolated from the snowgrass litter, and incorporated into kaolin pellets buried in the soil; and

3 Weight loss of IBP cellulose at the soil surface and at 5 cm depth.

The chemical composition of the plant components (green leaves, standing dead material, surface litter, and roots) indicates that protein and hemicellulose generally form about half, cellulose up to a third, and lignin and the water-soluble fraction most of the remainder; the amount of residual ash was small except for the roots (Table 5.4).

Table 5.4. Chemical composition of snowtussock (*Chionochloa rigida*) leaves, standing dead material, surface litter and roots (site climatic data Table 5.5) (L.F. Molloy, unpublished data).

	(% by weight)			
	Green Leaves	Standing Dead Material	Surface Litter	Roots
Hot water soluble	17·7	5·2	7·9	9·0
Protein/hemicellulose	49·7	52·8	53·0	44·8
Cellulose	27·5	34·3	25·7	19·9
Lignin	4·5	6·6	8·2	6·9
Ash	0·6	1·1	5·2	19·4

After a year, the loss of weight of the litter in the nylon bags ranged from 15–20% over most of the sites. Results from a typical site are given in Table 5.5. There appeared to be no significant loss of lignin, but much of the hot-water soluble fraction, and part of the protein, hemicellulose, and cellulose had been decomposed. These litter bag results are confirmed by data from the kaolin pellets which showed appreciable decomposition of cellulose and hemicellulose but no apparent decomposition of lignin.

The cellulose and hemicellulose remaining in the kaolin pellets was examined chemically to determine which monosaccharides had been decomposed; the original hemicellulose contained predominantly xylose (77%) and arabinose (18%). After a year, 80–90% of the polysaccharide had been decomposed, and the monosaccharide composition of the residue showed that there had been substantial microbial synthesis of mannose, galactose, and glucose.

Previous studies of some biochemical activities of soils in tussock grass-

Table 5.5. Decomposition of litter under tussock grass (*Chionochloa rigida*) in nylon net bags (about 3 mm mesh) in tussock grasslands, South Island, New Zealand (L.F. Molloy, unpublished data).

Mean annual precipitation (mm)	1,300
Mean monthly air temperature (°C)	
Warmest month (Jan/Feb)	8·6
Coldest month (July)	−0·8

Percent of individual litter components remaining after 1 yr (±SD)

Lignin	Cellulose	Protein and Hemicellulose	Hot water soluble
88±7	86±2	84±4	52±13

lands (Ross & Roberts 1968) had shown that the uppermost part of the soil profile appeared to be most affected by seasonal changes. Activities of glycoside hydrolase enzymes without added substrate, and of invertase (enzymes hydrolysing sucrose) and amylase (enzymes hydrolysing starch) generally increased with increasing altitude of the soil; the enzyme activities strongly correlated with organic C content and, to a lesser extent, with soil moisture content. The results suggested that the proportions of organic matter hydrolysable to reducing sugars by soil enzymes were greater in the lower horizons. The proportions of organic phosphorus compounds were also found to change with depth (Martin 1970). Oxygen uptakes and numbers of viable bacteria generally followed the same pattern.

Ten biochemical activities were measured throughout the present soil climosequence and the influence of ten environmental and soil chemical factors on these activities was assessed by principal components analysis (Ross *et al.* 1975a). Nearly 90% of the total variance in biochemical activities could be accounted for by only three components. The percentage of organic C, soil moisture regime and mean annual temperature accounted for much of the variance in all three components; soil total N was unimportant. As a result of this principal components analysis, the soils could be grouped biochemically into arrangements that were consistent with either the pedological classification of the soils or the influence of a dominant environmental factor. Respiratory activity was very much greater at 24° than at lower temperatures and temperature is obviously a key factor limiting the rate of decomposition (Ross *et al.* 1975b). Moisture may also be an important limiting factor, particularly at drier sites or at very wet sites (Fig. 5.13).

PASTURE LANDS

Much land in New Zealand is used as pasture. The use of native grasslands, and the conversion of forest land to pasture, has been the greatest single change of terrestrial ecosystems in New Zealand. The continued use of the land for pasture has been made possible by the introduction of grasses capable of sustaining continuous grazing, and particularly of clovers to ensure an adequate nitrogen economy. The levels of plant productivity attained by high-producing grass-legume associations, and the removal of large quantities of nutrients by grazing stock have induced deficiencies of major plant nutrients, such as phosphorus and potassium. This nutrient removal by stock has also highlighted deficiencies of minor and trace elements essential for stock health, such as copper and cobalt. The maintenance of adequate supplies of phosphorus, sulphur and molybdenum is essential for the establishment of an effective legume association, which has proved to be the decisive ecological factor in the introduced New Zealand grasslands. The correction of nutrient deficiencies necessary to establish a grass-clover association have resulted in increased nitrogen levels in the ecosystems; these are reflected in an increase in the levels of mull organic matter (Jackman 1964a, b), or in the gradual conversion of mor into mull organic matter (Walker 1968).

The principal grazing animals are sheep [58,000,000 (N.Z. Yearbook, 1974)], dairy cattle [3,000,000 (N.Z. Yearbook, 1974)], and beef cattle [5,000,000 (N.Z. Yearbook, 1974)]; but rabbits have acclimatized well to New Zealand conditions and have in the past reached pest numbers. There are a variety of insect pests, including the native scarab larvae, commonly known as grassgrubs, the larvae of the porina moth and numbers of introduced species. Nematodes are important parasites of the principal pasture species and are particularly damaging in pastures under moisture stress.

Estimates of pasture production have been based on direct measurement of herbage production, on butterfat production, or on the carrying capacity of the land. Production varies with season and soil type (Saunders & Metson 1971), and particularly with the application of fertilizers. Figures of 1,000–2,000 $g\ m^{-2}\ yr^{-1}$ dry matter of herbage have commonly been recorded, with even higher figures from pasture irrigated in the dry season. Root production is about 20–40% of herbage production (P.S. Evans, personal communication). An annual return of plant productivity as plant or animal debris to the soil or soil surface of at least up to 2,000 $g\ m^{-2}$ can be expected in a high-producing pasture.

Decomposer network

The principal agents of change in the conversion of plant residues to mull organic matter are the introduced earthworms, particularly *Lumbricus rubellus*, a surface-active species, and species of *Allolobophora*, principally subsurface species. The size of the earthworm population is closely related to the productivity of the pasture. There are approximately 170 kg live weight of earthworms for every 1,000 kg dry weight of herbage produced. The weight of earthworms is similar to the weight of the grazing stock (Sears *et al.* 1965; Waters 1955).

Most soil animals are either cosmopolitan or of European origin. The failure of the native earthworms to adapt to the exotic vegetation has not been explained; it may reflect nutritional or physical limitations. There are, however, a few instances of native species being associated with exotic forest. Conversely, the exotic earthworm species do not generally invade native forest, although one species, *Eisenia foetida*, is the dominant earthworm in the forests of Raoul Island.

The conversion of native forest or native grassland to exotic pasture in general has been associated with two major consequences. Firstly, the deeper rooting native vegetation which distributes organic matter more generally throughout the soil profile has been replaced by a more highly productive vegetation whose roots are largely confined to the top 5–10 cm. Secondly, the native soil fauna, including the native earthworms, has disappeared in the topsoils and been replaced in many cases by introduced lumbricids and other cosmopolitan or European soil animals. Native subsoil earthworms may persist after a forest to pasture conversion. The interrelated effects of earthworms on the soil organic cycle may be illustrated by the conversion of native tussock grassland in Otago (under *Chionochloa* and *Festuca*) to exotic pasture. The unmodified native tussock grassland soil is well structured and rich in organic matter. There is a diverse fauna both of arthropods and native earthworms, commonly with litter, topsoil and subsoil species, and the humus horizons tend to be mull-like (Barratt 1971a, b). This fauna disappears with the establishment of the exotic pasture, except possibly for the subsoil earthworms. The pasture is normally well limed, and topdressed with superphosphate and traces of molybdenum to promote legume growth and nitrogen fixation, and formerly was treated with an insecticide to control infestation by the scarab larvae (grassgrubs). The productivity of this pasture may be of the order of 7–800 g m^{-2} yr^{-1} of dry matter of herbage. However, without worms a peaty layer of dung and dead plant material accumulates at the surface. In addition, applied lime, superphosphate and insecticides tend to remain as a surface veneer. In short, the humus horizons are a laminated mor (Barratt 1966) with its characteristic shallow organic horizon, underneath which the mineral soil is compact, dry and of poor structure. Moisture penetration and plant root development are severely restricted. With the introduction of lumbricids, particularly *Allolobophora caliginosa*, dung and dead plant residues are incorporated into the topsoil and decompose much more rapidly. Lime, fertilizers and insecticides are mixed through the soil increasing their effectiveness. The effect on pasture production is marked, herbage dry matter yields increasing by 20–30% (i.e. about 200 g m^{-2} yr^{-1}). The structure of the humus horizons is an obvious mull. As the exotic earthworm population invades pasture, breaking down and burying the accumulated peaty layer of dung and plant material, the greatly stimulated rate of decomposition provides a rapid release of plant nutrients which promotes a greater pasture response during this phase than when a steady state is established. The relative mean pasture yields for two years for a pasture without worms, a pasture with invading worms; and a pasture with an established earthworm population are 100 : 121 : 119. (S.M.J. Stockdill, personal communication.)

The magnitude of the biological activity of the earthworms following their invasion is reflected in the physical and chemical changes of the soil, as well as in the greatly stimulated rate of decomposition and mineral recycling. Table 5.6 shows some physical and organic properties of an upland soil, where humid conditions occur, under tussock grasses and under in-

Table 5.6. Analyses of physical properties of an hygrous yellow-brown earth under native tussock grassland and introduced pasture with and without earthworms (S.J.M. Stockdill & Mr. G.G. Cossens, unpublished data).

Soil Properties	Soil Depth (cm)	Soil under pasture: With earthworms	Soil under pasture: Without earthworms	Soil under tussock
Moisture Properties				
Available moisture (cm)	0–10	3·1	1·8	2·4
	0–30	8·4	6·6	7·5
Field capacity (% by weight)	0–10	51·7	42·0	47·4
	0–30	42·7	36·6	40·8
Wilting Point (15 bar retention) (% by weight)	0–10	16·2	15·6	16·9
	0–30	13·7	13·5	14·2
Infiltration (cm)				
5-hr intake		61	29	42
Basic rate hr^{-1}		2·55	1·4	1·4
Macroporosity (%)	0–10	21·5	44·9	30·9
	0–30	22·7	29·3	26·2
Bulk Density (g ml^{-1}) (dry weight)	0–10	0·86	0·68	0·78
	0–30	0·96	0·96	0·94
Organic Carbon (% by volume)	0–10	4·46	3·54	4·62
	0–30	2·97	2·38	2·77

troduced pasture with and without earthworms. The measurements are of interrelated soil factors, but they all reflect the activity of the earthworms, or their absence. The field capacity of the soil with worms is greater than for either the tussock soil or the pasture without worms; for the top 30 cm of soil, the pasture soil with earthworms has a 17% greater capacity than the same soil without earthworms. The wilting point and dry bulk density have remained much the same. The soils with worms have a significantly greater available moisture because of their greater field capacity. Although total macroporosity has diminished, the fewer macropores in the soil with earthworms are of greater diameter, allowing freer movement of water through the profile, as indicated by the infiltration rate. The overall involvement of the earthworms in these changes is reflected by the distribution of organic matter throughout the profile, and particularly in the subsoil. Although no balance sheet of the energetics of a New Zealand pasture soil has been completed, it is clear that the earthworm populations must play the most substantial soil decomposer role in highly productive pasture land.

Where earthworms attain very large numbers, associated with high densities of grazing animals, the build up of cast material (e.g. up to 25 mm yr^{-1}) can cause a decrease in the bearing strength of the superficial soil layers, and give rise to 'poaching' (Lee 1959). Earthworms normally aestivate in the subsoil during the summer in the subhumid and humid New Zealand regions, but irrigation permits them to remain active in the topsoil. This is important as they are primarily responsible for incorporating dung, plant litter and dead roots into the soil organic matter, and normally, when the soil is dry, these accumulate and mineralization is inhibited.

Microbiologically, most pastures, in the milder regions at least, are dominated by endotrophic mycorrhizae of the vesicular-arbuscular type (Thornton, 1965; Jackson 1965a, b). There are, however, also extensive floras of yeasts (di Menna 1965) and bacteria (Stout 1973a) whose composition is related to soil conditions, and particularly to temperature and moisture regimes.

Rates of decomposition

The accumulation of litter occurs seasonally in many areas and is related to stocking practice. Under warm humid conditions, the growth of saphrophytes may be rapid and may include fungi producing animal toxins, such as *Pithomyces chartarum*. Normally, a superficial litter layer is absent from exotic pastures, although a thick mat layer of undecomposed roots accumulates in areas where earthworms are absent; this may be several centimetres thick, and comparable to the mor found in low-producing grasslands. Colonization by earthworms results in the breakdown of the mat and release of plant nutrients, and a consequent significant increase in the productivity and carrying capacity of the pasture (Stockdill 1959).

Rates of oxygen uptake and activities of invertase and amylase enzymes (hydrolysing sucrose and starch respectively) have been determined under a range of pastures of grasses and clovers (Ross 1965, 1966, 1968; Ross & Roberts 1970). Amounts of reducing sugars produced on incubation of soils without added substrate were related very closely to soil organic carbon content. Activities of invertase and amylase tended to decrease or increase respectively with increasing rainfall. The ratios of invertase to amylase activity were highest in the driest groups of soils. Although variations in soil organic carbon and annual rainfall together explained most of the variation of invertase activity, this activity appeared to be related more to the composition than the amount of organic matter in the soil. Variations in amylase activity could also be related to the content of soil organic carbon, clay content, and other soil factors. Amylase activity, compared with invertase activity, was relatively greater under legumes than under grasses. Oxygen uptakes were significantly correlated with organic carbon content, quantities of reducing sugars produced on incubation, and amylase activity in some soils, but significant relationships were not found in others. Uptake of oxygen generally increased two to threefold with the addition of glucose.

As previously stated, the pasture ecosystem is one of great biological activity, and has a high rate of turnover. However, organic matter steady states are reached at a high level of soil organic matter, because nutrient deficiencies, which may limit growth under natural conditions, tend to be corrected by fertilizers. Productivity consequently approaches the climatic maximum for grassland in some sites.

Modifications of the pasture ecosystem

Modifications of the pasture ecosystem by various land use practices include:

(i) Drainage and/or irrigation;

(ii) The disposal of agricultural or factory effluent, or sewage, by land irrigation;

(iii) The application of herbicides or pesticides;

(iv) Cultivation for market gardening, or agricultural crops as part of the pasture rotation.

(i) Drainage or irrigation dramatically alters the soil and herbage microclimate; it not only increases productivity, but also spreads it more evenly throughout the year. Consequently, the activity and size of the decomposer network is enhanced, and populations are generally higher in irrigated than control pastures (Ruscoe 1973). Amylase activity, though not invertase activity, was also increased by irrigation, although it may also be reduced by drainage. In a study on irrigated soils (Ross 1966), there was no increase in soil organic matter, the increase in decomposer activity compensating for increased plant productivity.

(ii) The disposal of agricultural effluent by spray-irrigation of pasture or other means is a standard practice in New Zealand. It may involve farm wastes, e.g. from milking sheds, wastes from dairy factories or abbatoirs, or the dispersal of refuse from feeding lots, and markedly increases the load placed on the decomposer network. Two alternative basic patterns can result: (1) The normal soil fauna and flora is able to deal with the effluent because of greatly heightened activity, particularly of earthworms and the bacterial and fungal microflora. (2) The effluent produces conditions inimical to the normal soil fauna and flora, which tend to be reduced or disappear, and may be replaced by a community of radically different composition. This latter pattern happens when whey, with lactose still present, is sprayed on pasture land; the earthworm and normal soil animal populations tend to be greatly reduced, or entirely absent, while micro-organisms that are facultative or obligate anaerobes, and able to fix atmospheric nitrogen, become key members of the decomposer network with an associated microflora and microfauna of yeasts and protozoa (di Menna 1966; Stout 1965). Here the metabolic pattern is one of dramatic increases in respiratory rates and population levels, following the addition of the effluent, and the production of locally anaerobic conditions and high fermentation rates. Where lactose is first removed from the whey, the pattern is more similar to that of a normal irrigated pasture, or of one irrigated with effluent from a milking shed, where the normal population shows a uniform and sustained increase of metabolic activity. With meat wastes effluent, containing nutrient-rich organic residues, there is similarly a more marked biological response (Tate 1973a).

(iii) The application of some insecticides and fungicides to soil may prove inhibitory or toxic to earthworms and thereby create problems of soil management (Keogh & Whitehead 1975). A study of the effects of four pesticides on a pasture ecosystem has shown that they had no detectable effect on soil properties, herbage yield or species composition (Martin 1974). Application of the pesticides at recommended field rates had no significant effect on respiratory activity, nor did they inhibit nitrogen mineralization or nitrification. However, higher concentrations in the laboratory proved inhibitory to respiration (Tate 1974) and had inhibitory or stimulatory effects on nitrogen mineralization or nitrification (Ross 1974).

(iv) Continued cultivation of pasture soil tends to reduce the content of organic matter, and radically alter the physical properties of the soil. Soil animals tend to be greatly reduced in numbers and variety, and the microbial population also tends to be reduced. Although respiratory rate tends to be reduced, this is primarily related to the reduction in the amount of organic matter (Stout & Dutch 1968).

Summary and conclusions

The broad theme of this contribution is the death cycle of a leaf in the terrestrial ecosystem: the transformation from plant to animal and microbial tissue, and, finally, to soil organic matter. This transformation is spread over hundreds, or even thousands of years, in which there is an endless diversity both in the rates of change and the pathways followed.

Decomposition of plant residues in soil is in large part the conversion of plant material to animal and microbial tissues, which may prove as resistant or even more resistant to degradation than the original plant structure. The organic material that accumulates in soil may therefore be divided into biomass and nonliving organic matter. The nonliving organic matter includes dead plant structures, or their immediate residues, and soil organic matter, which is the product of plant, animal and microbial decomposition.

Of this soil organic matter, only a fraction, the 'tip of the iceberg', is visible on the surface of the soil as decomposing plant litter, with the animal and microbial decomposers readily apparent. Generally, the greater part of soil organic matter is bound in intimate association with mineral particles. Much greater activity occurs, however, in the soil surface with its freshly-fallen plant debris and root activity, and its much greater heterogeneity, with plant, animal, microbial and mineral components still separately recognizable. The soil surface also has a different microclimate, a different decomposer network, and a different timescale of decomposition processes. With organic matter in the mineral soil there is a tendency toward superficial homogeneity of both mineral and organic components, and simplification of the decomposer network, while the timescale of decomposition processes stretches towards infinity.

This time sequence is clearly illustrated by consideration of the principal New Zealand terrestrial ecosystems, many of which, including pasture and exotic conifer forests, have been superimposed over the last century on much older native ecosystems of forest or tussock grassland. The litters and topsoils clearly represent the pattern of the current dynamics in the vegetation and decomposer network, whereas the subsoils may still preserve relics of the past ecosystem, such as subsoil native earthworms surviving in exotic pastures, or residues of forest lignins. In a sense, therefore, the ecological contrast is not dissimilar to that of the freshwater situation with its epilimnion, hypolimnion, and underlying sediment. It may not be unreal to speak of (1) an 'epiedaphon'—the predominantly animate surface region of a terrestrial ecosystem which may accumulate quite massive quantities of organic matter, such as the organic horizons of soils under North Auckland kauri forests or the pedestal of Macquarie Island tussock grasslands. However, organic matter is still essentially plant residues, albeit comminuted by soil animals, and is still comparatively young, generally less than 50 years old; and (2) a 'hypoedaphon'—the product of pedogenetic steady state which may extend long beyond the duration of the present ecosystem and where there may be much older accumulations of organic matter, as with many peats or subsoils, which are normally the product of a much slower process of decomposition. In analysing decomposition in a terrestrial ecosystem, it is essential to bear in mind the very different timescales of these two regions, their very different energy status; and the very different mechanisms by which the continuing transformation of organic matter into new living organisms and mineral material is effected. In this respect, carbon provides a unique and invaluable tool for the study of decomposition since it offers not only a measure of different mechanisms reflected by different fractionation rates but also, with radiocarbon, a measure of time. The data on the soils of the New Zealand area illustrate how important the factor of time may be, and how the diverse patterns of decomposition may be related to the climatic parameters which, in effect, define the different soil groups.

Acknowledgments

We are grateful to our colleagues who provided information, much of it unpublished, without which we could not have adequately covered this review. They include Dr D.J. Ross, Dr T.W. Speir, Mrs B.A. Bridger and Miss M.E. Dutch of Soil Bureau; Dr K.M. Goh of the Department of Soil Science, Lincoln College, Canterbury, (formerly of Soil Bureau D.S.I.R.); Dr R.M. Jackman and Mr P.S. Evans of Grasslands Division, D.S.I.R., Palmerston North; Dr N.A. Martin, Entomology Division, D.S.I.R., Auckland; Dr G.M. Will of Forest Research Institute, Whakarewarewa, Rotorua; and particularly Mr P.A. Williams of Botany Division, D.S.I.R. (formerly of Tussock Grasslands and Mountainlands Institute, Lincoln College) and Mr C. Meurk, Department of Botany, University of Otago, Dunedin, who provided information on the tussock grasslands. Data on radiocarbon was provided by the Institute of Nuclear Sciences, D.S.I.R., Lower Hutt, and we acknowledge the co-operation of the Director, Dr T.A. Rafter and his staff.

We acknowledge with thanks, the help of Mr W. Bick, Department of Soil Organic Chemistry, The Macaulay Institute for Soil Research, Craigiebuckler, Aberdeen, Scotland, with some amino acid determinations.

The scanning electron micrographs were prepared by the Physics and Engineering Laboratory, D.S.I.R., Lower Hutt, New Zealand.

We acknowledge the help of Miss D. Mangnall who drew the figures and Mr J.A. Small who helped generally in the preparation of the paper.

We finally acknowledge the helpful criticism of this paper by our colleagues Drs D.J. Ross and R.J. Furkert.

Dr V.M. Stout kindly presented the paper at the B.E.S. Symposium.

References

ANDERSON H.A. (1970) Some aspects of the chemistry of the Glentanar podzol profile. *Report Welsh Soils Discussion Group* **11**, 124–33.

ASHTON D.H. (1965) Regeneration pattern of *Poa foliosa* Hook F. on Macquarie Island. *Proc. R. Soc. Vict.* **29**, 215–33.

BARBER D.A. (1968) Microorganisms and the inorganic nutrition of higher plants. *A. Rev. Pl. Physiol.* **19**, 71–88.

BARRATT B.C. (1966) Factors of mull and mor development in temperate grasslands. *N.Z. Ecol. Soc. Proc.* **13**, 24–9.

BARRATT B.C. (1971a) Micromorphology of some zonal soils of New Zealand. *N.Z. Jl Sci.* **14**, 651–97.

BARRATT B.C. (1971b) Micromorphology of some intrazonal soils of New Zealand. *N.Z. Jl Sci.* **14**, 698–733.

BERGMAN A. (1963) *Organic Geochemistry*, ed. I. Breger, pp. 503–42. Pergamon Press, Oxford.

BIRCH J.W. & MELVILLE M. (1969) An electrolytic respirometer for measuring oxygen uptake in soils. *J. Soil Sc.* **20**, (1), 101–10.

BLISS L.C. & MARK A.F. (1974) High-alpine environments and primary production on the Rock and Pillar Ranges, Central Otago, New Zealand. *N.Z. Jl Bot.* **12**, 445–83.

BOWEN G.D. & ROVIRA A.D. (1969) The influence of microorganisms on growth and metabolism of plant roots. *In: Root Growth*, ed. E.J. Whittingdon, pp. 170–201. Butterworths, London.

BLAKEMORE L.C., COLLIE T.W. & METSON A.J. (1968) Sulphur. In: *Soils of New Zealand. Bull. Soil Bur. N.Z.* **26**, (2), 103–9.

BREMNER J.M. (1955) Studies on soil humic acids. 1. The chemical nature of humic nitrogen. *J. agric. Sci. Camb.* **46**, 247–56.

CARLES J. & DECAU J. (1960) Variations in the amino acids of soil hydrolysates. *Proc. R. Soc.*, Dublin Al, 177–8.

CHESHIRE M.V., MUNDIE C.M. & SHEPHERD H. (1971) The origin of the pentose fraction of soil polysaccharide. *J. Soil Sci.* **22**, 222–36.

DANIEL M.J. (1975) Preliminary account of litter production in a New Zealand lowland podocarp-rata-broadleaf forest. *N.Z. Jl Bot.* **13**, 173–87.

DUGDALE J.S. (1974) Arthropod species composition of the beech forest floor. *Beech Research News*, June 1974, 25–7.

DUTCH M.E. & STOUT J.D. (1968) The carbon cycle in a beech forest ecosystem in relation to microbial and animal populations. *Trans. 9th Int. Congr. Soil Sci.* **2**, 37–48.
DI MENNA M.E. (1965) Yeasts in New Zealand Soils. *N.Z. Jl Bot.* **3**, 194–203.
DI MENNA M.E. (1966) The incidence of *Azotobacter* and of nitrogen-fixing Clostridia in some New Zealand soils. *N.Z. Jl agric. Res.* **9**, 218–26.
EGUNJOBI J.K. (1969) Primary productivity and nutrient cycling in terrestrial ecosystems; a review with particular reference to New Zealand. *Tuatara* **17**, 49–66.
FLORENCE R.G. & LAMB D. (1970) Litter decomposition and nutrient release in *Pinus radiata* plantations. Paper I. 2nd Radiata pine Symposium Australian National University, Canberra, August, 1970.
FRENEY J.R. (1967) Sulphur-Containing Organics. In: *Soil Biochemistry*, Vol. 1, eds. A.D. McClaren & G.H. Peterson, pp. 229–59. Marcel Dekker, New York.
FORD G.W., GREENLAND D.J. & OADES J.M. (1969) Separation of the light fraction from soils by ultrasonic dispersion in halogenated hydrocarbons containing surfactant. *J. Soil Sci.* **20**, 291–6.
GADGIL R.L. & GADGIL P.D. (1971) Mycorrhiza and litter decomposition. *Nature, Lond.* **233**, 133.
GOH K.M. (1972) Amino-acid levels as indicators of paleosols. *Geoderma* **7**, 33–47.
GOH K.M. & STOUT J.D. (1973) Radiocarbon enrichment and the turnover of soil organic matter in a chronosequence of soils developed on wind-blown sand in New Zealand. *Proc. 8th Int. Radiocarbon Dating Conf., Lower Hutt, New Zealand.* **2**, 449–63.
GOH K.M., RAFTER T.A., STOUT J.D. & WALKER T.W. (1976) The accumulation of soil organic matter and its carbon isotope content in a chronosequence of soils developed on wind-blown sand in New Zealand. *J. Soil Sci.* (in press).
HOY J.M. (1962) Eriococcidae of New Zealand. *N.Z. Department of Scientific and Industrial Research Bull.* **146**, edited by N.Z. Government Printer, Wellington.
JACKMAN R.H. (1964a) Accumulation of organic matter in some New Zealand soils under permanent pasture. I. Patterns of change of organic carbon, nitrogen, sulphur and phosphorus. *N.Z. Jl agric. Res.* **7**, 445–71.
JACKMAN R.H. (1964b) Accumulation of organic matter in some New Zealand soils under permanent pasture. II. Rates of mineralization of organic matter and the supply of available nutrients. *N.Z. Jl agric. Res.* **7**, 472–9.
JACKSON R.M. (1965a) Studies in pasture soils. Part 2: Fungi associated with plant debris and fungal hyphae in soil. *N.Z. Jl agric. Res.* **8**, 865–77.
JACKSON R.M. (1965b) Studies of fungi in pasture soils. Part 3: Physiological studies on some fungal isolates from the root surface and from organic debris. *N.Z. Jl agric. Res.* **8**, 878–88.
JANSSON J.L. (1971) Use of ^{15}N in studies of soil nitrogen. In: *Soil Biochemistry*, Vol. 2, eds. A.D. McClaren & J. Skujins, pp. 129–30. Marcel Dekker, New York.
JENKIN J.F. & ASHTON D.H. (1970) Productivity studies on Macquarie Island vegetation. In: *Antarctic Ecology*, ed. M.W. Holdgate, pp. 851–63. Academic Press, London.
JENKINSON D.S. (1966) Studies on the decomposition of plant material in soil. 2. Partial sterilisation of soil and the soil biomass. *J. Soil Sci.* **17**, 280–302.
JENKINSON D.S. (1971) Studies on the decomposition of ^{14}C labelled organic matter in soil. *Soil Sci.* **111**, 64–70.
KEOGH R.G. & WHITEHEAD P.H. (1975) Observations on some effects of pasture spraying with benomyl and carbendazim on earthworm activity and litter removal from pasture. *N.Z. Jl expl. agric.* **3**, 103–4.
KNIGHT P.J. & WILL G.M. (1970) An appraisal of nutrient supplies available for tree growth in a pumice soil. *For. Sci. Jl* **4**, 1–15.

KONONOVA M.M. (1966) *Soil organic matter*, 2nd ed., Pergamon Press, New York.

LANCASTER MARGARET E. (1955) *Forest Fungi*, edited by N.Z. Government Printer, Wellington.

LEE K.E. (1959) The earthworm fauna of New Zealand *Bull. N.Z. Dep. Scient. ind. Res.* 130.

LEE K.E. (1968) A preliminary study of soil animals and their relationships to some New Zealand soils. In: *Soils of New Zealand. Bull. Soil Bur. N.Z.* **26**, (2), 168–82.

MARK A.F. (1965) The environment and growth rate of narrow-leaved snow tussock, *Chionochloa rigida*, in Otago. *N.Z. Jl Bot.* **3**, 73–103.

MARK A.F. (1975) Photosynthesis and dark respiration in three alpine snow tussocks (*Chionochloa* spp.) under controlled environments. *N.Z. Jl Bot.* **13**, 93–122.

MARTIN J.K. (1970) Soil Organic Phosphorus. V. Distribution of organic phosphorus in a sequence of soils under tussock grassland. *N.Z. Jl agric. Res.* **13**, 522–33.

MARTIN J.P. & HAIDER K. (1971) Microbes in soil humus formation. *Soil Sci.* **111**, (1), 54–63.

MARTIN N.A. (1974) Effect of four insecticides on the pasture ecosystem. I. Insecticide residues, soil moisture and bulk density, and herbage yield and composition. *N.Z. Jl agric. Res.* **17**, 485–94.

MCCOLL H.P. (1974) The arthropods of the floors of six forest types on the West Coast, South Island: a preliminary report. *N.Z. Ecol. Soc. Proc.* **21**, 11–16.

MCCOLL H.P. (1975) The invertebrate fauna of the litter surface of a *Nothofagus truncata* forest floor, and the effect of microclimate on activity. *N.Z. Jl Zool.* **2**, 15–34.

MEKHITEYA V.L. & PANKINA R.G. (1968) Isotopic composition of sulphur in aquatic plants and dissolved sulfates. *Geochem. International* **5**, 624–7.

MILLER R.B. (1963) Plant nutrients in hard beech. 3. The cycle of nutrients. *N.Z. Jl Sci.* **6**, 388–413.

MILLER R.B. (1968) Flows and cycles of macro- and micro-elements in a forest soil and its environment. *Trans. 9th Intern. Congr. Soil Sci., Adelaide* **4**, 323–31.

MILLER R.B. & BLAKEMORE L.C. (1968) Organic Matter. Soils of New Zealand. *Bull. Soil Bur. N.Z.* **26**, (2), 55–67.

MINDERMAN G. (1968) Addition, decomposition and accumulation of organic matter in forests. *J. Ecol.* **56**, 355–62.

MOLLOY L.F. & BLAKEMORE L.C. (1974) Studies on a climosequence of soils in tussock grasslands. I. Introduction, sites, and soils. *N.Z. Jl Sci.* **17**, 233–55.

N.Z. SOIL BUREAU (1968) *Soils of New Zealand. Bull. Soil Bur. N.Z.* **26**.

NEW ZEALAND OFFICIAL YEARBOOK (1974) *Authoritative reference on the resources, institutions, and social and economic conditions of New Zealand*, edited by N.Z. Government Printer, Wellington.

O'CONNOR K.F. (1959) Plant production in difficult environments. *New Zealand Institute of Agricultural Science, Proceedings*, pp. 86–94.

PIPER T.J. & POSNER A.M. (1972) Sodium amalgam reduction of humic acid. II. Application of the method. *Soil Biol. Biochem.* **4**, 525–31.

POOLE A.L. (1968) Forestry. *Soils of New Zealand. Bull. Soil Bur. N.Z.* **26**, (2), 96–8.

RAFTER T.A. & STOUT J.D. (1970) Radiocarbon measurements as an index of the rate of turnover of organic matter in forest and grassland ecosystems in New Zealand. In: *Radiocarbon Variations and Absolute Chronology*, ed. I.U. Olsson, pp. 401–15. Almquist & Wiksell, Stockholm.

ROSS D.J. (1965) A seasonal study of oxygen uptake of some pasture soils and activities of enzymes hydrolysing sucrose and starch. *J. Soil Sci.* **16**, 73–85.

Ross D.J. (1966) A survey of activities of enzymes hydrolysing sucrose and starch in soils under pasture. *J. Soil Sci.* **17**, 1–15.

Ross D.J. (1968) Activities of enzymes hydrolysing sucrose and starch in some grassland soils. *Trans. 9th Int. Congr. Soil Sci. Adelaide* **3**, 299–308.

Ross D.J. (1973) Biochemical activities in a soil profile under hard beech forest. 2. Some factors influencing oxygen uptakes and dehydrogenase activities. *N.Z. Jl Sci.* **16**, 225–40.

Ross D.J. (1974) Influence of four pesticide formulations on microbial processes in a New Zealand pasture soil. II. Nitrogen mineralization. *N.Z. Jl agric. Res.* **17**, 9–17.

Ross D.J. & McNeilly B.A. (1972a) Some influences of different soils and clay minerals on the activity of glucose oxidase. *Soil Biol. Biochem.* **4**, 9–18.

Ross D.J. & McNeilly B.A. (1972b) Effects of storage on oxygen uptakes and dehydrogenase activities of beech forest litter and soil. *N.Z. Jl Sci.* **15**, 453–62.

Ross D.J. & McNeilly B.A. (1973) Biochemical activities in a soil profile under hard beech forest. 3. Some factors influencing the activities of polyphenol-oxidising enzymes. *N.Z. Jl Sci.* **16**, 241–57.

Ross D.J. & Roberts H.S. (1968) A study of activities of enzymes hydrolysing sucrose and starch and of oxygen uptake in a sequence of soils under tussock grassland. *J. Soil Sci.* **19**, 186–96.

Ross D.J. & Roberts H.S. (1970) Enzyme activities and oxygen uptakes of soils under pasture in temperature and rainfall sequences. *J. Soil Sci.* **21**, 368–81.

Ross D.J. & Roberts H.S. (1973) Biochemical activities in a soil profile under hard beech forest. I. Invertase and amylase activities and relationships with other properties. *N.Z. Jl Sci.* **16**, 209–24.

Ross D.J., Speir T.W., Giltrap D.J., McNeilly B.A. & Molloy L.F. (1975a) A principal components analysis of some biochemical activities in a climosequence of soil. *Soil Biol. Biochem.* **7**, 349–55.

Ross D.J., McNeilly B.A. & Molloy L.F. (1975b) Studies on a climosequence of soils in tussock grasslands. 4. Respirometry activities and their relationships with temperature, moisture and soil properties. *N.Z. Jl Sci.* **18**, 377–89.

Ruscoe Q.W. (1971a) Mycoflora of living and dead leaves of *Nothofagus truncata. Brit. Mycological Soc. Trans.* **56**, 463–74.

Ruscoe Q.W. (1971b) The soil mycoflora of a hard beech forest. *N.Z. Jl Sci.* **14**, 554–67.

Ruscoe Q.W. (1973) Changes in the mycofloras of pasture soils after long-term irrigation. *N.Z. Jl Sci.* **16**, 9–20.

Saunders W.M.H. (1968) Phosphorus. In: *Soils of New Zealand. Bull. Soil Bur. N.Z.* **26**, (2), 95–103.

Saunders W.M.H. & Metson A.J. (1971) Seasonal variation of phosphorus in soil and pasture. *N.Z. Jl agric. Res.* **14**, 307–28.

Schnitzer M. & Ortiz de Serra M.I. (1973) The sodium-amalgam reduction of soil and fungal humic substances. *Geoderma* **9**, 119–28.

Sears P.D., Goodall V.C., Jackman R.H. & Robinson G.S. (1965) Pasture growth and soil fertility. 8. The influence of grasses, white clover, fertilisers and the return of herbage clippings on pasture production of an impoverished soil. *N.Z. Jl agric. Res.* **8**, 270–83.

Stevens P.R. & Walker T.W. (1970) The chronosequence concept and soil formation. *Quart. Rev. Biol.* **45**, 33–50.

Stockdill S.M.J. (1959) Earthworms improve pasture growth. *N.Z. Jl agric.* **98**, 227–33.

Stout J.D. (1958) Biological studies of some tussock grassland soils. 4. Bacteria. *N.Z. Jl agric. Res.* **1**, 943–57.

STOUT J.D. (1963) The terrestrial plankton. *Tuatara* **11**, 57–65.

STOUT J.D. (1965) The relation between Protozoan populations and biological activity in soils. In: *Progress in Protozoology*, p. 119. Excerpta Medica Foundation, London.

STOUT J.D. (1971) The distribution of soil bacteria in relation to biological activity and pedogenesis—Part 1. General introduction and factors affecting populations at Taita Experimental Station, New Zealand. *N.Z. Jl Sci.* **14**, 816–33.

STOUT J.D. (1973a) The distribution of soil bacteria in relation to biological activity and pedogenesis—Part 3. Soils of New Zealand *N.Z. Jl Sci.* **16**, 113–37.

STOUT J.D. (1973b) Soil Ecology. In: *The Natural History of New Zealand*, ed. G.R. Williams, pp. 131–54. A.H. & A.W. Reed, Wellington.

STOUT J.D. (1973c) The relationship between protozoan populations and biological activity in soils. *American Zoologist* **13**, (1), 193–201.

STOUT J.D. (1974) Protozoa. In: *The Biology of Plant Litter Decompoistion*, Vol. 2, eds. C.H. Dickinson & G.J.F. Pugh, pp. 385–420, Academic Press, London.

STOUT J.D. & DUTCH M.E. (1968) Rates of organic matter decomposition measured by Warburg respiratory experiments. *Trans. 9th Int. Congr. Soil Sci. Adelaide* **3**, 203–9.

STOUT J.D. & O'BRIEN B.J. (1973) Factors affecting radiocarbon enrichment in soil and the turnover of soil organic matter. *Proc. 8th Int. Radiocarbon Dating Conf., Lower Hutt, New Zealand*, 1972. **2**, 394–407.

STOUT J.D., RAFTER T.A. & TROUGHTON J. (1975) The possible significance of isotopic ratios in palaeoecology. *Roy. Soc. N.Z. Bull.* **13**, 279–86.

STYLES J.H. (1967) Decomposition of *Pinus radiata* litter on the forest floor. Part 2. Changes in microfauna population. *N.Z. Jl Sci.* **10**, 1045–60.

SYERS J.K., ADAMS J.A. & WALKER T.W. (1970) Accumulation of organic matter in a chronosequence of soils developed on wind-blown sand in New Zealand. *J. Soil Sci.* **22**, 146–53.

TATE K.R. (1973a) Respiratory activity of soils irrigated by water and by meatworks effluent: a note. *N.Z. Jl agric. Res.* **16**, 385–8.

TATE K.R. (1973b) Radiocarbon dating in studies of soil organic matter—vegetation relationships. *Proc. 8th Int. Radiocarbon Dating Conf., Lower Hutt, New Zealand*, 1972. **2**, 408–19.

TATE K.R. (1974) Influence of four pesticide formulations on microbial processes in a New Zealand pasture soil. I. Respiratory activity. *N.Z. Jl agric. Res.* **17**, 1–7.

TAYLOR N.H. & POHLEN I.J. (1968) Principles and criteria of genetic classification. *Soils of New Zealand. Bull. Soil Bur. N.Z.* **26**, (1), 20–8.

THORNTON R.H. (1965) Studies of fungi in pasture soils. I. Fungi associated with live roots. *N.Z. Jl agric. Res.* **8**, 417–40.

WALKER T.W. (1968) Changes induced in the soil by Pastoral Farming. In: *Soils of New Zealand. Bull. Soil Bur. N.Z.* **26**, (1), 104–9.

WARDLE P. (1973) Native Vegetation. In: *The Natural History of New Zealand*, ed. G.R. Williams, pp. 155–69, A.H. & A.W. Reed.

WATERS R.A.S. (1955) Numbers and weights of earthworms under a highly productive pasture. *N.Z. Jl Sci. Technol.* **A36**, 516–25.

WHITE E.G. (1975) A survey and assessment of grasshoppers as herbivores in the South Island alpine tussock grasslands of New Zealand. *N.Z. Jl agric. Res.* **18**, 73–85.

WIDDOWSON J.P. & BLAKEMORE L.C. (1974) Available sulphur in a New Zealand soil climosequence. *Sulphur Institute Journal* **10**, (3–4), 8–9.

WILL G.M. (1966) Root growth and dry matter production in a high-producing stand of *Pinus radiata*. *N.Z. Forestry Res. Notes.* No. 44.

WILL G.M. (1967) Decomposition of *Pinus radiata* litter on the forest floor. Part 1. Changes in dry matter and nutrient content. *N.Z. Jl Sci.* **10**, 1030–44.
WILSON A. (1959) Organic nitrogen in New Zealand snows. *Nature* **183**, (4657) 318–19.
YEATES G.W. (1974) Studies on a climosequence of soils in tussock grasslands. 2. Nematodes. *N.Z. Jl Zool.* **1**, 171–7.

6 The role of termites (Isoptera) in decomposition processes

T. G. WOOD *Centre for Overseas Pest Research, College House, Wrights Lane, London W8 5SJ, U.K.*

Introduction

Termites are entirely terrestrial and are typically inhabitants of tropical, subtropical, semi-arid and, to some extent, warm temperate regions where they are often dominant members of the soil fauna. In considering their role in decomposition processes it is pertinent to remember that the vast majority of work on decomposition processes has been done in cool temperate regions and although a few species of termites occur in these regions they have little impact on decomposition. The accepted pathway of decomposition of plant debris in soils, which is based very largely on work in cool temperate regions (Burges 1967), can be summarized as follows:

1 Development of phylloplane microflora
2 Colonization by saprophytic micro-organisms
3 Comminution and ingestion by invertebrates, resulting in incorporation of organic matter into the soil, a great increase in its surface area but little chemical change
4 Microbial colonization and utilization of faeces and comminuted litter resulting in chemical degradation of tissues and production of complex phenolic polymers (humic acids)
5 Formation of stable organo-mineral complexes (humus).

This is, of course, a much oversimplified picture and the processes, particularly 2, 3 and 4 are concurrent rather than sequential. Nevertheless, several stages can be recognized in the decomposition of plant debris; the relative rates of these stages are biological phenomena and account for the amount, composition and distribution of organic matter (i.e. the humus form) in different soils. Undoubtedly similar processes occur in tropical, subtropical and semi-arid soils but in this paper I hope to show that where termites are abundant their activities are sufficient to generate additional pathways of decomposition quite dissimilar to the one summarized above. In order to appreciate this, some account must be taken of their feeding habits, digestive processes and social behaviour.

Feeding habits

The food of termites consists of plant material of all kinds (wood, grasses, herbs, roots), from living plants to decomposed plant remains mixed with mineral soil (i.e. soil organic matter). Their food and feeding habits have been recently reviewed (Wood, in preparation); here it is only necessary to indicate their feeding habits in relation to the stage of decomposition of the food.

(1) LIVING PLANTS

Relatively few species have living plants as their principal source of food. Certain species of *Coptotermes* habitually feed on living trees but most termites that consume living woody tissues (including roots) also have dead wood as a principal source of food. In this category are included the 'dry wood' termites (Kalotermitidae) and many polyphagous species such as *Mastotermes darwiniensis* Froggatt in Australia and *Psammotermes hybostoma* Desneux and *Macrotermes bellicosus* (Smeathman) in Africa. Living grasses are consumed by several genera of grass-harvesting termites (*Hodotermes*, *Microhodotermes* and *Trinervitermes* in Africa, *Syntermes* in South America, *Drepanotermes* and *Schedorhinotermes* in Australia and *Anacanthotermes* in Asia) and by certain Macrotermitinae (fungus-growing termites). Generally, the dry parts of grasses are consumed in preference to fresh, green tissues and dead grasses and herbs are also consumed.

(2) DEAD (STANDING) PLANTS

Dead plant material is attacked before it has fallen to the ground by several species in addition to those (see above) feeding on living tissues. By 'wasteful' feeding some of these species (e.g. some of the grass-harvesters) actually produce plant litter which they often leave lying on the ground. Dead tissues of living plants (e.g. bark of trees) are an important source of food for many species (particularly Macrotermitinae such as *Macrotermes* and *Odontotermes*) which construct runways and sheets of soil cemented with saliva over the trunks and branches of trees.

(3) DEAD (FALLEN) PLANTS

Fresh woody litter is consumed by a great variety of species and in some areas it is difficult to find woody litter that has not been attacked by termites. In Africa the principal consumers of woody litter are Macrotermitinae but in Australia, where this subfamily does not occur, this food niche is largely exploited by *Nasutitermes*, *Coptotermes*, *Amitermes* and *Schedorhinotermes*. Macrotermitinae are also, along with the grass-harvesting species (see (1) above), the principal consumers of grass litter in Africa; some, such as *Macrotermes* and *Odontotermes* also consume leaf litter. In Australia this niche is exploited by the grass-harvesters (see (1) above) and a great diversity of debris feeders, principally species of *Amitermes*.

(4) DECOMPOSING LITTER

Woody litter in various stages of decay is the principal source of food for many lower termites (particularly Hodotermitidae and Rhinotermitidae) and for higher termites (Termitidae) such as certain Termitinae and Nasutitermitinae. In some cases there is a relationship between the stage or type of decay and feeding by specific termites but the relationship is complex as fungi may produce attractive substances, break down repellent substances, produce repellent substances or aid in digestion (Sands 1969). Dung, which can be regarded as partially decomposed plant material, is consumed by certain termites, in particular by Macrotermitinae (*Macrotermes* and *Odontotermes*) in Africa and Asia and by a wide variety of *Amitermes* species in Australia (Ferrar & Watson 1970).

(5) SOIL

Soil-feeding has been adopted as a specialized habit by many species of higher termites (except Macrotermitinae). The majority consume topsoil rich in organic matter and there appears to be a clear demarcation between soil-feeders and those feeding on decomposing vegetation which is associated with morphological specialization of the mandibles (Deligne 1966; Sands 1965a). All the Apicotermitinae are soil-feeders as are many Termitinae and Nasutitermitinae.

(6) SPECIALIZED FOOD

Specialization of feeding habits appears to be rare but does occur. Examples are *Hospitalitermes* of Indo-Malaya which feeds on lichens (Kalshoven 1958), *Anoplotermes pacificus* Fr. Muller of South America which feeds on root apices of certain forest trees (Kaiser 1953) and *Amitermes hastatus* (Haviland) which feeds preferentially on partially decomposed stems of Restionaceae (Skaife 1955).

Of particular importance in termite colonies are the practices of coprophagy, oophagy, necrophagy, cannibalism and trophallaxis (stomodeal and proctodeal feeding). These practices result in the recycling and efficient utilization of organic matter and nutrients within the colony and thus determine the pathways and rates of return to the ecosystem.

It is immediately apparent that termites have the potential for consuming plant material before it has been attacked by saprophytic micro-organisms (categories (1), (2) and (3) above). In contrast, the vast majority of the invertebrate decomposers only consume plant tissues which have undergone some degree of microbial decay.

Feeding groups in natural populations

It is obvious that not only do many termites have a wide range of food sources but that they can play a role in decomposition processes from the initial stages of the degradation of living tissues to the final end product, soil organic matter. This is illustrated in Table 6.1, which shows the abundance and feeding habits of termites in an area of tropical savanna woodland in Nigeria (Wood *et al.* in press). Over half of the population is made up of several species of Macrotermitinae which consume standing dead trees and fresh litter to the extent that it is difficult to find woody litter which has not been attacked by one of these species. Other species of wood-, grass- and fresh litter-feeders make up most of the remainder of the population. This may account for the virtual lack of litter in the 'decomposing' category (4) above. The only species which exploit decomposing wood are *Amitermes evuncifer* and *Coptotermes*, neither of which are very abundant. There are only two relatively abundant soil-feeding species.

The relative importance of the different feeding groups varies with the habitat. There are no population data to compare with those in Table 6.1, but an indication of likely differences is given by data in Table 6.2 which illustrate the number of species of the major feeding groups occurring in different regions of West Africa from the semi-arid Sahel savanna to lowland Tropical Rain Forest. Although the intensity of collecting and the area

Table 6.1 Population density and feeding habits of termites in an undisturbed area of Southern Guinea Savanna Woodland near Mokwa, Nigeria (data from Wood, Johnson & Ohiagu, in press)

		Living Plant Tissues			Dead (standing) Plants				Fresh Litter		Soil	
Species	Numbers per m²	Wood	Grass	Roots etc.	Bark	Wood	Grass	Roots	Wood	Grass Herbs	Topsoil	Subsoil
WOOD–FEEDERS												
Coptotermes	—	++				++		++	+			
Amitermes evuncifer Silvestri	37			++		++		++	++			
Microcerotermes spp.	1,084	+		+		++		++	++			
Fulleritermes	—	+			++	++			+			
WOOD AND LITTER FEEDERS												
Macrotermes bellicosus (Smeathman)	19* (both species)	+		+	++	++	+	+	++	+		
M. subhyalinus (Rambur)			+	+	+	+	++	+	++	++		
Odontotermes smeathmani (Fuller)	232 (both species)					++						
O. pauperans (Silvestri)						++						
Ancistrotermes	857			+	+	++		++	++	+		
Microtermes spp.	820			+		++		++	++			
GRASS AND LITTER FEEDERS												
Trinervitermes geminatus (Wasmann)	148* (all five species)		+				++			+		
T. trinervius (Rambur)			+				++			+		
T. togoensis (Sjostedt)			+				+			++		
T. occidentalis (Sjostedt)							+			++		
T. oeconomus (Tragardh)										++		

Table 6.1.—*continued*

Species	Numbers per m²	Living Plant Tissues			Dead (standing) Plants				Fresh Litter		Soil	
		Wood	Grass	Roots etc.	Bark	Wood	Grass	Roots	Wood	Grass Herbs	Topsoil	Subsoil
SOIL FEEDERS												
Cubitermes	201*										++	
Basidentitermes	r										++	
Pericapritermes	r										++	
Promirotermes	r										++	
Noditermes	r										++	
Adaiphrotermes	70										++	
Astalrotermes	r										+	++
Anenteotermes	r											++

++Primary source of food r =rare species
+Secondary source of food —=no population data

* Data from soil cores; these species are mound—inhabiting and total populations exceed these figures. Provisional estimates are 680/m² for *Trinervitermes* (C. E. Ohiagu, personal communication) and 180/m² for *Macrotermes bellicosus* (N. M. Collins, personal communication).

Table 6.2. Feeding habits of termites in different ecosystems in West Africa

Locality	Latitude	Total Number of Species	Living Wood and Fresh Woody Litter	Grass and Herbaceous Litter	Decomposing Wood	Soil
Sahel savanna, Senegal[1]	16·5°N	19	11	8	0	3
Northern Guinea savanna, Nigeria[2]	11°N	19	9	11	1	5
Southern Guinea savanna, Nigeria[3]	9°N	23	10	10	2	8
'Derived' savanna, Ivory Coast[4]	6°N	36	18	11	3	13
Rain forest, Cameroun[5]	4·5°N	43	8	2	8	31

1. Lepage (1973). Total area: 100 ha.
2. Sands (1965b). Area I. Total area: 4 ha.
3. Wood, Johnson and Ohiagu (in press). Including primary and secondary savanna woodland. Total area: 7 ha.
4. Josens (1972). Including sparsely wooded and densely wooded savanna woodland. Total area: 2700 ha., but not including other vegetation types within this area.
5. N.M. Collins (personal communication). Total area: 1·5 ha.

searched was not constant in the five localities there is an obvious north to south increase in the number of species feeding on soil and decomposing wood, and few species feeding on grass and herbaceous litter in the rain forest. Obviously the impact of termites on decomposition processes will be very different in these contrasting areas with their major influence being during the initial stages of decomposition in the savanna and in the later stages of decomposition in the rain forest.

Rates of consumption

Laboratory measurements of food consumption have been carried out largely on wood-feeding lower termites (i.e. families other than Termitidae) and the figures range from approximately 10–90 mg g termites^{-1} day^{-1} (dry weight of food/fresh weight of termites) and vary according to the species of wood and species of termite (see review by Wood, in preparation). There are few figures for Termitidae but two independent estimates for *Nasutitermes exitiosus* gave values of 10·6 and 12·2 mg g termites^{-1} day^{-1}. All these figures were obtained using nonbreeding colonies and would probably be exceeded by breeding colonies where workers collect food for themselves and also for the dependent castes. The only figure for grass-feeders is 49 mg g termites^{-1} day^{-1} for *Hodotermes mossambicus* (Hagen) (Nel, *et al.* 1970).

Field estimates of consumption by populations of known size are virtually nonexistent. Ohiagu and Wood (in press) showed that in the latter half of the dry season daily rates of grass-harvesting by a population of *Trinervitermes geminatus* at a density of 200 m^{-2} (1·5 g m^{-2}) varied from 7·3 to 55·3 mg g termites^{-1} day^{-1} although much of this grass is stored and consumed during periods when there is no foraging. Lepage (1973) estimated that a population of *Macrotermes subhyalinus* of 25–30 m^{-2} (0·5 g m^{-2}) consumed 7·2 g m^{-2} year^{-1} of litter, a rate of 3 mg g termites^{-1} day^{-1}. Wood (in prep.) suggested a rate of 60 mg g^{-1} day^{-1} for Macrotermitinae based on field measurements by Lepage (1974) and Collins & Wood (unpublished data). These figures, which were based on measurements of removal of naturally available food in the field, contrast with the extremely high rates of consumption obtained by extrapolation from 'baiting' techniques (see discussion in Wood, in preparation) which measure the consumption of weighed amounts of food placed in the field without alternative food sources. Josens (1972) estimated that subterranean Macrotermitinae at a population density of 500 m^{-2} (0·6 m^{-2}) consumed 130 g m^{-2} year^{-1}, a rate of 594 mg g termites^{-1} day^{-1}. Similar high rates of removal (540–1,508 g m^{-2} year^{-1}) by *Pseudacanthotermes militaris* (Hagen) (population density not known) were obtained by Williams (1973). It is likely that results obtained by these baiting methods overestimate consumption rates by not allowing for a functional response to the provision of food and for seasonal variation in rates of foraging. Satchell (1967) noted similar discrepancies among estimates of food consumption by lumbricid earthworms.

The above figures refer to species feeding on living and fresh dead tissues. There is no information on consumption by species feeding on decomposing tissues. Indirect estimates of consumption by soil-feeders (*Cubitermes*) have been made on the basis of metabolic rate and assumed values of assimilation efficiency and amounted to approximately 2,760 mg g termite^{-1} day^{-1} (Hébrant 1970). Annual consumption of soil was calculated at 11,300 g m^{-2} by a population of 1·15 g m^{-2}.

Digestion of food

Digestion determines the extent to which food is chemically altered during its passage through the alimentary canal. Noirot & Noirot-Timothée (1969) noted that various workers had recorded the presence of a number of enzymes capable of breaking down disaccharides (see also Retief & Hewitt 1973), simple polysaccharides (e.g. starch), complex polysaccharides of plant (hemicelluloses, cellulose) and insect (chitin) tissues, and proteins. Termites derive most of their energy from the digestion of complex plant polysaccharides and the evidence reviewed by Noirot & Noirot-Timothée (1969) indicated that cellulases were produced solely by symbiotic intestinal protozoa in lower termites and probably by bacteria in higher termites (Termitidae). However, Potts & Hewitt (1973) showed that *Trinervitermes trinervoides* (Sjostedt), a member of the Termitidae, produced its own cellulase. Seifert & Becker (1965) and Lee & Wood (1971a) demonstrated that lignin was also degraded but the mechanisms are unknown. However, it is the overall quantitative changes that are of interest in studies of decomposition.

(1) QUANTITATIVE CHANGES

Assimilation efficiency has been recorded for several wood-feeding species. Among lower termites there are figures for *Kalotermes flavicollis* of 61–62% (Seifert 1962), 54–64% (Seifert & Becker 1965) and 55–60% (Noirot & Noirot-Timothée 1969). Seifert & Becker (1965) recorded figures of 86–93% for *Reticulitermes lucifugus santonensis* de Feyteaud and 62–69% for *Heterotermes indicola* (Wasmann). Among the Termitidae Seifert & Becker (1965) recorded 75–85% for *Nasutitermes ephratae* (Holmgren) and Lee & Wood (1971a) recorded 54% for *Nasutitermes exitiosus*. Nel *et al.*'s (1970) figure of 61% in *Hodotermes mossambicus* is the only record for grass-eating termites. There is no comparable information for species feeding on herbaceous litter, decomposing wood or soil although Hébrant (1970) assumed that the soil-feeding *Cubitermes exiguus* Mathot had an assimilation efficiency of 15%. There are also no figures for Macrotermitinae but it appears that the termite-fungus system results in almost complete utilization of food (see Return via faeces, below).

(2) QUALITATIVE CHANGES

Studies of the chemical changes occurring during digestion in wood-feeding species have largely been concerned with cellulose, hemicelluloses and lignin. Cellulose appears to be intensively degraded to the extent of 74–91% by *Kalotermes flavicollis*, 78–89% by *Heterotermes indicola*, 96–99% by *Reticulitermes lucifugus santonensis* and 91–97% by *Nasutitermes ephratae* (Seifert & Becker 1965). Hemicelluloses are also intensively degraded to the extent of 60–70% by *Kalotermes flavicollis* (Seifert 1962) with preferential hydrolysis of pentosans in heartwood and hexosans in sapwood. There is conflicting evidence on the extent to which lignin is degraded. For *Kalotermes flavicollis* Seifert (1962) obtained values of 3–4% but values of 2–36% were obtained by Seifert & Becker (1965) for the same species. The latter authors obtained values of 15–41% for *Heterotermes indicola*, 70–83% for *Reticulitermes lucifugus santonensis* and 42–52% for *Nasutitermes ephratae*. Seifert (1962) fed *Kalotermes flavicollis* on sound wood and on wood which had been exposed to a brown rot fungus, *Coniophora cerebella* Pers., for 28 days and found that although the latter wood contained less cellulose and more hexosans than sound wood, faeces derived from both types of wood contained similar amounts of cellulose, hemicellulose and lignin. The implication that decomposition due to a combination of termites and fungus is similar to that due to termites alone may not apply to the many termites that feed on decomposing wood but are incapable of attacking sound wood. However, there is no information on the digestive capabilities of such species.

Nel *et al.* (1970) made a more detained study (Table 6.3) of chemical changes during digestion in the grass-eating *Hodotermes mossambicus*. There was intense degradation of hexosans (90·7%) and pentosans (84·0%) but little change in lignin (0·3%). There is no comparable information for soil-feeders and Macrotermitinae.

The intense degradation of plant material by the species discussed above results in the production of faeces with distinct chemical properties (Lee & Wood 1971a, 1971b) and the most significant of these are shown in Table 6.4. Nitrogen content is low and C/N ratio high (Lee & Wood (1971a) recorded a value of 107·0) particularly in faeces of wood-feeding species.

Table 6.3. Utilization of redgrass (*Themeda triandra* Forsk.) by *Hodotermes mossambicus* (after Nel *et al.* 1970).

	Food		Faeces		Coefficient of apparent digestibility
	Dry wt. (g)	%	Dry wt. (g)	%	
Ash	3·96	12·08	4·04	31·60	−0·022
Lignin	5·69	17·36	5·67	44·30	0·003
Soluble carbohydrates	0·55	1·68	0·76	5·91	−0·374
Hexosans	12·34	37·66	1·15	8·99	0·907
Pentosans	7·10	21·66	1·14	8·87	0·840
Other solubles	0·36	1·11	0·10	0·80	0·819
Nitrogen	0·25	0·75	0·13	0·99	0·486
Unaccounted for	2·52	7·70	—	—	—
Total		100·00		101·46	0·609

Table 6.4. Composition (as % loss on ignition) of structures made from faeces by various Australian termites (data from Lee & Wood, 1971a).

	% N	C/N	% Polyphenolic material*	% Lignin	% Total Carbohydrate
WOOD-EATERS					
Mastotermes darwiniensis (Froggatt)	0·53	92·0	5·6	64·1	11·9
Coptotermes acinaciformis (Froggatt)	0·60	89·3	12·2	40·0	17·0
Coptotermes lacteus (Froggatt)	1·07	50·0	8·1	46·1	19·2
Amitermes sp.	0·98	50·0	11·1	41·8	8·2
Microcerotermes nervosus (Hill)	1·11	32·4	3·9	44·2	10·8
Nasutitermes exitiosus (Hill)	0·83	57·8	6·6	53·1	14·5
GRASS-EATERS					
Nasutitermes magnus (Froggatt)	2·16	21·3	10·7	47·9	15·1
Nasutitermes triodiae (Froggatt)	2·10	22·9	15·9	40·3	11·2
Tumulitermes pastinator (Hill)	2·17	22·1	11·7	41·3	12·1

* Alkali—soluble, acid—insoluble fraction of the organic matter.

Lignin content is high and carbohydrate content low, the latter consisting both of cellulose structurally combined with lignin and of undigested fragments of plant tissue. The lignin/carbohydrate ratios are considerably greater than those in wood (which has a ratio of approximately 0·5) and grass (approximately 0·25). The polyphenolic material has properties similar to polyphenolic materials (humic acids) extracted from soils (Lee & Wood 1971a, 1971b).

Macrotermitinae are a special case as all their excreta is used to construct fungus combs (see 'Return via faeces', below). Analysis of these structures by Becker & Seifert (1962) indicated lignin/cellulose ratios of 0·2–0·6, i.e. similar to that of fresh plant material, which suggests that Macrotermitinae cannot digest structural polysaccharides. This is supported by microscopic examination of fungus combs (Wood, unpublished) which show that coarsely comminuted plant materials, with visible cellular structures, are a dominant constituent in contrast to the amorphous appearance of the faeces of other termites (Lee & Wood 1971b). The fungus combs support a monoculture of *Termitomyces*, the species of fungus varying with the species of termite (Sands 1969). The fungus is assumed to degrade structural polysaccharides and the termites assimilate the breakdown products when they ingest the old portions of the comb. As the combs are eaten away they are either simultaneously built up on the upper surface or replaced by new combs in the space beneath. Josens (1971) showed that the combs of the subterranean *Odontotermes pauperans* had a turnover time of 5–6 weeks and those of *Ancistrotermes cavithorax* (Sjostedt) and *Microtermes toumodiensis* Grassé of two months. In addition to consuming the mature comb the termites crop the white spherules of fungus on the surface of the comb. Matsumoto (1974) showed that in combs of two species of *Macrotermes* these spherules contained 7·4–7·6% nitrogen, thus providing a nitrogen-rich diet. Matsumoto (1976) also showed that the C/N ratio and nitrogen content of these *Macrotermes* fungus combs was 25·5–28·3% and 1·5–1·7%, respectively. Thus, in order to maintain the C/N ratio at this level and balance the loss of nitrogen, the fungus must metabolize considerable quantities of carbon. The termites, therefore, will have to consume more food than is required for their own maintenance requirements, a fact which may lend some credibility to the high weight-specific rates of consumption obtained for Macrotermitinae in some field 'baiting' experiments (see Rates of consumption, above). A further factor with some Macrotermitinae (e.g. *Macrotermes bellicosus*) is that food is not consumed directly but some (or all) is finely macerated and stored in special heaps ('food stores') in the nest and is consumed after a period of conditioning. It is not known what proportion of the food collected reaches the fungus comb via the food store nor what proportion of the food in the food store is metabolized by micro-organisms

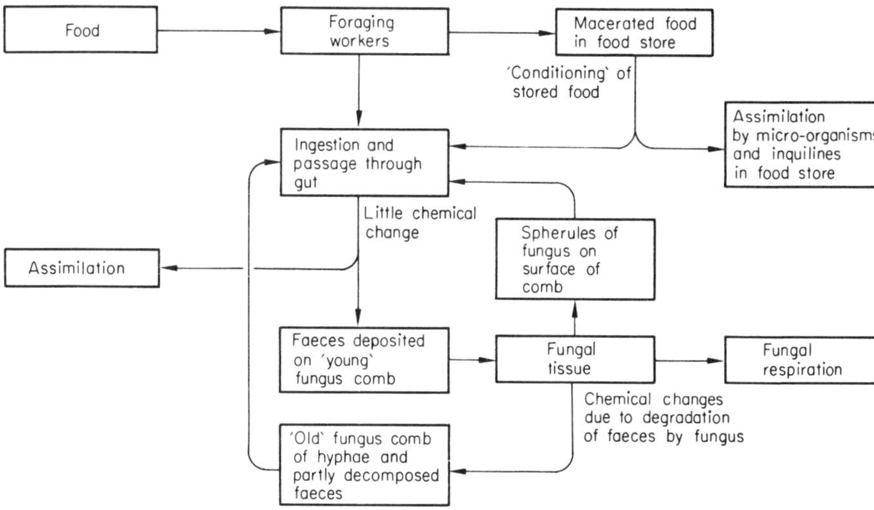

Figure 6.1. Utilization of food by Macrotermitinae; food stores are constructed only by certain species.

and invertebrate inquilines. These complex relationships involving utilization of food by Macrotermitinae are illustrated in Fig. 6.1.

Return of organic matter and nutrients to the ecosystem

Materials consumed by termites are subject to great physical and chemical changes in the alimentary system and are returned to the ecosystem either as faeces or as assimilated material, either salivary secretions (widely used for nest-building), termite corpses or via termite predators. The ways in which these returns are accomplished are largely dependent on the social behaviour of the termites themselves.

(1) RETURN VIA FAECES

Unlike other soil animals, which deposit their faeces within the soil where they are available to micro-organisms and coprophagous invertebrates, termites use their faeces for constructing certain parts of their nest-system. Before being used in this way the excreta may have passed through the alimentary system of several individuals by virtue of proctodeal feeding (lower termites) or reworking of structures made from faeces.

The central part of the nest-system (often referred to as the hive or habitacle) may be arboreal (within, or attached to the outside of, standing trees), epigeal (either in the form of mounds or within fallen trees or branches lying on the ground) or subterranean (the majority are in this category). From the hive a complex system of galleries and in some species epigeal runways lead to sources of food. All species, with the notable exception of the Macrotermitinae, use faecal material for lining the internal walls of their runways and galleries and many species also construct certain portions, or even the major portion, of the hive entirely from faeces or from a mixture of faeces and mineral soil. The Macrotermitinae appear to use all their faeces to construct fungus combs which may be located in the hive (e.g. *Macrotermes*, *Odontotermes*) or widely distributed throughout the soil (*Microtermes*, *Ancistrotermes*). The entire nest-system, the hive and associated galleries and runways, is maintained and protected against invasion of micro-organisms by intensive nest-sanitation and against other invertebrates (with the exception of inquilines) by defensive behaviour. Although Sands (1969) recorded the presence of certain specific saprophytic fungi in termite nests they probably have little effect on decomposition of faeces. Thus the faecal material is not available to other organisms until the nest-system, or parts of it, are abandoned. Peripheral galleries and runways are abandoned and new ones constructed as food supplies diminish and new ones are exploited, whereas the hive may be maintained for many years. Estimates of the longevity of colonies range from less than ten years for *Cubitermes fungifaber* (Sjostedt) (Noirot 1970), *Tumulitermes hastilis* (Froggatt) and *T. pastinator* (Williams 1968), 30 years for *Amitermes hastatus* (Skaife 1955), 50 years for *Nasutitermes exitiosus* (Ratcliffe *et al.* 1952) and more than 80 years for certain *Macrotermes* (Grassé 1949).

The only available information on decomposition of termite faeces is the work of Lee & Wood (1968, 1971a, 1971b) on the wood-eating *Nasutitermes exitiosus*. This species builds mounds which are constructed largely of faeces (the internal nursery) or a mixture of soil and faeces (the outer wall surrounding the nursery) which is capped with a thin external layer of soil. When the colony dies the soil capping is rapidly eroded by rain leaving the rest of the mound exposed. A comparison of some of the chemical properties of faecal material in occupied mounds with that in mounds which had been abandoned for 11 years showed that little decomposition of faeces had occurred (Table 6.5). The C/N ratio in the abandoned mound was high (>60), although slightly lower than in occupied mounds and the slightly greater methoxyl content in the abandoned mound indicated that some decomposition of less complex organic materials had occurred with consequent increase in the percentage of lignin. The structure of the outer wall and nursery of the abandoned mound was intact and there were some fungal hyphae, particularly in the basal portion in contact with the soil. It is possible

Table 6.5. Composition of occupied and abandoned mounds of *Nasutitermes exitiosus* near Canberra, Australia (adapted from Lee & Wood, 1968).

	% Organic matter	% Organic Carbon	% Nitrogen	C/N	% Alkali-soluble material[1]	% Methoxyl[1]
OCCUPIED MOUND						
Nursery	85	41	0·60	67·9	9·3	9·3
Outer wall	54	26	0·39	66·7	10·0	7·3
MOUND ABANDONED FOR 11 YEARS						
Nursery	70	36	0·58	61·5	16·9	10·5
Outer wall	62	30	0·49	60·9	11·1	9·6

1. Expressed as % of organic matter.

that the ultimate fate of the organic matter in these abandoned mounds is combustion in bush fires.

Further investigations (Lee & Wood 1971b) showed that the microbial population of the outer wall and nursery of occupied and abandoned mounds was very low. Total plate counts of bacteria gave populations ($\times 10^5$ g^{-1} dry weight) of 0·06 and 0·12 in the outer wall of occupied and abandoned mounds respectively and 0·005 and 4·07 for the nursery. There was some colonization of abandoned mounds by streptomyces ($\times 10^5$ g^{-1} dry weight: 0·001 in outer wall and nursery of occupied mounds, 0·35 and 7·0 in abandoned mound) and to some extent by spore-forming fungi ($\times 10^5$ g^{-1}: 0·02 and 0·006 in the outer wall and nursery of occupied mound and 0·08 and 1·16 in abandoned mounds). However, these organisms were largely inactive as respiration of samples from outer wall and nursery of both occupied and abandoned mounds was so low that rates could not be reliably measured. Lee & Wood (1971b) showed that inhibition of enzyme activity by compounds in faeces having properties similar to soil humic acids may partially explain the inhibition of microbial activity. Other factors possibly contributing to the slow rate of decomposition include low pH (3–5), high C/N ratio, high lignin and low carbohydrate content (see Table 6.4) and possibly concentration of resins, phenols and other resistant components of wood.

Many other termites (see Table 6.4) produce faeces with similar properties to those of *N. exitiosus* and also construct parts of their nest-system largely from faeces. Some of these structures can be expected to have similarly slow rates of decomposition. Others, such as those constructed by *Coptotermes lacteus* and *C. acinaciformis* have lower lignin/carbohydrate ratios (Table 6.4) and appear to decompose more rapidly (Lee & Wood, 1971b). There is no information on the decomposition of faecal material used to line walls of subterranean galleries. As indicated above, the faeces of Macrotermitinae appear to be used up entirely by the termite-fungus system. The only return via faeces to decomposer organisms in soil will be after colonies become moribund or parts of the nest-system containing fungus comb are abandoned.

(2) RETURN VIA SALIVARY SECRETIONS

All termites use salivary secretions to feed dependent castes (stomodeal feeding) and all (except possibly those nesting entirely within wood) use salivary secretions to cement soil particles together. The quantities used for the purposes of construction are unknown but must be considerable in the case of Macrotermitinae (particularly those species of *Macrotermes* and *Odonto-*

termes which build massive mounds with a net weight of several hundred kg ha^{-1}: Lee & Wood, 1971b, Table 6.3) which use saliva as the sole cementing medium. The incorporation of saliva contributes organic matter to the mounds and other structures built by Macrotermitinae and it is probably because of this that Boyer (1955, 1956) and Meikeljohn (1965) found that mounds of *Macrotermes* spp. contained considerably greater numbers of micro-organisms than surrounding soils. There is no information on the decomposition of saliva incorporated into mounds, covered runways and sheets or other structures.

(3) RETURN VIA CORPSES

The termite colony is virtually a closed system with unhealthy and dead individuals being consumed *in situ*. The major losses are via predation (see below) and the annual flights of secondary reproductives (alates) which are the potential founders of new colonies. The annual production of alates varies from low values of 1–5% of the population in *Trinervitermes geminatus* (Sands 1965c; Josens 1972) and 10% in *Cubitermes exiguus* (Bouillon 1970) to high values of 39% in *Zootermopsis laticeps* (Banks) (Nutting 1969) and 28–43% in *Odontotermes obesus* (Rambur) (Roonwall 1960). These numbers represent a proportionally higher biomass of the population. The figure of 10% for *C. exiguus* quoted above represents 22% by weight of the population. Lepage (1973) estimated that alates represented 56% of the biomass of the colony in *Macrotermes subhyalinus* and 59% in *Trinervitermes trinervius*. In addition the alates are rich in stored proteins and fats (alates of *Nasutitermes costalis* (Holmgren) have a calorific value of 6776 cal g^{-1} (1·6 J g^{-1}) (Weigert 1970). Alates therefore represent a considerable loss of energy to the colony. For instance, alates of *T. geminatus* representing 1% of the population contain 15% of the energy; net loss of energy in some species exceeds 50% (Josens 1972).

These vast numbers are released over very short intervals of time (flights often continue for no more than 1–2 hours) and only on certain favourable occasions during the year. Unlike faeces, which are deposited locally, alates are widely dispersed; some are consumed by predators (see below), others drop their wings, pair and burrow into the soil where, under conditions of stable population density, the majority perish. There is no information on the effect on soil micro-organisms of this sudden influx of winged termites.

(4) RETURN VIA PREDATORS

Winged termites are consumed by birds, lizards, ants and many other animals; this predation appears to be opportunistic (Nutting 1969) but has not been quantified. In contrast predation on colonies and foraging parties is often accomplished by specialized predators (Lee & Wood, 1971b, Chapter 4) and although there are no quantitative data large numbers of termites appear to be lost in this way. C. Longhurst's (personal communication) observations at Mokwa, Nigeria, on *Megaponera* sp. indicated heavy predation, largely on foraging parties of Macrotermitinae, on some days more than 1000 termites being collected by a party of 300 ants. Average rates of predation (mainly on *Odontotermes* spp.) were 0·8 termites m^{-2} day^{-1} over the period of several weeks. If this rate is continued throughout the year the annual rate of consumption (292 m^{-2} day^{-1}) would exceed the mean population of *Odontotermes* (Wood *et al.* in press) of 223 m^{-2}.

Discussion

There are no studies which provide reliable information on consumption, conversion and elimination of food by termite populations to enable the role of termites in decomposition processes to be quantitatively assessed. Nevertheless using some of the data presented in this paper and making several assumptions I have attempted to do this to provide a basis for considering the relevance of termites in decomposition processes. The population data are those of Wood *et al.* (in press) for Southern Guinea savanna woodland in Nigeria (see Table 6.1). Soil feeders have not been included as they are not numerous and their effects on decomposition are different to those species feeding on wood, grass and fresh litter. The calculated data and the basis for the calculations are shown in Table 6.6. The production/biomass ratios are from Lepage (1974) and Josens (1973) (see Wood & Sands, in prep.), but these are conservative estimates and may be exceeded in some situations: e.g. Josens (1973) estimated a value of 14 for *Ancistrotermes cavithorax* (Macrotermitinae). The figure for consumption from laboratory data (see 'Rates of consumption', above).

Total consumption of 168 g m^{-2} $year^{-1}$ is a significant fraction of the estimated grass production of 200–300 g m^{-2} (Wood & Ohiagu, in press) and litter fall from trees of 200–400 g m^{-2} (N. M. Collins, personal communication). It assumes even greater significance when one considers that much of the grass, leaf and fine woody litter is lost during annual bush fires. The figures indicate that returns to the ecosystem via alates and predation are approximately one fifth of the returns via faeces but could be considerably

Table 6.6. Food consumption, utilization and return to the ecosystem by termites feeding on wood, grass and fresh litter in Southern Guinea Savanna Woodland, Nigeria. (Population data from Table 6.1; biomass obtained by weighing individuals of each species.)

	Macrotermitinae	Others	Total
Numbers m^{-2}	1,928	1,540	3,468
$g\,m^{-2}$ (fresh wt.)	5·9	3·5	9·4
$g\,m^{-2}$ (dry wt.)	1·5	0·9	2·4
Consumption (g dry wt. m^{-2} $yr.^{-1}$)	129·2	38·3	167·5
Returns (g dry wt. m^{-2} $yr.^{-1}$) via:			
Faeces	nil	12·3	12·3
Alates and neuters	9·0	2·7	11·7

Calculation based on:
Consumption (laboratory data) = 30 mg (dry wt.) g (fresh wt.) $termites^{-1}$ day^{-1} for non-Macrotermitinae; 60 mg (d.w.) g (f.w.) $termites^{-1}$ day^{-1} for Macrotermitinae.
Assimilation efficiency (non-Macrotermitinae) = 68%.
Production/Biomass = 3·0 for non)Macrotermitinae, 6·0 for Macrotermitinae.

greater if higher values (which may have been more realistic) had been used for production/biomass of Macrotermitinae. Returns via faeces will be reduced by loss of faeces during bush fires.

There are no comparable figures for tropical forests where species feeding on decomposing plants and soil organic matter are relatively more abundant. However, soil-feeding termites may be unlike other termites but similar to other decomposers (e.g. earthworms, diplopods, insects etc.) in their effects on decomposition, as Hébrant (1970), on the basis of an *assumed* assimilation efficiency of 15%, indicated that *Cubitermes exiguus* (1·15 g m^{-2}) produced 9,600 g m^{-2} of faeces per year.

One is now in a position to compare decomposition as affected by termites with that in which termites are not involved (Fig. 6.2).

The first major difference between termites and other invertebrate decomposers is that considerable quantities of plant material are consumed by wood-, grass- and fresh litter-feeding termites before it has been attacked by saprophytic micro-organisms. This process is more significant in savanna regions, where these feeding groups are dominant, than in forests, where termites feeding on decomposing vegetation and soil become more significant.

The second major difference is that recycling via faeces is slow due to the fact that (1) the faeces are kept *within* the nest-system and only become available to the majority of saprophytic micro-organisms when part of or the entire nest-system is abandoned and (2) efficient digestion and assimilation (enhanced by coprophagy) produces partially humified faeces with certain properties (see Table 6.4) that inhibit microbial activity and result in

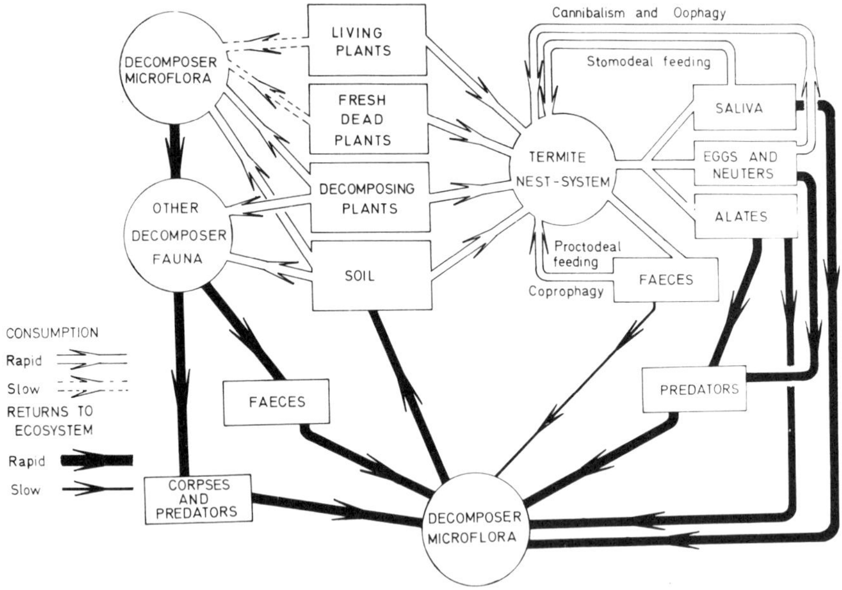

Figure 6.2. The influence of termites on decomposition processes contrasted with decomposition in the absence of termites. The diagram is designed for a population, such as the one enumerated in Table 6.1, where soil-feeding termites are scarce.

slow rates of decomposition. The ultimate fate of some faecal material, particularly that concentrated in structures made largely from faeces, may be combustion in bush fires. In the special case of Macrotermitinae there is little or no return via faeces. In contrast, other soil macrofauna and mesofauna return approximately 70–90% of their ingested food directly to the soil, immediately following passage through the alimentary system, where it is rapidly colonized and degraded by other decomposer organisms. Soil feeding termites may be similar to other soil macrofauna (see above) in having poor assimilation efficiencies but their faeces are, like those of other termites, used for construction purposes and nothing is known of their decomposition.

A third feature of termites is the extensive re-cycling (of dead, unhealthy and supernumerary individuals, salivary secretions, faeces) within the colony. This has often been attributed to the fact that termites need to make efficient use of the small amounts of nitrogen in their food. The net effect is that returns to the ecosystem via termite tissues are accomplished either by decomposition of alates which are released from the colony *en masse* over short periods of time or by predation on colonies, foraging workers or alates. Decomposition of alates has never been studied but can be expected to be relatively rapid, in comparison with decomposition of faeces, as can return via predation or salivary secretions. In the savanna ecosystem considered in Table 6.6, annual returns of dry weight through termite tissue and faeces are quantitatively similar but termite tissue, with a nitrogen content of 5·6–12·6% (Matsumoto 1975), is a more important pathway for nutrient re-cycling than faeces (nitrogen content 0·5–2·2%, Table 6.4).

It is not possible to say whether or not these processes have any significant role to play in determining the amounts and nature of organic matter in tropical soils but in savanna regions, where levels of organic matter are notoriously low (Jones 1973), they may be important determining factors. The effects of low returns via faeces are accentuated in situations where Macrotermitinae are the dominant group, as they are in many parts of Africa and Asia. The effects can also be expected to be greater in cultivated soils where accession of dead plant material (via crop residues) is low. Wood *et al.* (in press) showed that under continuous cultivation the majority of species present in undisturbed savanna woodland (Table 6.1) disappeared and in fact only two species survived in any numbers. The most abundant species, *Microtermes*, not only survived but increased in abundance to 4,000–4,500 m^{-2} (approximate dry weight biomass 1·4–1·6 $g\,m^{-2}$). As a member of the Macrotermitinae returns via faeces would be expected to near zero.

It may be possible to boost the levels, or at least to slow down the decline, of organic matter in cultivated tropical savanna soils by shifting the balance between Macrotermitinae and other termites in favour of the latter, particularly those feeding on decomposing vegetation and soil. This is also good pest management strategy as the majority of termites damaging crops in tropical savannas of Africa and Asia belong to the Macrotermitinae.

Summary

Termites are principally inhabitants of tropical, sub-tropical and semi-arid regions. They feed on all forms of plant material from living tissues to humidified material in soil. In savannas species feeding on living and fresh dead tissues are the dominant feeding group; they are less dominant in rain forests where there are many species feeding on decomposing plants and soil.

Species feeding on living and fresh dead tissues consume significant quantities of these substrates before colonization by saprophytic micro-organisms. They digest a large proportion of structural polysaccharides, assimilate 60–80% of the ingested food and use their excreta within the nest-system to build structures and line walls of galleries. In this form the faeces are maintained free of all but a few saprophytic micro-organisms and only become available to other decomposers when the structures are abandoned.

Certain structures may be occupied for several decades. After they are abandoned subsequent decomposition of faeces may be slow with relatively little chemical change being apparent after ten years; contributory factors are high C/N ratio, high lignin/carbohydrate ratio and the presence of 'humic acids' which inhibit enzyme activity. The fungus-termite symbiotic system in Macrotermitinae appears to result in complete utilization of ingested food with little or no return to the ecosystem via faeces. These processes are in sharp contrast to those in which terminates are not involved where invertebrate faeces, derived largely from the consumption of partially decomposed plant material which undergoes little chemical change during passage through the alimentary system, are deposited in the soil where they act as microsites of intense microbial activity.

The return to the ecosystem of organic matter and nutrients consumed by termites feeding on living and fresh dead plant tissues may be quantitatively as significant via salivary secretions, corpses and predation as via faeces. Although there is no information on these pathways of decomposition they are likely to be more rapid than returns via faeces.

Soil-feeding termites consume large quantities of humified organic matter mixed with mineral soil. Nothing is known of their assimilation efficiency but it is assumed to be low as they produce large quantities of faeces. In this respect they resemble other invertebrate decomposers although it is not known whether or not their excreta stimulate or inhibit microbial activity and therefore the subsequent rate of decomposition.

References

BECKER G. & SEIFERT K. (1962) Über die chemische zusammensetzung des Nest- und Galeriemateriels von Termiten. *Insectes soc.* **9**, 273–87.

BOUILLON A. (1970) Termites of the Ethiopian Region. In: *Biology of Termites*, eds. K. Krishna & F.M. Weesner, Vol. 2, pp. 153–280. Academic Press, New York and London.

BOYER P. (1955) Premières études pedologiques et bacteriologiques des termitières. *C.r.hebd. Séanc. Acad. Sci. Paris* **240**, 569–71.

BOYER P. (1956) Action des termites constructeurs sur certains sols d'Afrique Tropicale. *Proc. 6th Int. Congr. Soil Sci. Paris 1956* **3**, 95–103.

BURGES A. (1967). The decomposition of organic matter in the soil. In: *Soil Biology*, eds. A. Burges & F. Raw, pp. 479–92. Academic Press, New York and London.

DELIGNE J. (1966) Charactères adaptifs au régime alimentaire dans la mandibule des termites (Insectes, Isoptères). *C.r. hebd. Séanc. Acad. Sci. Paris* **263**, 1323–5.

FERRAR P. & WATSON J.A.L. (1970) Termites associated with dung in Australia. *J. Aust. ent. Soc.* **9**, 100–2.

GRASSÉ P.P. (1949) Ordre des Isoptères ou Termites. In: *Traité de Zoologie*, Vol. 9, ed. P.P. Grassé, pp. 408–544. Masson, Paris.

HÉBRANT F. (1970) Etude du flux energetique chez deux especes du genre *Cubitermes* Wasmann (Isoptera, Termitinae), termites humivores des savanes tropicales de le region Ethiopienne. These de Docteur en Sciences, Univ. Catholique de Louvain.

JONES M.G. (1973) The organic matter content of the savanna soils of West Africa. *J. Soil. Sci.* **24**, 42–53.

JOSENS G. (1971) Le renouvellement des meules à champignons construit par quatre Macrotermitinae (Isoptères) des savanes de Lamto-Pakobo (Côte d'Ivoire). *C. r. hebd. Séanc. Acad. Sci. Paris* **272**, 3329–32.

JOSENS G. (1972) *Etudes biologique et écologique des termites (Isoptera) de la savane de Lamto-Pakobo (Côte d'Ivoire).* These de Doctorat d'Etat, Faculte des Sciences, Bruxelles.

JOSENS G. (1973) Observation sur bilans energetiques dans deux populations de Termites à Lamto (Côte d'Ivoire). *Annls Soc. r. zool. Belg.* **103**, 169–76.

KAISER P. (1953) *Anoplotermes pacificus* einemit Pflanzenwurzeln vergesellschaftet lebende Termite. *Mitt. zool. Mus. Hamb.* **52**, 77–92.

KALSHOVEN L.G.E. (1958) Observations on the black termites, *Hospitalitermes* spp., of Java and Sumatra. *Insectes soc.* **5**, 9–30.

LEE K.E. & WOOD T.G. (1968) Preliminary studies of the role of *Nasutitermes exitiosus* (Hill) in the cycling of organic matter in a yellow podzolic soil under dry sclerophyll forest in South Australia. *Trans. 9th Int. Congr. Soil Sci. Adelaide, 1968* **2**, 11–18.

LEE K.E. & WOOD T.G. (1971a) Physical and chemical effects on soils of some Australian termites, and their pedological significance. *Pedobiologia* **11**, 376–409.

LEE K.E. & WOOD T.G. (1971b) *Termites and Soils.* Academic Press, New York and London.

LEPAGE M. (1972) Recherches ecologiques sur une savane sahèlienne du ferlo septentrional, Senegal: donnes preliminaires sur l'ecologie des termites. *Terre Vie* **26**, 383–409.

LEPAGE M. (1974) *Les Termites d'une savanne sahélienne (Ferlo Septentrional, Sénégal): peuplement, populations, consommation, rôle dans e'écosystèm.* These de Docteur en Sciences, Univ. de Dijon (France).

MATSUMOTO T. (1976) The role of termites in an Equatorial Rain Forest Ecosystem of West Malaysia. I. Population density, biomass nitrogen and calorific content, and respiratory rate. *Oecologia* **22**, 153–78.

MEIKELJOHN J. (1965) Microbiological studies on large termite mounds. *Rhod. Zambia Malawi J. agric. Res.* **3**, 67–79.

NEL J.J.C., HEWITT P.H. & JOUBERT L. (1970) The collection and utilisation of redgrass (*Themeda triandra* Forsk.) by laboratory colonies of the harvester termite, *Hodotermes mossambicus* (Hagen) and its relation to population density. *J. ent. Soc. sth. Afr.* **33**, 331–40.

NOIROT C. (1970) The nests of termites. In: *Biology of termites*, eds K. Krishna & F.M. Weesner, Vol. 2, pp. 73–125. Academic Press, New York and London.

NOIROT C. & NOIROT-TIMOTHEE C. (1969) The digestive system. In: *Biology of termites*, eds K. Khrisna & F.M. Weesner, Vol. 1, pp. 49–88. Academic Press, New York and London.

NUTTING W.L. (1969) Flight and colony foundation. In: *Biology of termites*, eds K. Krishna & F.M. Weesner, Vol. 1, pp. 233–82. Academic Press, New York and London.

OHIAGU C.E. & WOOD T.G. (in press) A method for measuring the rate of grass-harvesting by *Trinervitermes geminatus* (Wasmann) (Isoptera, Termitidae) and observations on its foraging behaviour in Southern Guinea. Savanna, Nigeria. *J. appl. Ecol.*

POTTS R.C. & HEWITT P.H. (1973) The distribution of intestinal bacteria and cellulase activity in the harvester termite *Trinervitermes trinervoides* (Nasutitermitinae). *Insectes soc.* **20**, 215–20.

RATCLIFFE F.N., GAY F.J. & GREAVES T. (1952) *Australian Termites.* C.S.I.R.O. Aust., Melbourne.

RETIEF L.W. & HEWITT P.H. (1973) Digestive carbohydrases of the harvester termite *Hodotermes mossambicus*: ∝-glycosides. *J. Insect Physiol.* **19**, 105–13.

ROONWALL M.L. (1960) Biology and ecology of Oriental termites. No. 5. Mound structure, nest and moisture content of fungus combs in *Odontotermes obesus*, with a discussion on the association of fungus with termites. *Rec. Indian Mus.* **58**, 131–50.

SANDS W.A. (1965a) A revision of the termite subfamily Nasutitermitinae (Isoptera: Termitidae), from the Ethiopian region. *Bull. Br. Mus. nat. Hist. Ent. Suppl.* **4**, 1–172.

SANDS W.A. (1965b) Termite distribution in man—modified habitats in West Africa, with special reference to species segregation in the genus *Trinervitermes* (Isoptera, Termitidae, Nasutitermitinae), *J. Anim. Ecol.* **34**, 557–71.

SANDS W.A. (1965c) Mound population movements and fluctuations in *Trinervitermes ebenerianus* Sjostedt (Isoptera, Termitidae, Nasutitermitinae). *Insectes soc.* **12**, 49–58.

SANDS W.A. (1969) The association of termites and fungi. In: *Biology of Termites* eds K. Krishna & F.M. Weesner, Vol. 1, pp. 495–524. Academic Press, New York and London.

SATCHELL J.E. (1967) Lumbricidae. In: *Soil Biology*, eds A. Burges & F. Raw, pp. 259–322. Academic Press, New York and London.

SEIFERT K. (1962) Die chemische veranderung der Holzzellwand-komponenten unter dem Einfluss pflanzlicher und tierischer Schadlinge. 4, Mitteilung: Die Verdauung von Kiefern- und Rotbuchenholz durch die termite *Kalotermes flavicollis* Fabr. *Holzforschung*, **16**, 161–8.

SEIFERT K. & BECKER G. (1965) Der chemische Abbau von Laub- und Nadelholzarten durch verschiedene Termiten. *Holzforschung*, **19**, 105–111.

SKAIFE S.H. (1955) *Dwellers in darkness.* Longmans Green, London.

WEIGERT R.G. (1970) Energetics of the nest-building termite, *Nasutitermes costalis* (Holmgren), in a Puerto Rican Forest. In: *A Tropical Rain Forest. A study of irradiation and ecology at El Verde, Puerto Rico.* ed H.T. Odum. Division of Technical Information, U.S. Atomic Energy Commission.

WILLIAMS M.A.J. (1968) Termites and soil development near Brock's Creek, Northern Territory, *Aust. J. Sci.* **31**, 153–4.

WILLIAMS R.M.C. (1973) Evaluation of field and laboratory methods for testing termite resistance of timber and building materials in Ghana, with relevant biological studies. *Tropical Pest Bulletin*, No. 3, Centre for Overseas Pest Research, London.

WOOD T.G. (in preparation) Food and feeding in Termites. In: *Production Biology of Ants and Termites*, ed. M.V. Brian, I.B.P. Synthesis Volume, Cambridge University Press.

WOOD T.G., JOHNSON R.A.J. & OHIAGU C.E. (in press) Populations of subterranean termites (Isoptera) in natural and agricultural ecosystems in Southern Guinea Savanna, Nigeria.

WOOD T.G. & OHIAGU C.E. (in press) A preliminary assessment of the significance of grass-eating termites (Isoptera) in pastures in Northern Nigeria. *Samaru Agric. Newsletter.*

WOOD T.G. & SANDS W.A. (in preparation). The rôle of Termites in Ecosystems. In: *Production Biology of Ants and Termites*, ed. M.V. Brian, I.B.P. Synthesis Volume, Cambridge University Press.

7 Allocation of resources. nutrient uptake and release by decomposer organisms.

P. DOWDING *Department of Botany, Trinity College, University of Dublin, Ireland.*

Introduction

The activities of decomposer organisms are thought to be a major pathway by which mineral nutrients elaborated into plant and animal material are returned to the soil/root system available for uptake by plants. During decomposition a large proportion of the plant carbohydrate is lost as carbon dioxide by microbial respiration.

The composition of the initial substrate has a profound effect on the type of colonization (Hudson 1968). Particular attention has been paid to carbohydrate status (Garrett 1962, Merrill & Cowling 1966). Garrett (1956) and Hudson (1968) both emphasize that colonization of plant parts begins long before they die. Early colonists must either be parasitic or 'commensal' to obtain sufficient carbohydrate and mineral nutrients. The 'commensal' phylloplane/rhizosphere floras depend on materials diffusing in solution, as do the balanced biotrophic parasites. The latter group have much more exacting nutrient requirements than the epiphytic organisms, if behaviour on artificial media forms an adequate basis for judgement. Two groups of colonists of dead plant material are thought to depend largely on 'soluble substrates'—primary and secondary sugar saprophytes. The remainder are capable of utilizing complex carbohydrates and as such are responsible for the disposal of the bulk of plant debris.

The study of nutrient requirements of those organisms, normally dependant on soluble sources of nutrient, on liquid or semi-solid media is a reasonable model of the natural situation. This is not the case with organisms which normally depend largely on cellulose, lignin or other complex substances. The behaviour of these organisms in a medium composed of soluble, evenly distributed, and diffusible nutrients may well be different from that on natural substrates.

There are further limitations to the applicability of results from liquid/agar cultures to the situation in plant litter or soil. The most important is the spatial concentration of each nutrient. Even if the concentration of the liquid medium is adjusted to the average, say, of wood (Park, Chapter 3), there is no way of producing local pockets of nutrients such as are found in medullary ray cells. Such local concentrations may well allow organisms to grow in habitats in which the *average* concentration of nutrients would be insufficient. The spatial localization in faecal pellets is discussed by Hissett & Gray, Chapter 2, and Hargrave, Chapter 13 (this volume), in soil and aquatic systems respectively.

The following account deals first with the natural substrate, and with the changes in nutrient amounts and concentration observed during decay. The second part deals with the behaviour of selected organisms in pure culture with regard to allocation of nutrients during growth in and 'decay' of artificial substrate. The gulf between the two is in part due to the basic differences in substrate type outlined above. In part it is due to the lack of widely accepted methods for separating microbial components from substrate components in opaque and physically heterogeneous media.

The substrate

Plant materials contain a variety of mineral nutrient sources. Nitrogen is mainly in protein form either as protoplasts or in the cell walls. After cell death it is usually bound up with oxidized polyphenols (Wood 1964). Smaller amounts of nitrogen are found in nucleic acids, membrane lipo-proteins, amino-sugars and lignin. During decay increasing amounts become incorporated into microbial protein, and into amino-polysaccharide cell walls.

All forms of nitrogen have to be digested by enzymes before they are available to saprophytic micro-organisms.

Phosphorus is present in membrane phospholipid, in nucleic acids, and as polyphosphates, particularly of inositol (phytic acid). Very little is present as inorganic phosphate (P_i) ion. All the organic forms (P_o) have to be degraded before assimilation.

Potassium and calcium contrast with N and P, in that both are mostly present as simple inorganic ions in the living plant. Potassium (K^+) is present in cell vacuoles, and calcium as insoluble calcium pectate in cell walls. A large proportion of the nutrients, particularly phosphorus, present in the live leaf may be back translocated (Rodin & Basilevitch 1967) before death to overwintering rhizomes or roots. The residue fraction in the leaf is unlikely to be readily available to micro-organisms.

The nutrients present at death will be competed for by micro-organisms, but soluble ionic forms will also be depleted by leaching. This type of loss must affect soluble ionic nutrients much more than those in insoluble or organic forms. At present it has not been quantified and, indeed, is extremely difficult to assess, since the amount of leaching will vary with both intensity and duration of rainfall as well as with the rate of drying and the nature and position of the materials subjected to leaching.

MOBILIZATION OF MINERAL NUTRIENTS

The nutrients sequestered by micro-organisms will at first be indistinguishable from substrate nutrients in total analyses (Fig. 7.1). However, microbial

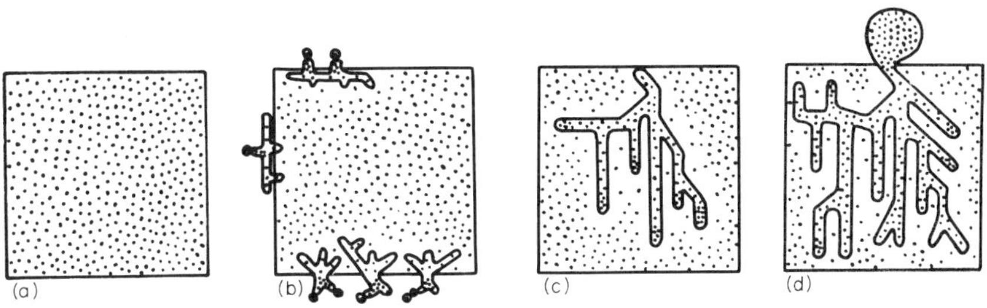

Figure 7.1. Changes in nutrient location with time in decomposing substrates.
(a) Original, uncolonized substrate with nutrient units as dots.
(b) Extraction of nutrients near surface only by rapidly sporulating surface colonists.
(c) Colonization by pervasive fungus with nutrient absorption by hyphae, but no overall change.
(d) Mobilization by pervasive fungus to massive exterior fruit body, with resultant drop in nutrient concentration in substrate.

activity is accompanied by respiratory loss of carbon as CO_2 and often consequent narrowing of carbon/nutrient ratios (Table 7.1). When nutrients sequestered are translocated outside the substrate via hyphae (e.g. mycorrhizal fungi) or by mycelial aggregations, or to fruiting structures on the surface, then the carbon/nutrient ratio will widen (Table 7.1). It is because these ratios are the resultant of two independent processes that they are un-

Table 7.1. Changes in Carbon/Nitrogen and Carbon/Phosphorus ratios with time in leaves of *Eriophorum* spp. at Barrow, Alaska and Moor House, U.K., and soil ratios at the same sites.

Year		0	1	2	3	4	soil
E. angustifolium	C/N	17·2	29·4	42	42		21
(Barrow)	C/P	500	417	454	999		5,520
E. vaginatum	C/N	50	42	33	28	25	53
(Moor House)	C/P	877	767	635	556	504	450

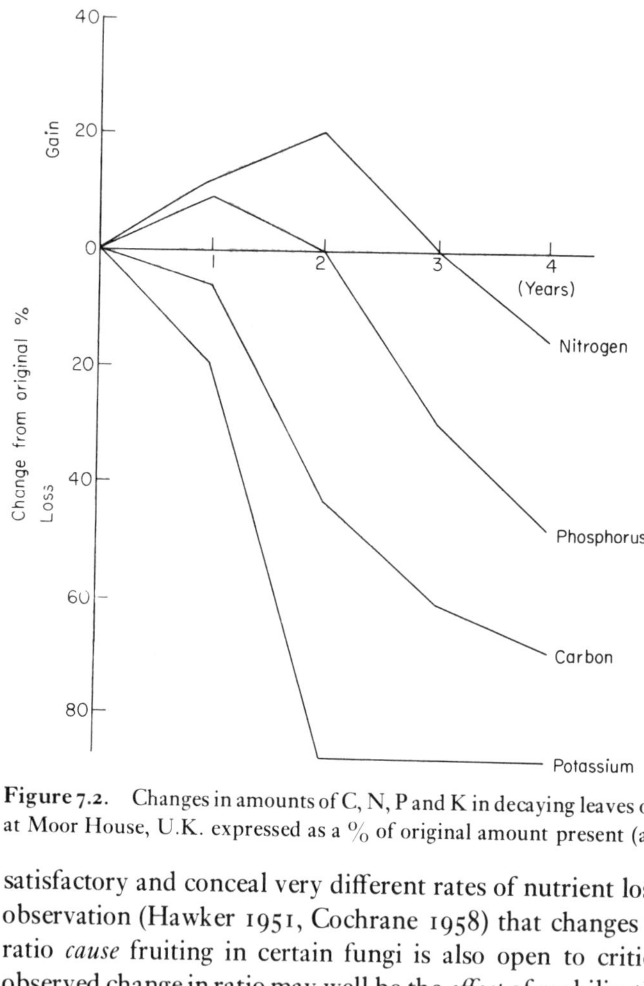

Figure 7.2. Changes in amounts of C, N, P and K in decaying leaves of *Rubus chamaemorus* at Moor House, U.K. expressed as a % of original amount present (after Dowding 1974).

satisfactory and conceal very different rates of nutrient loss (Fig. 7.2). The observation (Hawker 1951, Cochrane 1958) that changes in substrate C/N ratio *cause* fruiting in certain fungi is also open to criticism, in that the observed change in ratio may well be the *effect* of mobilization of protoplasm and stored nutrients to the (external) fruiting site.

Gotz *et al.* (1973) observed that in a mixed hardwood forest losses of nitrogen exceeded carbon losses when initial litter C/N ratios were below 20–30 ('critical point') and that litter C/nutrient ratios tended to stabilize near those of the underlying A horizon. The 'critical point' appears to be site specific but is not directly related to soil/nutrient ratio (Table 7.1) at Tundra IBP sites. Such equilibration of *ratios* is a result of so many different factors that its use creates more problems than it solves.

It has frequently been observed (e.g. Kendrick & Burges 1962) that the comminution phase of decomposition is delayed until some fraction of dry weight has been lost by microbial respiration. Increased digestibility is thought to be caused by higher nutrient concentrations, or to microbial metabolism of toxic or indigestible compounds. A coincident change in physical environment (the overlying layer of fresh litter) which allowed greater microfaunal activity could also be the cause.

Another effect of micro-organisms, that of concentrating the mineral nutrients from a relatively large volume of indigestible material into a digestible form (as spores/fruit bodies), has been noted by Leach (1940), Franke-Grossmann (1963) and Dowding (1973) among others. The destruction of hymenomycete fruit bodies by slugs and dipteran larvae is an example of the same phenomenon. In addition to concentrating the nutrients and presenting them in a more palatable form to saprovores, micro-organisms may also synthesize growth factors (Baumberger 1935; Leach 1940, Dowding 1967). This is certainly one of the advantages of the gut flora to the ruminant (Weller *et al.* 1958). Most spores that are not consumed before they are dispersed do not land on suitable host or substrate, and are washed out or fall out of the air on to the soil. They are there consumed by animals or other micro-organisms, so contributing to the biological cycle of nutrients.

Pure culture studies

Nutritional studies in pure culture have been largely conducted with liquid media (Burnett 1969; Cochrane 1958). The system has a number of advantages when compared with natural substrates (Table 7.2) but, with the

Table 7.2. A comparison of the attributes of liquid and natural solid media.

	Liquid	Solid
Spatially uniform	✓	
Separable from micro-organism	✓	
Definable	✓	
Substrate loss quantifiable	✓	✓
Sporing structures separable		✓
Sporing of higher fungi		✓

exception of commensal organisms is one rarely met with in nature. Hybrids, such as cellulose in mineral solution (e.g. Chang & Hudson 1967; Bhargava 1972) are frequently used, but suffer particularly from the inseparability of substrate and micro-organism. Another disadvantage shared by most artificial media, is their insufficiency as far as Basidiomycete fruiting is concerned.

YEASTS

Some of the earliest organisms investigated were yeasts (Prescott & Dunn 1959), both as sources of protein and of fat (Fig. 7.3), where either carbohydrates or nitrogen (as protein) were stored by the yeast cells if they were

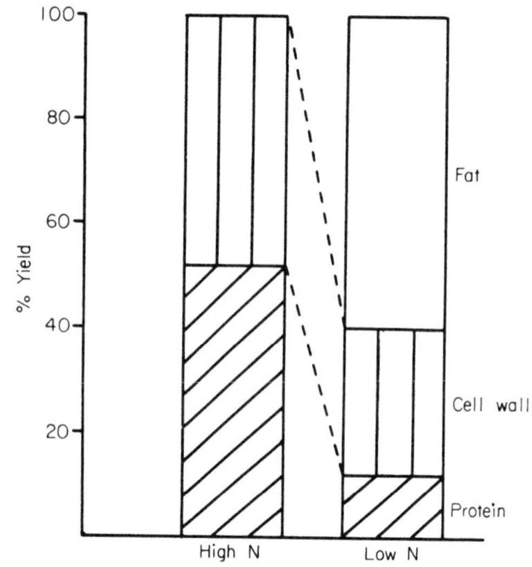

Figure 7.3. Crude fractions of yeast dry weight in protein, fat, and cell wall on media at two levels of nitrogen (after Prescott & Dunn 1959).

in excess of the other. Large amounts of nitrogen caused much greater utilization of the P and K in the medium than when nitrogen was limiting (Table 7.3). Yeast cells appeared to be able to modify nutrient ratios (Table 7.3), primarily by accumulation of mobilizable carbohydrate. Phosphorus and potassium were 'spread' more thinly when they were limiting growth (Table 7.3).

Table 7.3. Percent of nutrients in media taken up by yeast cells, and the final carbon/nutrient ratios of the cells, under two nitrogen regimes.

Nutrient	N	P	K
High nitrogen 73 g l^{-1}			
Percent uptake	66	100	100
Cell carbon/nutrient	6	26	24
Low nitrogen 10 g l^{-1}			
Percent uptake	45	53	25
Cell carbon/nutrient	25	56	48
Cell carbon/nutrient (if fat removed)	6	14	12

MYCELIAL FUNGI

Various forms of mycelial fungi have been investigated, and a network of tubes within which protoplasts could move freely would seem an ideal means of allocating scarce nutrient resources first to exploitation of the substrate, then to reproduction (Burnett 1969).

Among the fungi imperfecti, *Botrytis cinerea* Pers. Ex. Fr. was shown to produce sclerotia when nitrogen was limiting and conidia when N was freely available (Pieris 1947). The form of nitrogen was not important (Leach & Moore 1966). Sclerotial formation would appear to demand excess carbohydrate as it was not observed under limiting N conditions by Dowding & Royle (1971). They found that under low N (C/N=250) (Table 7.4) most of the nitrogen was assimilated in the first three days, but under high N assimilation was more gradual and was reduced if P was limiting. Phosphate uptake was far less efficient, but under limiting conditions little more was taken up by cultures over three days old. Phosphorus uptake was limited by low N levels. When either N or P was limiting the mycelium continued to assimilate carbon, though at a lower rate than non-limited cultures (Table 7.5). The carbon/nutrient ratios attained are strikingly higher for *Botrytis* than for yeast cells (Tables 7.3 and 7.5).

Table 7.4. Percent of nutrients in medium taken up by *Botrytis cinerea* after 3 and 14 days (amounts in mg/flask).

Nutrient regime		N		P	
		3 days	14 days	3 days	14 days
Low N (1 mg)	High P (2·5 mg)	80%	80%	2%	5%
High N (10 mg)	Low P (0·25 mg)	7%	30%	10%	11%
High N (10 mg)	High P (0·25 mg)	8%	80%	2%	70%

Table 7.5. Final (14-day-old cultures) carbon/nutrient ratios of *Botrytis cinerea* mycelia. (Nutrient regimes as in Table 7.4.)

Nutrient régime		C/N	C/P
High N	High P	20	200
High N	Low P	50	5,000
Low N	High P	170	1,750
Low N	Low P	160	5,100

Chalal & Gray (1968) grew a number of fungi on wood pulp amended with soluble nutrients, and measured nitrogen immobilization after five days' growth. *Myrothecium verrucosum* and *Rhizoctonia solani* took up 60% of the available N and *Polystictus versicolor*, *Chaetomium globosum* and a number of Fungi Imperfecti 40–50%. The amount of dry matter in the fungi was not determined. The net weight loss only gives an estimate of growth, assuming a metabolic efficiency of 50% (Burnett 1969) (Table 7.6). The estimated

Table 7.6. Weight loss of wood pulp (orig. 750 mg) after 5 days' decay by various fungi, estimated weight of fungus (assuming carbon assimilated = 50% carbon absorbed), and weight of nitrogen immobilized.

Fungus	Weight Loss (mg)	Estimated fungal mycelium (mg)	Nitrogen (mg)	C/N /(fungal)
Trichoderma sp.	298	150	8·6	27
Chaetomium sp.	190	95	10	9·5
Polystictus versicolor	260	130	8·6	15

C/N ratios are low compared with Levi and Cowling's (1968) for *Polystictus versicolor* of 73–278. The variation recorded by these authors is ascribed to adaption to extremely low substrate nitrogen. The age of the colony also has a great effect (Levi & Cowling 1969; Dowding & Royle 1971).

The commercial mushroom has also been studied by Treschow (1944) and Watson (1974). Watson (1974) found that mycelium in defined liquid culture could vary its N and P content adaptively (Table 7.7). Efficiency

Table 7.7. Carbon/nutrient ratios of 6-week-old *Agaricus bisporus* cultures on liquid media (n = 3), and efficiency of uptake from the medium.

		Mycelial ratio		Uptake efficiency/%	
	Medium ratio	C/N	C/P	N	P
Not limiting	C/N 20–100; C/P 50–200	15	541	31	2·2
P—limited	(C/P 400)	60	250	1	6·0
N—limited	(C/N 200)	40	417	37	1·0

of phosphorus uptake was much lower than for nitrogen uptake (Table 7.7) and nitrogen uptake was more affected by limiting P than vice versa. Most of the phosphate uptake occurred early with the exception of cultures given organic phosphates as a phosphorus source.

Experiments on mycelial cultures as food (Prescott & Dunn 1959) gave mycelia with C/N ratios of 60 and C/P ratios of 40 after a short growth period.

During the vegetative phase some of the nitrogen assimilated by the micro-organism has to be re-exported as enzymes, particularly if insoluble carbohydrates forming the carbon source. Levi & Cowling (1968) found that cellulase production by *P. versicolor* was reduced at low N concentrations, but not proportionately. They also found that there was accumulation of fat under these conditions (cf. yeast cells) and that almost all the available nitrogen had been taken up after 14 days' growth. The relative proportion of nitrogen in nucleic acids increased with decreased nitrogen concentration until 25% of nitrogen assimilated in nucleic acids at medium C/N ratios ⩾1,600.

ALLOCATION TO REPRODUCTION

The degree to which reproduction, or sporophore and spore production, acts as a drain on the mycelium increases with the complexity of the fruiting structures. This hypothesis has been used to explain the succession of fruit bodies on herbivore dung (Webster 1970) but does not explain succession on other substrates, except perhaps on wood (Chesters 1950). Even if the mycelium of fungi can adapt its N and P content to the exigencies of the medium it is unlikely that spore nutrient content can vary as widely. It would be far more plausible to adopt Salisbury's (1942) hypothesis, that for any species the size and content of sexually reproductive propagules are largely genetically fixed, and that the only adaptation possible in starvation conditions is to reduce the number produced by each individual. Even though they are asexual propagules, the majority of conidia would fall into this principle, since their size is constant enough for taxonomic purposes.

The nutrient content of spores is always high. Merill & Cowling (1966) report spore C/N ratios of 12 for *Fomes pinicola* growing on wood with C/N ratios >400. The C/N ratio of the fruit body was 50. C/N ratios for mushroom fruit bodies are reported as 8 (Prescott & Dunn 1955 and Watson 1974). Merill & Cowling (1966) estimated that to provide the nitrogen in the fruit body and in the spores produced each year, *F. pinicola* would need to extract *all* the nitrogen from 0·2 cubic metres of wood. Such total extraction is probably impossible, if only because some N will be left in hyphal walls.

Both *Agaricus bisporus* on compost (Watson 1974) and *B. cinerea* on agar (Dowding & Royle 1971) seem to have low efficiencies as far as turning substrate N and P into sporophore or spore N and P are concerned (Table 7.8). If the sporophores of *A. bisporus* had been allowed to age until spore production ceased and the spores had been trapped, the utilization might have been higher. It is remarkable that two very different fungi have a similar range of efficiencies.

The reproductive allocation of unicellular organisms, is of course identical to their vegetative allocation so long as the cell numbers are still increasing. These organisms do not have to invest capital in an exploitative

Table 7.8. The efficiencies of utilization (%) of nutrients in the medium in the production of sporophores by *A. bisporus* (on compost) and of conidia by *B. cinerea* (on agar media).

Nutrient status	*A. bisporus*		*B. cinerea*	
	N	P	N	P
Non limiting	15	8	11·2	5
N—limited	21	6	8·0	1·5
P—limited	12	10	3·3	4
N+P limited	—	—	5·6	4

transport network, and at the same time cannot capitalize on solid plant remains, as they cannot penetrate plant cell walls. Some unicellular organisms do manage to exploit solid substrates with the aid of animals to break open new surfaces, e.g. maggots in carrion (Dowding 1967), or bark beetle larvae in wood (Leach 1940).

Discussion

So far there is information on changes in carbon and nutrient contents of complex solid substrates inextricably infested with micro-organisms. The allocation by these organisms to reproductive structures appearing outside the substrate can be measured, if they can be easily detached. Reliable estimates of the allocation to vegetative growth, in solid substrates, either in pure culture or in competition are more difficult to obtain. Organisms which can 'stretch' their reserves furthest are obviously in a position to gain new resources from a greater volume of substrate than are organisms (such as *Penicillium*) which cannot allocate their nutrients this way. The inability to 'recycle' protoplast to growing tips and to translocate nutrients over considerable distances may both be additional features of primary sugar saprophytes (sensu Garrett 1956).

There are several possible solutions to these complex problems. All are necessarily incomplete, but, after the observational phase of investigation some more definite questions (hypotheses) need to be answered:

(a) 'During vegetative growth of wood rotting fungi nutrients taken up by the mycelium are translocated to the growing tips:'

Analyses of uncolonized, just colonized (ca 2–3 days) and long colonized wood, will show whether nutrient concentration on a **volume** basis changes with zone. It is to be expected that gravimetric concentration will increase as the carbohydrates are in part lost as CO_2 (Fig. 7.4) during decay.

(b) 'Extraction of substrate nutrients into fruit body and spores is never complete:'

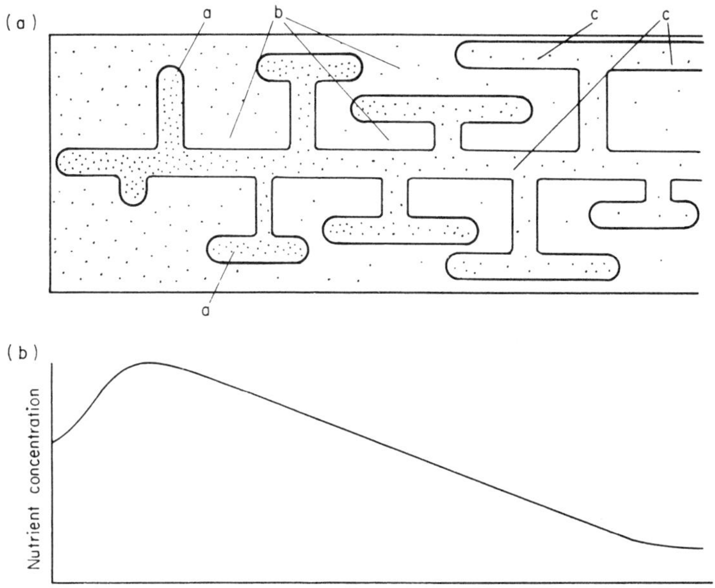

Figure 7.4. (a) Vegetative exploitation of solid substrate by a fungus showing:
a concentration of nutrients in the growing tips of hyphae, primarily in the advancing margin.
b absorption of substrate nutrients by the fungus.
c mobilization of nutrients within the fungus from older areas to the younger margin.
(b) The probable changes in nutrient concentration observed across the advancing margin of the colony shown in a.

Analysis of wood after fruit bodies have been matured (Hungate 1940) as well as during decay and before colonization will yield indices of extraction efficiency (e.g. Merrill & Cowling 1966). Cross-checks can be made by estimating the nutrients in the sporophore and in spores, if the production rate of spores is known.

(c) 'Competition lowers the extraction efficiency of individual species, but leads to more complete extraction of the whole:'

Herbivore dung appears to be the best substrate as the successions are well defined, as are the boundaries of the substrate. In addition there are sufficient nutrients for the production of basidiomycete fruit bodies. The hypothesis is supported by Webster's (1970) observation that the fruiting period of each component is shortened by competition, but not by Garrett's (1956) general hypothesis of nutrient limitation as one of the chief regulators of succession.

Nutrient loss from the substrate to fruit bodies could be estimated by either or both 'cropped' substrate and fruit body analysis. Pure cultures of components on sterilized dung would yield information on maximum extractive efficiency of each species, and of the rate of depletion.

Wood would also be a suitable substrate, but the units would have to be larger, necessitating subsampling. The size of unit would also be rather arbitrary and the time course of events much longer than on dung.

(d) '"Grazing" (by microfauna) (a) keeps the vegetative mycelium in an active state, (b) leads to a more rapid mobilization of nutrients and (c) slows decay (carbon removal) by fungi:'

These are all examples of effects the arthropod consumption of external fungal structures (usually fruit bodies) is supposed to have. The last is incompatible with the more widely held idea that microbial activity removes dry matter and possibly unpalatable compounds from freshly-dead material until it becomes palatable to the microfauna.

Simulation of grazing can be obtained by physical removal of external hyphae and fruiting structures. Depletion of nutrients and loss in dry weight of 'grazed' are then directly comparable with 'ungrazed' substrates.

More carefully controlled substrates can be built on an alpha-cellulose base. Known concentrations of more soluble carbohydrates and of mineral nutrient sources can easily be incorporated with reliable reproducibility. Hot alkali treatment removes fungal carbohydrates and leaves alpha-cellulose largely intact (Table 7.9). At early stages of growth, however, weight gains

Table 7.9. Weight loss of filter papers infected with *Chaetomium globosum* Kuntze ex Fries, hydrolyzed with 23M KOH.

Age of culture			
Nett weight change due to fungal activity (orig. wt. 300 mg)	+10	−26	−100
Subsequent weight loss on hydrolysis (e.g. fungal mycelium + non cellulose)	−16	−66	−104
Gross change (cellulose removal)	−6	−94	−204

of fungal material are larger than 'gross decay', and likely that either (i) the inoculum is exerting a sizeable effect or, (ii) enzymic activity has resulted in some short chain carbohydrates in the substrate which are soluble in hot alkali or, (iii) are caused by the rapid accretion of nutrients (10 mgN and 2·5 mg P per flask).

After four weeks the metabolic efficiency (weight gain/carbohydrate consumed) drops from 66% to 50%. The latter figure is more in line with published data (Burnett 1969).

Nitrogen and phosphorus analyses will yield data on the rate and extent of uptake of these two nutrients after the (original) soluble component has been washed out.

Insoluble sources N and P can be localized at particular spots on a cellulose matrix, and export from them traced by ordinary analysis, or by the use of radio tracers (cf. Watkinson 1971, for work with P_{32} in agar cultures). This type of system can be used to test the same hypothesis as was tested by analysis of wood before and during decay, only in a more controlled fashion.

The extent to which microbial decay of natural substrates is accompanied by the synthesis of e.g. essential amino acids (Weller *et al.* 1958) (Table 7.10) may be as important for subsequent microfaunal activity, as loss in dry weight or removal of unpalatable compounds. Techniques for amino acid analysis are now routine, but are confined to the nutritional/biochemical fields of study. Investigators concerned with the flow of nutrients through the soil biota may well gain some insight to the parts played by various

Table 7.10. Proportions (%) of food protein intake in various fractions in the rumen of sheep (Weller *et al.* 1958).

In bacteria or protozoa	63–83
In plant residue	10–27
As soluble N	5–10

Table 7.11. Weight of mycelium and weight loss of cellulose substrate after 6 weeks' growth of *C. globosum* at 2 levels of N and P (mg. dw).

Nutrient status		Weight mycelium	Cellulose removed
C/N 17	C/P 68	104	204
C/N 17	C/P 680	60	80
C/N 170	C/P 68	16	33
C/N 170	C/P 680	17	33

Table 7.12. Fungal taxa growing as well on sodium phytate as on K_2HPO_4

Alternaria
Aspergillus niger
Botrytis cinerea
Botryodiplodia
Chaetomium globosum
Epicoccum purpurascens
Fusarium oxysporum
Trichoderma spp.
Stemphylium
Agaricus bisporus (Watson 1974)
Aspergillus spp. (Casida 1959)

groups of organisms by the use of methods at present confined largely to dietary analyses of food for farm animals. Similarly, plant nutritionists are paying increasing attention to phosphate as a limiting mineral in ecosystems. It has already been demonstrated that phosphate status can have a profound influence on the nitrogen uptake and growth of fungi (Dowding & Royle 1971; Watson 1974) (Table 7.11). Weller *et al.* (1958) note that phytic acid restricts phosphorus uptake by the sheep and that microbial transformation could have a considerable effect on the type of phosphorus compounds entering the intestine. A wide range of fungi can utilize sodium phytate as sole phosphorus source (Table 7.12) and may, therefore, not only move phosphorus outside the substrate, but also present it to the microfauna in a more digestible form.

References

BAUMBERGER R. (1935) A nutritional study of insects with special reference to micro-organisms and their substrate. *J. exp. Zool.* **28**, 1–81.

BHARGAVE S. (1972) Cellulolysis of filter paper and wheat straw by foot rot fungi. *Trans. Br. mycol. Soc.* **58**, 348–50.

BURNETT J.H. (1969) *Fundamentals of Mycology*. Arnold, London.

CHALAL D.S. & GRAY W.D. (1968) The growth of selected cellulolytic fungi on wood pulp. In: *Biodeterioration of Materials*, eds A.H. Walters & J.J. Elphick, Elsevier, Amsterdam. pp. 584–93.

CHANG Y. & HUDSON H.J. (1967) The fungi of wheat straw compost I. Ecological studies. *Trans. Br. mycol. Soc.* **50**, 667–77.

CHESTERS C.G.C. (1950) On the succession of microfungi associated with the decay of logs and branches. *Trans. Lincolnshire Naturalists Union* **12**, 129–35.

COCHRANE V.W. (1958) *The physiology of fungi*. J. Wiley, New York.

DOWDING P. (1973) Effects of felling time and insecticide treatment on the inter-relationships of fungi and arthropods in pine logs. *Oikos* **24**, 422–9.

DOWDING P. (1974) Nutrient losses from litter on IBP Tundra sites. In: *Soil organisms and Decomposition in Tundra*, eds A.J. Holding, O.W. Heal, S.F.Jr. Maclean and P.W. Flanagan, Tundra Biome Steering Committee, Stockholm.

DOWDING P. & ROYLE M.C.I. (1971) The uptake and partitioning of nitrate and phosphate by cultures of *Botrytis cinerea*. *Trans. Br. mycol. Soc.* **59**, 193–208.

DOWDING V.M. (1967) The function and ecological significance of the pharyngeal ridges occurring in some cyclorhaphous Diptera. *Parasitology* **57**, 371–88.

FRANKE-GROSSMANN H. (1963) Some new aspects in forest entomology. *Ann. Rev. Ent.* **8**, 415–66.

GARRETT S.D. (1956) *The biology of root infecting fungi*. Cambridge University Press.

GARRETT S.D. (1962) Decomposition of cellulose in soil by Rhizoctonia solani Kuhn. *Trans. Br. mycol. Soc.* **45**, 115–20.

GOTZ J.R., LIKENS G.E. & BORMANN F.M. (1973) Nutrient release from decomposing leaf and branch litter in the Hubbard Brook Forest, New Hampshire. *Ecol. Monog.* **43**, 173–91.

HAWKER L.E. (1951) *The physiology of fungi*. Methuen, London.

HUDSON J.H. (1968) The ecology of fungi on plant remains above the soil. *New Phytologist* **67**, 837–74.

HUNGATE R.E. (1940) The nitrogen content of sound and decayed coniferous wood and its relation to loss in weight during decay. *Botanical Gazette* **102**, 382–92.

KENDRICK W.B. & BURGES A.A. (1962) Biological aspects of the decay of *Pinus sylvestris* leaf litter. *Nova Hedwigia* **4**, 313–42.

LEACH J.G. (1940) *The insect transmission of plant disease.* McGraw-Hill, New York.

LEACH R. & MOORE K.G. (1966) Sporulation of *Botrytis fabae* on agar cultures. *Trans. Br. mycol. Soc.* **49**, 593–601.

LEVI M.P. & COWLING E.B. (1968) The effect of C/N ratio of the substrate on the growth, composition, cellulose production and wood destroying capacity of *Polystictus versicolor.* In: *Biodeterioration of Materials*, eds J.H. Walter & J.J. Elphick, Elsevier, Amsterdam. pp. 573–83.

MERRILL W. & COWLING E.B. (1966) The role of nitrogen in wood deterioration. Amount and distribution of nitrogen in fungi. *Phytopathology* **56**, 1083–90.

PIERIS G. (1947) *The Botrytis disease of Gladiolus together with a physiological study of certain Botrytis species.* Ph.D. Thesis, University of London.

PRESCOTT S.C. & DUNN R.G. (1959) *Industrial Microbiology*, 3rd edn. McGraw-Hill, New York.

RODIN L.E. & BASILEVITCH N.I. (1967) *Production and Mineral cycling in terrestrial vegetation.* Oliver & Boyd, Edinburgh.

SALISBURY E.J. (1942) *The reproductive capacity of flowering plants.* G. Bell & Sons, London.

TRESCHOW C. (1944) Nutrition of the cultivated mushroom. *Dansk Botanisk Arkiv* **11**, (6), 1–180.

WATKINSON S.C. (1971) Phosphorus translocation in stranded and unstranded mycelium of *Serpula lachrymans. Trans. Br. mycol. Soc.* **57**, 535–9.

WATSON J.M. (1974) *The nitrogen and phosphorus nutrition of* Agaricus bisporus. M.Sc. Thesis, University of Dublin.

WEBSTER J. (1970) Coprophilous fungi. *Trans. Br. mycol. Soc.* **54**, 161–80.

WELLER R.A., GRAY F.J. & PILGRIM A.F. (1958) The conversion of plant nitrogen into microbial Nitrogen in the rumen of the sheep. *Br. J. Nutr.* **12**, 421–9.

WOOD R.K.S. (1964) *Physiological plant pathology.* Blackwell Scientific Publications, Oxford.

8 Species diversity and the structure of microbial communities in terrestrial habitats

M. J. SWIFT *Department of Plant Biology and Microbiology, Queen Mary College, University of London, U.K.*

Introduction

The last decade has seen an increasing involvement of microbiologists in integrated ecological research programmes such as those promoted by the IBP. As a result of this much firmer statements can be made than previously concerning the quantitative role of micro-organisms within natural ecosystems. The role of microbes in the energy transformations and nutrient cycling of terrestrial and aquatic systems are described elsewhere in this volume (Hissett, Ausmus, Dowding, Saunders, Fenchell). In addition it has been possible to establish more clearly the trophic relationships between micro-organisms and other components of the decomposer biota (Wiegert *et al.* 1970; Heal & MacLean in press).

In the majority of these studies, however, it has only proved possible to deal with the microflora as a single entity, or at best to subdivide it into its major components (fungi, bacteria, actinomycetes). Yet the microflora associated with dead plant or animal materials may be extremely diverse. For instance Novack & Whittingham (1968) described 161 species of fungi on the mixed deciduous leaf litter of a single woodland, Hayes (1965) found 120 species on needles of *Pinus sylvestris* and Frankland (1966) found 113 on petioles of *Pteridium aquilinum*. Comparable numbers of species have been recorded for other microbial groups and on other types of organic material.

The interactions between these microbial species and their patterns of distribution may be important factors in determining rates and patterns of decomposition. It is the purpose of this paper to review some of the current concepts of the functional roles of different components of the microbial communities associated with the decomposition of plant matter in terrestrial habitats and to propose some aspects of the structure of such communities. In doing this I shall draw mainly on information derived from patterns of distribution of species in space and time. Questions of species abundance, trophic relationships between species or their symbiotic or competitive interaction will only be touched on. Thus 'diversity' in the context of this paper relates usually to species number rather than carrying any implications of equitability or of abundance. In carrying out such an analysis a number of fundamental problems of microbial ecology must be considered.

THE DETERMINATION OF SPECIES COMPOSITION

Any attempt to discuss the species composition of a microbial community is fraught with dangers of many kinds. The methods of detection and determination of microbial species are all highly selective. The species list derived for any habitat will have two characteristics which are a consequence of these problems of method. It will be incomplete, and it will be biased in respect of some physiological feature selected for by the method. In addition most isolation techniques do not distinguish between the actively growing decomposer or the resting-spore of an 'invader'. These difficulties have been discussed many times and nowhere more informatively than by Warcup (1965). Qualification must thus be made to any generalization made in terms of a species list of micro-organisms.

In the bacteria and the actinomycetes the concept of what is a species is itself by no means a matter of general acceptance (e.g. see Gray 1969, Williams *et al.* 1969, Szabo 1974). The lack of readily recognizable morphological diversity has led to the utilization of a series of biochemical and physiological features to distinguish taxa. This leads to a merging of taxonomic status and ecological function which can be confusing in some circumstances, enlightening in others. The fungi have a clearer taxonomic structure—(but see Hudson 1970)—and readily gain major prominence in this discussion. Most of the examples are drawn from woodland habitats of temperate regions which lends potential consistency to the argument but introduces inevitable bias.

Despite these reservations and qualifications it is the thesis of this review that the patterns of distribution of microbial species in space and time are informative and provide a framework within which the functioning of individual species populations of decomposers can be instructively investigated.

THE QUESTION OF SCALE IN TIME AND SPACE

Analysis of the structure within communities depends on the recognition of pattern in the distribution of the component species in space and time.

It is a feature of all communities, but particularly so for microbial communities, that the scale at which its features are observed determines the patterns that are deduced. Burges (1960) emphasized how easy it was to be misled if techniques of investigation were used in a scale of dimension for which they were not devised. The integrated studies of the IBP have also exposed the dangers of losing sight of scale differences. The multiplication of data derived by microbiologists from a few grammes of soil, to derive data on a per hectare basis ignores all potential scales of pattern between these two and runs the risk of producing absurd results. Patterns of microbial distribution and activity may also be expressed over widely differing scales in time, ranging from the evolutionary development of the microflora characteristic of a particular soil, to the responses of populations to environmental fluctuations with scales in hours or minutes.

It is clearly of particular importance to the understanding of the role of micro-organisms at the ecosystem level that the scale in space and time of any described microbial process be fully appreciated. A structure for microbial communities can be obtained by describing the spatial patterns of species distribution and the ways in which they change with time.

Patterns of distribution in space

The spatial structure of decomposer communities is primarily determined by the nature of the food resources on which they depend. The heterogeneous nature of the primary food resource, the dead organic material of plant origin, accounts for a large proportion of the observed diversity in microbial species. This resource may be categorized in terms of the plant organ from which it is derived (leaves, stems, roots, fruits etc.) each with differing physical and chemical structures and distributions in time and space. These structures are sometimes described as 'substrates'; this usage however conflicts with the biochemical definition of a chemical entity that interacts in a specific manner with an enzyme. I would, therefore, propose that the term '*primary resource*' be used as the generic term at the morphological level. In the same way *secondary resources* may be used for materials originating from secondary production which become food resources for decomposer organisms. These include all dead animal and microbial tissues. Decaying vegetation quickly becomes a mixture of both primary and secondary resources; much of the faecal material of soil animals is an intimate mixture of both. Thus the distinction may quickly lose its practical usefulness in the face of an inability to distinguish the interwoven patterns of distribution. However, I would suggest that the distinction is useful conceptually and that many of the arguments applied below to the primary resources and their selective influence on the microflora may also be widened to include the secondary resources.

DISTRIBUTION OF MICROBIAL SPECIES IN RELATION TO PRIMARY RESOURCES

A comprehensive analysis of the microflora of decaying plant debris will yield many hundreds of species. An initial partitioning of this community can be readily made on the basis of specificity for the different types of primary resource. Thus there is a recognizably different fungal flora of leaves and branch-wood about which some taxonomic generalizations can be made; for instance the Aphyllophorales and Xylariaceous Ascomycetes commonly found on decaying wood are absent from leaf litter. The nature of this specialization is unclear; one possible relationship may be the size of the resource in relation to the energy demand for fruit-body production in these particular groups. This is emphasized further by the observation that the microflora of twigs may be very different from that of large branch-wood. However, whilst size may be a selection factor, other features may be of greater importance.

The selective effect of primary resource types on microfungi is illustrated by the work of Caldwell (1963); when segments of root and twig of beech of approximately similar dimensions were buried in the same soil they were colonized by different components of the mycoflora. Of the fourteen genera and seven species of microfungi identified in one experiment, nine genera and six species were resource specific. However, some overlap between different types of resources is common and many microfungi may be found on a variety of different plant tissues. This distinction between *resource-specific* (RS) and *nonspecific* (RN) species is also seen when any one type of resource is examined for microbial distribution patterns in relation to plant species of origin. Hering (1965) isolated or identified eighty-two species of fungi associated with mixed deciduous leaf litter. Details were given for the distribution of fifty of the species and these data are summarized in Table 8.1. The fungi were partitioned between the species of leaves in such a way that about half the recorded species were only found on one of the four leaf types. Conversely over 40% occurred on at least three of the four species of leaf. On this basis the fungal flora can be subdivided into RS and RN groups. This behaviour is not confined to the fungi; for instance Hissett & Gray (1974) found that of twelve species—groups of bacteria isolated from leaf litter in a mixed deciduous woodland, five were found on ash alone, two on oak alone and five were common to both.

Table 8.1. The distribution of species of fungi between four species of leaf (=four different primary resources) in the litter layer of an oak wood. Data recalculated from Hering (1965).

	Numbers of Species			
1. Total Species	50			
	Ash	*Birch*	*Hazel*	*Oak*
2. Distribution between resources	28	28	23	31
3. Resource specific (RS)	8	6	2	10
4. Total RS	26			
5. Resource non-specific (RN)				
(a) All resources	15			
(b) Three resources	6*			

*A further three species were found on two of the leaf types.

The basis of such resource-specificity is not well understood but is generally attributed to the presence of specific chemicals (= modifiers, see p. 193). For instance the distribution of wood decay fungi between different types of timber has been attributed to the presence of a variety of organic chemicals including terpenes and polyphenols (Hintikka 1970; Scheffer & Cowling 1966). Comparable evidence is not available for decomposer microbes on other materials. However, the selective effect of roots on the soil microflora is well documented and is attributed to the presence of specific substances in root exudates (Chan *et al.* 1963). As the organisms from the rhizosphere are among the primary colonisers and decomposers of roots (Dix 1964; Waid 1974) the selective activity of the living root may well determine the nature of the subsequent decomposer microflora. This effect may also be seen in other resources where the initial decomposer microflora often spans both parasitic and saprophytic phases.

The composition of the RN microflora deserves further mention. Many of these organisms have an extremely wide resource distribution. For instance of the fifteen species of fungi* found on all four species of leaf litter by Hering (see Table 8.1), seven were found on beech branch wood by myself,[1] eleven on *Pteridium* petioles by Frankland (1966),[2] six on needles of *Pinus sylvestris* by Hayes (1965)[3] and seven on the roots of various crop

*The list of fungal species found by Hering on all four types of leaf litter are as follows (the numbers refer to the references given above); *Botrytis cinerea* (1, 2, 3, 4); *Paecilomyces farinosus* (2); *Gliomastix murorum* (2); *Cephalosporium* sp; *Verticillium lecanii*; *Aureobasidium pullulans* (2, 3, 4); *Cladosporium herbarum* (1, 2, 4); *Epicoccum nigrum* (1, 2, 4); *Penicillium frequentans* (2, 3); *P. thomii* (1, 2); *P. coeruleoviride*; *Trichoderma viride* (1, 2, 3, 4); *Mucor hiemalis* (1, 2, 3, 4); *M. rammanianus* (1, 2, 3, 4); Sterile 1785.

plants by Parkinson *et al.* (1963).[4] Among the commonest of these species are various Mucoraceae such as *Mucor hiemalis* or Hyphomycetes such as *Aureobasidium pullulans*, *Epicoccum nigrum*, *Cladosporium herbarum* and various species of *Trichoderma* and *Penicillium*. As Hudson (1968) pointed out, the definition of the role played by these 'generalist' micro-organisms within decomposer communities is a problem of particular interest.

The 'unit-community'

Plant resources occur as a series of discrete 'units' of variable size and origin. In the case of woody materials, for instance, the units may vary considerably in size from small twigs to large branches or stems (Whittaker & Woodwell 1968). The distribution of the twigs and branches in the litter layer of a woodland is nonrandom in space and time, dependent on the spatial and age distribution of the vegetation from which it originates (Healey & Swift 1971). The decomposer organisms associated specifically with wood decay will therefore have spatial and temporal distribution patterns determined partially by the resource distribution. A different type of resource, such as leaf litter, may, in temperate woodlands with only one or a few tree species, and a much more restricted season of fall, show much less variation in size and more homogeneity in distribution. In other areas, tropical forest for instance, the variation may be just as great as for wood. Within this resource determined pattern there is also a considerable extent of secondary spatial heterogeneity in the distribution of the microbes. Fig. 8.1 shows data derived from an isolation study of fungi from 10 cm long samples of wood taken from beech branches of variable length and between 2 and 5 cm in diameter. The branches were selected from the litter layer of a woodland by a 'nearest neighbour' sampling technique (Healey & Swift 1971). 56 species were isolated from the total sample of 32 branches. 15 isolation plates were made for each branch. The distribution of species between branches shows a modal species number of four with a range from one to nine (Fig. 8.1a).

Thus each individual branch has a community of micro-organisms of varying diversity, comprising several species of fungi plus bacterial and other microbial components. Such partitioning could presumably be demonstrated for other resource types such as roots, fruits, or leaves. Such communities may be termed '*unit-communities*' and regarded as having a functional integrity. As such the unit-community resembles an 'association'. However, as the following discussion will show, the degree of constancy of the unit community in species composition may be low, which is in direct contrast to the implication of the 'association'. It should rather be regarded as a spatial unit within the whole microbial community structure which is de-

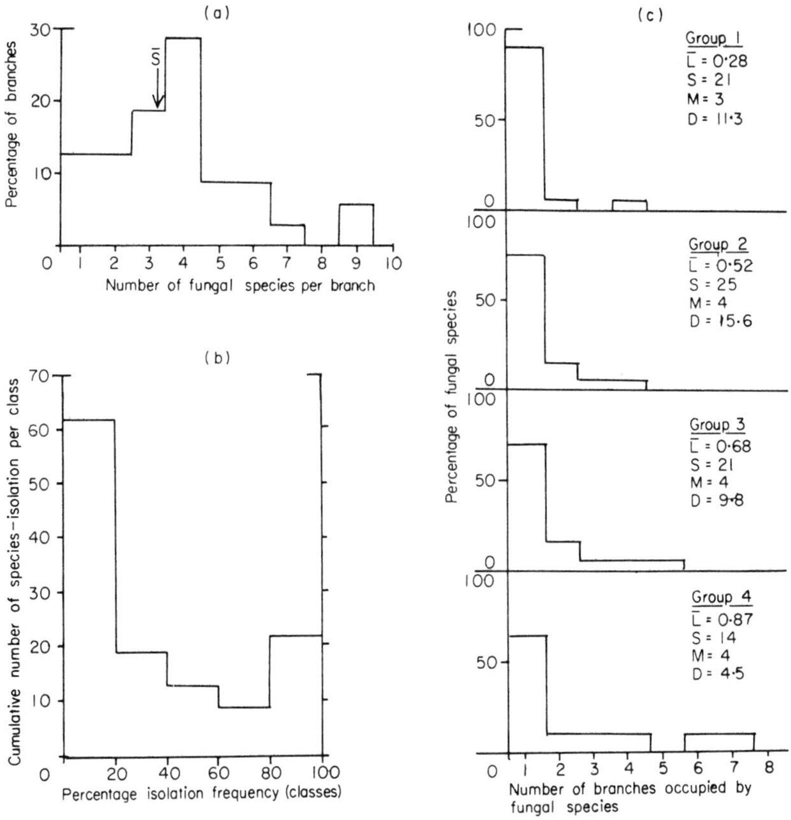

Figure 8.1. Distribution of species of fungi within a sample of branches of beech (*Fagus sylvatica* L.) taken from the litter layer of coppiced woodland. (a) Numbers of species isolated from each branch (Total number of branches, N=32; $\bar{S}$=Mean number of species per branch=3·9). (b) Distribution of species within branches (within unit-communities). For each of the 32 branches the frequency of isolation of each species was determined as the percentage occurrence on the (15) isolation plates for each branch. The species were grouped into frequency classes (X axis) and totalled for the full sample (Y axis). Thus for the full sample a total of 125 species isolations were made (=$\bar{S}$.N). The histogram thus gives a 'profile' of the species abundance pattern within an 'average' unit community branch. (c) Changes in the mycoflora during decomposition. Each histogram contains data for N/4=8 branches as sub-samples of the samples of (a) and (b). $\bar{L}$=Mean fraction of weight lost (compared with living beech wood on the basis of RD—see text); S=total number of fungal species from subsample; M=Modal number of fungal species per branch; D= diversity index=$1/\Sigma p_i^2$, where p_i=relative frequency of isolation of any species i (Simpson 1949; MacArthur 1972) in this case summed for all the species within each of the subsamples.

termined solely by the spatial, temporal and size class distribution of resource units. The unit-community provides a convenient basis for the analysis of community structure and function and will be the main concern of the remainder of this paper.

The number of microbial species within a unit-community may be relatively small. The data of Fig. 8.1a show a modal fungal species number of four. However, further examination of this wood decay data shows that a very considerable degree of variation exists *between* unit-communities. Two-thirds of the fungal species were isolated from only one of the branches. Only five fungi were isolated from five or more of the 32 branches; these included such RN types as *M. hiemalis* and *T. viride* and specialized wood decay (though not species specific) Basidiomycetes as *Stereum hirsutum*. The occurrence of regular associations between species was almost totally confined to that of the former two species. The major component of diversity in the fungal community of branch-wood is explained by partitioning between different species of resource or between units of the same resource. The features which may determine the partitioning of species between units are discussed in detail below; the question of the factors which determine the internal structure of the unit community are discussed in the next section.

THE INTERNAL STRUCTURE OF THE UNIT-COMMUNITY

Fig. 8.1b shows some aspects of the structure within the unit-community. The majority of species were isolated only infrequently and the second largest group of species were those that were isolated with high frequency (i.e. occurred on more than twelve out of fifteen plates) from the logs they inhabited. This distribution suggests that most of the microcommunities have single dominants associated with two or three rare species. Exceptions to this occurred in six of the 32 branches in which the two commonest RN species, *T. viride* and *M. hiemalis* were both isolated with high ($>50\%$) frequency. Thus in these cases alone something of a constant association can be said to occur. More will be said in relation to the role of these particular species later but on this evidence for wood-decay specific associations seem relatively rare. Only further work will reveal whether this is general, both in wood decay and for other resource types.

The co-existence of microbial species within unit-communities is dependent on the presence of a variety of niches which are determined by either chemical (substrate) factors or by factors of the physical environment. These factors determine how the capacities of microbes for decomposition are deployed.

Substrate affinities among decomposer micro-organisms

Plant materials, the primary resources of decomposer organisms, are complex both in their internal structure and their chemical composition. The structural complexity is due to the presence of different cell types and tissues within the resource. This may affect the distribution of the microflora in three ways: (1) The differing chemical composition of different cell types may have a selective effect on the microflora. (2) Different types of plant tissue create different physical environments again leading to differential effects on the microflora. (3) The differing chemical and physical nature of different tissues affect their palatibility to decomposer animals; the activity of these organisms may exert a strong selective and/or regulatory effect on the microflora. Only the first of these effects will be considered in this section.

The organic components of plant litter can conveniently be placed in three categories of differing significance to decomposer organisms; as sources of carbon and energy (*energy molecules*); as sources of elements other than carbon (*nutrient molecules*); or as molecules that have effects at relatively low concentrations either because they are essential to the growth of auxotrophic microbes or because they have an effect, stimulatory or inhibitory, on the activity of the microbes (*modifier molecules*). All three of these components may be distributed heterogeneously within primary resources and thus lead to spatial partitioning of the microflora.

Modifier molecules may have marked effects of this kind; many tannins and polyphenols which have an inhibitory effect on the activity of microbes may be localized in their distribution. For instance in wood they are in the highest concentration in the bark (periderm) and in the heartwood. This may be one of the features determining the existence of a distinct bark microflora (Bier 1963; Grosjean 1950). The suberins and cutins of external tissues also have selective effects on an invading microflora although ultimately they are metabolised as energy molecules (Martin & Juniper 1970). The distinction of bark and wood microfloras in woody tissues is also enhanced by the distribution of energy molecules such as sucrose or starch which have a storage or translocatory function in the plant. The distribution of these molecules is strongly correlated with that of living (parenchymatous) cells (Priestley 1962). The distribution of nutrient molecules, such as those containing N, also follows a similar pattern (Merrill & Cowling 1966). It is a feature of colonization by many wood decay organisms that they grow initially in the cambial and ray parenchyma. In some cases, e.g. the blue-stain fungi, growth may be limited to this area.

In most resource types the bulk of the energy molecules are found in the components of the cell wall. The structure of cell walls in different tissues

Table 8.2. Chemical composition of the cell walls of sycamore stem tissue, with particular reference to the polysaccharide components. Data re-calculated from Thornber & Northcote (1961a, b, 1962).

(a) Composition of different plant tissues (% dry weight)

	Outer Phloem	Inner Phloem	Cambium	Sapwood	Heartwood
Pectic substances	3·1	6·6	15·0	3·8	1·3
Hemicelluloses	21·3	25·8	45·0	31·5	31·4
α-cellulose	29·0	45·0	36·7	40·0	43·0
Lignin	46·9	22·6	2·5	25·3	24·3
Mean cell weight (mg)	—	—	15·3	157	171
Mean cell length (mm)	—	—	0·52	0·72	0·68

(b) Composition of different polysaccharide fractions in sapwood (% dry weight)

	Hemicellulose fractions				
	Xylan	Glucomannan	Hemicellulose I	Hemicellulose II	α-cellulose
Glucose	18·7	20·4	8·9	30·4	95·0
Galactose	0·0	0·0	2·8	0·0	0·0
Arabinose	Trace	0·0	1·6	0·0	0·1
Xylose	69·2	33·2	73·8	15·0	1·2
Mannose	0·0	39·8	0·0	51·0	3·5
Uronic anhydride	12·6	3·9	11·4	3·2	0·0
Methylgluronic anh.	12·2	0·0	0·0	0·0	0·0
Percentage of total fraction	48	2	42	8	—

of plant resources may differ considerably (Table 8.2a). Moreover, modern studies on the composition and structure of plant cell walls have revealed an extremely high degree of complexity at the molecular level. This exists not only in the internal complexity of many of the individual polysaccharides (Table 8.2b) but also in the physical arrangement and intercomponent binding of the polymers in the wall (Northcote 1972). Cellular differentiation and the deposition of additional polysaccharide and lignin, with a corresponding increase in the density of the wall (Table 8.2a) adds further elements of complexity.

The heterogeneity of distribution of cell wall types within plant resources creates the potential for a considerable extent of partitioning of the microflora on the basis of differing nutritional capacities and preferences. To what extent may this be realized among the decomposer microflora? Only a small group of organisms, the white-rot wood decay Hymenomycetes is capable of totally degrading this complex cell-wall material (e.g. see Findlay 1940). In order to do this they must possess a formidable array of enzymes; the polysaccharases which catalyse the depolymerization of cell wall components are highly specific in their action. So much so indeed that Albersheim *et al.* (1969) in a review of the enzymatic breakdown of cell wall components remarked that 'in all biology there is no molecular interaction more specific ... than the recognition of saccharides by proteins'. This specificity relates to the constituent sugars, to the position of the glycosidic bond(s) (e.g. 1—2, 1—3, 1—4, 1—6 etc.), to the steric relationship of the monomers (α or β bonds) and finally to the presence or absence of side branches or substituted groups (Reese 1968). This is illustrated in Table 8.3 for a brown-rot fungus *Coniophora cerebella* for which more depolymerizing enzymes have been positively identified than for any other decomposer microbe (King 1966). It should be noted that the extent of definition of the enzyme system

Table 8.3. Depolymerizing enzymes detected in cultural filtrates of *Coniophora cerebella* (King 1966; King & Fuller 1968).

Polysaccharide	Bond attacked	Polysaccharase	Glycosidase
Cellulose	Gβ1—4G	+	+
Laminarin	Gβ1—3G	+	
Xylan	Xβ1—4X	+	+
Glucomannan	Gβ1—4G	+	
	Gβ1—4M	+	
	Mβ1—4G	+	
Arabinogalactan	Galβ1—3Gal	+	
	Galβ1—6Gal	+	
Pectin	GalUα1—4GalU	+	

G=Glucose; X=Xylose; M=Mannose; Gal=Galactose; Gal U=Galacturonic Acid.

is dependent on the availability of substrate molecules to determine their action and it is the absence of the latter that is probably the main limitation to our description of the enzymatic capacities of decomposer organisms.

The most intensively studied of the polysaccharide-degrading systems, the cellulose enzymes of fungi, has revealed further levels of complexity. The catabolism of cellulase involves the participation of at least four different enzymes C1 ('hydrocellulase'), endo- and exo-β-glucanase (Cx components), and β-glycosidase, each catalysing a different step in the depolymerization of the molecule and its assimilation into the metabolic pool of the fungus. It is probable that most of the other polysaccharides require at least two of these enzyme types, the endo-polysaccharase and the glycosidase, for their extracellular catabolism although this has been demonstrated for only a few polymers (Table 8.3). The enzymatic degradation of lignin is not well understood but it is clear that several enzymic steps must be implicated in view of the variety of monomers and inter-monomeric bonds (Kirk 1971).

Most of the available data on the subject of the enzymatic capacity of decomposer micro-organisms relates to only a limited range of activities commonly including some measure of cellulolytic, pectinolytic and ligninolytic ability. Furthermore, the usual methods of testing do not necessarily correlate with the abilities of the organism in relation to the decomposition of plant tissues. For instance *Trichoderma viride* has been used in the laboratory as a producer of some of the most active of cellulase systems (Selby 1968). However, when inoculated on to plant materials such as oak leaves (Hering 1967) no cellulose breakdown was demonstrable although breakdown of hemicellulose components was detected. Three possible reasons exist for this type of conflicting evidence:

(1) The interactive effects of the cell wall components. It is known that the presence of lignin in the cell wall can have a 'masking' effect and slow the rate at which cellulose and other polysaccharides are degraded (Bailey *et al.* 1968). This emphasizes the multiple-enzyme nature of the degradative systems; the possession of a single enzyme, even one as fundamental as a cellulase, does not guarantee the decomposition capacity of the organism.

(2) The use of misleading tests in the laboratory. In the case of *T. viride* the existence of the C1 component of the cellulase complex is well established. However, much of the laboratory screening for cellulolytic ability in micro-organisms by ecologists has employed the use of degraded forms of cellulose such as filter paper. Whilst this establishes the capacity for polysaccharase production, the activity of the organism on intact cell walls may depend on the production of C1 fractions. On the other hand it should be noted that the absence of C1 does not necessarily mean that an organism will be unable to benefit from the capacity to produce Cx. Selby (1968) has demonstrated how the C1 and Cx components from two different organisms

may combine in the degradation of crystalline cellulose. Similar criticisms may apply to the use of the Bavendamm Reaction as a test for ligninolytic activity (Henderson 1960). This test involves distinguishing the capacity to colourize by oxidative action phenolic compounds such as gallic or tannic acids. The polyphenol oxidase enzymes involved in the oxidative reactions probably do not play any direct role in lignin depolymerization.

(3) The existence of genetic variation. The same species of fungus may embrace variants differing in one or more of these enzyme capacities. This feature may well explain some of the apparently contradictory reports on enzymic capacities found in the literature (e.g. see Hudson 1968).

For these reasons it is difficult at this stage to make any substantial generalizations concerning the distribution of enzymic capacities within the decomposer microflora. Two features do emerge from recent work however; firstly that as more investigation is carried out the more widely distributed these enzymic capacities are shown to be. For instance the earlier suggestions that there was a general taxonomic correlation between the absence of cellulolytic ability in the Phycomycetes and the confining of ligninolytic activity to the Basidiomycetes has now been broken down by an increasing number of exceptions (e.g. Park, Chapter 3; Henderson 1960). Secondly that there is considerable variation in the combination of capacities exhibited by any

Table 8.4. Degradative capacities of fungi isolated from soil or from debris of *Eriophorum angustifolium* or *Carex aquatilis* at Point Barrow, Alaska (Flanagan & Scarborough 1974).

Total species tested	35
(1) Individual Capacities	Number of species
Starch Hydrolysis	30
Pectin Hydrolysis	27
Cellulose Hydrolysis	7
Gallic Acid Coloration	14
Humic Acid Clearing	8
(2) Combinations of Capacities	
Total possible combinations	32
Number combinations observed	13

one organism. Table 8.4 illustrates this for a sample of litter and soil fungi from Tundra (Flanagan & Scarborough 1974). It was possible to subdivide the 35 species tested into 13 groups varying from those combining all activities to those with none. This implies a very considerable potential difference in the 'biochemical strategy' for the obtaining of energy and carbon on the part of decomposer fungi.

Table 8.5. The utilization of polymeric or simple C sources by bacterial isolates from two species of leaf litter (Hissett & Gray 1974).

	Percentage of isolates utilizing C source	
Source of Isolates	Oak	Ash
Pectin	1	0
Xylan	6	5
Cellobiose	36	45
Glucose	72	63
Xylose	76	69
Chitin	0	2
Glucosamine	3	36
Vanillic Acid	0	1

The capacity for depolymerization is probably less widespread in the bacteria and actinomycetes than in the fungi (Table 8.5). However, considerable variation does exist in the combinations of molecules that can be used by these decomposer microbes as is shown in the patterns of utilization of C and N sources by an actinomycete flora illustrated in Fig. 8.2. The number of different combinations of utilization was again considerable in these examples.

The potential for interaction of a synergistic kind between decomposers of differing biochemical potential has been mentioned above in connection

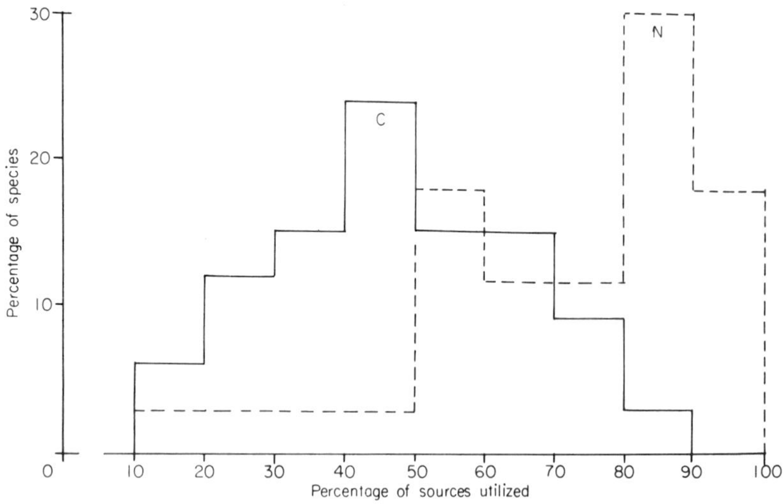

Figure 8.2. The proportion of simple carbon sources (n=22) and nitrogen sources (n=25) utilized by 33 species-groups of actinomycetes isolated from a woodland humus soil (Szabo 1974).

with depolymerizing reactions. A considerable extent of this may also exist through other biochemical functions. It is now accepted that the regulation of many depolymerizing enzymes probably operates through a system of catabolite repression (Lewis 1973). This has been demonstrated clearly for the cellulase system of *Myrothecium verrucaria* by Hulme & Stranks (1971) who demonstrated that concentrations above any but negligible levels of glucose, cellobiose, or other simple carbohydrates will repress the production of cellulase enzymes.

The activity of extracellular enzymes are outside the normal regulatory mechanisms of the cell. Thus it is entirely possible for the production of sugars to exceed the rate at which a cellulolytic organism can absorb and use them. This would be regulated in a mono-culture situation by the switching on and off of the cellulase production by the catabolite repression mechanism. However, the presence of other organisms utilizing the sugars may result in the continuing production of the polysaccharase. Garrett (1963) defined noncellulolytic organisms which grow at the expense of cellulosic substrates in this way as 'secondary sugar fungi'. There have been a number of demonstrations of the capacity of these organisms to benefit from the activity of cellulose decomposers (Tribe 1966; Hedger & Hudson 1974) and of the stimulation of the rate of cellulose breakdown by their synergistic interaction (Enebo 1963).

Clearly the potential for co-existence in a noncompetitive manner of microbial species by the adoption of differing patterns of substrate utilisation is considerable. It is also possible that within different unit-communities a species may adopt a different biochemical role, i.e. an organism that is cellulolytic in one community may act as a 'sugar-fungus' in another. It is probable that the interaction between species within unit-communities are symbiotic rather than antagonistic in nature, at least during the early stages of decomposition before nutrient sources become limiting. This may contrast with the low nutrient situation of soil environments where antagonistic interactions probably predominate (Park 1960). The potential for niche diversity created by substrate factors is broader than the above discussion implies for it has been confined to the carbon and energy obtainable from cell walls. The other constituents of primary resources (including nutrient molecules), add further substrate diversity as do the components of the secondary resources which become available during the later stages of decomposition. The dead tissues of fungi, bacteria and soil animals contain a variety of energy molecules in bulk which are not found in plant cells. For instance chitin is found in the cell walls of fungi and in the exoskeleton of many arthropods; the cell walls of all the microbes contain a range of other polysaccharides not found in plants; the storage polysaccharides of both animals and microbes are different from those of plants; the extracellular products of decomposition, including the components of the humus, also provide a potential food resource for the decomposer community.

THE INFLUENCE OF THE PHYSICAL ENVIRONMENT

The complexity of the features of the physical environment at the scale of the microhabitat and the selective effect that many of the features have on microbes in the soil have been reviewed in detail by Griffin (1972). Much less is known of the micro-environments of plant litter, and very little indeed in relation to the distribution of species components of the decomposer microflora (Williams & Gray 1974). Webster (1956) found that the diversity and composition of the fungal flora on stems of cocksfoot showed significant differences between internodes. Whilst nutritional factors played some role in this the moisture regime was demonstrated to be one of the major determinants (Webster & Dix 1960). Similar stratification of moisture levels within decomposing plant tissues, and associated differences in the partial pressures of O_2 and CO_2, may result in the spatial segregation of different components of the microflora. For instance bacteria may dominate waterlogged parts whereas fungi predominate in the more aerobic zones. This could occur where litter is in contact with very wet soils.

The definition of micro-environmental features of this kind within plant litters has been given little attention by microbiologists. The application of direct observation methods such as soil sections to the study of the litter horizons (Anderson 1971), has revealed the topographic complexity of these layers and indicates the potential diversity of physical environments within them. In particular the distinction between 'surface colonizers' and colonizers of the interior of plant resources which has been made in a number of studies (e.g. Kendrick & Burges 1962; Hogg & Hudson 1966) may well relate to adaptations to different physical environments.

Niche diversity and species diversity

In the preceeding discussion I have suggested that the microflora of decomposer habitats may exist in a mosaic structure, exhibiting patterns of distribution at varying scales which are determined by the spatial distribution of the primary resources and their components. Most of this pattern is undescribed and these suggestions are as yet hypothetical. The structure of experimental investigation has not been designed to describe it and such evidence as exists is incidental in nature (e.g. Fig. 8.1). In Fig. 8.3 the proposed structure is summarized in diagrammatic form. Features of structural

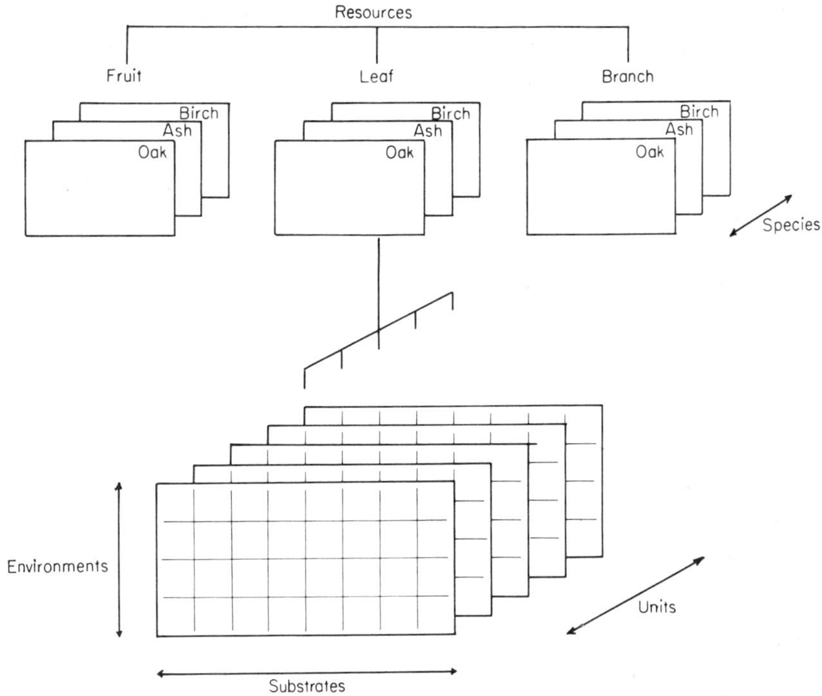

Figure 8.3. Diagrammatic illustration of the spatial partitioning of the decomposer microflora. Organisms occupying different boxes are regarded as existing noncompetitively (except possibly during colonization, see p. 213) those separated by lines may enter into competitive interaction if there is any significant reduction in the within-box area and heterogeneity (see also Fig. 8.6). Note that there is amplification of scale in the diagram between the resource and unit levels.

complexity additional to those illustrated are derived from primary resources such as roots and dead standing material which are spatially separated from the litter horizon, and from size variation in the primary resources. Any increase in the variety of the primary resources increases the heterogeneity, as does the variety of secondary resources. A considerable proportion of the partitioning of the microbial community is pictured as noncompetitive in nature. Organisms in different unit-communities are spatially separated (Fig. 8.3) and often may be separated in time as well, particularly in resources such as branch wood, where colonization may occur throughout the year (see below).

Within the unit-community the ability to co-exist depends on the continuing variety of available substrates and the complexity of micro-environments. The fundamental niches are largely defined by these two sets of parameters. Any decrease in the variety, quantity and accessibility of substrates and in the variety of micro-environments will lead to overlap in realized niches and an increase in competition between the micro-organisms occupying them. The development of this situation during the decomposition of a resource unit is discussed below.

A further prediction may be made concerning the niche structure of microbial communities—the existence of a large number of *equivalent species*. Within different units of the same resource the potential exists, because of their spatial and temporal discreteness, for extremely similar functional niches to be occupied by different species. The described structure of the community of beech wood is suggestive of this. It is possible to hypothesize that a Basidiomycete such as *Coriolus versicolor* in branch A and *Stereum hirsutum* in branch B occupy niches which are essentially definable by the same parameters e.g. cellulose, hemicellulose and lignin degrading capacity, tolerance of low O_2 and high CO_2 concentrations, demand for moisture contents above fibre saturation point. On large branches the two species may be found within the same unit; the similarity of niches is underlined by their spatial separation, often with the formation of interactive 'zone-lines' (Swift 1976). The explanation for the coexistence of such ecologically equivalent species probably lies in the circumstances of colonization of the resources. This may be a highly competitive process (e.g. see Rishbeth 1951) or the variety of successful species may be explained by the variety of colonization patterns available to micro-organisms.

I have suggested that a repeating mosaic of spatial heterogeneity determines the nature of the microbial community which is characteristic of any particular macrohabitat. Differences in the structure of this niche space are the factors which determine the differences in the diversity, composition, and structure of the microbial communities of different vegetation and soil types. Differences of this kind have been demonstrated in soil types both of close relationship (Christensen 1969) and of extreme difference (Sundman 1970). Attempts have only rarely been made to correlate these differences in microfloras with anything other than gross features of difference at the macrohabitat level and this is rarely informative. However, a factor which may prove useful in exploring the basis of such contrasts is the nature and distribution patterns of the primary resources within the habitat. The pattern of distribution of unit-communities may be a fundamental feature of community structure that is determined by the resource distribution.

Changes in community structure with time

The composition and structure of biological communities may be subject to change during the course of time. Such changes may be directional and permanent or they may be cyclical, involving not permanent change in one direction but a regeneration of structure in response to a fluctuating, and often seasonal, environment (Park 1968). Changes of both types are discernible within the decomposer microflora and it is worth considering what factors are involved in directing these changes in the light of the discussion above.

SUCCESSIONAL CHANGE

Successional changes in the vegetation of an area towards the establishment of a climax community bring about progressive changes in the physical and chemical environments which in their turn exert selective effects on the heterotrophic component of the community. Changes in the composition of the plant community have direct effects on the component of the microflora which is symbiotic with the plants. The RS component of the saprotrophic microflora will be similarly subjected to changing selection pressures. Thus increasing species diversity in the vegetation will lead to increased diversity in the plant-symbiotic and RS saprotrophic microflora. The increasing variety of the primary resources entering the decomposer system and the progressive accumulation of organic matter within the system will increase the amount and variety of substrates and the heterogeneity of the micro-environments available to the microflora. Thus it can be predicted that the RN microflora will also increase in diversity during the development of the primary macrosere.

Pugh (1962) has shown changes in fungal species composition and increases in species diversity during the development of vegetation on coastal muds and Brown (1958) has demonstrated the same for sand dunes but in general very little information is available on microbial successions of this kind. Hayes (1965) commented on the striking differences in the number of species isolated in his study of pine needle fungi in comparison to those found by other workers. He suggested that the differences in species number might reflect the differing ages and maturity of the forests in question. The effects on microbial diversity of the diversion of the primary sere by factors such as fire and cultivation has yet to be explored in a comprehensive manner by microbiologists but the indications are that such peturbations result in a decline in species diversity and a change in species composition (e.g. see Widden & Parkinson 1974).

Whilst the seral development of the decomposer microflora has not received a great deal of attention by microbial ecologists the cyclical changes in microbial community structure have been a favoured subject for study and speculation.

CYCLIC CHANGE

The progressive decomposition of a resource unit breaks down the niche structure that I have suggested exists within it. Microbial catabolism depletes the primary resource of both the amounts and diversity of substrates, although the production of secondary resources may to a certain extent compensate for the latter. The physical character of the material is altered by the 'skeletonizing' effect of microbial decomposition and even more drastically by the comminutive activities of soil animals. This results in the complete restructuring of the physical environment at the microhabitat level. It is probable that the nett effect is to increase the diversity of such micro-environments as the soil condition is approached. These changes in niche structure are not confined to the within-unit level; as decay procedes so both the physical and the chemical distinction between different primary resources declines so that within the humic materials (the *terminal resources* of decomposition) the origin of the components among the primary resources may be difficult to determine.

It has long been realized that such changes in the resources would have a profound effect on the inhabitant microflora. Garrett (1956, 1963) has pictured a succession of microbes progressively adapted to the changing chemical composition of the resource. Changes in the composition of the microflora during decomposition have been demonstrated for a variety of resource types and have been well reviewed by Hudson (1968) and others. The community change involved is essentially a cyclic one when considered from the scale of the decomposer habitat as a whole. A species which becomes inactive in a particular microhabitat survives by the re-colonization of a fresh resource; these microbes therefore exhibit behaviour typical of 'fugitive species' as characterized by MacArthur & Wilson (1967). The highly efficient dispersal methods shown by many of them is also characteristic of this pattern of colonization. As long as the supply of that resource is maintained, without change in its nature or amount, then the 'cycle' of each species can be maintained. Such changes are thus not directional or permanent except at the scale of individual units or microhabitats. Such a cycle will involve successive phases of colonization, exploitation, dispersal and re-colonization of the resource (Fig. 8.6). If a change in the resource supply does take place then the cycles will be broken and permanent directional change may occur. A number of the features of this type of cyclic change are reviewed here in relation to the illumination they cast on the postulated structure of microbial communities.

Changes in species diversity

An illustration of change in species diversity during progressive decomposition is given by the sample of decayed beech branches referred to earlier. The extent of decomposition of branch wood can be determined by measuring its relative density (RD) and comparing it with that of living wood (Healey & Swift 1971). The beech sample was found to show decay states ranging from 81% to 10% of the RD of undecayed branches. For further analysis the branches were grouped in four classes of equal size on the basis of consistency of state of decay. The data in Fig. 8.1c show that some quite marked differences in the structure of the microbial communities could be observed as decay progressed. The diversity of the macrocommunity (that of the total of branches in each of the four samples) decreased markedly in the terminal stages of decay. However, the size of the unit-communities remained constant at an average of four species per branch. The explanation for this apparent contradiction lies in the histograms of Fig. 8.1c. It is in the terminal stages of decay that the more frequent species, such as the *T. viride*/*M. hiemalis* association appear. Thus the effect of progressive decay is to bring about a decrease in diversity and an increase in homogeneity of species composition between different unit-communities.

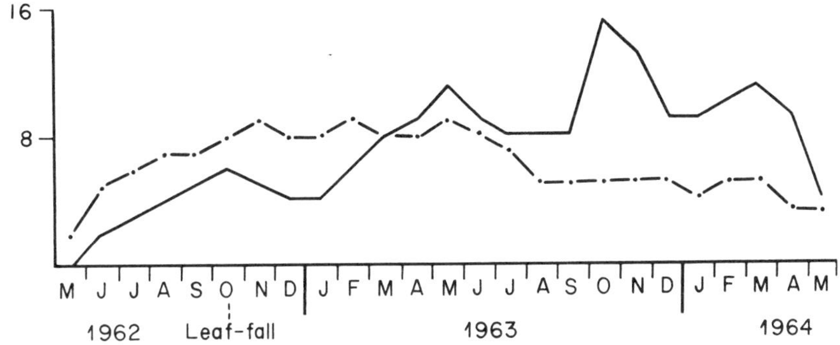

Figure 8.4. Variation in the number of species of micro-fungi recorded from monthly samples of twenty leaves of beech (*Fagus sylvatica L*). The leaves were placed in wire cages in the litter layer under a beech tree in 1962. The numbers of species recorded by direct observation are shown by the continuous line; the numbers of species isolated onto Potato-Dextrose Agar are shown by the broken line. Reprinted with permission from Hogg & Hudson (1966).

A similar decline in diversity is illustrated in Fig. 8.4 which is taken from a study by Hogg & Hudson (1966) of the succession of fungi on beech leaves. The number of species as determined by isolation techniques again shows a decline in the latter stages of decay. Note, however, the discrepancy between these results and those found by direct observation of fruiting.

If this decline in species diversity is a general feature then we should

Table 8.6. The numbers of species of micro-organisms isolated from leaf litter and different horizons of soil.*[1] In the case of bacteria and actinomycetes 'species' are groups determined as defined in the respective papers.*[2] The definition of soil horizons and the methods used to isolate organisms differed in the different studies so no vertical comparison of the data should be made.

Organisms	Dominant Litter Types	*[2]Litter	Humus	Soil	Reference
Fungi	*Eucalyptus maculata*	33	25	75	Eicker (1973)
Fungi	*Pinus strobus,*				
Fungi	*Acer saccharum,*				
	Ulmus americana	59	30	30	Widden & Parkinson (1973)
	Fraxinus pennsylvanica,				
	Quercus bicolor	161	96	—	Novack & Whittingham (1968)
Bacteria*[1]	*Quercus robur*				
	Fraxinus excelsior	12	11	10	Hissett & Gray (1974)
Actinomycetes*[1]	*Quercus robur*				
	Fraxinus excelsior	91	126	130	Williams *et al.* (1969)

expect to find significant differences between the number of species of fungi found in the successive horizons of soil. Table 8.6 records the results of a number of studies contrasting the species of micro-organisms isolated from leaf litter, humus, and mineral horizons of soil. The examples were chosen on the basis of internal consistency, i.e. the same workers using comparable methods at the same time on the same site. The decline in fungal diversity between litter and humus is confirmed although the increase in species number in the mineral horizon in the first example shows an interesting deviation from the general pattern. Further examples of declining diversity with depth have been shown for a variety of soils by Wicklow & Whittingham (1974).

An interesting feature of these results is the contrast in the behaviour of the fungi and the other microbial groups. Whilst the fungal diversity is generally lower in the organic and mineral horizons than in the litter layer that of the actinomycetes is considerably enhanced. The same may apply to the bacteria for Hissett & Gray comment that the heterogeneity of the humus layer bacteria was even greater than their results indicate because of a large proportion (56%) of isolates which did not associate with any of their defined groups. This is probably related to the much smaller size unit of the bacterial colony than the functional unit of the fungi which is the apical region of the hyphae. This probably confers great responsiveness to the developing complexity of the physical environment at the micro scale during the establishment of the particulate structure of soil.

Changes in species composition

The fact of change in species diversity and in particular that of change in the number of species implies that the species composition of decomposer communities varies as decomposition proceeds. This 'succession' involves the addition of new species to the community at specific times during the process of decomposition and the extinction of others at subsequent times. The demonstration of these events is by no means simple as the isolation data for two selected species from Hering's (1965) study of Ash leaf litter illustrates (Fig. 8.5). Both species are shown to be present throughout the decay period but show considerable fluctuation in the frequency of their isolation. The periodic absence of the species from the isolation plates can be attributed to a number of possible causes. (1) The sample taken at that particular time lacked any leaves with that particular fungus present, i.e. this is an effect of the distribution of species between 'units' of the resource. (2) Environmental factors may have differentially eliminated the species over this period but subsequent reinoculation re-established it. (3) The fungus

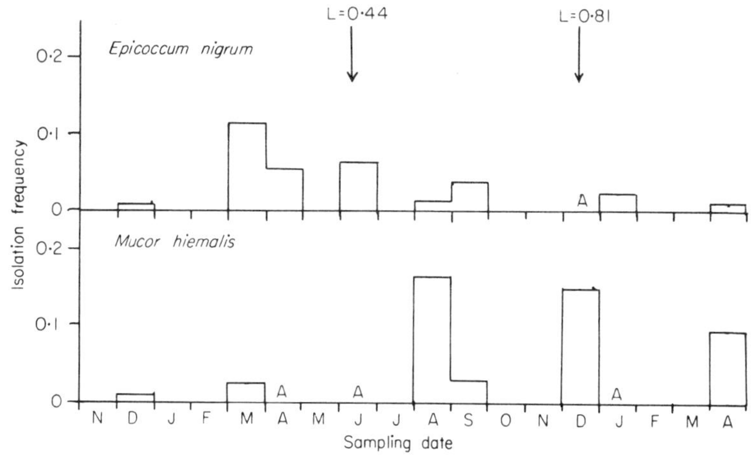

Figure 8.5. Contrasted patterns of change in relative abundance of two species of fungi during the decomposition of Ash leaves. Data redrawn from Hering (1965), Table 3, p. 396. Isolation frequency = Ci/Ct, where Ci = number of colonies of species *i* and Ct = total number of colonies for all species; A = species absent; L = fraction of weight lost by decomposition.

was present but failed to appear on the isolation plates, possibly because of competitive effects of other fungi on the agar medium.

However, despite the fluctuations of abundance from month to month a contrast in pattern between these fungi can be discerned. *Epicoccum nigrum* shows an early maximum in relative abundance which then declines whilst *Mucor hiemalis* shows a complementary pattern. The periods of high isolation frequency may be interpreted as representing periods of exploitation of the resource. The periods of low relative abundance may represent stages of either inoculation and colonization or terminal resting phases such as chlamydospores or sclerotia. In most recent work washing techniques have been employed to try and rid the resource of spores prior to plating-out so that it is probably reasonable to assume that isolation frequencies are proportional to present or past mycelial abundance. The direct observation of the organism fruiting on the plant material (Fig. 8.4) may confirm previous exploitative activity but it cannot establish the period or extent for which this activity has taken place.

Despite these difficulties of interpretation it has proved possible to build up a picture of the 'successions' of fungal presence and activity on a number of primary resources. Hudson (1968) has reviewed these studies and suggested that for many types of leaf and herbaceous litters some general correspondence of pattern could be discerned in the observed changes. The primary flora, originating on the senescent standing leaf, was comprised of weak parasites which are of course highly resource specific, and a number of RN '*primary saprophytes*'. Some of these originate from a persistent phase in the phylloplane of the living leaf and could therefore also be symbiotic. After litter fall this RN component, which includes such species as *Cladosporium herbarum* and *Epicoccum nigrum*, persists and increases in abundance. The RS component is added to by the appearance of saprophytic species whilst the parasitic component disappears. Maximum diversity coincides with the upsurge of a group of '*secondary saprophytes*' most of which are RN components of the 'typical soil flora'. These include such genera as *Penicillium*, *Trichoderma* and *Mucor* as well as the litter inhabiting Basidiomycetes. This community eventually replaces the earlier 'primary' flora. These changes are illustrated in Fig. 8.6.

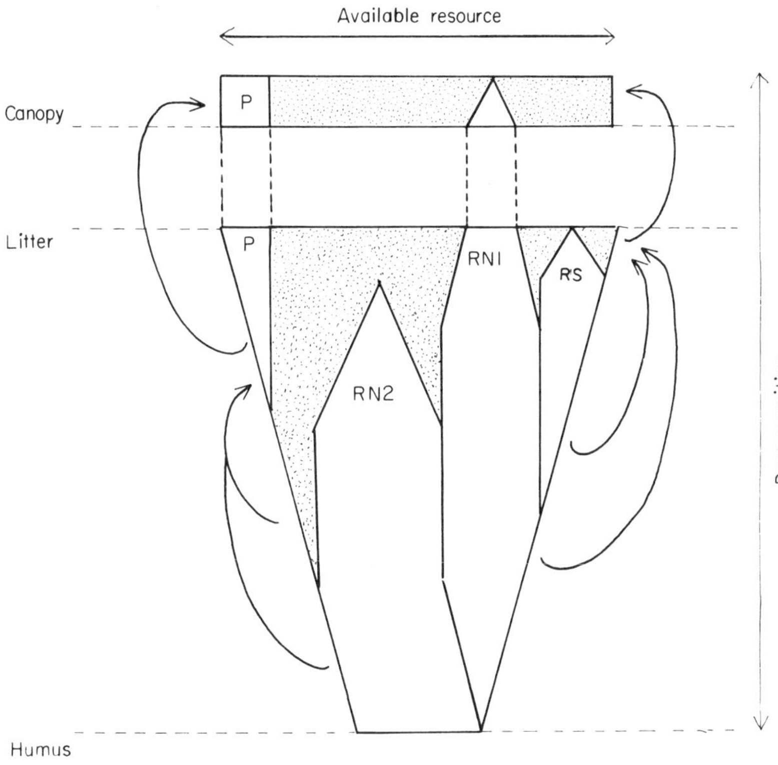

Figure 8.6. Diagrammatic representation of the cyclic changes in fungal community structure on an 'average' resource unit. Unstippled areas represent the extent of occupation of available resource by fungi. The arrows represent the dispersal of fungal propagules. P = parasitic species; RS = resource-specific saprophytes; RN_I = primary resource-nonspecific species; RN_2 = secondary resource-nonspecific species. Note that this 'succession' may be diverted if the resource is consumed by soil animals with consequent changes in pattern of colonization, chemical composition of the resource and particle size.

INTERPRETATION OF THE CHANGES IN DIVERSITY AND COMPOSITION

Succession in relation to the concept of the unit-community

In all the studies reviewed by Hudson, as with the beech-wood 'succession' reported here, the described changes in species composition relate to multiple-unit samples taken from the resource type in question; in no case is it possible to describe the changes that may occur *within* individual units or of the possible differences in pattern that may exist *between* different units. If it is accepted that it is at the unit-community level that the most significant interaction of species occurs, then it is clear that attempts to categorize species ecologically in terms of the times at which they occur in 'successions' which are based on analysis is of the macrocommunity may result in the comparison of organisms which have a markedly different significance in terms of the development of their own unit-community. To put it another way; whilst the 'succession' is conceived at a microhabitat level, its description is usually based on observations at a much greater scale. The one major exception to this has been the coprophilous 'succession' where the fruiting sequence on the same faecal pellet has commonly been observed through protracted periods of time. In this case it has clearly been demonstrated by Harper & Webster (1964) that all the member species of the unit-community are present from the initiation of the decay period; the sequence of fruiting observed is explicable in terms of differing physiological behaviour among the members of the unit community. Whilst the coprophilous 'succession' may be regarded as a special case—it is in effect a secondary or divergent succession of the sequence on the primary resource—the findings of Harper & Webster do urge caution in assuming the reality of sequences based on comparisons of samples of greater than single-unit size.

The nature of the communities at different stages of decomposition

Changes in species composition and diversity may imply changes in the functional attributes of the community and its component members. Garrett (1951, 1956, 1963) proposed that there was a sequence corresponding with the dominance of different 'substrate groups' of fungi; sugar fungi followed by cellulolytic fungi followed by ligninolytic fungi. He also recognized that sugar organisms might coexist with cellulose decomposers by benefiting from the excess sugars produced by extra-cellular enzyme activity. As mentioned earlier it has become increasingly difficult to define these simple substrate groups with any clarity. The correlation of organism sequence with

Table 8.7. Pattern of change in chemical composition of various resources during the initial stages of decomposition. (a) Comparison of the fractional loss in the components. (b) Comparison of the fractional contribution of each component to the total weight loss. + = apparent increase in component weight in comparison to original. Initial decay periods, 1 = 27 days, 2 = 66 days. Data re-calculated from Tenney & Waksman (1929).

	(a) Fraction of original component lost				(b) Fraction of total weight loss			
	Corn stalk[1]	Rye straw[2]	Oak leaf[2]	Alfalfa[1]	Corn stalk[1]	Rye straw[2]	Oak leaf[2]	Alfalfa[1]
Dry matter	0·36	0·17	0·22	0·30	—	—	—	—
Ether-soluble	0·30	0·53	0·26	—	0·02	0·06	0·04	—
Cold H_2O soluble	0·80	0·61	0·81	0·30	0·23	0·16	0·30	0·12
Hot H_2O soluble	0·55	+	0·46	0·26	0·05	+	0·12	0·04
Hemicellulose	0·31	0·22	0·27	0·49	0·20	0·27	0·16	0·14
Cellulose	0·44	0·24	0·34	0·41	0·35	0·54	0·21	0·36
Lignin	+	+	+	0·02	+	+	+	0·01

functional sequence is also unclear. Recently Pugh *et al.* (1972) have pointed out that the majority of Hudson's 'primary saprophytes' (RN1, Fig. 8.6), that are present on the leaf from early senescence possess some measure of cellulolytic and other depolymerizing activities. The most telling evidence against a nutritional interpretation, however, rests on the absence of correlation with the pattern of chemical change in the resource. One of the features that led to the derivation of the substrate group hypothesis was the observation that the initial rates of loss of different components could be clearly ranked; sugar>hemicellulose>cellulose>lignin. This feature, which has been discussed in some detail by Minderman (1968), can however be misleading in the present context. There is surprisingly little detailed data from which the comparison can be made but it does generally hold. This is shown in Table 8.7a for the classical studies of Tenney & Waksman (1929). Here three out of the four types of resource show a much higher fractional rate loss in the water-soluble (sugar) components than in the cell wall polysaccharides; the breakdown of lignin in all cases is negligible or undetectable. However, if the same data are recalculated (Table 8.7b) to show the proportional contribution to total weight loss of the weight lost in each chemical fraction, then it can be seen that the absolute quantities of cell-wall polysaccharides metabolized during this initial period exceeds that of the sugars. Putting it another way, the amount of initial microbial production attributable to energy from cell wall polymers is greater than that from sugars, particularly when it is remembered that a considerable fraction of the latter is probably lost due to physical leaching rather than microbial metabolism.

Thus there seems no reason to suppose that the functional interrelationships within the unit-community, or indeed the macrocommunity, change significantly with time, at least during the early stages of decay. The predominance of some niche determinants may change and with it the community structure but the internal functional structure will remain fundamentally similar and the interaction of polysaccharolytic and sugar-utilizing functions will be a common feature rather than the predominance of one or other feature.

The factors determining the appearance of new species

If the 'successional' features that have been described for the macrocommunity apply to the unit-community, namely that there is a significant time sequence in the addition of new species to the unit-community, then the elucidation of the factors that determine these differences in time of colonization becomes a feature of considerable interest. In considering this question it is useful to distinguish between factors concerned with the process of *inoculation* and those determining *colonization.*

Success in occupation will depend in the first instance on the presence of propagules of the organism at the surface of the resource. It is noteworthy that the RN component of Hudson's primary microflora contains many organisms, such as *Cladosporium herbarum*, that are common members of the phylloplane flora of the living leaf, or in some cases may even be mildly parasitic in relation to senescing leaves, e.g. *Botrytis cinerea*, *Aureobasidium pullulans* (Pugh & Buckley 1971). The ability to persist in the specialized habitat of the phylloplane thus contributes to the decomposer potential of these organisms. They may be contrasted with Hudson's secondary RN fungi, which are less commonly found in the phylloplane or as major components of the air-spora than the primary saprophytes (Gregory 1973). It is possible to also contrast these secondary species with the other component of the primary group, the RS species, in terms of their differing modes of dispersal. The Ascomycete members of the latter group discharge their spores by violent means in to the air; most of the former group, *Penicillium* spp. and the Mucoraceae for instance, are reliant on passive means of dispersal. In many cases this may most commonly mean the carriage of the spores in soil water and on (or in) the bodies of soil animals. Thus the dispersal of this group will be over distances on a microscale largely within the litter and soil milieu as contrasted with the wider-scale aerial dispersal of the primary saprophytes. It is noteworthy that the commonest time of fruiting in many of the leaf-inhabiting ascomycetes is in the late summer and autumn, coinciding with the time of leaf senescence and litter fall.

It is clear that the pattern of spore discharge and dispersal and the timing and location of spore deposition may be an important factor in the establishment stage of community structure. Little can be said in detail at present but this is an area readily accessible to experimental investigation.

The presence of spores or other propagules on the surface of a resource does not necessarily ensure success. Selective effects may be imposed on the microflora at the penetration stage of colonization. On senescent or dead tissues these effects will not be as manifold and complex as they are known to be in the resistance of living plants to invasion by potential pathogens but some features in common may be expected. Thus the presence of cuticular materials resistant to enzymic breakdown or containing fractions that have fungistatic properties (Martin & Juniper 1972) may delay the penetration of some species. Further restriction may be imposed by selectively toxic 'plant protection' chemicals in the senescent tissue (Harrison 1971). In short the same type of mechanism that may operate to select specific components on a resource level may also determine some of the time sequence of colonization. It is worth noting, in view of the probable significance of soil animals in spore dispersal of the later colonizers, that modifier compounds of this kind have also been regarded as important in establishing the palatibility of the leaves to components of the decomposer fauna (Heath & King 1964, Anderson 1973). One of the significant actions of the primary saprophytes may be in the metabolism of these modifier compounds resulting in increased accessibility to fungal colonization and palatability to soil animals.

The soil animals may affect the colonization process by micro-organisms in another important fashion—by their comminutive action. This is discussed more fully by Hargraves in this volume but three possible points of significance should be emphasized here. (1) The addition of specific nutrients of which nitrogenous materials such as urea or uric acid may be the most significant (Hudson 1968). (2) A process of inoculation, homogenizing and localizing the microfloral components. (3) Increasing the surface area of the resources by the reduction of particle size thereby creating a wider range of micro-environments for the microflora. This latter may have particular significance for the bacteria and actinomycetes which lack the capacity of the mycelial fungi to directly penetrate plant materials and are characterized more by surface habitats. But it may also be a feature which is of selective value to the secondary RN fungi. The nett effect of the intervention of the soil animals is to divert the primary resource succession on to a typical coprophilous path (Nicholson *et al.* 1966).

The factors determining the disappearance of species

The decline in the relative abundance of individual species, its disappearance from membership of the active microflora and the general decrease in fungal species diversity within the macrocommunity can all be related to a progressive increase in competition between the constituent species of the community. The depletion of the energy and carbon resource leads to increasing overlap in substrate-determined niches and thus to increased competition for the remaining substrates, notably cellulose and lignin. This is illustrated in Fig. 8.6. It is no doubt significant that the most persistent fungal flora includes components that have the highest capacity for the production of antibiotics (the penicillia) and the widest range of enzymatic diversity (the basidiomycetes). Webster and his co-workers have demonstrated the way in which antagonistic phenomena may increase during the decomposition of herbivore faeces (Webster 1970) and this feature is probably widespread on a variety of resources.

It is interesting to speculate briefly on the differing patterns of response that may be observed in different groups of fungi to the impact of increasing competition.

The phylloplane members of the early leaf colonizers appear to have a 'tight' cycle between inoculation and sporulation (see Fig. 8.6). By means either of resistant chlamydospores persisting on twigs and buds or by the timing of sporulation on leaf litter to coincide with bud opening they are rapidly re-established in the position on the leaf surface which ensures their early colonizing role. The life-cycle of the RS component is longer and typically seems based on a more seasonal pattern correlated with leaf litter fall as mentioned above. It is probable therefore that fruiting in these species will be environmentally controlled rather than endogenous. Interesting variants in this type of life-cycle may occur between species which occupy either rapid or slow decomposing litter, e.g. ash versus beech. In the case of rapidly decaying leaves there may be a resting resistant stage between the formation of ascocarps or pycnidia (resource determined) and the release of spores (which will be determined by the seasonal physical environment). Again this seems a promising area for investigation.

An endogenously regulated tight life-cycle, resembling that described above for the phylloplane organisms, has been described by Harper and Webster (1964) for the mucoraceous species that are pioneers on the coprophilous succession. It is possible that this group may exhibit a similar pattern of behaviour on primary resources even though they are associated with the terminal rather than the pioneer stages of decay. A repeated pattern of short time span between spore and spore will effectively 'avoid' competition. It is also likely that their increased abundance late in decomposition is correlated with the intervention of soil animals and the constant opening of 'pioneer' microhabitats. The type of behaviour exhibited by these species may fall within the category of r-selected species as defined for other groups of organisms (Pianka 1970) and also represent most clearly the behaviour of 'fugitive species' as mentioned earlier. The basidiomycetes have a contrasting behaviour more reminiscent of K-selection. They are probably the most persistent of all the fungi, colonizing soon after leaf fall and remaining of importance into the terminal and even humic stages of decay. Fruiting in the basidiomycetes requires the diversion of very considerable resources. Perennial fruit bodies or repetitive fruiting at the same site is also common in these fungi. Again the seasonal pattern of fruiting is probably associated with rapid colonization after litter fall although most isolation studies fail to confirm this. Basidiomycete mycelia are probably larger in size and more perennial in nature than those of most other fungi, which is further indicative of their competitive success. A considerable potential exists for examining the ecological role of different micro-organisms within the type of life-cycle strategies they exhibit.

The origin of unit-communities

The problem was posed earlier as to what was the basis of the hypothesized variety of unit-communities found associated with a resource. One answer may be sought in the potential for variation in the circumstances of colonization coupled with the enormous abundance of resource available to the microflora. This can be seen by considering the example of wood decay. The decomposition of branch wood is initiated in the canopy and in contrast to the deciduous leaf litter of temperate climates the input of dead branches to the decomposer system occurs continuously through the year. Potential colonization courts therefore become exposed at random points in a very extensive spatial and temporal volume. Variations in the temporal and spatial distribution of potential colonizers in the air spora could therefore lead to considerable diversity in the initiating microflora, and thus to a diversity of unit-communities by the time the decaying branches have reached the litter layer—by which time some 40% decomposition may have occurred. Whether such diversity occurs in other resources is less certain. The greater seasonality in leaf and root death and the select nature of the phylloplane and rhizoplane microfloras suggest that a more homogenous distribution may be characteristic of these resources. However, the same potential for 'opportunistic' colonization may exist once senescence and death have occurred and before substrate-limitation imposes high competition on the process of occupation.

It is probable that the features determining the variation in unit-community composition should be sought on a scale between that of the microhabitat and the macrohabitat. The demonstration, for instance, of pattern at the 10–20 cm scale in the distribution of nutrients in soils (Usher, this volume, Chapter 4) may be indicative of the types of variation that could influence the pattern of colonization of different units of the same resource.

Conclusion; the structure of microbial communities

The main elements of the structure of the microbial community of terrestrial decomposer habitats are pictured in Fig. 8.3. The fundamental unit of primary decomposition is an integrated community partitioned into niches determined by features of the chemical and physical make-up of the occupied resources. The passage of time and the progress of decay result in changes in composition and integrative structure of these unit-communities. Fig. 8.7 summarizes the possible nature of the shifts in niche structure that occur during decay. The declining availability and variety of substrates is to some measure compensated by the increase in physical diversity of the system

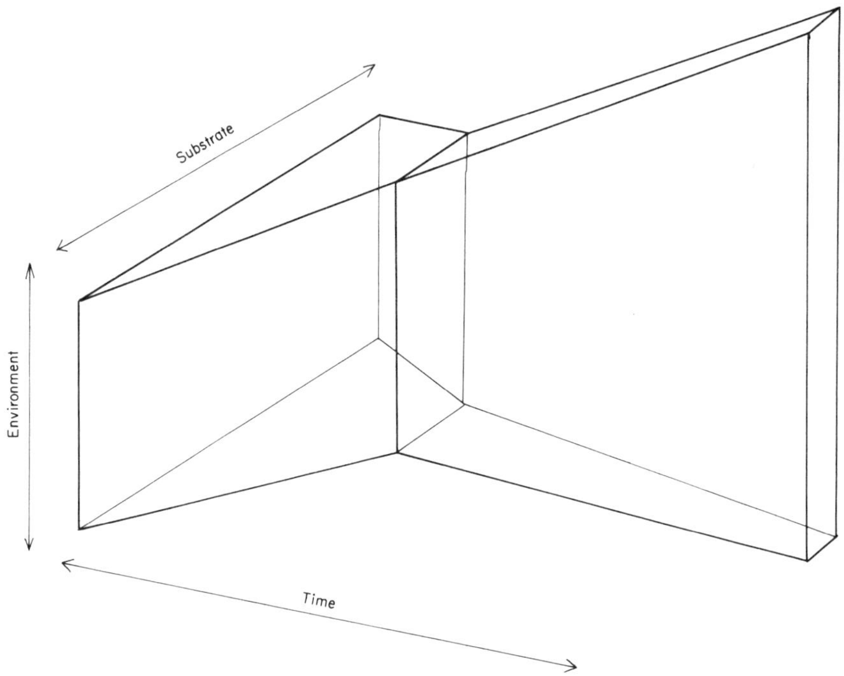

Figure 8.7. Changes in the potential variety of fundamental microbial niches within primary resources during decomposition. A cross section of the volume at right angles to any point in time represents the available niche space in relation to the two major determinants. The timescale is simplified to represent two components in decomposition; the initial rapid decomposition phase characteristic of primary resources, and the much slower decomposition of the terminal resources. These two phases are not represented in scale with regard to time.

and allows the development of a secondary microflora, that characteristic of soil rather than plant litter. In this flora bacteria and actinomycetes to a considerable extent replace the fungi which dominate litter habitats.

The proposed structure must at present be largely hypothetical. The analysis of spatial patterns of distribution of microbes in plant litter environments has been given little attention in comparison with the interest paid to time sequences or to micro-environments in the soil. However, the former would seem to be if anything the easier of experimental access.

The status of micro-organisms in decomposer food webs would seem to be determined primarily by their metabolic capacity. This is allied to a versatility in response to rapidly changing chemical and physical environments which is reflected in their enzymic abilities, their patterns of growth and their methods of reproduction, and dispersal. This versatility is no doubt made possible by the variety of genetic mechanisms with which both the prokaryotic and eukaryotic members are endowed. However, the very versatility imposes caution on attempts to categorize micro-organisms into functional ecological groups. As with symbiotic microbes, fluctuating environments, both physical and biotic, may bring about shifts in the functional role of any one species (Lewis 1973). Thus the erection of ecological groups must be recognized as specific to the circumstance under investigation. Greater understanding of the nature of microbial community structure is likely to come from seeing it as a dynamic rather than a static concept.

Acknowledgment

I should like to acknowledge the debt that I owe to the late Ian Healey for first awakening my interest in the ecological analysis of microbial communities. I have also benefited from discussions with many colleagues of whom I should like to thank in particular Alan Hildrew, Jo Anderson and Juliet Frankland who read the paper in manuscript at various stages of preparation.

References

ALBERSHEIM P., JONES T.M. & ENGLISH P.D. (1969) Biochemistry of the cell wall in relation to infective processes. *Ann. Rev. Phytopath.* 7, 171–94.

ANDERSON J.M. (1971) Observations on the vertical distribution of Oribatei (Acarina) in two woodland soils. In: *IV Colloquium Pedobiologiae, C.R. 4eme Coll. Int. Zool. Soc. Ed. INRA*, pp. 257–72. Paris.

ANDERSON J.M. (1973) The breakdown and decomposition of sweet chestnut (*Castanea sativa* Mill.) and beech (*Fagus sylvatica* L.) leaf litter in two deciduous woodland soils. II. Changes in the carbon, hydrogen, nitrogen and polyphenol content. *Oecologia* **12**, 275–88.

BAILEY P.J., LIESE W. & ROSCH R. (1968) Some aspects of cellulose degradation in lignified cell walls. In: *Biodeterioration of materials.* eds. A.H. Walters & J.J. Elphick, pp. 546–57. Elsevier, Amsterdam.

BIER J.E. (1963) Tissue saprophytes and the possibility of control of some tree diseases. *Forestry. Chron.* **39**, 82–4.

BURGES A. (1960) Time and size as factors in ecology. *J. Ecol.* **48**, 273–85.

BROWN J.C. (1958) Soil fungi of some British sand dunes in relation to soil type and succession. *J. Ecol.* **46**, 641–64.

CALDWELL R. (1963) Observations on the fungal flora of decomposing beech litter in soil. *Trans. Br. mycol. Soc.* **46**, 249–61.

CHAN E.C.S., KATZNELSON H. & ROUATT J.W. (1963) The influence of soil and root extracts on the associative growth of selected soil bacteria. *Can. J. Microbiol.* **9**, 187–97.

CHRISTENSEN M. (1969) Soil microfungi of dry to mesic conifer-hardwood forests in Northern Wisconsin. *Ecology* **50**, 9–27.

DIX N.J. (1964) Colonisation and decay of bean roots. *Trans. Br. mycol. Soc.* **47**, 285–92.

EICKER A. (1973) The mycoflora of *Eucalyptus maculata* leaf litter. *Soil Biol. Biochem.* **5**, 441–8.

ENEBO L. (1963) Symbiosis in thermophilic cellulose fermentation. *Nature, Lond.* **163**, 805.

FINDLAY W.P.K. (1940) Studies in the physiology of wood-destroying fungi. III. Progress of decay under natural and controlled conditions. *Ann. Bot. NS.* **4**, 701–12.

FLANAGAN P.W. & SCARBOROUGH A.M. (1974) Physiological groups of decomposer fungi on Tundra plant remains. In: *Soil Organisms and decomposition in tundra*, eds. A.J. Holding, O.W. Heal, S.F. Maclean Jnr., P.W. Flanagan, pp. 159–81. Tundra Biome Steering Committee. Stockholm.

FRANKLAND J.C. (1966) Succession of fungi on decaying petioles of *Pteridium aquilinum*. *J. Ecol.* **54**, 41–63.

GARRETT S.D. (1951) Ecological groups of soil fungi: a survey of substrate relationships. *New Phytol.* **50**, 149–66.

GARRETT S.D. (1956) *Biology of the root-infecting fungi.* Cambridge University Press.

GARRETT S.D. (1963) *Soil fungi and soil fertility.* Pergamon, Oxford.

GRAY T.R.G. (1969) The identification of soil bacteria. In: *The Soil Ecosystem*, ed. J.G. Sheals, pp. 73–82. The Systematics Association, London.

GREGORY P.H. (1973) *The microbiology of the atmosphere.* 2nd edn. Leonard Hill, London.

GRIFFIN D.M. (1972) *Ecology of Soil Fungi.* Chapman and Hall, London.

GROSJEAN J. (1950) Substances with fungicidal activity in the bark of deciduous trees. *Nature, Lond.* **165**, 853–4.

HARPER J.E. & WEBSTER J. (1964) An experimental analysis of the coprophilous fungus succession. *Trans. Br. mycol. Soc.* **47**, 511–30.

HARRISON A.F. (1971) The inhibitory effect of oak leaf litter tannins on the growth of fungi; in relation to litter decomposition. *Soil. Biol. Biochem.* **3**, 167–72.

HAYES A.J. (1965) Some microfungi from Scots Pine litter. *Trans. Br. mycol. Soc.* **48**, 179–85.

HEAL O.W. & MACLEAN S.F. Jnr. (1975) *Comparative productivity in ecosystems—secondary productivity.* In: *Unifying concepts in ecology*, eds. W.H. Van Dobben & R.H. Lowe-McConnell, Junk, Hague and Purdo, Wageningen (in press).

HEALEY I.N. & SWIFT M.J. (1971) Aspects of the accumulation and decomposition of wood in the litter of a coppiced Beech–Oak woodland. In: *IV Colloquium Pedobiologiae, C.R. 4eme Coll. Int. Zool. Soc.* Ed. INRA, pp. 417–30. Paris.

HEATH G.W. & KING H.G.C. (1964) Litter breakdown in deciduous forest soils. *8th Int. Congr. Soil Sci.* **3**, 979–87.

HEDGER J.N. & HUDSON H.J. (1974) Nutritional studies of *Thermomyces lanuginosus* from wheat straw compost. *Trans. Br. mycol. Soc.* **62**, 129–43.

HENDERSON M.E.K. (1960) Studies on the physiology of lignin decomposition by soil fungi. In: *The ecology of soil fungi.* eds. D. Parkinson & J.S. Waid, pp. 286–96. Liverpool University Press.

HERING T.F. (1965) Succession of fungi in the litter of a Lake District oakwood. *Trans. Br. mycol. Soc.* **48**, 391–408.

HERING T.F. (1967) Fungal decomposition of oak leaf litter. *Trans. Br. mycol. Soc.* **50**, 267–73.

HINTIKKA V. (1970) Selective effects of terpenes on wood-decomposing Hymenomycetes. *Karstenia* **11**, 28–32.

HISSETT R. & GRAY T.R.G. (1974) Bacterial populations of litter and soil in a deciduous woodland. I. Qualitative studies. *Rev. Ecol. Biol. Sol.* **4**, 495–508.

HOGG B.M. & HUDSON H.J. (1966) Micro-fungi on leaves of *Fagus sylvatica*. 1. The micro-fungal succession. *Trans. Br. mycol. Soc.* **49**, 185–92.

HUDSON H.J. (1968) The ecology of fungi on plant remains above the soil. *New Phytol.* **67**, 837–74.

HUDSON H.J. (1970) Infraspecific categories in fungi. *Biol. J. Linn. Soc.* **2**, 211–19.

HULME M.A. & STRANKS D.W. (1971) Regulation of cellulase production by *Myrothecium verrucaria* grown on non-cellulosic substrates. *J. gen. Microbiol.* **69**, 145–55.

KENDRICK W.B. & BURGES A. (1962) Biological aspects of the decay of *Pinus sylvestris* leaf litter. *Nova. Hedwig.* **4**, 313–42.

KING N.J. (1966) The extracellular enzymes of *Coniophora cerebella*. *Biochem. J.* **100**, 784–92.

KING N.J. & FULLER D.B. (1968) The xylanase system of *Coniophora cerebella*. *Biochem. J.* **108**, 571–6.

KIRK T.K. (1971) The effects of micro-organisms on lignin. *Ann. Rev. Phytopath.* **9**, 185–210.

LEWIS D.H. (1973) Concepts in fungal nutrition and the origin of biotrophy. *Biol. Rev.* **48**, 261–78.

MACARTHUR R.H. (1972) *Geographical ecology; patterns in the distribution of species*. Harper & Row, London.

MACARTHUR R.H. & WILSON E.O. (1967) *The theory of Island Biogeography*. Princeton University Press, New Jersey.

MARTIN J.T. & JUNIPER B.E. (1970) *The cuticles of plants*. Edward Arnold, London.

MERRILL W. & COWLING E.B. (1966) Role of nitrogen in wood deterioration: amounts and distribution of nitrogen in tree stems. *Can. J. Bot.* **4**, 1555–80.

MINDERMAN G. (1968) Addition, decomposition and accumulation of organic matter in forests. *J. Ecol.* **56**, 355–62.

NICHOLSON P.B., BOCOCK K.L. & HEAL O.W. (1966) Studies on the decomposition of the faecal pellets of a millipede (*Glomeris marginata* Villers). *J. Ecol.* **54**, 755–66.

NORTHCOTE D.H. (1972) Chemistry of the plant cell wall. *Ann. Rev. Plant. Physiol.* **23**, 113–32.

NOVACK R.D. & WHITTINGHAM W.F. (1968) Soil and litter microfungi of a maple-elm-ash floodplain community. *Mycologia.* **60**, 776–87.

PARK D. (1960) Antagonism—the background to soil fungi. In: *The ecology of soil fungi*, eds. D. Parkinson & J.S. Waid, pp. 148–59. Liverpool University Press.

PARK D. (1968) The ecology of terrestrial fungi. In: *The Fungi*, Vol. III, eds. G.C. Ainsworth & A.S. Sussman, pp. 5–39. Academic Press, London.

PARKINSON D., TAYLOR G.S. & PEARSON R. (1963) Studies on fungi in the root region. I. The development of fungi on young roots. *Plant and Soil* **19**, 332–49.

PIANKA E.R. (1970) On r- and K-selection. *Am. Naturalist.* **104**, 592–7.

PRIESTLEY C.A. (1962) Carbohydrate resources within the perennial plant. Commonwealth Bureau of Horticulture and Plantation Crops. *Tech. Comm.* **27**, 116 pp.

PUGH G.J.F. (1962) Studies on fungi in coastal soils. II. Fungal ecology in a developing salt marsh. *Trans. Br. mycol. Soc.* **45**, 560–6.

PUGH G.J.F. & BUCKLEY N.G. (1971) *Aureobasideum pullulans:* an endophyte in sycamore and other trees. *Trans. Br. mycol. Soc.* **57**, 227–31.

PUGH G.J.F., BUCKLEY N.G. & MULDER J. (1972) The role of phylloplane fungi in the early colonisation of leaves. *Symp. Biol. Hung.* **11**, 329–33.

REESE E.T. (1968) Microbial transformation of soil polysaccharides. In: *Organic Matter and Soil Fertility*, pp. 535–77. North Holland, Amsterdam.

RISHBETH J. (1951) Observations on the biology of *Fomes annosus* with particular reference to East Anglian pine plantations. II. Spore production, stump infection and saprophytic activity in stumps. *Ann. Bot., Lond. N.S.* **15**, 221–76.

SCHEFFER T.C. & COWLING E.B. (1966) Natural resistance of wood to microbial deterioration. *Ann. Rev. Phytopathology* **4**, 147–70.

SELBY K. (1968) Mechanism of biodegradation of cellulose. In: *Biodeterioration of materials*, eds. A.H. Walters & J.J. Elphick, pp. 62–78. Elsevier, Amsterdam.

SIMPSON E.H. (1949) Measurement of diversity. *Nature Lond.* **163**, 688.

SUNDMAN V. (1970) Four bacterial soil populations characterised and compared by a factor analytical method. *Can. J. Microbiol.* **16**, 455–64.

SWIFT M.J. (1976) The ecology of wood decomposition. *Sci. Progress* (Oxford) (in press).

SZABO I.M. (1974) *Microbial communities in a forest-rendzina ecosystem; the pattern of microbial communities.* Akademiai Kiado, Budapest.

TENNEY F.G. & WAKSMAN S.A. (1929) Composition of natural organic materials and their decomposition in the soil. IV. The nature and rapidity of decomposition of the various organic complexes in different plant materials, under aerobic conditions. *Soil Sci.* **28**, 55–84.

THORNBER J.B. & NORTHCOTE D.H. (1961a) Changes in the chemical composition of a cambial cell during its differentiation into xylem and phloem tissue in trees. 1. Main components, *Biochem. J.* **81**, 449–54.

THORNBER J.B. & NORTHCOTE D.H. (1961b) Changes in the chemical composition of a cambial cell during its differentiation into xylem and phloem tissue in trees. 2. Carbohydrate constituents of each main component. *Biochem. J.* **81**, 455–64.

THORNBER J.B. & NORTHCOTE D.H. (1962) Changes in the chemical composition of a cambial cell during its differentiation into xylem and phloem tissue in trees. 3. Xylan, glucomannan and α-cellulose fractions. *Biochem. J.* **82**, 340–6.

TRIBE H. (1966) Interactions of soil fungi on cellulose film. *Trans. Br. mycol. Soc.* **49**, 457–66.

WAID J.S. (1974) Decomposition of roots. In: *Biology of plant litter decomposition*, eds. C.H. Dickinson & G.J.F. Pugh, pp. 175–211. Academic Press, London.

WARCUP J.H. (1965) Growth and reproduction of soil micro-organisms in relation to substrate. In: *Ecology of soil-borne plant pathogens*, eds. K.F. Baker & W.C. Snyder, pp. 52–68. John Murray, London.

WEBSTER J. (1956) Succession of fungi on decaying cocksfoot culms. (Part 1.) *J. Ecol.* **44**, 517–44.

WEBSTER J. (1970) Coprophilous fungi. *Trans. Br. mycol. Soc.* **54**, 161–80.

WEBSTER J. & DIX N.J. (1960) Succession of fungi on decaying cocksfoot culms. III. A comparison of the sporulation and growth of some primary saprophytes on stem, leaf blade and leaf sheath. *Trans. Br. mycol. Soc.* **43**, 85–99.

WHITTAKER R.H. & WOODWELL G.M. (1968) Dimension and production relations of trees and shrubs in the Brookhaven Forest, New York. *J. Ecol.* **56**, 1–25.

WICKLOW D.T. & WHITTINGHAM W.F. (1974) Soil microfungal changes among the profiles of disturbed conifer-hardwood forests. *Ecology* **55**, 3–16.

WIDDEN P. & PARKINSON D. (1973) Fungi from Canadian coniferous forest soils. *Can. J. Bot.* **51**, 2275–90.

WIDDEN P. & PARKINSON D. (1974) The effects of a forest fire on soil microfungi. *Soil Biol. Biochem.* 7, 125–38.

WIEGERT R.G., COLEMAN D.C. & ODUM E.P. (1970) Energetics of the litter-soil subsystem. Methods of study in soil ecology. *Proc. Paris Symp.*, pp. 93–8. UNESCO, Paris.

WILLIAMS S.T., DAVIES F.L. & HALL D.M. (1969) A practical approach to the taxonomy of Actinomycetes isolated from soil. In: *The soil ecosystem*, ed. J.G. Sheals, pp. 107–17. The Systematics Association, London.

WILLIAMS S.T. & GRAY T.R.G. (1974) Decomposition of litter on the soil surface. In: *Biology of plant litter decomposition*, eds. C.H. Dickinson & G.J.F. Pugh, pp. 611–32. Academic Press, London.

9 Decomposition processes in the sea with special reference to man-made waste.

G. D. FLOODGATE *Marine Science Laboratories, University College of N. Wales, Menai Bridge, Gwynedd, U.K.*

I. Organic carbon cycle in the sea

The organic matter in the world's oceans, whether in the form of dissolved organic carbon (DOC), or as particulate organic carbon (POC) is derived from one of two sources. These are firstly the organic carbon fixed *in situ* by primary production due to algal photosynthesis, and secondly additions made by rain, rivers and human pollution. All this organic carbon either moves through the food web to higher trophic levels to appear in one of several possible forms, or is degraded by microbial action to become incorporated into microbial tissue along with the formation of chemically simpler substances. In some instances complete mineralization to carbon dioxide will occur. A small proportion of the annual production of organic carbon is highly resistant to degradation, and has a turnover time which may extend to several thousand years, so that biologically it is effectively removed from the system. The portion that moves into the food web appears partly as the biomass of higher organisms, and partly as plant exudates or the excretory products of animals. All these fractions are eventually subjected to biodegradation to microbial tissue and simple by-products of metabolism.

The most important plant exudate is glycollate as far as the micro-algae are concerned. The amount of fixed carbon which appears in this form is very varied. The details are further confused by the fact that glycollate can be reabsorbed by algal cells themselves as well as providing a suitable substrate for bacteria (Fogg 1963). The contribution from animal waste products is higher than might be supposed. Work by Steele & Baird (1972) on a Scottish loch showed that the carbon settling out into sediment traps was largely made up of faecal pellets. According to Butler *et al.* (1970) about 20% of the phosphorus and 40% of the nitrogen ingested by grazers is excreted as faecal pellets which settle out to form a major fraction of the organic input into the bottom. The pellets become infested by bacteria which may then become food for coprophagic organisms, the pellets being discharged again for recolonization.

An unknown fraction of the dissolved organic carbon will be absorbed on to various surfaces particularly clays. The process is selective depending upon the molecular structure of the organic compound and the surface structure of the inorganic material involved. The rate of turnover of absorbed substrates as compared with the same substrate in solution is not easy to estimate. Stotzky & Rem (1966) noted that montmorillonite stimulated bacterial respiration probably by maintaining the pH at an optimum value. Both montmorillonite and kaolin served as a source of minerals for bacterial nutrition.

The annual amount of marine primary production is estimated at $3{\cdot}6 \times 10^{16}$ g yr^{-1} (Williams 1971) though it varies from area to area and seasonally (Ryther 1963). Further amounts of organic carbon are added by land drainage and from rain but these are smaller than the amount formed by primary production by several orders of magnitude (Williams 1971). Estimates of the DOC average around $1{\cdot}0$ mg C l^{-1} in the upper 300 m of water and approximately half this concentration below this depth. The POC averages about $0{\cdot}1$ mg C l^{-1} in the upper layers, and $0{\cdot}01$ mg C l^{-1} in deep water. The DOC is in much greater supply than the POC though the concentration of both vary a great deal from place to place and from time to time. The greater part of both DOC and POC consists of humic type material which is refractory to chemical and microbiological degradation. Only about 10% of organic carbon in a sea water sample is labile and available for microbial attack. These proportions are due more to the relative rates of turnover rather than relative amounts of refractory and degradable materials produced by photosynthesis.

The variety of chemical species among the naturally formed organics is very considerable. However, largely because of problems associated with sampling and analysis the number of substances which have been identified and studied in detail is very small. Lipids are found in the surface slicks and include fatty esters, free fatty acids, fatty alcohols and hydrocarbons (Wagner 1969). Even carbon fatty acids from C_{12} to C_{22} were present in all samples. Fatty acids with C_{15}, C_{17} and C_{19} chain lengths were present in some samples with C_{17} being found in high concentration on occasion. Low molecular weight organic acids such as formic, acetic, lactic and glycollic acids were reported from Pacific waters by Koyama & Thompson (1959). The presence of sterols such as cholesterol, stigmasterol and β-sitosterol have been demonstrated (Matthews & Smith 1968).

Carbohydrates, mainly of phytoplankton origin, are known to be present in sea water both as free sugars and as polymers. Starch was found as suspended particles along with dissolved free sugars including glucose, mannose and galactose, in coastal waters off California. Pentose was also found (Degens *et al.* 1964). Vaccaro and his colleagues (1968) showed that glucose concentration was closely connected to phytoplankton density. A similar range of hexoses has been identified in the hydrolysates of POC by Handa & Tommaga (1969).

Amino acids are mostly present condensed to polypeptides. The most abundant are aspartic, threonine, serine, glutamic acid, alanine, valine, isoleucine and leucine (Hood 1963). Many other amino acids have been identified both as constituents of polypeptides and as free acids (Duursma 1965), and in sediments (Litchfield *et al.* 1974). Many complex ring structures have also been reported. These include partially degraded chlorophylls, cobalamine and other vitamin B_{12} analogues, carotenoids and phenolic compounds.

The second major source of organic carbon in the sea includes that derived from human activity and consists of domestic and industrial sewage. It is difficult to obtain reliable data as to the quantities involved, but taken globally the total organic carbon input due to pollution is very much less than that fixed annually by photosynthesis or derived via run off and rain. Pollution, however, tends to be local because industry is concentrated in certain parts of the world. These areas too tend to have the maximum concentration of humans each producing their quota to add to the pollution burden. With certain important exceptions the impact of pollutants is confined mainly to fairly localized areas. The variety of types of chemicals that is found in polluting effluents is enormous. Many of the chemicals belong to the same chemical series as naturally produced organics. Polysaccharides, polypeptides, phenolics, lipids and hydrocarbons are all included. There are, however, many novel substances which appear in the effluent particularly of the chemical and manufacturing industries. These include phenoxy acids herbicides, organic phosphorous and carbamate insecticides, quinones and diotins from the paper pulp industry, nitriles and organic phosphoresters from plastic manufacture (Widmark *et al.* 1972). Others such as the polycyclic aromatic hydrocarbons although found in nature can give rise for concern because of their toxicity, in this case as carcinogens, and their slow rate of degradation. Other toxins such as cyanide are known to be biodegraded, though the rate of decomposition is unknown. Novel synthetic materials present a problem to the degradative capacity of the oceans and it might have been expected that a long period of evolutionary change would be required in order that the natural environment could develop adequate means of dealing with them. However, such is the chemical versatility of fungi and bacteria that many, though not all these substrates are apparently degraded in a measurable time. The essential point is that from the degradative point of view there is no intrinsic difference between an organic substance whose origin is in the photosynthetic activity of phytoplankton and one manufactured by man and dumped in the oceans. The same degradative processes will operate. In addition, the sea areas which are mainly used for dumping domestic and industrial waste are also those which are noted for their generally high productivity. Coastal areas and the continental shelf often receive a high pollution loading even though they are the areas which are most productive of fish and hence economically important. Estuaries too receive a good deal of sewage and are also biologically active. In considering the capacity of a sea area to deal with a given pollutant loading, the natural as well as the added substrates must be taken into consideration. As yet too little is known about the effects of pollutants on the natural degradative and mineralization processes for confident predictions to be made.

II. Factors governing the rate of biodegradation

Although the main agents of oxidation of organic carbon in the marine environment are micro-organisms there are at least two other mechanisms which may operate though their significance is not certain. The first of these is oxidation due to abiological chemical processes, especially photochemical processes. For example, the photochemical changes produced in oil may be important in its removal from the sea (Burwood & Speers 1974). The second mechanism is due to the intervention of macrobiological agents. Particulate organic carbon is ingested by a wide variety of animals who considerably alter its chemical composition (Seki 1972). Similarly pollutants may be directly ingested as in the case of the removal of microbial pollutants by the red beard sponge (Claus *et al.* 1967) and the uptake of oil droplets by protozoa (Andrews & Floodgate 1974).

In order that a substance may be degraded by micro-organisms, several conditions must be fulfilled.

(a) BIODEGRADABILITY OF THE SUBSTRATE

Firstly the substrate itself must be of such a chemical configuration that it will fit into one of the biochemical pathways that have been evolved. In addition, the energy cost to the organism must not be greater than it receives from the oxidation. Most of the naturally formed metabolites are readily degraded though some polymers such as cellulose and chitin disappear from the environment rather slowly. By and large as the complexity of the molecule increases so does the difficulty in biodegradation. Ring structures are more difficult to break down than alkyl chains, and aromatic polymers such as lignin degrade at a very slow rate.

Nevertheless, such is the versatility of the degradative abilities of micro-organisms, especially bacteria and fungi, that very many pollutants are degraded at a greater or lesser rate. Unfortunately this ability of heterotrophs to destroy such a wide variety of substrates has lead to the 'fallacy of bacteriological infallibility' that is to the notion that all pollutants are biodegradable (Alexander 1965). The chlorinated hydrocarbons are a classic example of the folly of putting biologically active nondegradable material into natural water systems without first ascertaining the rate at which biodegradation proceeds and the chemical structure of the byproducts which are formed.

(b) THE PRESENCE OF SUITABLE MICRO-ORGANISMS

A second requirement is the presence of an organism in the environment which has the catabolic potential to deal with the substrate. Typically a natural microbial community will contain a diversity of physiological types, so that the spectrum of enzymic potential within the population is very wide. Many of the enzymes required to carry out the degradative processes are induced; that is the presence of the substrate is required in order to produce the enzyme in quantity within the organism. The presence of the substrate will have two effects on the local microbial population. There will be a large and possibly rapid increase in the physiological types able to break down the substrates and there will be a marked increase in the appropriate enzyme concentration in each bacterial cell.

(c) MICROBIAL INTERACTION

However, the mechanism of bio-oxidation may be more complex than the above would suggest. Micro-organisms in culture are known to interact in several ways (Bungay & Bungay 1968), although it is as yet impossible to assess the importance of these interactions in nature. Kuentzer (1969) has pointed out that the formation of carbon dioxide by bacterial mineralization of organics can trigger off a bloom of blue-green algae which will in turn produce more organic material. Again the production of antibiotics by algae will tend to limit bacterial growth (Sieburth 1968). It has been frequently observed in this department that the bloom of the colonial flagellate *Phaeocystes* is accompanied by an increase in the total heterotrophic count and at the same time a diminution of the number of bacterial species present in the Menai Straits. The organic compounds themselves can also limit the growth and activity of micro-organisms. Phenolic compounds which are produced by macro-algae, and also appear in effluents, prevent the reproduction of some bacterial strains but are excellent substrates for the growth of others.

(d) SUBSTRATE INTERACTIONS

The presence of two substances together can affect their degradation rate. The more easily degraded will be oxidized first thereby sparing the more recalcitrant material. Such sparing effects emphasize the care that needs to be taken in predicting the period of time required to remove a pollutant by microbiological means.

Another important substrate interaction is known as co-oxidation. This involves the breakdown of a recalcitrant material only in the presence of another substrate. For example *Mycobacterium rhodocroes* grown on n-decane is capable of oxidizing a number of methyl substituted aromatic compounds which it will not attack when provided alone (Duros & Frankenfeld 1968).

(e) ENVIRONMENTAL CONDITIONS

It is important to realize that the presence of a biodegradable compound and the presence of one or more organisms able to metabolize it will not guarantee that the substrate will in fact degrade. The environment will have an important effect on the efficiency of degradation. Andrews & Williams (1971) found a seasonal variation in the degradative rate of glucose and amino acids. During the winter the oxidation rate of glucose was below the limit of detection, but increased to around 20%–50% per day during April to October with occasional days of very high rates. Amino acids showed a lower rate of oxidation than the sugars but the same pattern of seasonal variation. Wright & Hobbie (1965) found a similar pattern in turnover time for the uptake of acetate and glucose in the waters of Lake Eiken. A seasonal change of this sort would suggest that temperature variation may be responsible. However, Nedwell & Floodgate (1971) working with inshore muds concluded that the bacterial population adapted to changing temperatures, that is psychrophilic organisms become predominant in the cold months. These experiments however, were primarily concerned with rates of division rather than with heterotrophic oxidation. It is also interesting to note that

Redmond (1972) found higher rates of oxidation of plant carbohydrates in Southampton Water as compared with the Vellar estuary in South East India, possibly due to the higher nutrient levels due to pollution in the cooler water.

Another factor which will govern the rate of heterotrophic degradation is the concentration of the dissolved organics. A great deal of trouble is taken when disposing of a pollutant to ensure that it is diluted to the highest possible degree. Experiments with pure cultures of bacteria of various origins have shown that under artificial cultivation there is a limit of concentration below which the organism will not grow (van Uden 1969). Extrapolation of laboratory threshold figures to the marine environment could lead to unreliable conclusions, but equally they suggest that from the point of view of maximum degradation great dilution may not be beneficial.

Another environmental factor which can govern the rate of biodegradation is the concentration of other essential nutrients notably nitrogen and phosphorus. As set out below the nitrogen supply is a major factor limiting the rate of breakdown of oil in the marine environment. Some pollutants of course add nitrogen either organically bound as in dairy and food wastes or inorganic as in ammoniacal waste liquors from coaking processes. Phosphate too is provided in sewage, but this is likely to be less limiting in the sea than nitrogen. Oxygen lack which can be an important factor in fresh water systems rarely becomes a limiting factor in marine biodegradation. This is partly due to the high degree of dilution that is achieved in the sea and partly to the turbulence that is a characteristic of the shallow waters into which pollutants are discharged.

Many micro-organisms in the aquatic environment adhere to the surface of solid objects which may be particles of plant materials and sand grains, or experimental objects such as glass slides. Bacteria are also to be found at interfaces such as the sea–air interface, water sediment interfaces and water interfaces where density differences are found as in the thermocline. The former case, that is adhesion to solid surfaces, has been studied to a considerable extent in sewage disposal units (Charackliss 1973) and some rates of biodegradation established. However, although bacterial adhesion has been observed on many surface exposed to sea water, it is still difficult to assess its significance in biodegradation, though it may be considerable. Kriss (1963) found the phenomenon to be very widespread in oceanic as well as coastal waters.

Finally, hydrostatic pressure exerts an effect on the rate of destruction of organic matter. Seki *et al.* (1974) produced evidence of high heterotrophic potentiality of bacteria taken from depths of more than 5,000 m. The rod-shaped barophilic organisms they observed had a mean generation time of only ten hours which represents a fast growth rate for the marine environment. This is in contrast to the conclusions of Sorokin (1970) who examined the heterotrophic activity of microflora of seawater using ^{14}C labelled algal hydrolysate. Activity was confined to the upper 800 m of water with a maximum around 400–600 m. Below this depth both heterotrophic numbers and activity were low. Jannasch (1971) also considers that heterotrophic activity at great depths is minimal. The problem has a bearing on pollution problems since many noxious substances are dumped into very deep water in containers of various types. Should the containers leak there is a danger of the toxin not being biodegraded and its concentration increasing.

III. Problems in measuring the rate of biodegradation

There are very considerable technical difficulties in measuring the rate at which substances will degrade in the marine environment. Work in the field is beset with analytical and sampling difficulties, and by the fact that the many variables which influence the rate of removal are not controlled by the investigator. Laboratory-based investigations are often misleading. The kinetics of growth in batch culture appear to be very different from those found in the sea although continuous culture appears to be closer. One technique which has proved of value was originally devised by Parson & Strickland (1962) and recently reviewed by Wright (1973). The solutes suitably labelled with ^{14}C are added to seawater and the release of carbon dioxide followed over a short period. The method assumes that Michaelis–Menten enzymic relationships apply, but this is not always the case. The mixed nature of natural populations means that uptake rates are complex because of different rates of transport across cell membranes. Nevertheless some useful data has been obtained with small molecules such as glucose, glycollate, acetate and lactate.

Unfortunately in the natural situation organics, whether pollutants or not, rarely appear singly, but usually as complex mixtures of substances. Since they are in high dilution the analytical difficulties in following changes in concentration are formidable. During recent years, a new armoury of analytical techniques has been developed, such as mass spectrophotometry and atomic absorption spectrophotometry. But, even with these powerful weapons, making direct measurements in the changes of concentration of a pollutant is often difficult to do, and once achieved, difficult to interpret. The concentration at any moment will depend upon input rates and mixing factors as well as biochemical oxidation so that the problem of sampling becomes paramount. A large number of samples, while statistically desirable, can become unwieldy from the analyst's point of view. The tendency is therefore to devise techniques which give an integrated value for the rate of degradation of a mixture of organics over a stated period. The demand for oxygen or the increase in bacterial heterotrophic numbers are examples of these integrated techniques.

The oxygen demand is estimated chemically (COD) or biochemically (BOD), the latter being a measure of the microbial activity. The principle is simple; a sample of water containing the organic matter is stored tightly stoppered at a suitable temperature for five days. During this period the natural microflora will increase, oxidize the organic matter and reduce the oxygen content of the water. The BOD value is then quoted in terms of the weight of oxygen removed. The method is widely used, sometimes with modifications but there are difficulties and limitations to be considered when applying the technique to the marine environment (Gaudy 1972).

Respirometric methods using the Warburg or Gilson apparatus have been used. However, they have not been very successful because of the small volumes of gas involved and the long duration of the experiments. In addition, sea water is largely buffered by its carbonate-bicarbonate system, so that the measurement of the carbon dioxide involved can present some problems. Other respirometers have been developed in order to study a particular pollutant. Thus Gibbs (1972) developed a novel respirometer to investigate the uptake of oxygen by crude oil.

Oxygen electrodes have been used to monitor the removal of oxygen, partly because they provide a continuous record and partly because they eliminate the practical difficulties involved in performing a large number of Winkler titrations. Pomeroy & Johannes (1968) devised an apparatus in which the microplankton below the 10 µm size range, whether algae, protozoa or bacteria, was concentrated, so enabling rapid measurements of the rate of oxygen uptake to be obtained. Other pieces of apparatus have been devised to measure the rate of oxygen uptake in bottom sediments over a given period (e.g. Edwards & Rolley 1965).

Another group of techniques have been devised to estimate the electron transport activity. These methods rely on changes in colour between the oxidized and reduced state of a number of dye substances. These reactions can be used in anaerobic as well as aerobic systems. Essentially the dyes act as electron acceptors for the co-enzymes nicotinamide-adenine-dinucleotide (NAD) and flavine adenine dinucleotide (FAD). The commonest dye used is 2,3,5-triphenyl-tetrazolium chloride, which is reduced to formazan, a red compound whose concentration is measured spectrophotometrically. Hobbie and his colleagues (1972) used the method in conjunction with respiratory rate estimates, and adenosine-triphosphate measurements of bacterial biomass, and found good agreement between the methods for surface waters, but they diverged widely when the water was taken from the deeps.

The disadvantage of all these techniques which rely on oxygen uptake is that there is no indication of which substrate is being oxidized in the complex mixture, nor is there any information as to what its products are.

Besides these chemical methods, various biological estimations have been devised as well. These centre on an estimate of the number of bacteria present, or on a chemical determination of the biomass. The enumeration of 'viable counting units' is an unreliable technique both because the methods underestimate the number of bacteria present and because of the difficulties inherent in the statistical interpretation of the results (Postgate 1969). It is impossible to obtain a medium that will support the growth of all the different species which can be expected to be found in any sample of sea water. Direct microscopic counts invariably give a higher value than viable counts. But since only small volumes of the sample are examined at a time, the sampling error can be large. Again statistically satisfactory data can only be obtained after a great deal of tedious counting. Jannasch & Jones (1959) examined a number of sea water samples by five cultural methods and two direct counting techniques. Very considerable variations were found between techniques which in some cases were as high almost as four orders of magnitude.

Biomass can also be estimated on the basis of the amount of a cell constituent that is present. Total nitrogen following Kjeldhal digestion has been used, but can only be meaningful if the bacteria are first separated from other sources of nitrogen. Furthermore, the concentration of organic nitrogen per cell is not constant, nor is it directly related to the metabolic activity, a drawback which is true of most of the constituents of the microbial cell. The possible exception to the rule is adenosine triphosphate (ATP). This compound, though rather labile, is rapidly and easily measured even at high dilution by the light emitted from the luciferin-luciferase reaction. The light is measured directly by a photomultiplier and suitable ancillary circuits. The ATP is broadly related to the cell activity. In a mixed population of protozoa, algae, bacteria and fungi, all will contribute to the ATP value, so that again difficulties in interpreting the results arise (Holm-Hansen & Paerl 1972).

In spite of such difficulties bacterial counts can be useful as indicators of biodegradation (Prochazka & Payne 1965). Although the actual values are of only limited value the trends they show from time to time imply that heterotrophic activity is changing and so give some overall picture of the situation.

IV. Rate of decomposition of certain pollutants

(a) DOMESTIC SEWAGE

Domestic sewage is a complicated mixture of human excreta, detergent, soap, fats, food particles, paper and other materials. Most of these materials are known to be biodegradable, though there is very little information as to the rate of breakdown in the sea. Between 30% and 50% of the organic matter is soluble, around 9–14% colloidal, approximately 20% supra colloidal particles and 23–30% settleable particles (Hunter 1971).

A major nitrogenous dissolved compound is urea which is present in urine at concentrations of the order of 20 g l^{-1}. Two microbial degradative pathways are known, one involving the classical urease process by which urea is hydrolysed to carbon dioxide and ammonia. This system is present in the bacteria and in some algae which grow on urea heterotrophically. Other urea degrading algae do so by the urea amidolase (UAL-ase) pathway.

Remsen and his colleagues (1972) observed the most rapid rate of decomposition of 19·4 nM urea l^{-1} h^{-1} in the Savannah River. Slowest rate were found in the North Atlantic oceanic waters of 0·31 nM l^{-1} h^{-1}. In general, rivers gave the highest rates of decomposition followed by coastal waters and finally the open oceans. Algae were found to be the major decomposers in rivers, brackish water and coastal waters whereas the bacteria were the major decomposing agents in the open ocean. In view of the high bacterial loading of rivers and coastal waters this result is rather surprising.

Another nitrogenous source in domestic sewage is the amino acids which appear both in the free and bound form and in both the dissolved and particulate phase. Hunter (1971) quotes figures from a number of sources to show that the amino acid content in waste water varies from 40–140 mg amino acid l^{-1} sewage. Particulates in one instance contained approximately 40 mg l^{-1} of which about one third was in the settleable fraction. Using pure solutions of amino acid Williams & Gray (1970) noted a rapid response by the microbial population of the English Channel to an increase in amino acid concentration so that within a few days the ambient mineralization rate rose from 10–50 μg amino acid l^{-1} d^{-1} to 1,000–2,000 μg amino acid l^{-1} d^{-1}.

Carbohydrates are also found in sewage both as free sugars and as polymers, the most important of which is cellulose. The soluble carbohydrates were present at 10 mg l^{-1} in American domestic waste with much of it as free hexose. In British domestic waters pentoses accounted for as high a fraction as the exxoses (Hunter 1971). The rate of degradation of glucose in the English Channel was found by Andrews & Williams (1971) to vary seasonally and to oscillate between zero and 50% per day. Wright

& Hobbie (1965) showed that algae could be involved in glucose uptake in lake water, but when the substrate concentration was low the bacteria were the major utilizers. They attributed this difference to the fact that algae take up the sugar by diffusion only, whereas the bacteria have a specific transport mechanism. Of the polysaccharides, cellulose is the more abundant and is of course a natural product as well as a pollutant. Indeed this is one example where the input to the sea from human sources must be very much smaller than that synthesized naturally. Its rate of degradation in several forms of fibre has been studied by Hofsten & Edberg (1972) in a sewage treatment plant, lake and river water as well as the Baltic. Newly-processed fibres were rapidly degraded in aerobic and nutrient rich water, whereas previously dried pulp and cotton fibres went more slowly. Lignin containing material did not change significantly. Decomposition was slow in unpolluted sea water and in heavily polluted anaerobic water, but relatively rapid both in unpolluted anaerobic water and in lake sediment. Rapid in this context refers to 40–50 days.

A great deal of human faecal matter is made up of bacterial cells, notably *Escherischia coli*. It has been estimated that as much as 50% by weight of faeces consists of bacterial tissue (Lovatt Evans 1947). Many of these cells are still living when voided but they die off in sea water in a matter of hours or days, depending upon the conditions. Some of the bacterial biomass will be taken into the food web by way of detritus feeders. Some protozoans are known to predate *E. coli* causing a rapid loss of cells from a culture within a few days. Others will be devoured by organisms belonging to the genus *Bdellovibrio*. These small bacteria parasitize other gram negative bacteria and are found in considerable numbers in the sea area around a sewage outfall. It is possible that they are responsible for the transformation of a sizeable proportion of faecal coliforms (Mitchell 1972). Other cells which escape predation will be mineralized by other bacteria adhering to the particles and their carbon recycled.

The increasing use of detergents for domestic purposes has lead to the development of a family of soft, that is biodegradable, products appearing on the market. The concentration of detergent that reaches the marine environment is probably very small, so that its effect is not great. There does not appear to be any data on the rates of disappearance in sea water. The earlier tetrapropylene benzyl-sulphonates have been replaced by the alkyl benzyl-sulphonates. The short side chains of the former material are an effective obstacle slowing down degradation whereas the linear form of the latter will aid breakdown until all surface activity is destroyed, so preventing the foaming which at one time threatened to overwhelm sewage plants and rivers. The remaining ring structure left after the loss of surface activity will degrade more slowly (McCutcheon 1966).

(b) CRUDE OIL AND REFINED OIL PRODUCTS

Approximately 1,355 million tons of oil are transported every year by sea. The annual input to the marine environment of both crude oil and oil products is estimated at close to 6 million metric tons (Anon 1975). The greatest sources are tankers which do not use the 'load on top' method of operating, and river run off which supply 0·77 mta and 1·6 mta respectively. The vigorous cleaning of all ships which is essential for safety during a dry dock inspection and refitting accounts for 0·25 mta. Refinery losses account for 0·2 mta as do tanker accidents. Oil terminal operations account for only 0·003 mta. Furthermore, fossil hydrocarbons seep naturally into the oceans at a rate of about 0·6 mta. In addition the microalgae add a considerable amount of hydrocarbons mostly as alkanes every year (Blumer *et al.* 1971; Youngblood *et al.* 1971, 1973). The fossil oil itself is a very complex mixture consisting mainly of several types of hydrocarbon. The simplest are the n-alkanes with carbon chain lengths up to C_{60} together with homogenous branched chain alkanes. The cyclic-paraffins are represented by cyclohexane and cyclopentane and their analogues, some with alkyl side chains. There are many ring structures such as benzene, alkyl benzenes, naphthalenes and polyaromatic hydrocarbons. At the high boiling point end there are asphaltic materials and resins. One reference oil is known to contain 234 different n-alkanes with individual concentrations between 2·3% and 0·0004% (Mair 1964). Moreover, the mixture differs between oil fields often quite considerably and even between wells within the one oilfield. The problems concerned with the biodegradation of oil are thus very complex. When solubilization and photoxidation are taken into account the problem becomes formidable indeed. Hence the literature abounds with inconsistencies and contradictory generalizations can be made by selecting the appropriate data (Floodgate 1973).

When crude oil, or its refined products are spilled into the sea, about 40% of it is lost in a few days by evaporation. The rate of loss of these 'light ends' will depend upon the temperature. The volatile materials also tend to be the toxic ones. A few components dissolve to a very limited extent but in some cases to a sufficient extent to cause toxicity (e.g. Takahashi & Kittredge 1973). The oil will spread over the surface to form slicks which eventually may be only a few molecules thick. Alternatively it may be churned up by wind and wave action to form a stable water in oil emulsion known as 'chocolate mousse'. Degradation does not proceed very far in such a state; and may last with little evidence of bacteriologically induced changes for at least two years (Davis & Gibbs 1975). Wave action will also break up the slick into small pieces which can accumulate detritus and tend to form pellets. These pellets are usually very small having a diameter of only a few millimetres. However, when rolled along a shallow bottom they can become considerably enlarged and develop into tar balls of up to 30 cm in diameter. These tar balls, which have been found in many parts of the ocean, have an encrusted degraded outer surface of asphaltic material and a centre which consists of virtually unchanged oil. This is because the outer layer cuts off the inner material from oxygen, bacteria and nutrients so inhibiting further destruction. Many species of marine bacteria are hydrocarbonoclastic and their distribution in the oceans is very wide (Floodgate 1972). There are also many species of yeasts (Ahearn & Myers 1972) and filamentous fungi (Perry & Cerniglia 1973) capable of attacking oil so that organisms with oil degrading ability are found in all marine environments.

In order to degrade the hydrocarbon, the micro-organism adds molecular oxygen to the substrate. Hence oil biodegradation is aerobic. There are a few reports of anaerobic degradation, but since a strictly anaerobic biochemical pathway is unknown and since the Michaelis constant of the oxygenase concerned is such that it indicates that the enzyme has a high affinity to oxygen, these reports are likely to be untrue. Furthermore, degradation is a surface phenomenon with the bacteria being found at the interface between the water and the oil. It follows that finely divided oil will present a larger interfacial area than undivided oil, and so will degrade more quickly.

It is now well established that the first fraction to disappear is the alkanes, the more linear the chain, the greater the rate of degradation. The processes involves the oxidation of the terminal methyl group to form a fatty acid via alcohol and aldehydes. Diacids formed by the oxidation of the terminal methyl radical at both ends are also known. Depending upon conditions, these substrates are removed in periods ranging from weeks to months. Other fractions are removed more slowly. Cyclanes are oxidized to diols before breaking the carbon ring. Aromatics are degraded slowly in sea water probably by way of ring opening to form muconic acids or muconic hydroxyaldehyde derivatives. Hence the intermediate by-products of oil biodegradation is a variety of organic acids, fatty acids, aldehydes and ketones some of which will be soluble in water. These intermediates will be mineralized to carbon dioxide and water, while some will be assimulated to form bacterial tissue, which will form the food supply of predators. In this way the carbon from the fossil fuel will be recycled into the biosphere. However, much of the oil is degraded very slowly, so that most laboratory experiments are stopped when a considerable proportion of the original oil is left. The polyaromatic hydrocarbon are known to be degraded under laboratory conditions and it is reasonable to suppose that they will also degrade in nature. The resinous and asphaltic compounds are also very resistant and it is reasonable to suppose that they will require decades before finally being mineralized. Since degraded oil tends to become heavier than sea water, due to the incorporation of oxygen into the molecule and because sand, silt and other detritus adhere to the surface, particles of oil which have been at sea for some time will slowly sink to the bottom. Hence most of the recalcitrant material will degrade slowly in sediments. If oil becomes covered, as for example, by burial on a beach, there is very little change in the oil and it can be expected to persist indefinitely.

The effects of organic matter also present in the sea water can be expected to have an effect on oil degradation. Peptone and sugars have a sparing effect, but co-oxidation may also occur. In addition essential nitrogen and phosphorus may be provided (Gatelier *et al.* 1973). It follows that discharge of oil along with other degradable material could produce changes in the fate of the oil.

The supply of inorganic nutrients is known to be one of the important rate limiting factors (Atlas & Bartha 1972). This was confirmed by Gibbs (1975) using a semiclosed system with Kuwait crude oil as substrate. Various amounts of inorganic nitrogen, usually nitrate, and phosphorus were added periodically; the doses being calculated so that the upper values of concentration were not greatly in excess of those found in British coastal waters. He found that there was no uptake of nitrate in the absence of oil and that the nitrogen accumulated. When oil was present the nitrate disappeared from the solution having presumably been taken up by the bacteria. Oxygen uptake was taken as a measure of oil degradation, and was stimulated by the addition of nitrate. The oxygen uptake rate fell off only slowly even when the nitrogen had disappeared from the solution, thus suggesting that the bacteria were storing the nitrogen in some form. At the summer temperature (14°C) used in this experiment approximately 4μ mols available nitrogen were required to oxidize 1 mg oil calculated on a BOD for oil of 3·5 mg oxygen required to mineralize 1 mg oil. When the temperature was lowered to a winter temperature (4°C), the uptake of nitrogen and phosphorus was not markedly reduced. It is interesting to note 11μ mols available nitrogen are required to mineralize 1 mg oil at this lower temperature. Again the nitrogen supply rate is considered to be the rate limiting factor. On the basis of these figures it has been calculated (Gibbs *et al.* 1975) that the possible oil degradation rate for the Irish Sea would be around 30 g m^{-3} a^{-1} in summer and 11 g m^{-3} a^{-1} in winter. These figures apply only to that portion of the oil which degraded over a period of 9–12 months.

(c) INDUSTRIAL EFFLUENTS

Industrial effluents are complex mixtures of many substances which vary in composition and dilution. The majority of them are degraded during treatment in a normal sewage works. There are of course some very important exceptions such as the polychlorinated biphenyls (PCB). Toxicity is not itself a good guide to biodegradability. Cyanide for example can be broken down in a plant which has been suitably adapted biologically (Pettit & Mills 1954). Similarly aromatic compounds are mineralized under suitable conditions (Gibson *et al.* 1967; Evans 1969). It is, therefore, reasonable to assume that degradation will take place in the sea. However, conditions are markedly different in terms of the concentrations of the pollutant, the density of the degrading microbial population and the availability of other essential nutrients. Almost nothing is known of the disappearance rates of industrial effluents under marine conditions. The best that can be said is that the rates found in sewage works are maximum possible. Marine rates will be much lower.

Some of the pesticides, herbicides and fungicides used in agriculture are known to be persistant and therefore give cause for concern. Alexander (1965) has summarized much of the earlier work on microbial degradation of biocides in soils. Patel and his colleagues (1972) incubated radio labelled DDT and cyclodiene insecticides for 30 days in water from open oceans and found no change. However, there were indications of activity in surface films, plankton and algae. Similarly Moore & Dorward (1968) have demonstrated that oceanic cultures of green and blue-green algae apparently accumulated and metabolized a number of pesticides such as lindane, DDT and malathion providing the concentration was sufficiently low so that growth was not impaired. Bousch & Matsumura (1973) also implicate algae in pesticide degradation but suggest that the process may be non-metabolic.

Although the greatest part of sewage from industry and domestic use is treated in sewage works, considerable quantities of sludge from the major conurbations are dumped daily at various points around the coast of the UK. In spite of this long established practice, and despite the fact that primary sludge contains 70–80% of organic matter, there is very little data concerning the rate of degradation in the sea. Recently Barrett and his colleagues (1972) in a few preliminary experiments incubated both suspended and settled sludge in sea water at 20°C. This temperature is about 5°C higher than the maximum found in British coastal waters. The suspended solids oxidized rapidly for the first ten days and then slowed down. During the period between the 20th and 30th day of the experiment ammonia was oxidized to nitrate probably due to nitrifying bacteria. After 110 days the concentrations of organic carbon and organic nitrogen had decreased by around 60% to 70%. The results for the settled solids were less satisfactory so that the authors could only conclude that biological oxidation takes place, though the rate depends on the thickness of the deposit.

(d) PLASTICS AND OTHER POLYMERS

The rapidly increasing use of plastics has lead perhaps inevitably to a great increase in the amount of these materials that eventually litter the marine environment. Carpenter and his co-workers (Carpenter & Smith 1972; Carpenter *et al.* 1972) found an abundance of polystyrene granules (0·1–2 mm) in coastal areas with a maximum density of 0·1 mg m^{-2}. Somewhat higher concentrations were found offshore compared with the immediate coastal strip. Plastic pieces were also found in plankton hauls taken in the Sargasso sea at an average concentration of 0·29 mg m^{-2}. Diatoms and hydroids were attached to these particles.

Some plastic films are potentially biodegradable by either bacterial or fungal action (Evans & Levisohn 1968; Booth *et al.* 1968; Sharpe & Woodrow 1971). However, rates of destruction in the marine environment can be expected to be slower than in the laboratory. Polyvinyl chlorides did not show evidence of microbial attack when left in the sea though there was evidence that under favourable laboratory conditions some oxidation took place depending upon the formulation. A few PVC formulations were degraded anaerobically (Coscarelli 1964). Polyurethanes which were submerged in the Mediterranean sea for several years were found to support the growth of several fungi and of an associated pink bacterium (Jones & le Campion—Alsumard 1970). Algae too utilize certain vinyl additives which can cause problems in swimming baths lined with vinyl (Yeager 1970). The problem may be overcome by adding a suitable biocide. Polyethylenes are not attacked by micro-organisms either in the sea or under laboratory conditions.

A large-scale investigation into the biodegradation of many different synthetic and natural polymers was carried out by Muroaka (1965). For almost 200 days 2,385 specimens of 603 different materials were exposed on the floor of the Pacific. Untreated wood, jute, cotton and manilla rope were attacked by micro-organisms to a significant degree. Glass, plastic, laminates, plastic rope and synthetic rubber were not attacked although they were covered with bacterial slimes. Boring animals were in fact responsible for greater damage than micro-organisms. The same author tested a number of materials in sea water containing nutrients at 15°C with bacteria from deep sediments. Neoprene rubber was susceptible to bacterial action; silicone rubber, GRS-Rubber and polyvinyl chlorides were fairly resistant while polyethylene was not attacked at all.

(f) METAL TRANSFORMATIONS

Obviously metals cannot be degraded. Nevertheless micro-organisms modify the form in which they are found in the marine environment to a much greater extent than is generally realized. Many metals such as gold, uranium, zinc, vanadium, cobalt and manganese are involved in microbial geochemical processes (Zajic 1969).

Some substrates which are prone to microbial oxidation also act as chelating agents and remove the metal in effect from the solutions (Siegel 1971). The microbial intervention may be direct or indirect. The most important indirect transformation is the precipitation of iron as sulphide formed by the reduction of sulphate by the anaerobic *Desulphovibrio desulphuricans* in the presence of organic matter. However, from the point of view of pollution the bacterial intervention in the cycle of mercury is currently the most interesting, and possibly the most important because of the high toxicity of this element. Much of the man-made pollution of the sea by mercury arises from the use of agricultural fungicides, and from spent catalysts used in industrial processes. It is also found in fossil fuels (coal, oil, lignite) which accounts for an annual release into the environment of around 5×10^3 mt at the present rate of energy utilization. Some of this atmospheric mercury eventually will find its way into the natural water systems. It has been calculated that at least $1{\cdot}2 \times 10^5$ metric tons of mercury have entered the environment by human activity (Gavis & Ferguson 1972). The mercury in the aquatic environment is associated very largely with the particulate phase, either in suspension or as bottom sediment (Jernelov 1972). It is known that some bacteria develop a high resistance to mercury. *Staphylococcus aureus* which is common in the human body, develops a high resistance in patients treated with mercurial drugs. Similarly factory workers exposed to mercury carried a higher percentage of resistant nasal staphylococci than people not so exposed (Hall 1970). In contrast some bacteria are killed and this property was used by Hansen & Bonde (1969) as a means of determining low-level concentrations of the metal.

Transformations of mercury by bacteria in the marine environment was first noted by Jansen & Jernalov (1969), when they found that sediment from an aquarium methylated mercury from mercuric chloride. This methylation did not occur with heat sterilized sediments. The sediments from Swedish lakes and rivers was found to form both methyl and dimethyl mercury from the ionic species (Jernelov 1969). The methylation did not proceed under anaerobic conditions, presumably because the presence of hydrogen sulphide caused the precipitation of the mercury as sulphide. A detailed study of the biochemical mechanism of methylation was made by Wood *et al.* (1968) using extracts from an anaerobic fresh water methanogenic bacterium. They found that both dimethyl and methyl mercury was formed and that trivalent cobalt was implicated in the methylation process as methyl donor. Mercury thus will circulate in the aquatic biosphere in the methylated form, but in the sediments it is more likely to be inorganically bound either as the sulphide or in iron complexes. With time it will be covered to form the mineralogenic layers and so be removed from interference with biological entities.

Conclusions

It is only since World War II that the microbiologist has begun to examine the role of micro-organisms in the natural environment as compared with their role in disease or as biochemical tools. As a result very little is really known of the role they play in the cold dilute medium that is the sea. Sufficient information has been discovered to reveal the enormous complexities of the problem and the technical difficulties that have yet to be overcome. Nevertheless some insight is being gained into the processes of degradation especially of politically sensitive pollutants such as oil. Probably the major advances of the near future will be made with less complex substrates. However, it is now clear that no pollutant should be added to the sea until it is reasonably certain that it will biograde. To do so is to invite further problems such as are now being encountered with chlorinated hydrocarbons, PCB and plastics. Furthermore, since marine rates of decomposition are likely to be much slower than in the laboratory, and the situation much more complex, all additions to the sea by human action should be made with circumspection and caution.

References

AHEARN D.G. & MYERS S.P. (1972) The role of fungi in the decomposition of hydrocarbons in the marine environment. In: *Biodeterioration of materials*, eds. A.H. Walters & Hueck van der Plas **2**, pp. 12–18. Applied Science Publishers, London.

ALEXANDER M. (1965) Biodegradation problems of molecular recalcitrance and microbial infallibility. *Adv. appl. Microbiol.* **7**, 35–80.

ANDREWS A.R. & FLOODGATE G.D. (1974) Some observations on the interactions of marine protozoa and crude oils. *Mar. Biol.* **25**, 7–12.

ANDREWS P. & WILLIAMS P.J. LE B. (1971) Heterotrophic utilisation of dissolved organic compounds in the sea. III. Measurement of oxidation rates and concentrations of glucose and amino acids in sea water. *J. mar. biol. Ass. UK* **51**, 111–25.

ANONYMOUS (1975) *Petroleum in the marine environment.* Workshop at Airlie, Virginia, U.S.A. May 1974, published by National Academy of Science, Washington.

ATLAS R.M. & BARTHA R. (1972) Degradation and mineralisation of petroleum in sea water; limitations of nitrogen and phosphorus. *Biotech. Bioeng.* **14**, 309–18.

Barratt M.J., Gilligan R.M. & Mollowney B.M. (1972) Degradation of sludge. In: *Out of sight, out of mind. Report of Working Party on disposal of sludge in Liverpool Bay*, Vol. 2, Appendix 3. HMSO, London.

Blumer M., Guillard R.R.L. & Chase T. (1971) Hydrocarbons in marine phytoplankton. *Mar. Biol.* **8**, 183–8.

Booth G.H., Cooper A.W. & Robb J.A. (1968) Bacterial degradation of plasticised P.V.C. *J. appl. Bact.* **31**, 305–10.

Bungay, H.R. & Bungay M.L. (1968) Microbial interactions in continuous culture. *Adv. appl. Microbiol.* **10**, 269–89.

Bousch G.M. & Matsumura F. (1973) Pesticide degradation by marine algae. *Government Reports Announcement* **73**, 50.

Burwood R. & Speers G.C. (1974) Photo-oxidation as a factor in the environmental dispersal of crude oil. *Estuarine and Coastal mar. Sci.* **2**, 117–36.

Butler E.I., Corner E.D.S. & Marshall S.M. (1970) On the nutrition and metabolism of zooplankton. VII. Seasonal survey of nitrogen and phosphorus excretion by Calanus in the Clyde Sea. *J. mar. biol. Ass. UK* **50**, 525–520.

Carpenter E.J. & Smith K.L. (1972) Plastics in Sargasso sea surface. *Science* **175**, 1240–2 (erratum in *Science* **177**, 85).

Carpenter E.J., Anderson S.J., Harvey G.R., Miklas H.P. & Peck B.B. (1972) Polystyrene spherules in coastal waters. *Science* **178**, 749–50.

Characklis W.G. (1973) Attached Microbial Growth. I. Attachment and growth. *Wat. Res.* **7**, 1113–27.

Claus G., Madri P. & Kimen S. (1967) Removal of microbial pollutants from waste effluents by the red beard sponge. *Nature, Lond.* **216**, 712–14.

Coscarelli W. (1964) Deterioration of organic materials by marine organisms. In: *Principles and application in aquatic microbiology*, eds. H. Heukelekian & N.C. Dondero, pp. 113–47. Wiley & Sons, London.

Davis S.J. & Gibbs C.F. (1975) The effects of weathering on a crude oil residue exposed at sea. *Wat. Res.* **9**, 275–88.

Degens E.T., Reuter J.H. & Shaw K.N.F. (1964) Biochemical compounds in offshore California. Sediments and Sea water. *Geochim cosmochim Acta* **28**, 45–66.

Douros J.D. & Frankenfeld J.W. (1968) Effects of culture conditions in production of *trans* cinnamic acid from alkyl benzenes by soil micro-organisms. *Appl. Microbiol.* **16**, 320–5.

Duursma E.K. (1965) The dissolved organic constituents of sea water. In: *Chemical Oceanography*, Vol. I, eds. J.P. Riley & Skirrow, pp. 433–75. Academic Press, London.

Edwards R.W. & Rolley H.L.J. (1965) Oxygen consumption in river muds. *J. Ecol.* **53**, 1–19.

Evans W.C. (1969) Microbial transformation of aromatic compounds. *Ferm. Adv.* 1969, 649–87.

Evans D.M.E. & Levisohn I. (1968) Biodeterioration of polyester based polyurethanes. *Int. Biodeter. Bull.* **4**, 89–92.

Floodgate G.D. (1972) Biodegradation of hydrocarbons in the sea. In: *Water Pollution Microbiology*, eds. R. Mitchell, pp. 153–72. Wiley & Sons, London.

Floodgate G.D. (1973) A threnody concerning the biodegradation of oil in natural waters. In: *Microbial degradation of oil pollutants*, eds. D.G. Ahearn & S.P. Meyers, pp. 17–24. Louisiana State University No. LSU-SG-7301.

Fogg G.E. (1963) The role of algae in organic production in aquatic environments. *Brit. Phyc. Bull.* **2**, 195–205.

Gatelier C.R., Oudin J.L., Fusey P., Lacaze J.C. & Priou M.L. (1973) Experimental ecosystems to measure fate of oil spills dispersed by surface active products. *Proc. Int. Conf. on prevention and control of oil spills*. pp. 497–504. API-EPA-USGC, Washington.

GAUDY A.F. (1972) Biochemical oxygen demand in water. In: *Water Pollution Microbiology*, ed. R. Mitchell, pp. 305–31. Wiley & Sons, London.

GAVIS J. & FERGUSON J.F. (1972) The cycling of mercury through the environment. *Wat. Res.* **6**, 989–1008.

GIBBS C.F. (1972) A new approach to the measurement of rate of oxidation of crude oil in sea water. *Chemosphere* **1**, 119–24.

GIBBS C.F. (1975) Quantitative studies on marine biodegradation of oil. I. Nutrient limitation at 14°C. *Proc. Roy Soc. Lond.* **B188**, 61–82.

GIBBS C.F., PUGH K.B. & ANDREWS A.R. (1975) Quantitative studies on marine biodegradation of oil. II. Effects of temperature. *Proc. Roy Soc. Lond.* **B188**, 83–94.

GIBSON D.T., WOOD J.M., CHAPMAN P.J. & DAGLEY K.S. (1967) Bacterial degradation of aromatic compounds. *Biotech. Bioeng.* **9**, 33–44.

HALL B.M. (1970) Mercury resistance of *Staphylococcus aureus*. *J. Hyg.* **68**, 121–9.

HANDA N. & TOMINAGA H. (1969) A detailed analysis of carbohydrates in marine particulate matter. *Mar. Biol.* **2**, 228–35.

HANSEN J.C. & BONDE G.J. (1969) Microbiological determination of mercury in trace amounts including marine sediments. *Rev. int. oceanogr. med.* **15–16**, 5–28.

HOBBIE J.E., HOLM-HANSEN O., PACKARD T.T., POMEROY L.R., SHELDON R.W., THOMAS J.P. & WIEBE W.J. (1972) The study of the distribution and activity of micro-organisms in ocean water. *Limnol Oceanogr.* **17**, 544–55.

HOFSTEN B.V. & EDBERG N. (1972) Estimating the rate of degradation of cellulose fibres in water. *Oikos* **23**, 29–34.

HOLM-HANSEN O. & PAERL H.W. (1972) The applicability of ATP determinations for estimation of microbial biomass and metabolic activity. *Mem. Inst. Ital. Idrobiol.* **29**, Supplement, 149–68.

HOOD D.W. (1963) Amino acids in the marine environment. In: *Oceanography and Marine Biology*, ed. H. Barnes, Vol. I, pp. 129–56. Allen & Unwin, London.

HUNTER J.V. (1971) Origin of organics from artificial contamination. In: *Organic compounds in aquatic environments*, eds. S.D. Faust & J.V. Hunter, pp. 51–94. Marcel Dekker, New York.

JANNASCH H.W., EIMHJELLEN K., WIRSENCO & FARMANFARMAIAN A. (1971) Microbial degradation of organic matter in the deep sea. *Science* **171**, 672–5.

JANNASCH H.W. & JONES G.E. (1959) Bacterial population in sea water as determined by different methods of enumeration. *Limnol Oceanogr.* **4**, 128–39.

JENSEN S. & JERNELOV A. (1969) Biological methylation of mercury in aquatic organisms. *Nature, Lond.* **223**, 753–4.

JERNELOV A. (1972) Mercury—a case study of marine pollution. In: *The changing chemistry of the oceans. Proceeding of 20th Nobel Symposium*, eds. D. Dyrssen & D. Jagner, pp. 161–9. Wiley Interscience, London.

JONES E.G.B. & LE CAMPION-ALSUMARD T. (1970) The biodeterioration of polyurethane by marine fungi. *Internat. Biodeter. Bull.* **6**, 119–24.

KUENTZER L.E. (1969) Bacteria, carbon dioxide and algal blooms. *J. wat. poll. contr. Fed.* **41**, 1737–47.

KRISS A.E. (1963) Marine Microbiology (Deep Sea). Translated by J.M. Shewan & Z. Kabata. Oliver & Boyd, London.

KOYAMA T. & THOMPSON T.D. (1959) Organic acids in sea water. *A.A.A.S. preprints of International Oceanographic Congress 1959*, 925–6.

LITCHFIELD C.D., MUNRO A.L.S., MASSIE L.C. & FLOODGATE G.D. (1974) Biochemistry and microbiology of some Irish sea sediments. I. Amino acid analyses. *Mar. Biol.* **26**, 249–60.

LOVATT EVANS C. (1947) *Principles of human physiology*, 9th Edition, p. 865. J.A. Churchill Ltd., London.

MCCUTCHEON J.W. (1966) Chemical structure in biodegradability. *Proceeding annual meeting Chemical Specialities Manufacturers Association* **53**, 62–4.

MAIR B.J. (1964) Hydrocarbons isolated from petroleum. *Oil gas. J.*, Sept. 14, 1964, 130–4.

MATTHEW W.S. & SMITH L.L. (1968) Sterol metabolism. III. Sterols in marine waters. *Lipids* **3**, 239–46.

MITCHELL R. (1972) Ecological control of microbial imbalances. In: *Water Pollution Microbiology*, ed. R. Mitchell, pp. 273–88. Wiley & Sons, London.

MOORE R.B. & DORWARD D.A. (1968) Accumulation and metabolism of pesticides by algae. *J. physol.* **4**, Supplement, p. 7.

MURAOKA J.S. (1965) Deep sea biodeterioration of materials. II. Six months at 2340 feet. *Technical Report U.S. Naval Civil Engineering Laboratory*, R. 393.

MURAOKA J.S. (1966) Effects of deep sea micro-organisms on rubber and plastic insulation. *Mat. Protect.* **5**, 35–7.

NEDWELL D.B. & FLOODGATE G.D. (1971) The seasonal selection by temperature of heterotrophic bacteria in an intertidal sediment. *Mar. Biol.* **11**, 306–10.

PARSONS T.R. & STRICKLAND J.D.H. (1962) On the production of particulate organic carbon by heterotrophic processes in sea water. *Deep Sea Res.* **8**, 211–22.

PATEL K.C., MATSUMURA F. & BOUSCH G.M. (1972) Microbial transformation of DDT, dieldrin, aldrin and endrin by marine micro-organisms. *Environ. Sci. Technol.* **6**, 629–32.

PERRY J.J. & CERNIGLIA C.E. (1973) Studies on the degradation of petroleum by filamentous fungi. In: *Microbial degradation of oil pollutants*, eds. D.G. Ahearn & S.P. Myers, pp. 89–94. Louisiana State University No. LSU-S9-7301.

PETTET A.E.J. & MILLS E.V. (1954) Biological treatment of cyanides, with and without sewage. *J. appl. chem.* **4**, 434–44.

POMEROY L.R. & JOHANNES R.E. (1968) Occurrence and respiration of ultra plankton in the upper 500 m of the ocean. *Deep Sea Res.* **15**, 381–91.

POSTGATE J.R. (1969) Viable counts and viability. In: *Methods in Microbiology*, Vol. I, eds. J.R. Norris & D.W. Ribbons, pp. 611–28. Academic Press, London.

REDMOND J.D. (1972) Mineralisation of dissolved carbohydrates in tropical and temperate waters. *Mar. poll. Bull.* **3**, 45–7.

REMSEN C.C., CARPENTER E.J. & SCHROEDER B.W. (1972) Competition for urea among estuarine micro-organisms. *Ecology* **53**, 921–6.

PROCHAZKA G.J. & PAYNE W.J. (1965) Bacterial growth as a practical indicator of extensive biodegradability of organic compounds. *Appl. microbiol.* **13**, 702–5.

RYTHER J.H. (1963) Geographic variations in productivity. In: *The Sea*, Vol 2, ed. M.N. Hill, pp. 347–80. Wiley Interscience, London.

SHARPE A.N. & WOODROW M.N. (1971) A rapid test for the biodegradability of PVC film by pseudomonads. *J. appl. Bact.* **34**, 485–9.

SEKI H. (1972) The role of micro-organisms in the marine food chain with reference to organic aggregates. *Mem. inst. Ital. Idrobiol.* **29**, Supplement, pp. 245–60.

SEKI H., WADA E., KOIKE I. & HATTORI A. (1974) Evidence of high organotrophic potentiality of bacteria in the deep ocean. *Mar. Biol.* **26**, 1–4.

SIEBURTH J. MCN. (1968) The influence of algal antibiosis on the ecology of marine micro-organisms. In: *Advances in the Microbiology of the sea*, Vol. I, eds. M.R. Droop & E.J.F. Woods, pp. 63–94. Academic Press, London and New York.

SIEGEL A. (1971) Metal organic interactions in the marine environment. In: *Organic Compounds in aquatic environments*, eds. S.J. Faust & J.V. Hunter, pp. 265–96. Marcel Dekker, New York.

SOROKIN YU I. (1970) Determination of activity of heterotrophic microflora in the oceans by means of C^{14} labelled organic substances. *Mikrobiologiya* **39**, 149–56.

STEELE J.H. & BAIRD I.E. (1972) Sedimentation of organic matter in a Scottish Loch. *Mem. inst. Ital. Idrobiol.* **29**, Supplement, pp. 73–88.

STOTZKY G. & REM L.T. (1966) Influence of clay minerals on micro-organisms. I. Montmorillonite and kaolin in bacteria. *Canad. J. Microbiol.* **12**, 547–63.

TAKAHASHI F.T. & KITTREDGE J.S. (1973) Sublethal effects of the water soluble component of oil; chemical communication in the marine environment. In: *The microbial degradation of oil pollutants*, eds. D.G. Ahearn & F.P. Myers, pp. 259–64. Louisiana State University, LSU-SG-7301.

UDEN N. VAN (1969) Kinetics of nutrient limited growth. *Ann. Rev. Microbiol.* **23**, 473–86.

VACARRO R.F., HICKS S.E., JANNASCH H.W. & CAREY F.G. (1968) Occurrence and role of glucose in sea water. *Limnol Oceanogr.* **13**, 356–60.

WAGNER F.S. (1969) Composition of the dissolved organic compounds in sea water: a review. *Contrib. mar. sci. Univ. Texas* **14**, 115–53.

WIDMARK J.G., GARRETT W.D. & PALMORK K.H. (1972) Organic chemicals in Report of the Seminar on methods of detection, measurement and monitoring of pollutants in the marine environment. *Supplement to the report of the Technical conference on marine pollution and its effects on living resources and fishing*, pp. 53–64. F.A.O. Rome. Dec. 4–10, 1970.

WILLIAMS P.M. (1971) The distribution and cycling of organic matter in the oceans. In: *Organic compounds in aquatic environments*, eds. S.J. Faust & J.V. Hunter, pp. 145–63. Marcel Dekker, New York.

WILLIAMS P. LE B. & GRAY R.W. (1970) Heterotrophic utilisation of dissolved organic compounds in the sea. II. Observations on the responses of heterotrophic marine populations to abrupt increases in amino acid concentrations. *J. mar. biol. Ass. UK.* **50**, 871–81.

WOOD J.M., KENNEDY F.S. & ROSEN G.G. (1968) Synthesis of mercurial compounds by extracts of a methogenic bacterium. *Nature, Lond.* **220**, 173.

WRIGHT R.T. (1973) Some difficulties of using C^{14} organic solutes to measure heterotrophic bacterial activity. In: *Estuarine Microbial Ecology*, eds. L.H. Stevenson & R.R. Colwell, pp. 199–217.

WRIGHT R.T. & HOBBIE J.E. (1965) The uptake of organic solutes in lake water. *Limnol Oceanogr.* **10**, 22–8.

YEAGER C.C. (1970) Biodegradation of plastics. ASTM, meeting Atlantic City Abstracted by IBRIS **6**, abstract No. 70/2-38-0974.

YOUNGBLOOD W.W., BLUMER M., GUILLARD R.L. & FIORE R. (1971) Saturated and unsaturated hydrocarbon in marine benthic algae. *Mar. Biol.* **8**, 130–201.

YOUNGBLOOD W.W. & BLUMER M. (1973) Alkanes and alkenes in marine benthic algae. *Mar. biol.* **21**, 163–72.

ZAJIC J.E. (1969) *Microbial biogeochemistry*. Academic Press, New York.

10 Decomposition of marine macrophytes

K. H. MANN *Department of Biology, Dalhousie University, Halifax, N.S., Canada*

In the coastal marine environment, productivity at all trophic levels is normally higher than offshore. In part, this is due to the high productivity of marine macrophytes such as seaweeds, seagrasses and marsh grasses which flourish at the edge of the sea. These plants are seldom grazed extensively by herbivores. Instead, they die, decay and enter detritus food chains. An understanding of decomposition processes is essential to an understanding of the dynamics of coastal marine productivity.

Evidence for the foregoing statements was reviewed (Mann 1972a) in the IBP symposium 'Detritus and its ecological role in aquatic ecosystems'. This paper reports further investigations into the decomposition process, as it occurs on the plant or after deposition in the sediments. The work was all performed on material from inlets on the east coast of Nova Scotia, Canada, but the results should have general application.

Bacteria and seaweeds

Laminaria longicruris, a common kelp on the east coast of Canada, is attached to a solid surface by a holdfast, which in turn leads to a stalk, or stipe. Beyond this is a ribbon-shaped blade, which may be several metres long. Growth occurs at the meristem at the base of the blade while older, often dead and decaying material is eroded from the tip of the blade. Hence the fronds are in effect moving belts of tissue, continually being added to at the base and eroded at the tip. In the course of a year the tissue of a blade may be replaced several times, so that a large amount of organic detritus is liberated into the water, and eventually sinks to the sediments (Mann 1972b). Growth is rapid in winter and early spring, and slows down in summer and early fall, when reserves of laminaran are accumulated (Black 1950).

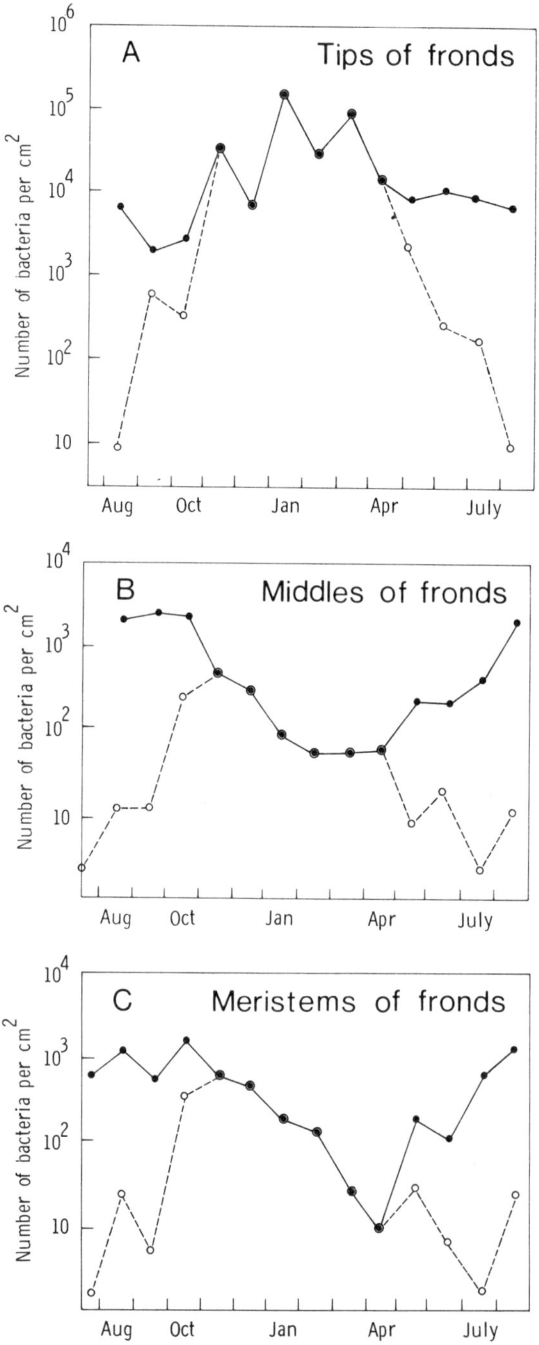

Figure 10.1. Numbers of bacteria per cm² of blade of *Laminaria longicruris*, as determined by incubation of plates at ambient temperature (solid line) and 3°C (broken line), in various months of the year. A, from tips of fronds; B, from middles and C from meristems at bases of fronds. Observations made from March 1972 to March 1973, but rearranged to emphasize mid-winter peak. (After Laycock 1974.)

In some sheltered habitats, at this time, the region of decaying tissue is seen to spread from the tip of the blade down towards the meristem.

Laycock (1974) studied the bacterial flora of different parts of the blade of *L. longicruris* through an annual cycle of plant growth and decay. Each month she took from five different plants four samples of tissue from each of the three zones: tip, middle and meristem. The samples were blended in an appropriate medium, diluted, plated, and incubated for 15 days at ambient seawater temperature, and at two of the temperatures 3°, 10° and 18°C most different from ambient. Plates having between 20 and 200 colonies were counted, and numbers of bacteria per cm^2 of blade were calculated. Bacteria were then isolated from the plates and tested for their ability to utilize protein, mannitol, alginate and laminaran.

Fig. 10.1 shows the total numbers of bacteria per cm^2 of blade from plates incubated at ambient temperature, using material from the three regions of the blades. Numbers were greatest at the decomposing tips, and reached their maximum in January. At this time, bacteria were 1,000 times more abundant on the tips than elsewhere on the blade. The broken line shows the results of incubation at 3°C, at times when the ambient seawater temperature was above this level. These counts may be regarded as representing the psychrophilic bacteria. They were comparatively scarce in summer, when they were replaced by a mesophilic population, but they dominated in winter, and were particularly abundant on the decomposing tips. Table 10.1 shows that forms capable of hydrolysing laminaran were particularly abundant on the decomposing tips between November and March. Hence, the psychrophilic forms were characterized by the laminaran hydrolysers.

Fig. 10.2 shows the numbers of organisms capable of hydrolysing mannitol, alginate, and protein, as determined by incubations at ambient temperature. They were reasonably abundant on the decomposing tips at all times of year, but on other parts of the blades their numbers increased greatly during summer and early fall. They may be regarded as mesophilic forms, which built up their numbers as the zone of decaying tissue spread toward the base of the blade in late summer.

Table 10.1. *Laminaria longicruris.* Numbers of laminaran hydrolysers per cm^2. (From Laycock 1974.)

	Month						
Frond region	Oct.	Nov.	Dec.	Jan.	Feb.	March	April
Decomposing tips	1,000	2,500	1,000	6,500	1,000	21,000	310
Middles	1,000	1,000	130	1·2	15	1·2	<0·6
Meristems	12	1,300	4·7	11	1·7	<0·19	<1·4
Healthy tips					240	<3·2	<1·4

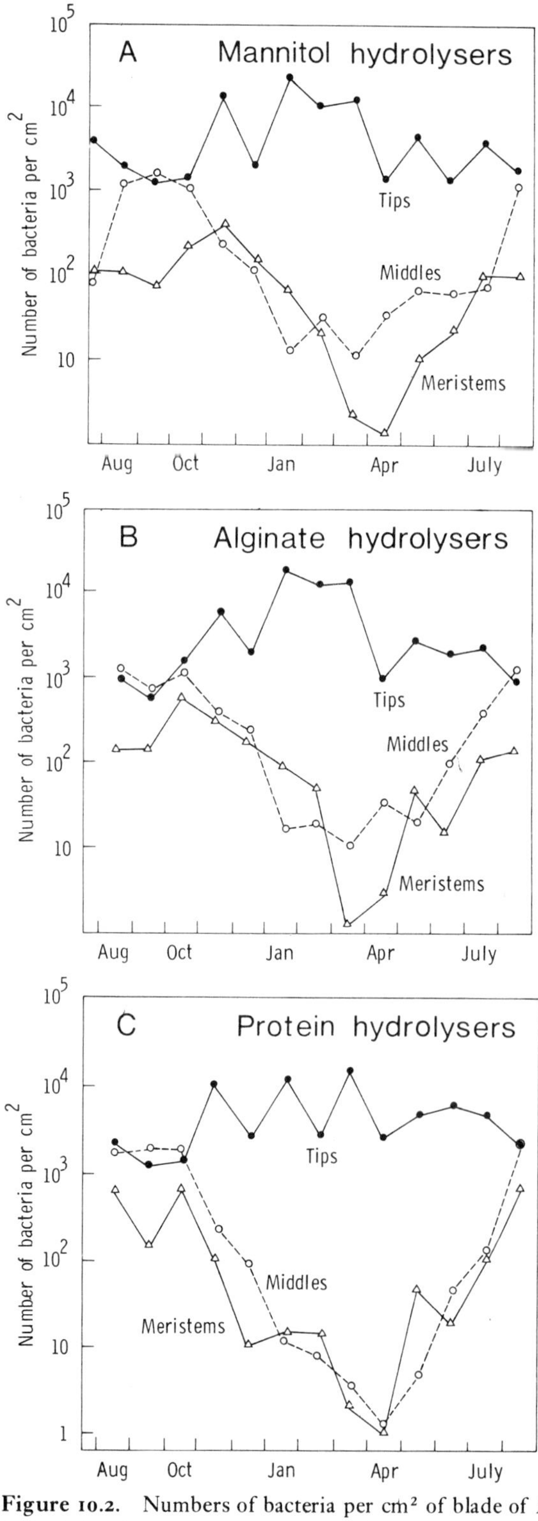

Figure 10.2. Numbers of bacteria per cm² of blade of *Laminaria longicruris* capable of hydrolysing A, mannitol; B, alginate; C, protein. Data from tips, middles and meristems of blades displayed separately. (After Laycock 1974.)

Thus, there was a meaningful pattern in the annual cycle of growth and decay of plants and the numbers and functional characteristics of the epiphytic bacteria. In winter the plants grew rapidly. Most of the blade was strong and healthy, and laminaran content was declining from its autumn peak. At this time, bacteria were most abundant near the eroding tips, and those capable of metabolizing laminaran reached their seasonal peak. Bacteria capable of metabolizing mannitol (the characteristic product of photosynthesis) and alginates and proteins, were abundant at the tips of the blade at all times of year, but became much more abundant on the remainder of the blade during the warm summer months.

This study was carried out in a sheltered inlet with relatively high summer temperatures. On the open coast, where wave action (and hence nutrient renewal) is more vigorous, and where summer warming is less pronounced, growth continues at a faster rate in summer, and the decomposing tissue is more or less confined to the tip of the blade. This condition is perhaps more common than that described in detail earlier. Nevertheless, when spot checks were made on the distribution of bacterial types on plants from an exposed area, they were found to give results qualitatively and quantitatively similar to those from the sheltered area.

The implications of these findings are that a great deal of the productivity of *Laminaria longicruris* is taken up by epiphytic bacteria. In winter and early spring the tip of the blade is apparently healthy, but is eroded by wave action. The majority of organisms isolated during the winter maximum hydrolysed neither mannitol, alginate or protein (cf. Figs. 10.1 and 10.2). It seems probably that the dense population of bacteria was taking up the contents of ruptured, healthy cells, e.g. amino acids, and stored laminaran.

However, with increasing senescence of the algal tissues during the summer months, the number of proteolytic isolates increased very rapidly, and there is a strong presumption that the plant tissues were being attacked and disrupted by the bacteria.

Decomposition of eelgrass

While seaweeds colonize areas of strong water movement and rocky substrates, the areas of quiet water and soft sediments tend to be colonized by seagrasses such as *Zostera marina*. In the classic work of Petersen (1918) it was proposed that the food chains leading to fish in the inshore waters of Denmark depended on eelgrass as the primary producer. In the 1930's the eelgrass stock on both sides of the Atlantic was drastically reduced; the explanation is still being debated (Burkholder & Doheny 1968; Rasmussen 1973). However, the expected decline in commercial fish stocks did not occur, and the ecological role of eelgrass was questioned (Nelson 1947; Thorson 1971).

Harrison (1974; Harrison & Mann 1975) studied the decomposition of *Zostera marina* leaves under controlled laboratory conditions, with a view to determining rates of decomposition, and the relative importance of microbial decay, as opposed to autolysis and leaching. Leaf material of various ages and sizes was incubated in flasks with Millipore HA (0·45 μm) filtered seawater. The leaf material was weighed at the start of each experiment, and was recovered at the end by filtering the contents of each flask through a 0·8 μm Millipore membrane. Changes in total dry weight, organic matter and carbon and nitrogen content were monitored under various experimental conditions. It was found that at 20°C, fresh material lost organic matter more rapidly than dead leaves, and dead leaves that had been dried decomposed more rapidly than those that had not. The experiments with dried, dead leaves gave the most reproducible results, and were adopted as standard.

When eelgrass detritus of three different particle sizes was incubated in seawater to which a natural microbial flora had been added, it was found that particles of less than 1 mm diameter lost organic matter about twice as fast as particles averaging 12 mm in diameter (Fig. 10.3).

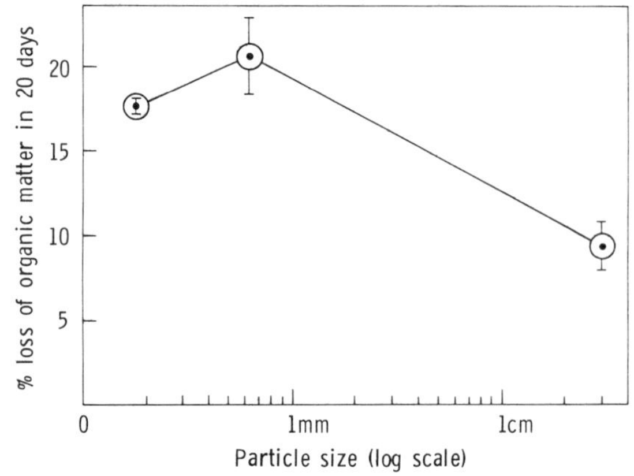

Figure 10.3. Loss of organic matter when dried, dead eelgrass detritus of various particle sizes was incubated for 20 days at 20°C in seawater to which a natural microflora had been added. (After Harrison & Mann 1975.)

To investigate the relative importance of microbial decay, as opposed to leaching, incubations were carried out under sterile and unsterile conditions. Sterilization was achieved either by incubating in a 1/1,000 (w/v) solution of potassium cyanide, or by subjecting the material to a dry heat of 100°C for 60 hours. The sterile flasks were checked for bacteria by taking

Table 10.2. Losses of dry weight and organic matter from dried eelgrass detritus of various size fractions, to show differences between sterile material and material incubated with micro-organisms. Values are means ± 1 S.D. Experiments in triplicate. Significance determined by t-test. (From Harrison 1974.)

Particle size	Method of sterilization	% loss of dry wt.		Signif. of difference	% loss of organic matter		Signif. of difference
		Sterile	Unsterile		Sterile	Unsterile	
Whole leaf	KCN	21·4 ± 2·9	22·6 ± 1·2	None	8·6 ± 3·4	9·3 ± 1·4	None
0·25–1·0 mm	KCN	27·0 ± 3·5	34·5 ± 1·7	95%	15·0 ± 2·9	20·5 ± 2·2	None
Under 0·42 mm	Heat	30·5 ± 0·3	32·5 ± 1·2	None	15·4 ± 0·5	20·1 ± 1·5	99%
Under 0·25 mm	KCN	24·7 ± 0·7	26·0 ± 0·4	None	15·0 ± 0·9	17·6 ± 0·4	95%

subsamples, diluting appropriately and incubating on plates of Zobell's 2216e medium. Some results, after incubating 20 days at 20°C, are shown in Table 10.2. In general, the sterilized material lost 90% as much weight and 80% as much organic matter as the unsterilized material, presumably by leaching of soluble matter. This surprising result caused us to question conventional ideas on the role of micro-organisms in the decomposition of eelgrass.

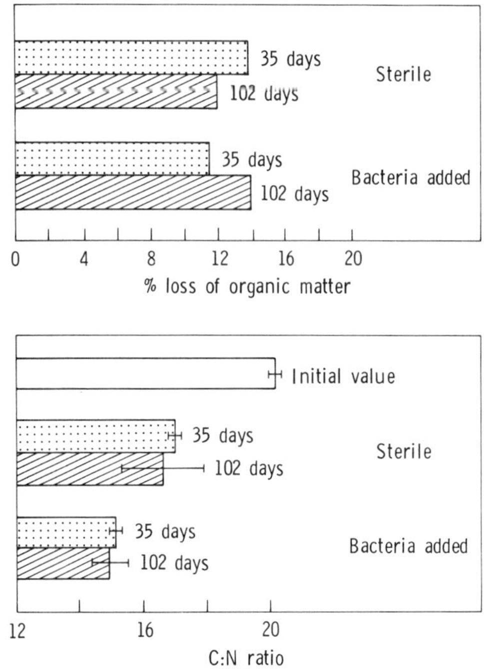

Figure 10.4. Results of incubating finely ground eelgrass detritus in sterile seawater, or with bacteria added, at 20°C for 35 days or 102 days. Upper figure shows percent loss of organic matter; lower figure the change in carbon : nitrogen ratio. (After Harrison 1974.)

Fig. 10.4 shows the result of long-term incubation of finely ground detritus. 102 days at 20°C would be expected to produce as much decomposition as a whole year at sea temperatures. The loss of organic matter was no greater with bacteria present than under sterile conditions. However, there was a shift in C/N ratio suggesting that a loss of organic carbon may have been compensated by uptake of nitrogen by micro-organisms.

It might be thought that incubation in closed flasks, with no opportunity for renewal of the water or the nutrients, was inhibiting microbial decomposition. Experiments were repeated with continuous renewal of the medium, so that it was replaced every 12 hours, and in other cases with enrichment of the medium with ammonia, nitrate or phosphate. In no case was there a significant increase in the rate of loss of organic matter, and in several cases there was a decrease. One can only conclude that the results given in Fig. 10.4 are representative of natural events. Harrison has shown (see Fenchel & Harrison Chapter 12) that the presence of protozoa produces some increase in bacterial decomposition. However, starting with dry eelgrass leaves (as would be the case with those tossed up on the shore) the rate of decomposition is very slow, involving a loss of no more than 20% of the organic matter in 100 days at 20°C. Similar experiments starting with fresh eelgrass leaves gave more variable results, but in general there was only slightly more loss of organic matter in a given set of experiments.

Hence, in contrast to the kelps, eelgrass seems to be resistant to decay and can be expected to accumulate in the sediments. Such accumulations

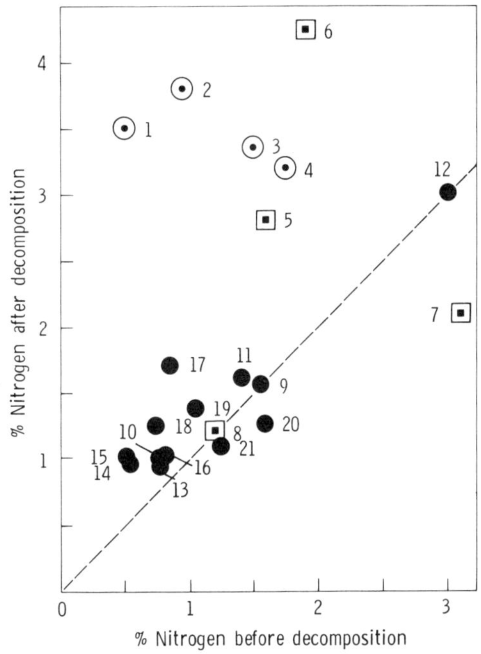

Figure 10.5. Changes in the nitrogen content (as % of remaining dry weight) during decomposition of detritus derived from eelgrass, and from various other marine, freshwater and terrestrial macrophytes. Key: 1, Mangrove leaves in field (Heald 1969); 2, marsh grass in field (Odum & de la Cruz 1967); 3, eelgrass leaves in field (Jensen 1914); 4, eelgrass leaves in lab. (Fenchel 1973); 5–8 eelgrass leaves in lab. (Harrison 1974); 9–17, terrestrial leaf litter decomposed six months on forest floor: 9, ash; 10, oak; 11, hazel; 12, alder (Bocock 1964); 13, chestnut; 14, beech (Anderson 1973); 15, maple; 16, beech; 17, birch (Gosz *et al.* 1973); 18–21 terrestrial leaf litter decomposed in freshwater for six months (18) or 50 days (19–21); 18, beech (Iversen 1973); 19, elm; 20, alder; 21, oak (Kaushik & Hynes 1971). Dashed line indicates no change in percentage nitrogen. (From Harrison 1974.)

are commonly seen in the vicinity of eelgrass beds. Coastal ecosystems may thus be expected to have a large reserve of slowly decomposing eelgrass detritus. Any food chains based on this detritus will be expected to continue to function for many years after a catastrophic decline in eelgrass production. This may be the explanation of the events referred to earlier.

The carbon to nitrogen ratio is often taken as a valuable index of the nutritional value of detritus (Russell-Hunter 1970). As Fig. 10.4 shows, over a period of 102 days the presence of micro-organisms is associated with a decrease in the C/N ratio of eelgrass detritus and hence an improvement in its nutritional value. Fig. 10.5 summarizes data from a number of sources on change in C/N ratios during decomposition, and shows that in a large number of studies there is evidence of an improvement in C/N ratio. Those in the eelgrass detritus (open circles and squares) are more marked than in terrestrial leaf litter.

THE FATE OF ORGANIC DETRITUS

A. Sedimentation

The environment of seaweeds and seagrasses is characterized by strong wave action and tidal currents. Hence, the decomposing plant detritus does not stay long in the vicinity of the plants, but is taken up in suspension and moved considerable distances. In St. Margaret's Bay, Nova Scotia, Sutcliffe (1972) monitored total particulate carbon in suspension in the water column. He distinguished the living from the dead by ATP determinations, and sorted the material into three size fractions. He recognized the spring bloom of phytoplankton by the peaks of living particulate carbon of large particle size. However, there were peaks of nonliving particulate carbon about one month before the spring bloom, at a time when the seaweeds were growing most rapidly. Concentrations were highest after strong winds, and he attributed the late winter peaks to material eroding from the kelps. There were similar peaks in June, July and August, at a time when phytoplankton production was low, but the seaweeds were showing signs of decomposition and would be expected to yield detritus.

Several years later, in the same bay, Webster *et al.* (1975) monitored the settlement of organic matter just above the sea bed, by means of sediment traps. One station was about 0·5 km offshore from a dense kelp bed, where the bottom was at a depth of 70 m, and the traps were suspended at 60 and 65 m. Another station was near to a seagrass bed, where the water was 10 m deep and the traps were placed at six depths between 0·5 and 2·0 m from the bottom. At the shallow station, the traps usually caught between 2 and

6 g per square metre per day of sediment having carbon content of about 7%. There were several peaks of sediment catch which were probably caused by wind-induced resuspension of bottom materials. At the deep station, the rate of sedimentation and the carbon content were about the same as at the shallow station but there was a clearly defined seasonal pattern (Fig. 10.6) with high catches in the period January to April. Since this coincides with the period of most rapid growth of the seaweeds, there is a strong presumption that the rain of organic materials is seaweed detritus.

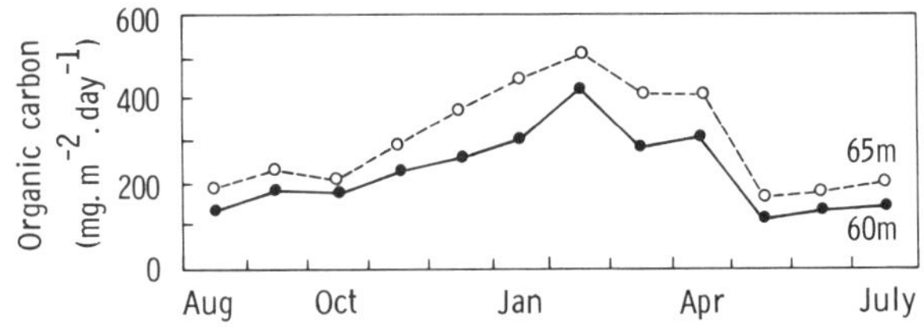

Figure 10.6. Daily rates of sedimentation of organic carbon in St. Margaret's Bay, Nova Scotia, in traps placed at 60 m and 65 m depth, over a bottom at 70 m depth.

An alternative explanation would be that winter is a time of storms, and that, even at a depth of 70 m, there is sufficient turbulence to resuspend bottom material and deposit it in the traps. This argument was checked by comparing the organic matter content of material in the traps with that in nearby sediments. For example, at the shallow station, the organic content of the material in the traps was 10·77%, while in nearby sediments it was only 4·52%. There was a similar difference at the deep station. Clearly, it is unlikely that resuspended bottom sediment accounts for all the material in the traps.

Other investigators have found that in sites remote from beds of macrophytes, it is usual to find that sedimenting material consists primarily of phytoplankton, and zooplankton faeces (e.g. Steele & Baird 1972). However, since the peak of fallout into our traps did not coincide with the bloom in the plankton, it is reasonably clear that the bottom deposits of St. Margaret's Bay receive a rain of decomposing organic detritus derived mainly from seaweeds and sea grasses.

B. Decomposition in the sediments

In St. Margaret's Bay, our studies indicated that seaweed productivity, averaged over the bay, amounted to about 600 g C m^{-2} yr^{-1} (Mann 1972b) while phytoplankton production was about 200 g C m^{-2} yr^{-1} (Platt 1971). Total annual catch in the sediment traps was 118 g C m^{-2} at the deep station and 134 g C m^{-2} at the shallow station. Hence, about 15% of the original plant production appeared to reach the sediments. However, there is evidence (Breen & Mann 1976) that seaweed beds were being destroyed by sea urchins in the years between the original measurement of seaweed productivity (1968) and the measurement of sedimentation (1971–2). Hence, the difference between the two may be less than we have indicated. Hargrave (1973) found that oxygen consumption of cores from sediments in St. Margaret's Bay followed the same seasonal pattern as the sedimentation of organic matter described above, and consumed 298 l of oxygen per m^2 per year. Assuming a respiratory quotient of 0·85, this would be enough oxygen to metabolize about 125 g C m^{-2} yr^{-1}, a figure which agrees well with the estimates of sedimentation given above.

It is of interest to know what proportion of this organic matter is metabolized by bacteria, and what proportion by benthic invertebrates. With this question in mind, Dale (1974) studied the biomass and distribution of bacteria in marine sediments. For convenience of sampling, he chose an area of extensive intertidal flats, including part of an eelgrass bed, and having a range of sediment types. Seven sampling stations were set up, with stations three, four and five within a stand of *Zostera marina*. Core samples were taken on three occasions at each station, using a Plexiglass tube of 55 mm diameter. The core was transported to the laboratory packed in crushed ice. Previous studies have tended to use methods which enumerate only aerobic bacteria. Zobell (1938) and Tsernoglou & Anthony (1971) used plate counts. Hargrave (1972) and Fenchel (1970) used oxygen uptake as an index of bacterial activity. In this study, the bacteria were stained with acridine orange and counted directly by fluorescence microscopy. Samples were taken from both aerobic and anaerobic portions of the core.

A number of parameters were measured to see which ones best described the distribution of aerobic and anaerobic bacteria. They included Eh, particle size distribution (as determined with a Coulter counter), organic carbon content, and total nitrogen content. The last two were determined with an automatic C:H:N analyzer. Table 10.3 shows the correlations between variables. There was a strong negative correlation between bacterial numbers and mean grain diameter, and good positive correlations with C and N content. However, there were also strong correlations between mean grain size and carbon and nitrogen values. In an attempt to disentangle these relationships, partial correlation analysis was used, with one variable being controlled while other correlations were being tested. Table 10.4 gives the relevant results, abstracted from a more complete analysis. When grain size was controlled, partial correlations with carbon and nitrogen were greatly reduced, but the partial correlation coefficients were still significant. Hence, although part of the correlation between bacteria and carbon or nitrogen

Table 10.3. Pearson correlation matrix for sediment properties: B—bacterial numbers; G—mean grain size; A—area index; c/s—clay: silt ratio; %C—% clay by volume; S—sorting coefficient; C—organic carbon content; N—total nitrogen. Correlations involving bacterial numbers, area index, % clay, C and N, are based on log transformed data. For 50 samples, critical r values are 0·28 and 0·36 for error probabilities of 0·05 and 0·01 respectively. (From Dale 1974.)

	B	G	A	c/s	%C	S	*Eh*	C	N
B	1·00	−·91	·89	·57	·88	·72	·07	·90	·91
G		1·00	−·98	−·66	−·98	−·80	−·03	−·93	−·93
A			1·00	·77	·99	·74	−·03	·94	·95
c/s				1·00	·76	·40	−·13	·69	·70
%C					1·00	·77	−·03	·94	·94
S						1·00	·27	·67	·68
Eh							1·00	−·07	−·04
C								1·00	·99
N									1·00

Table 10.4. Correlation between bacterial numbers and carbon and nitrogen, with and without statistical control for grain size. (From Dale 1974.)

	Organic carbon	Total nitrogen	Critical r value at 0·01 level
Correlation with bacterial numbers (Pearson r)	0·90	0·91	0·36 (48 d.f.)
Partial correlation with bacterial numbers, controlling for grain size	0·40	0·42	0·36 (47 d.f.)

was attributable to the relation between bacteria and grain size, not all of it could be explained in this way. Newell (1965) suggested that the carbon and nitrogen content of the bacteria themselves might account for the correlation between bacteria and sediment carbon and nitrogen. Calculations, using bacterial numbers in this study together with data from Luria (1960) on dry weight and composition of microbial cells showed that the bacteria would account for only 1% to 3% of the carbon and nitrogen present.

It turns out that grain size, organic carbon and total nitrogen are about equally good as predictors of bacterial numbers in these sediments. It is probable that both organic matter and inorganic particles are sorted by wave and tidal action so that fine particles of sediment and buoyant particles of organic matter tend to be deposited where wave and tidal action is low. Fine sediments have a large surface area per unit volume, and thus present favourable conditions both for adsorption of organic matter and colonization by bacteria. Hence, this study which embraced both aerobic and anaerobic bacteria, in a wide range of Eh values, confirms and extends the suggestion of Hargrave (1972) and others, that there is a basically simple relationship between micro-organisms and the surface area and organic content of sediments.

Table 10.5 gives some calculations of bacterial biomass observed in this study, together with other figures on bacteria and macrofauna abstracted from the literature. In general, bacterial biomass and macrofaunal biomass are of the same order of magnitude. From the well-known relationships between size and metabolic rate we may conclude that the bacteria are responsible for metabolizing the major portion of the organic matter in the sediment. Hargrave's (1973) figures suggest that in St. Margaret's Bay the rate of metabolism is about equal to the rate of sedimentation of organic matter, leaving very little to accumulate over time.

On the other hand, Keen & Piper (1976) took cores from the sediments

Table 10.5. Comparison of the biomass per unit area of bacteria and other benthic organisms.

Group	Habitat & Location	Dry wt $g\,m^{-2}$	Source
Bacteria	Fine silt Petpeswick	21·5	This study
Bacteria	Coarse silt Petpeswick	26·5	This study
Bacteria	Fine sand Petpeswick	5·5	This study
Bacteria	Foram ooze Pacific	0·3	Sorokin (1971)
Bacteria	Lagoon sand Pacific	9·7	Sorokin (1971)
Bacteria	Bottom sand Black Sea	15–30	Kriss *et al.* (1950)*
Bacteria	Shallow mud Azov Sea	17–57	Zhukova & Fedosov (1963)
Molluscs	Eelgrass bed Petpeswick	16·2	Burke (unpublished)
Molluscs	Sand flat Petpeswick	6·6	Burke (unpublished)
Macrofauna	Sand beach	9·2	McIntyre (1970)
Macrofauna	Mud flat Florida	30	Moore *et al.* (1968)
Meiofauna	Sublittoral mud Scotland	0·7–1·3	McIntyre (1964)

*Cited in Sieburth (1960).

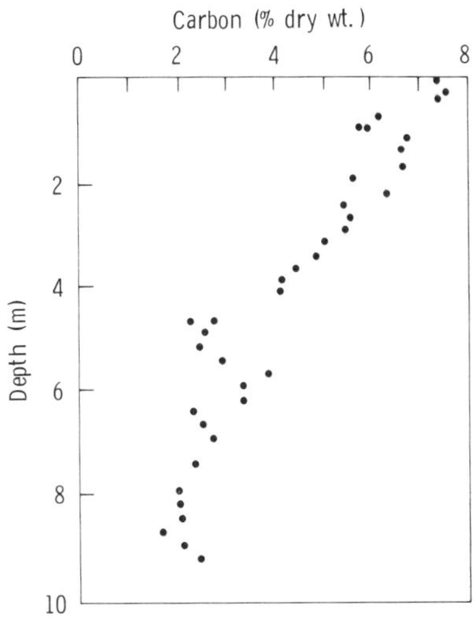

Figure 10.7. Organic carbon content (percent dry weight) of a core taken from the deepest part of St. Margaret's Bay, Nova Scotia. (From Keen & Piper 1976.)

in the deep part of the bay and found that the carbon content decreased from over 7% in the first half-metre to about 2% at 8–10 metres (Fig. 10.7). The presumed explanation of this change is slow bacterial decomposition of the organic matter over long periods of time. A radiocarbon date gave 3,970±260 BP at a depth of 6·25 m. The deposits are anaerobic after the first few cm and the authors found evidence of considerable quantities of methane being generated. The rate of sedimentation works out at about 0·15 cm per year. At a density of two for the solids and a roughly 50% water content, this is 1,500 $g.m^{-2}.yr^{-1}$. At 7% carbon content this is 105 $g\,C\,m^{-2}\,yr^{-1}$, close to the figure of 118 $g\,C\,m^{-2}\,yr^{-1}$ determined by Webster *et al.* (1975). However, it should be noted that the calculation of rate of sedimentation is very tentative, being dependent on a single radiocarbon date.

C. *Fate of material not sedimenting in the bay*

The proportion of primary production reaching the sediments has been shown to be quite low, perhaps as low as 15%, although this figure is open to question. What happens to the remainder? The bay is not a closed system and it has been estimated (Sharaf el Din *et al.* 1970) that the water exchanges with the water on the continental shelf about ten times per year. Hence, a proportion of the detritus generated by primary producers may be exported from the bay before being utilized by consumers. J. Foulds (unpublished data) has shown that *Mysis stenolepis* in St. Margaret's Bay filters eelgrass detritus from suspension in the water column, and Heinle *et al.* (1974) showed that the copepod *Eurytemora affinis* can grow and reproduce on a diet of plant detritus. Gerber & Marshall (1974) showed that the gut of a calanoid copepod in a coral reef lagoon had about 95% detritus in its gut contents. This is one of the areas currently in need of research. The role of benthic invertebrates in reducing particle size and generally aiding the process of decomposition, on and in the sediments, is relatively well documented (for review, see Mann, 1972a). However, the extent to which analogous processes occur in the water column is not yet clear.

Sorokin *et al.* (1970) showed that some marine planktonic crustaceans can effectively filter bacteria from the water, and that the concentrations of bacteria required for maximum feeding efficiency would be found near the thermocline even in oligotrophic marine waters. Later, Sorokin (1972) showed that in a coral reef community there are many kinds of invertebrates, including bivalve molluscs, ascidians, sponges, and tubicolous polychaetes, capable of filtering bacteria from the water column. If we assume that decomposition of abundant marine macrophyte detritus leads to high productivity of bacteria, many of which are held in suspension in the water column, then it seems likely that the filter feeding groups mentioned are supported largely by marine macrophyte decomposition.

Discussion

We have traced the fate of organic matter in a coastal marine environment from its production by seaweeds and seagrasses, through various stages of decomposition to the stage where a small proportion of the original production is buried deep in the sediments. The process of decomposition is vastly different for the seaweed, *Laminaria*, and for the seagrass, *Zostera*. *Laminaria* is an alga which grows rapidly, exudes soluble organic compounds (Sieburth & Jensen 1969; Khailov & Burkalova 1969) and erodes at the tips of the blades to produce a copious supply of particulate organic matter. Its surface is coated with an abundant bacterial flora which varies in characteristic ways between one part and another and from season to season. There is every reason to suppose that the abundant organic particles in the water column, which reached their peak of abundance before the onset of the spring bloom of phytoplankton, consisted largely of seaweed detritus and bacteria associated with seaweeds. It also seems probable that a large proportion of the organic material sedimenting out at about this time was derived from the same source.

The fate of organic matter synthesized by eelgrass seems to be quite different. Although the corresponding bacteriological study of living plants has not been carried out, it appears that eelgrass leaves are very resistant to decay, and may remain intact on the sea floor for many months. As is well known for the leaves of terrestrial plants, mechanical reduction of particle size speeds up decomposition. However, in laboratory experiments finely ground eelgrass leaves lost most of their organic matter by leaching rather than by bacterial action, and under the best conditions that could be devised, eelgrass detritus lost no more than 20% of its organic matter in 100 days. Hence, the two kinds of plants appear to contribute pools of particulate organic matter with quite different turnover times: seaweed organic matter decomposes rapidly, eelgrass organic matter decomposes slowly.

There is a third major category of marine macrophyte contributing to the productivity of the coastal zone: the marsh grasses such as *Spartina*. Odum & de la Cruz (1967) showed that in a series comprising successively smaller particle sizes of *Spartina* detritus, protein content (estimated as $6{\cdot}25 \times$ total N) increased from 6% in the dead leaves entering the water to about 24% in the smallest particles, and this was attributed to buildup of microbial biomass. Gosselink & Kirby (1974) studied decomposition of *Spartina alterniflora* in the laboratory. They claimed that for the smallest particle size the biomass of *Spartina* decreased by 90% during 30 days' incubation at 30°C, while microbial biomass increased with a conversion efficiency around 50%. These figures were not observed directly, but were calculated indirectly. For example, in experiments with the smallest particles (67 µm), total biomass decreased about one third; oxidizable C decreased about 5%, but particulate nitrogen increased over ten times. The nitrogen content of the *Spartina* was determined at the start of incubation, and the nitrogen content of the micro-organisms was assumed to be 13% of the ashfree dry weight. The relative changes in proportions of *Spartina* and micro-organisms required to give the above analyses, could then be calculated. If these results are correct, *Spartina* tissue is converted rather rapidly and completely to microbial biomass, especially after being finely divided.

We know very little about the fate of macrophyte detritus while it is in suspension in the water column, but we now know a great deal about what happens near the sediment surface. Large particles are attacked by detritivores such as amphipods, which digest the colonizing micro-organisms and produce faecal pellets rich in indigestible carbon compounds but poor in protein (Heald 1969; Fenchel 1970). The faecal pellets are quickly recolonized by micro-organisms, their proportional nitrogen content increases, and they are incorporated in the sediments, to be ingested by sediment feeding organisms such as gastropods, lamellibranchs and annelids (Newell 1965; Rhoads 1963, 1967; Nichols 1974). In many situations, the organisms in the upper 5–10 cm are thought to ingest the whole of the upper layer, passing it through their guts and entering it as faeces, once every few years (Rhoads 1967, 1973). This kind of activity results in the intimate association of inorganic particles, organic matter and bacteria observed by Dale (1974).

Once the organic matter has been buried below the layer that is turned over and aerated by the invertebrates, the environment is anaerobic. Here we find what Fenchel & Reidl (1970) have called 'the sulfide system: a new biotic community'. In this zone, bacteria continue to oxidize organic matter, while reducing sulfate to sulfide instead of utilizing molecular oxygen. Jorgensen and Fenchel (1974) measured rates of sulfate reduction in a laboratory model system and estimated that more than 50% of the total mineralization of eelgrass detritus was due to sulfate reduction.

In systems approaching steady state, it appears that 90% or more of marine macrophyte production enters detritus food chains rather than grazer food chains (Teal 1962; Heald 1969; Miller *et al.* 1971). However, it is also clear that few organisms are capable of digesting the plant material, and that they are dependent on the micro-organisms to convert the plant tissue to microbial biomass, which they then ingest. It is of interest to know what is the efficiency of this intermediate step. If most of the energy of marine macrophyte production is released in microbial metabolism, the role of the macrophytes in supporting productivity of invertebrates and fish could be considerably less than one would otherwise expect. In this connection, the calculations of Gosselink & Kirby (1974) on the conversion efficiency of micro-organisms growing on *Spartina* detritus is important. They found that the efficiency was very dependent on the mean particle size of the detritus, being highest (66%) on the smallest particles and lowest (24%) on the largest particles. This was mainly because the low ratio of surface area to volume in the large particles reduced the accessibility of the materials to microbial decomposition, and in turn reduced the metabolic rate of the microbial communities. Hence, the decomposition process was prolonged in time, and more of the energy in the *Spartina* was used to offset metabolic maintenance costs. Another factor contributing to the difference may have been the presence of a greater proportion of refractory organic material in the larger particles.

More work of this type is needed, but the range of efficiencies reported is in fair agreement with values quoted for a variety of microbial systems (Payne 1970). They suggest that the additional trophic level interposed between macrophytes and macroscopic consumers in the sea may lead to the loss of one-half to two-thirds of the energy as a metabolic cost, but certainly not the 90% that conventional trophic dynamics might have suggested.

Acknowledgements

The author is grateful for permission to quote from the published and unpublished work of R.A. Laycock, P.G. Harrison, N.G. Dale, T.J.M. Webster, J.B. Foulds, B.T. Hargrave, W.H. Sutcliffe, M.J. Keen and J.W. Piper, all of whom have been colleagues in either the Bedford Institute of Oceanography or Dalhousie University. Most of them kindly read and criticized drafts of the paper. Much of the work was supported by the Marine Ecology Laboratory, Bedford Institute of Oceanography, or by the National Research Council of Canada.

References

ANDERSON J.M. (1973) The breakdown and decomposition of sweet chestnut (*Castanea sativa* Mill.) and beech (*Fagus sylvatica* L.) leaf litter in two deciduous woodland soils. II. Changes in the carbon, hydrogen, nitrogen and polyphenol content. *Oecologia* **12**, 275–88.

BOCOCK K.L. (1964) Changes in the amounts of dry matter, nitrogen, carbon and energy in decomposing woodland leaf litter in relation to the activities of the soil fauna. *J. Ecol.* **52**, 273–84.

BLACK W.A.P. (1950) The seasonal variation in weight and chemical composition of the common British Laminariaceae. *J. Mar. Biol. Ass. U. K.* **29**, 45–72.

BREEN P.A. & MANN K.H. (1976) Changing lobster abundance and the destruction of kelp beds by sea urchins. *Mar. Biol.* **34**, 137–42.

BURKHOLDER P.R. & DOHENY T.E. (1968) *The biology of eelgrass.* Dept. of Conservation and Waterways, Town of Hempstead, Long Island, New York. 120 pp.

DALE N.G. (1974) Bacteria in intertidal sediments: Factors related to their distribution. *Limnol. Oceanogr.* **19**, 509–18.

FENCHEL T. (1973) *Aspects of the decomposition of seagrasses.* International Seagrass Workshop, Leiden, Netherlands.

FENCHEL T. (1970) Studies on the decomposition of organic detritus derived from turtle grass *Thalassia testudinum. Limnol. Oceanogr.* **15**, 14–20.

FENCHEL T. & REIDL R.J. (1970) The sulfide system: a new biotic community underneath the oxidized layer of marine sand bottoms. *Mar. Biol.* **7**, 255–68.

GERBER R.P. & MARSHALL N. (1974) Ingestion of detritus by the lagoon pelagic community at Eniwetok Atoll. *Limnol. Oceanogr.* **19**, 815–24.

GOSSELINK J.G. & KIRBY C.J. (1974) Decomposition of salt marsh grass, *Spartina alterniflora* Loisel. *Limnol. Oceanogr.* **19**, 825–32.

GOSZ J.R., LIKENS G.E. & BORMANN F.H. (1973) Nutrient release from decomposing leaf and branch litter in the Hubbard Brook Forest, New Hampshire *Ecol. Monogr.* **43**, 173–91.

HARGRAVE B.T. (1972) Aerobic decomposition of sediment and detritus as a function of particle surface area and organic content. *Limnol. Oceanogr.* **17**, 583–96.

HARGRAVE B.T. (1973) Coupling carbon flow through some pelagic and benthic communities. *J. Fish. Res. Board Can.* **30**, 1317–26.

HARRISON P.G. (1974) *Growth and detritus formation in a temperate seagrass,* Zostera marina. Unpublished Ph.D. Thesis, Dalhousie University, Halifax, N.S. 190 pp.

HARRISON P.G. & MANN K.H. (1975) Detritus formation from eelgrass (*Zostera marina L.*): the relative effects of fragmentation leaching and decay. *Limnol. Oceanogr.* **20**, 924–34.

HEALD E.J. (1969) *The production of organic detritus in a south Florida estuary.* Doctoral dissertation, University of Miami, 110 pp.

HEINLE D.R., FLEMER D.A., USTACH J.F., MURTAGH R.A. & HARRIS R.S. (1974) The role of organic debris and associated micro-organisms in pelagic estuarine food chains. *Final report to O.W.R.R.*, U.S. Dept. Interior, N.R.I. Ref. No. 74–29.

IVERSEN T.M. (1973) Decomposition of autumn-shed beech leaves in a spring brook and its significance to the fauna. *Arch. Hydrobiol.* **72**, 305–12.

JORGENSEN B.B. & FENCHEL T. (1974) The sulfur cycle of a marine sediment model system. *Mar. Biol.* **24**, 189–201.

KAUSHIK N.K. & HYNES H.N.B. (1971) The fate of leaves that fall into streams. *Arch. Hydrobiol.* **68**, 465–515.

KEEN M.J. & PIPER D.J.W. (1976) Kelp, methane and an impenetrable reflector in a temperate bay. *Can. J. Earth Sci.* **13**, 312–18.

KHAILOV K.M. & BURLAKOVA Z.P. (1969) Release of dissolved organic matter by marine seaweeds and distribution of their total organic production to inshore communities. *Limnol. Oceanogr.* **14**, 521–7.

LAYCOCK R.A. (1974) The detrital food chain based on seaweeds. I. Bacteria associated with the surface of *Laminaria* fronds. *Mar. Biol.* **25**, 223–31.

LURIA S.E. (1960) The bacterial protoplasm: Composition and organization, pp. 1–34. In: *The Bacteria*, eds. I.C. Gunsalus & R.Y. Stanier, vol. 1. Academic Press, New York.

MACINTYRE A.D. (1970) The range of biomass in intertidal sand with special reference to the bivalve *Tellina tenuis. J. Mar. Biol. Assoc. U.K.* **50**, 561–75.

MANN K.H. (1972a) Macrophyte production and detritus food chains in coastal waters. *Mem. Ist. Ital. Idrobiol.* **29**, Suppl. 353–83.

MANN K.H. (1972b) Ecological energetics of the seaweed zone in a marine bay on the Atlantic coast of Canada. II. Productivity of the seaweeds. *Mar. Biol.* **14**, 199–209.

MILLER R.J., MANN K.H. & SCARRATT D.J. (1971) Production potential of a seaweed-lobster community in Eastern Canada. *J. Fish. Res. Bd. Canada* **28**, 1733–8.

MOORE H.B. *et al.* (1968) Some biomass figures from a tidal flat in Biscayne Bay, Fla. *Bull. Mar. Sci. Gulf Caribb.* **18**, 261–329.

NELSON T.C. (1947) Some contributions from the land in determining conditions of life in the sea. *Ecol. Monogr.* **17**, 337–46.

NEWELL R. (1965) The role of detritus in the nutrition of two marine deposit feeders, the prosobranch *Hydrobia ulvae* and the bivalve *Macoma balthica. Proc. Zool. Soc. Lond.* **144**, 25–45.

NICHOLS F.H. (1974) Sediment turnover by a deposit-feeding polychaete. *Limnol. Oceanogr.* **19**, 945–50.

ODUM E.P. & DE LA CRUZ A.A. (1967) Particulate organic detritus in a Georgia salt marsh-estuarine ecosystem. In: *Estuaries*, ed. by G.H. Lanff, 383–8. Amer. Ass. Adv. Sci. Publ., 83. Washington, D.C.

PAYNE W.J. (1970) Energy yields and growth of heterotrophs. *Ann. Rev. Microbiol.* **24**, 17–52.

PETERSEN C.G.J. (1918) The sea bottom and its production of fish-food: a survey of the work done in connection with valuation of the Danish waters from 1883–1917. *Rep. Dan. Biol. Stn.* **25**, 1–62.

PLATT T. (1971) The annual production by phytoplankton in St. Margaret's Bay, Nova Scotia. *J. Cons. Int. Explor. Mer.* **33**, 324–33.

RASMUSSEN E. (1973) Systematics and ecology of the Iselfjord marine fauna (Denmark) with a survey of the eelgrass (*Zostera*) vegetation and its communities. *Ophelia* **11**, 1–495.

RHOADS D.C. (1963) Rates of sediment reworking by *Yoldia limatula* in Buzzard's Bay, Massachusetts, and Long Island Sound. *J. Sediment. Petrol.* **33**, 723–7.

RHOADS D.C. (1967) Biogenic reworking of intertidal and subtidal sediments in Barnstaple Harbor and Buzzards Bay, Massachusetts. *J. Geol.* **75**, 461–76.

RUSSELL-HUNTER W.D. (1970) *Aquatic productivity*. MacMillan, N. Y.; Collier-MacMillan, Toronto.

SHARAF EL DIN S.H., HASSAN E.M. & TRITES R.W. (1970) The physical oceanography of St. Margaret's Bay. *Fish. Res. Board Canada Tech. Rep.* **219**, 242 pp.

SIEBURTH J.M. (1960) Soviet aquatic bacteriology: a review of the past decade. *Quart. Rev. Biol.* **35**, 179–205.

SIEBURTH J. McN. & JENSEN A. (1969) Studies on algal substances in the sea. III. The production of extracellular organic matter by littoral marine algae. *J. exp. Mar. Biol.* **3**, 290–309.

SOROKIN Y.I. (1971) Population activity, and production of bacteria in bottom sediments of the Central Pacific. *Oceanology* **10**, 853–63.

SOROKIN Y.I. (1972) Bacteria as food for coral reef fauna. *Oceanology* **12**, 169–77.

SOROKIN Y.I., PETIPA T.S. & PAVLOVA Y.V. (1970) Quantitative estimate of marine bacterioplankton as a source of food. *Oceanology* **10**, 253–60.

STEELE J.H. & BAIRD I.E. (1972) Sedimentation of organic matter in a Scottish sea loch. *Mem. Ist. Ital. Idrobiol.* **29** Suppl. 74–88.

SUTCLIFFE W.H. Jr. (1972) Some relations of land drainage, nutrients, particulate materials and fish catch in two eastern Canadian bays. *J. Fish. Res. Bd. Canada* **29**, 357–62.

TEAL J.M. (1962) Energy flow in a salt-marsh ecosystem of Georgia. *Ecology* **43**, 614–24.

THORSON G. (1971) *Life in the sea*. McGraw-Hill Book Co. New York, Toronto.

TSERNOGLOU D. & ANTHONY E.H. (1971) Particle size, water-stable aggregates, and bacterial populations in lake sediments. *Can. J. Microbiol.* **17**, 217–27.

WEBSTER T.J.M., PARANJAPE M.A. & MANN K.H. (1975) Sedimentation of organic matter in St. Margaret's Bay, N. S. *J. Fish. Res. Bd Can.* **32**, 1399–1407.

ZOBELL C.E. (1938) Studies on the bacterial flora of marine bottom sediments. *J. Sediment. Petrol.* **8**, 10–18.

ZHUKOVA A.I. & FEDOSOV M.V. (1963) Significance of microorganisms of upper sediment layer of shallow water basin. In: *Transformation of organic matter*, ed. C.H. Oppenheimer pp. 711–19. Marine microbiology. Thomas.

11 Nitrogen regeneration and the metabolism of coastal marine bottom communities

S. W. NIXON, C. A. OVIATT and S. S. HALE
Graduate School of Oceanography, University of Rhode Island, Kingston, Rhode Island 02881, U.S.A.

Abstract

The rates of oxygen uptake and nutrient regeneration by subtidal bottom communities in Narragansett Bay, Rhode Island, were measured *in situ* over an annual cycle. Oxygen uptake ranged from 10–150 mg m^{-2} hr^{-1}, with an annual total of about 360 g O_2 m^{-2}. Inorganic nitrogen was released to the water column almost totally as ammonia at rates up to 400 µM m^{-2} hr^{-1}. The total annual flux of inorganic nitrogen (~870 mM m^{-2}) was only about half that expected on the basis of oxygen uptake or inorganic phosphate regeneration. It appeared that the anomalously low inorganic nitrogen flux may be balanced by high rates of dissolved organic nitrogen release from the bottom. The rapid regeneration of inorganic nitrogen by bottom communities plays a major role in maintaining the high rates of primary production in coastal waters, but the incomplete metabolism of nitrogen on the bottom may be responsible for the low N/P ratios characteristic of shallow waters and for the importance of nitrogen as a limiting factor in these areas.

Introduction

The study of marine bottom communities has largely been devoted to the description and analysis of community structure. In spite of the impressive efforts often involved in such work, we still know very little about most aspects of community function in the benthos, and almost nothing at all about how the structure of the community influences its role in nutrient cycling or energy transfer. Recent papers by Rowe (1971) and Hargrave (1973) have emphasized the coupling of pelagic and benthic communities

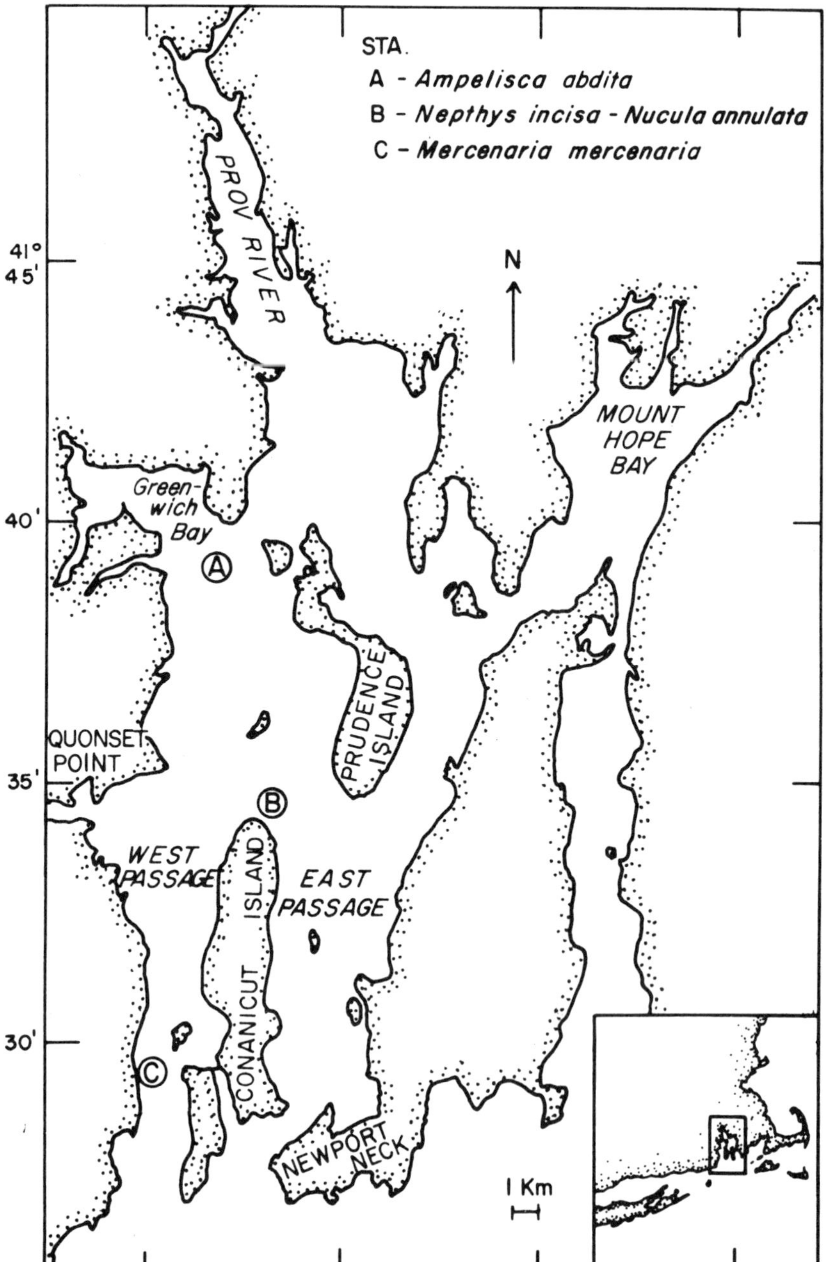

Figure 11.1. Location of the sample bottom communities in Narragansett Bay, Rhode Island. (A) An amphipod mat with *Ampelisca*, (B) a polychaete-bivalve community with *Nepthys*, *Nucula* and *Pitar* that was taken over by the amphipod *Leptocheirus*, (C) a hard clam bed with *Mercenaria*.

in terms of organic matter input to the bottom, and there have been numerous measurements of benthic oxygen uptake that document the substantial metabolic rates of a variety of marine bottom communities (for example: Carey 1967; Teal & Kanwisher 1961; Hargrave 1969; Pamatmat 1971a, b, 1973; Smith *et al.* 1972, 1973; Nixon & Oviatt 1973; Smith 1973; Wells *et al.* 1973). As a consequence of the metabolism of organic matter by the benthos, significant amounts of nitrogen, phosphorus, and other nutrients may be 'regenerated' and made available to the water column in inorganic form. Alternatively, the sediments may take up and bind some nutrients from the overlying water through sorption reactions.

Beginning with the classic studies of English lakes by Mortimer (1941, 1942), the subject of sediment-water nutrient flux has received much more attention in freshwater systems than it has in the marine environment (Lee 1970; Keeney 1973). To some extent this is understandable, since the benthos play only a minor role in open ocean nutrient cycles. However, the importance of benthic communities may be expected to increase dramatically in coastal waters, and it is surprising that so little is known about their role in nutrient cycling or in maintaining the characteristically high levels of productivity found in shallower marine and estuarine regions. In order to study the magnitude and chemical composition of benthic nutrient regeneration, as well as its relationship to community oxygen uptake and taxonomic structure, we have made some of the first reported *in situ* measurements of net nutrient flux between marine bottom communities and the overlying water. The measurements were made over an annual cycle at three stations dominated by different assemblages of animals in the West Passage of Narragansett Bay, Rhode Island (Fig. 11.1). While dissolved organic carbon, reactive and organic phosphorus, and silica fluxes were also measured, this paper is devoted to a discussion of the oxygen uptake and nitrogen flux from these communities and their role in the nitrogen cycle in the Bay.

THE STUDY AREA AND BOTTOM COMMUNITIES

The physical oceanography of Narragansett Bay has been described by Hicks (1959). In general, it is a well-mixed estuary with a small fresh water input that results in only weak vertical stratification. Salinities range from 20 to 30‰ in the West Passage, with an annual temperature range from about 0 to 24°C. The mean depth is 8·8 m, with 75% of the bay being less than 12 m deep. The sediments of the bay have been described by McMaster (1960). There is relatively little salt marsh or seaweed in the area, and almost all of the organic input to the bottom communities must come from the very abundant phytoplankton and zooplankton populations in the water column (Smayda 1957; Pratt 1965; Martin 1968).

The three subtidal bottom areas chosen for study included an assemblage of macrofauna (>0·75 mm) dominated by the hard clam, *Mercenaria mercenaria* in the lower West Passage, a polychaete-bivalve community with *Nepthys incisa* and *Nucula annulata* in the middle of the bay, and an amphipod mat of *Ampelisca abdita* in the upper West Passage. General characteristics of the three stations during the annual study are summarized in Table 11.1. However, there were also seasonal changes in the relative abundance and composition of species, particularly at the mid-bay station during the

Table 11.1. Characteristics of the sampling stations in Narragansett Bay.

	Upper Bay, *Ampelisca*	Mid-Bay, *Nepthys-Nucula*	Lower Bay, *Mercenaria*
Depth (m)	5·8	7·3	6·1
Maximum bottom currents* (cm sec^{-1})	22	21	30
Sediments†			
Silt-Clay (%)	7	73	58
Clay (%)	3	19	9
Organic Matter (%)	1·5	4·5	4·1
Organic Nitrogen (%)	0·01	0·08	0·10
Organic Carbon (%)	—	0·61	0·68
Macrofauna (>0·75 mm)‡			
Numbers			
November	1,664±1,708	768±972	315±115
March	1,328±547	206±96	165±70
July	2,728±1,517	1,643±190	441±110
Biomass (gm dry wt)			
November	0·61±0·33	0·78±0·39	2·65±2·29
March	0·63±0·43	1·16±0·73	6·92±10·54
July	2·62±3·42	4·87±1·72	3·91±2·85

*Haight (1938).
†Sampled in November.
‡Means $\pm\sigma$ for 0·1 m^2 samples. N=5 replicates at each station in November, 3 in March and July.

second summer, when the large amphipod, *Leptocheirus pinguis* and a bivalve, *Pitar morrhuana*, exceeded *Nepthys* and *Nucula* by weight and numbers.

Principal coordinates analysis (Blackith & Reyment 1971) and correspondence analysis (David *et al.* 1974), were used to test for similarities and differences in the taxonomic composition of the three areas on the basis of three sets of quantitative bottom samples taken during the year. While some species were common among all of the stations at all times, the multivariate analysis clearly distinguished all of the communities as separate taxonomic entities. The upper bay station (*Ampelisca*) also did not change significantly in relative composition over the year, while the mid and lower bay stations showed substantial changes during the second summer. Thus, the three stations represented at least five separate taxonomic groups of macrofauna during the year of the study.

Methods

Nutrient fluxes and oxygen uptake at the three stations were measured *in situ* using opaque PVC chambers that covered 0·24 m^2 of bottom and contained 27 l of water. The small water volume to sediment surface ratio of the chambers produced substantial concentration changes with short-term, three- to four-hour incubations. Three chambers were carefully placed on each community by SCUBA divers at each sampling date and anchored to the bottom with long pins or a heavy metal collar so that the edges of the chambers extended 4 cm below the sediment surface. Corrections for plankton metabolism and nutrient flux in the entrapped water in the chambers were made by incubating bottom water samples separately adjacent to the chambers. The details of chamber design and sampling procedure have been given elsewhere (Hale 1974). Initial and final samples were withdrawn from the chambers using a specially constructed 700 ml PVC syringe and brought to the surface.

On deck, oxygen samples were withdrawn from the syringe by siphon and fixed for later laboratory titration (Strickland & Parsons 1968). The nutrient samples were filtered with Gelman Type A glass fibre filters and preserved (ammonia—Degobbis 1973; phosphate—Gilmartin 1967) for laboratory analysis of ammonia (Solorzano 1969), nitrate and nitrite (Technicon Autoanalyzer), and inorganic phosphate (Strickland & Parsons 1968). Dissolved organic nitrogen was measured as ammonia after Kjeldahl digestion and the addition of base. Oxygen and ammonia determinations were made the same day the samples were taken; samples for the other analyses were frozen in polyethylene bottles. The analytical precisions and the resulting effects on the fluxes for a three-hour run were:

O_2	$\pm$0·032 mg l^{-1}	$\pm$1·2 mg m^{-2} hr^{-1}
NH_3	$\pm$0·07 μmoles l^{-1}	$\pm$2·6 μmoles m^{-2} hr^{-1}
NO_3	$\pm$0·15 μmoles l^{-1}	$\pm$5·6 μmoles m^{-2} hr^{-1}
NO_2	$\pm$0·15 μmoles l^{-1}	$\pm$5·6 μmoles m^{-2} hr^{-1}
PO_4	$\pm$0·014 μmoles l^{-1}	$\pm$0·5 μmoles m^{-2} hr^{-1}

Results and discussion

OXYGEN UPTAKE

Oxygen uptake rates for the sub-tidal bottom communities in the bay ranged from a low of about 10 mg m^{-2} hr^{-1} in mid-winter to over 150 mg m^{-2} hr^{-1} in summer (Fig. 11.2). A seasonal effect appeared to be present, with higher

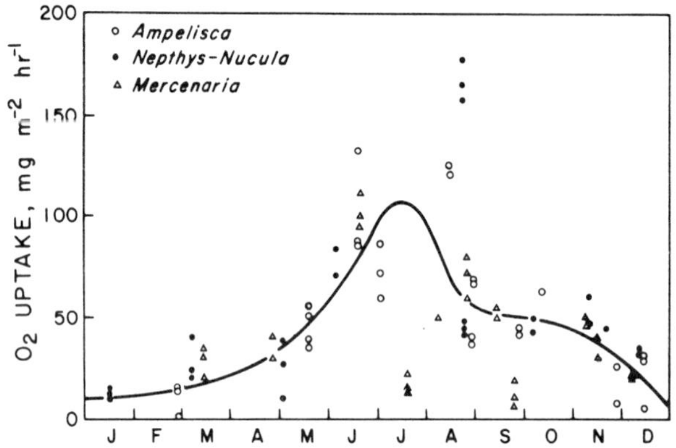

Figure 11.2. Oxygen uptake by the bottom communities from July 1973 to August 1974. The curve was drawn by eye for estimating the total annual uptake.

rates of metabolism in spring and early summer than found during late summer and autumn. This effect was apparent in plots of oxygen uptake as a function of bottom water temperature (Fig. 11.3). While measurements taken during the winter and spring when temperatures were rising showed

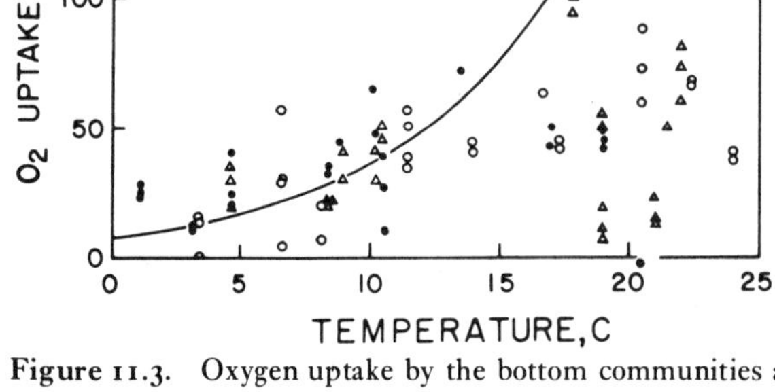

Figure 11.3. Oxygen uptake by the bottom communities as a function of bottom water temperature.

oxygen uptake to be a strong exponential function of temperature, virtually all of the measurements made towards the end of the summer and autumn showed uptake rates that were lower than those found at similar temperatures in spring. To some extent, this may reflect a suppression of metabolism by lower oxygen concentrations in the water at higher summer temperatures. An additional explanation may be that Narragansett Bay is characterized by an intense ($>20 \times 10^6$ cells l^{-1}) winter–spring phytoplankton bloom (Smayda 1957; Pratt 1965), and an associated early spring zooplankton pulse. Much of this organic production may have been deposited on the bottom while temperatures were low and benthic metabolism was suppressed. As temperatures rose in the spring, the metabolic potential of the benthos increased and there was a rapid oxidation of the accumulated organic matter. Later in the summer, the metabolic potential was high, but metabolism was limited by the supply of organic matter. Of course, some organic matter is always present in the sediments, even during late summer and autumn, but it probably represents what Marshall (1972) has described as 'the ashes of the fire', a refractory background level that is virtually unavailable to the benthos. Oxygen uptake remained relatively stable throughout the autumn, then declined sharply in December and remained low throughout the winter (Fig. 11.2).

For all communities, the total annual oxygen consumption was on the order of $370\,g\,O_2\,m^{-2}\,yr^{-1}$. Assuming a respiratory quotent of 1, this was equivalent to about $140\,g\,C\,m^{-2}\,yr^{-1}$, or 64% of the estimated net primary production (Smayda 1957). In fact, it is probably a larger fraction than that, since Smayda's primary production estimate was based on surface incubations rather than an integrated measurement of the water column.

NITROGEN REGENERATION

Almost all of the inorganic nitrogen released from the bottom was in the form of ammonia. The amphipod mat community dominated by *Ampelisca* did show a substantial release of nitrate on several occasions (Fig. 11.4), though the general pattern for all of the communities was a variable uptake or release of nitrate at rates less than $40\,\mu M\,m^{-2}\,hr^{-1}$. Nitrite fluxes were virtually always below the level of detection. The seasonal pattern of ammonia flux was similar to that of oxygen uptake, except that ammonia was almost always released by the communities (Fig. 11.5). The influence of season was not as apparent, with summer and autumn measurements remaining in line with the winter–spring data showing an exponential relationship between ammonia flux and temperature (Fig. 11.6).

Considering the low rates and erratic behaviour of benthic nitrate flux,

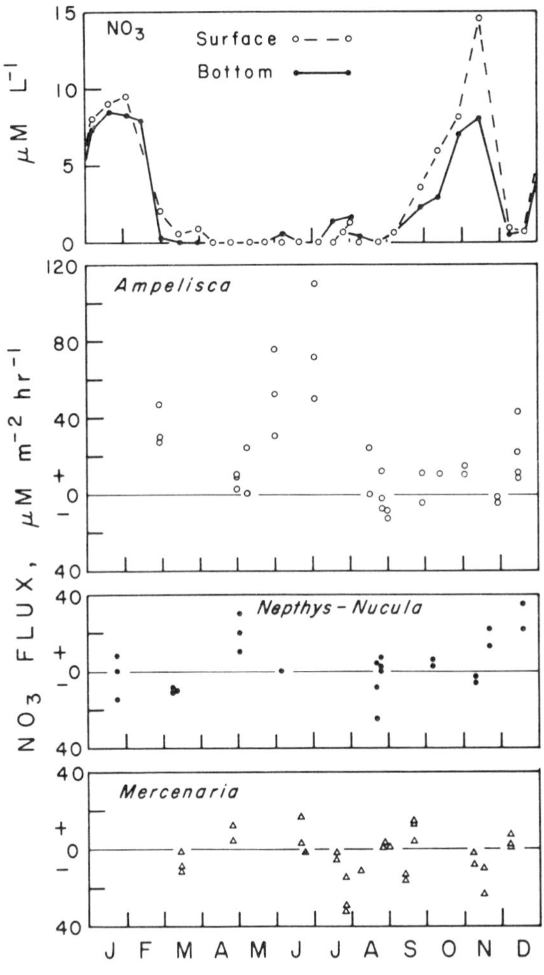

Figure 11.4. Seasonal cycle of nitrate in the surface and bottom waters of the West Passage of Narragansett Bay, and the flux of nitrate at the sediment-water interface at three stations in the West Passage.

it is not surprising that there was no correspondence with the seasonal nitrate pattern in the water column (Fig. 11.4). It seems clear that essentially all of the nitrate in the water is derived from pelagic processes, and that the benthic communities do not participate directly in the seasonal nitrate cycle in the water. This is not the case for ammonia, where the seasonal cycle of input by the benthos precedes the period of maximum abundance in the water column (Fig. 11.5). In order to see if benthic regeneration of ammonia could account for the levels found in the water, the regression of ammonia flux as a function of temperature (Fig. 11.6) was coupled with observed bottom water temperatures and a numerical hydrodynamic simulation

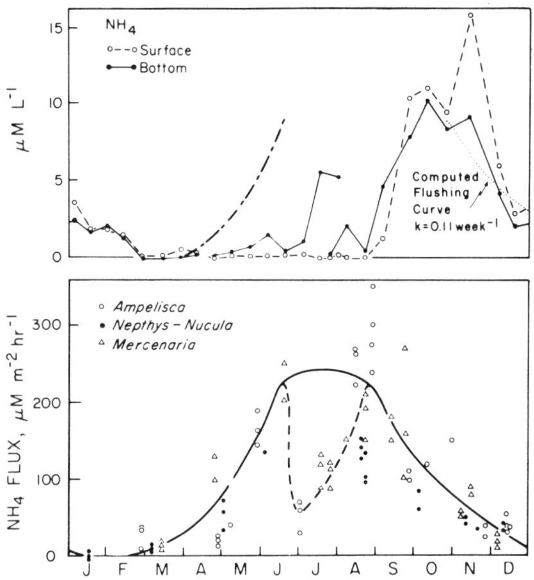

Figure 11.5. Seasonal cycle of ammonia in the surface and bottom waters of the West Passage of Narragansett Bay and the flux of ammonia at the sediment-water interface at three stations in the West Passage. The dotted line represents the computed ammonia concentrations using a numerical mixing model without any biological or chemical processes. The broken line represents the potential ammonia concentrations in the water column on the basis of benthic flux as a function of temperature (Fig. 11.6) and hydrodynamic mixing. The smooth curve at the bottom was drawn by eye for estimating the total annual ammonia flux.

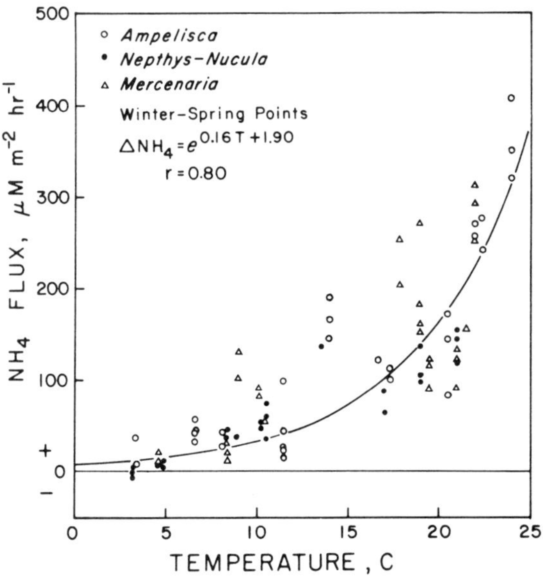

Figure 11.6. Ammonia flux at the sediment-water interface as a function of bottom water temperature.

model of flushing and circulation in the bay (Kremer & Nixon 1975; Hess & White 1974). The resulting computed rate of increase in ammonia concentration in the water agreed very well with observed rates (Fig. 11.5), and suggests that benthic regeneration is largely responsible for the basic seasonal pattern of abundance. Of course, once the ammonia is put into the water, biological events in the plankton exert a strong influence on the final levels observed. Thus, high rates of summer phytoplankton growth take up virtually all of the ammonia released by the bottom during the early summer, and concentrations in the water do not increase until July or August. Even though the benthic input does not accumulate in the water in spring, the rapid increase of nitrogen in the water column from benthic regeneration makes the high summer primary production possible. During the autumn and winter, as regeneration by the benthos declines and biological activity in the water column is reduced, the ammonia is lost through flushing and oxidation to nitrate (Fig. 11.5).

ANOMALOUS NUTRIENT RATIOS

Over the annual cycle, some 870 mM m^{-2} of ammonia were released by the three bottom communities. If the mid-summer dip in release rates suggested in Fig. 11.5 is correct, this value would be reduced to about 675 mM m^{-2} yr^{-1}. In either case, the amount of inorganic nitrogen regenerated was low in terms of the amount of oxygen consumed, giving O/N atomic ratios of 27 : 1 or 33 : 1, respectively. This represents a substantial departure from

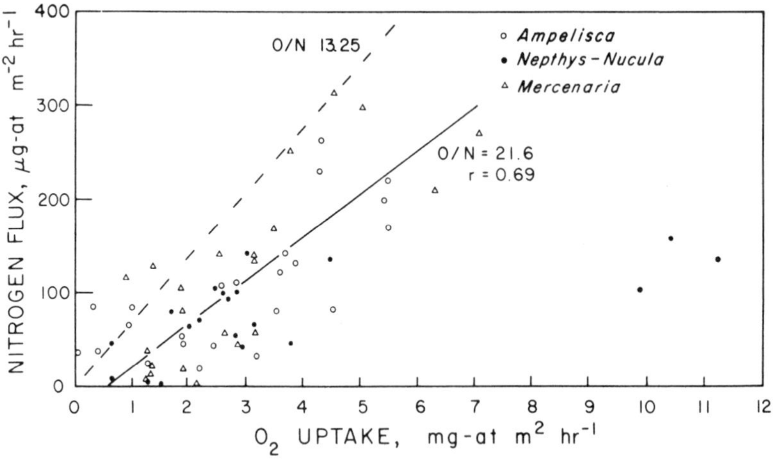

Figure 11.7. Oxygen uptake and inorganic nitrogen flux at the sediment-water interface at three stations in the West Passage of Narragansett Bay. A functional regression (Ricker 1973) was used to calculate the slope for comparison with the classic relationship given by Redfield (1934), Fleming (1940), and others.

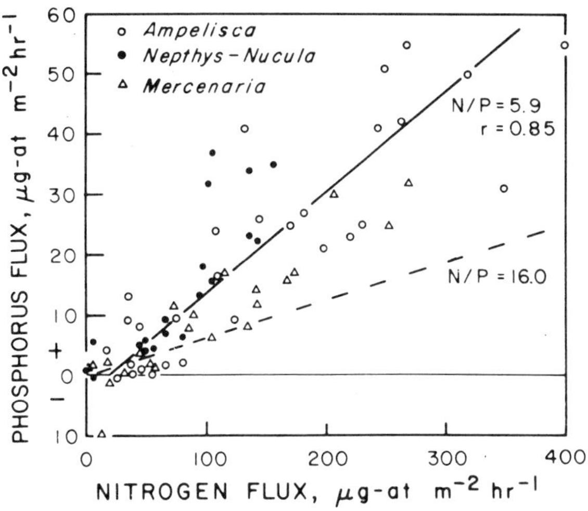

Figure 11.8. The flux of inorganic phosphate and nitrogen at the sediment-water interface at three stations in the West Passage of Narragansett Bay. A functional regression (Ricker 1973) was used to calculate the slope.

the O/N ratio of 13·25 : 1 commonly found associated with the production or decomposition of organic matter in the sea (Redfield 1934; Fleming 1940). The fact that the departure is based on annual totals indicates that the discrepancy was not caused by transient changes in the nitrogen content of phytoplankton caused by nutritional conditions, senescence, etc., but represents a more fundamental process. A functional regression (Ricker 1973) of total inorganic nitrogen flux with oxygen uptake for the individual measurements on all communities also showed a marked nitrogen deficit (Fig. 11.7), and a comparison of phosphate release with nitrogen release (Fig. 11.8) confirmed that inorganic nitrogen flux was less than half that expected on the basis of phosphorus regeneration and the N/P ratio of about 16 : 1 in organic matter (Redfield 1934; Fleming 1940). The evidence seems to indicate that not only is benthic nitrogen regeneration a dominant influence on the seasonal cycles of nitrogen and primary production in the water column, but that even with the high rates of regeneration involved, somewhere around half of the nitrogen that may be input to the bottom is not being recycled as nitrate, nitrite, or ammonia. This low nitrogen anomaly may well be responsible for the characteristically low N/P ratios found in Narragansett Bay (annual mean 3·5 : 1) and many other shallow marine waters (Riley 1941a, b; Jeffries 1962), though the uptake of nitrogen by vascular plant detritus may also be locally important in estuaries with large inputs of *Spartina* (Thayer 1974).

DISSOLVED ORGANIC FLUXES

At least two explanations for the apparent nitrogen deficit are possible, including a flux of large amounts of nitrogen from the bottom in dissolved organic form, and very high rates of denitrification resulting in a large flux of nitrogen gas. On the basis of a preliminary analysis from the summer period, it appears that the former may be the more important. Summer flux rates of dissolved organic nitrogen were similar to those of ammonia, and appear to lie between 100–400 μg-at N m^{-2} hr^{-1}. This half organic/half inorganic fractionation in nitrogen regeneration by the marine bottom communities is compatible with observations that much of the nitrogen in the water column is also in organic form, with 10–20 μg-at N l^{-1} of DON found in the lower West Passage (P. Kremer, personal communication). In contrast, dissolved organic phosphorus flux was usually low, with an erratic pattern of release and uptake. The evidence is very suggestive that this partitioning of nitrogen and not of phosphorus is the mechanism by which bottom communities produce the low reactive N/P ratios in the overlying water discussed earlier. And the low ratio of available N/P in the water column may be responsible for recent findings that nitrogen, rather than phosphorus, is the major nutrient limiting primary production in coastal waters (Ryther & Dunstan 1971; Goldman *et al.* 1973). The large amount of dissolved organic nitrogen input to the water may, however, serve as an important nitrogen source for pelagic bacteria and the microzooplankton that feed on them. John Sieburth has made *in situ* measurements of the growth of natural pelagic bacteria populations in Narragansett Bay that indicate summer doubling times of 6–8 hours. Some of the dissolved organic nitrogen may also be taken up by phytoplankton. Measurements of the composition of the organic nitrogen pool and of its turnover time have become critical next steps in following the nitrogen cycle in coastal waters. It should be pointed out, however, that different sediments and benthic communities may behave differently. A similar study by Hartwig (1974) on a sandy bottom off California found a dissolved organic nitrogen flux of only 15% of the total. For *in situ* measurements in nearby Buzzards Bay, Massachusetts, Rowe *et al.* (1975) report that inorganic nitrogen flux was about 80% of that expected on the basis of oxygen uptake. However, their N/P ratios were as low as those found in this study, and organic nitrogen was not measured. Until more measurements are available, it is probably not valid to attempt to use the existing literature data on oxygen uptake to calculate nutrient regeneration, especially for nitrogen.

THE EFFECT OF COMMUNITY STRUCTURE

While there were no large differences in oxygen uptake or nutrient regeneration among the three stations, statistically significant variations were found. Multivariate discriminant analysis (Blackith & Reyment 1971) was used to calculate a generalized distance function to separate the stations on the basis of oxygen uptake and ammonia, nitrite, nitrate, phosphate, and silicate flux. The results of this analysis indicated that the metabolism of the lower bay station, where *Mercenaria* was abundant, was significantly different at the 95% confidence level from the metabolism of the bottom at the other two stations. It is clear, however, that differences in community function were substantially smaller than differences in community species composition, and it was not possible to show significant differences in metabolism or nutrient regeneration between the two upper bay stations.

Acknowled

ırbin, Eric Klos and Stephen
ınd laboratory analysis. Björn
y Smith assisted with statisti-
rt for the project came from
. Department of Commerce,
Agency.

phometrics, Academic Press, Lon-

Sound. I. Oxygen utilization of
niversity **19**, 136–44.
ses in R- and Q-Mode Analysis:
Study of Geological Processes.

ammonia determination. *Limnol.*

units for reporting populations
, *1939*, *Proc.* **3**, 535–40.
centration occurring during sea-

ıorganic nitrogen removal from
wastewater: effect on phytoplankton growth in coastal marine waters. *Science* **180**, 955–6.

HAIGHT F.J. (1938) Currents in Narragansett Bay, Buzzards Bay, and Nantucket and Vineyard Sounds. U.S. Coast and Geodetic Survey. *Spec. Publ.* No. 208, 101 pp.

HALE S.S. (1974) *The role of benthic communities in the nutrient cycles of Narragansett Bay.* Unpublished M.Sc. Thesis, University of Rhode Island, Kingston, 123 pp.

HARGRAVE B.T. (1969) Similarity of oxygen uptake by benthic communities. *Limnol. Oceanogr.* **14**, 801–5.

HARGRAVE B.T. (1973) Coupling carbon flow through some pelagic and benthic communities. *J. Fish. Res. Board Can.* **30**, 1317–26.

HARTWIG E.O. (1974) *Physical, chemical and biological aspects of nutrient exchange between the marine benthos and the overlying water.* Unpublished Ph.D. Thesis, University of California, San Diego, 189 pp.

HESS K. & WHITE F.M. (1974) A numerical tidal model of Narragansett Bay. *University of Rhode Island Marine Tech. Rep. No. 20*, Kingston R.I., 141 pp.

HICKS S.D. (1959) The physical oceanography of Narragansett Bay. *Limnol. Oceanogr.* **4**, 316–27.

JEFFRIES H.P. (1962) Environmental characteristics of Raritan Bay, a polluted estuary. *Limnol. Oceanogr.* **7**, 21–31.

KEENEY D.R. (1973) The nitrogen cycle in sediment-water systems. *J. Environ. Qual.* **2**, 15–29.

KREMER J.N. & NIXON S.W. (1975) An ecological simulation model of Narragansett Bay—the plankton community. In: *Recent Advances in Estuarine Research*, 672–90, ed. L.E. Cronin, Proc. 2nd Intl. Est. Res. Conf., Myrtle Beach, S.C. 1973. Academic Press, N.Y. Vol. 1.

LEE G.F. (1970) Factors affecting the transfer of materials between water and sediments. *Literature review no. 1.* Univ. of Wisconsin Water Resources Center Eutrophication Program. Madison, Wis. 50 pp.

MARSHALL N. (1972) Interstitial community and sediments of shoal benthic environments. In: *Environmental Framework of Coastal Plain Estuaries*, ed. B.W. Nelson, pp. 409–15. Geological Society of America, Memphis **133**.

MARTIN J.H. (1968) Phytoplankton–zooplankton relationships in Narragansett Bay. III. Seasonal changes in zooplankton excretion rates in relation to phytoplankton abundance. *Limnol. Oceanogr.* **13**, 63–71.

MCMASTER R.L. (1960) Sediments of Narragansett Bay system and Rhode Island Sound, Rhode Island. *J. Sediment Petrol.* **30**, 249–74.

MORTIMER C.H. (1941) The exchange of dissolved substances between mud and water in lakes. *J. Ecol.* **29**, 280–329.

MORTIMER C.H. (1942) The exchange of dissolved substances between mud and water in lakes. *J. Ecol.* **30**, 147–201.

NIXON S.W. & OVIATT C.A. (1973) Ecology of a New England saltmarsh. *Ecol. Monogr.* **43**, 463–98.

PAMATMAT M.M. (1971a) Oxygen consumption by the seabed. IV. Shipboard and laboratory experiments. *Limnol. Oceanogr.* **16**, 536–50.

PAMATMAT M.M. (1971b) Oxygen consumption by the seabed. VI. Seasonal cycle of chemical oxidation and respiration in Puget Sound. *Int. Rev. gesamten Hydrobiol.* **56**, 769–93.

PAMATMAT, M.M. (1973) Benthic community metabolism on the continental terrace and in the deep sea in the North Pacific. *Int. Rev. gesamten Hydrobiol.* **58**, 345–68.

PRATT D.M. (1965) The winter–spring diatom flowering in Narragansett Bay. *Limnol. Oceanogr.* **10**, 173–84.

REDFIELD A.C. (1934) On the proportion of organic derivatives in sea water and their relation to the composition of plankton, pp. 176–92. In: *James Johnson Memorial Volume*, University Press, Liverpool, 348 pp.

RICKER W.E. (1973) Linear regressions in fishery research. *J. Fish. Res. Bd. Can.* **30** (3), 409–34.

RILEY G.A. (1941a) Plankton studies. III. Long Island Sound. *Bull. Bingham Oceanogr. Coll.* 7 (3), 1–93.

RILEY G.A. (1942b) Plankton studies. IV. Georges Bank. *Bull. Bingham Oceanogr. Coll.* 7 (4), 1–73.

ROWE G.T. (1971) Benthic biomass and surface productivity. In: *Fertility of the Sea*, ed. J.D. Costlow, Jr., pp. 441–54. Gordon and Breach, New York.

ROWE G.T., CLIFFORD C.H., SMITH JR. K.L. & HAMILTON P.L. (1975) Benthic nutrient regeneration and its coupling to primary productivity in coastal waters. *Nature Lond.* **255**, 215–17.

RYTHER J.H. & DUNSTAN W.M. (1971) Nitrogen, phosphorus and eutrophication in the coastal marine environment. *Science* **171**, 1008–13.

SMAYDA T.J. (1957) Phytoplankton studies in lower Narragansett Bay. *Limnol. Oceanogr.* **2**, 342–58.

SMITH JR. K.L. (1973) Respiration of a sublittoral community. *Ecology* **54**, 1065–75.

SMITH JR. K.L., BURNS K.A. & TEAL J.M. (1972) *In situ* respiration of benthic communities in Castle Harbor, Bermuda, *Mar. Biol.* (Berl.) **12**, 196–9.

SMITH JR. K.L., ROWE G.T. & NICHOLS J.A. (1973) Benthic community respiration near the Woods Hole sewage outfall. *Est. Coastal Mar. Sci.* **1**, 65–70.

SOLORZANO L. (1969) Determination of ammonia in natural waters by the phenolhypochlorite method. *Limnol. Oceanogr.* **14**, 799–801.

STRICKLAND J.D.H. & PARSONS T.R. (1968) A practical handbook of seawater analysis. *Fish. Res. Bd. Can. Bull.* **167**, 1–311.

TEAL J.M. & KANWISHER J. (1961) Gas exchange in a Georgia salt marsh. *Limnol. Oceanogr.* **6**, 388–99.

THAYER G.W. (1974) Identity and regulation of nutrients limiting phytoplankton production in the shallow estuaries near Beaufort, N.C. *Oecologia* (Berl.) **14**, 75–92.

WELLS J.M., WELLS A.H. & VANDERWALKER J.G. (1973) *In situ* studies of metabolism in benthic reef communities. *Helgol. wiss. Meeresunters.* **24**, 78–81.

12 The significance of bacterial grazing and mineral cycling for the decomposition of particulate detritus

T. FENCHEL and P. HARRISON *Department of Ecology, University of Aarhus, DK–8000 Aarhus C, Denmark*

Introduction

The classical concepts of decomposition and mineralization were straightforward and simple: during decomposition of organic debris mineral nutrients were regenerated to be used by the primary producers. In addition, a food chain was recognized which channelled dead organic material through bacteria, fungi, and 'detritus feeders'. However, this pathway for organic carbon was usually considered of little importance in the aquatic environment as evidenced by most graphical textbook representations of ecosystems.

More recent work has shown that a large part, at times most, of the primary production of shallow water, macrophyte-dominated systems is utilized as particulate detritus in decomposer food chains (Harrison & Mann 1975a; Mann 1972 and papers cited therein). Furthermore, it has been shown that essentially only bacteria and fungi act as primary decomposers and that organisms higher in the food chain (protozoa and metazoa) utilize mainly the microbial biomass. At the same time these microbial grazers, or detritus feeders, may stimulate bacterial growth and thus the rate of decomposition (Fenchel 1970; Hargrave 1970; Johannes 1965; Newell 1965). Finally, it is recognized that the quantitatively most important substrate for the primary decomposers, i.e. the remains of macrophyte tissue, may be poor in essential mineral nutrients. The bacteria, therefore, while mineralizing the organic carbon in their substrate must also assimilate dissolved minerals from the water. Thus while decomposers may enrich the detritus with nitrogen and phosphorus they may also compete with primary producers for these elements (Johannes 1965, 1968; Pomeroy 1970). These authors also assigned the stimulatory effect of bacterivores to their regeneration of nutrients through excretion; according to this theory nutrients would become tied up in bacterial biomass in the absence of grazers and the rate of decomposition would consequently decrease.

The present authors have independently carried out experiments on 'microcosm' systems, i.e. flasks containing particulate organic material such as detritus derived from various vascular plants or pure structural polysaccharides, water with known contents of mineral nutrients, and with different microbial inocula (Barsdate *et al.* 1974; Fenchel 1970, 1972, 1973, in press; Fenchel & Rahn unpublished results; Harrison & Mann 1975b). Our objectives have been to study (1) the dynamics of microbial populations associated with the breakdown of detritus and factors which control microbial numbers; (2) factors which influence the rates of decomposition, e.g., nutrient levels and the presence or absence of grazing; and (3) mineral cycling in decomposer systems.

It is acknowledged that the laboratory microcosm approach has limitations when one wishes to relate the results to the conditions in nature. The approach does allow, however, the manipulation and measurement of factors difficult to control in the field: e.g., accurate estimation of sizes of microbial populations, maintenance of systems with special compositions of the microbial communities and with special nutrient levels in the water, continuous measurements of the rates of decomposition (as oxygen consumption or the production of ^{14}C labelled carbon dioxide from homogeneously labelled plant materials), and the use of radiotracers for following mineral cycles. The use of such techniques in relatively simple microcosms has made possible the analysis of several fundamental properties of decomposer systems.

In the following will be described our main results concerning the dynamics of microbial populations, the effect of grazers, and the cycling of nutrients. Finally, the significance of the results for gaining an insight into natural systems as well as some evolutionary aspects will be discussed.

The composition and succession of microbial populations on detrital particles

The composition of the microbial communities of detrital particles is relatively constant with respect to the main groups of organisms. This holds true for detrital particles from a variety of sources (seagrasses, mangrove leaves, macroalgae, barley hay, and even pure cellulose particles). Also, the composition of artificial laboratory systems inoculated with a natural source does not differ in any evident way from samples of detritus collected in nature.

There are very few data on the qualitative composition of the bacterial flora of detritus. Pure isolates of *Pseudomonas* and *Cytophaga* have been obtained on cellulose-agar plates, but judging from microscopical observations the diversity of forms is higher; e.g., *Nocardia* sp. and filamentous forms are often seen (Fenchel 1973). Fungi are often represented by chytrids in marine systems and these have been isolated on cellulose-agar (Fenchel unpublished observations); however, they have not as yet been enumerated.

Using fluorescence microscopy Fenchel (1970, 1972, 1973, in press) found between 2 and 15 bacteria per 100 μm^2 detrital surface, depending somewhat on nutrient levels and other factors. This means that only from 2 to 15% of the detrital surface is covered by bacteria. Depending on the median particle size (and thus the total surface area) the numbers observed correspond to between 10^9 and 10^{10} bacteria per gram dry weight of detritus. Harrison & Mann (1975b) using plate counts found about one order of magnitude fewer bacteria (Fig. 12.2).

The consumers of bacteria are mainly small zooflagellates (genera such as *Bodo*, *Monas*, *Rhynchomonas* and choanoflagellates and colourless euglenoids) and ciliates (e.g., *Euplotes*, *Holosticha*, *Uronema*, and *Cyclidium*). Whereas the zooflagellates most probably feed exclusively on bacteria, the different kinds of ciliates may also utilize flagellates, microalgae, and other ciliates. Typically between 5×10^7 and 5×10^8 flagellates are found per gram of detritus and ciliates usually number between 10^4 and 10^5 per gram (Fenchel 1970, 1972, 1973, in press; Harrison & Mann 1975b).

If the average wet weight of a bacterium is estimated to be 8×10^{-13} g and that of an average zooflagellate and an average ciliate to be 5×10^{-11} g and 10^{-8} g respectively, the biomass of the three groups in one gram of detritus (with a median particle size of about 200 μm) will be 4, 5, and 0·5 mg respectively. Thus the biomass of the two major groups of protozoan bacterivores is typically of the same magnitude as that of the bacteria.

Other types of protozoa (heliozoans, naked amoebae), small metazoans (nematodes, turbellarians, rotifers, ostracods, harpacticoids), and microalgae (diatoms, blue-green algae) are frequent in detritus. Their numbers and significance, however, have not been studied in detail. Fenchel (1970) found about 2×10^7 diatoms and about 3×10^3 nematodes per gram natural detritus derived from the seagrass *Thalassia*.

When previously sterilized detrital particles are placed in sea- or freshwater and inoculated with a natural source of detritus a characteristic succession may be followed (Fig. 12.1). Bacteria are found in small numbers on the particles within 6–8 hours; they then increase with doubling times of 10–20 hrs and reach their maximum after from 50 to 150 hours. Thereafter their numbers decline and maintain a relatively stable level after about 200 hours. Flagellates usually appear about 20 hours after inoculation and typically show a maximum in numbers after around 200 hours. Finally, the ciliates appear after 50–100 hours and often reach maximum numbers after 200–

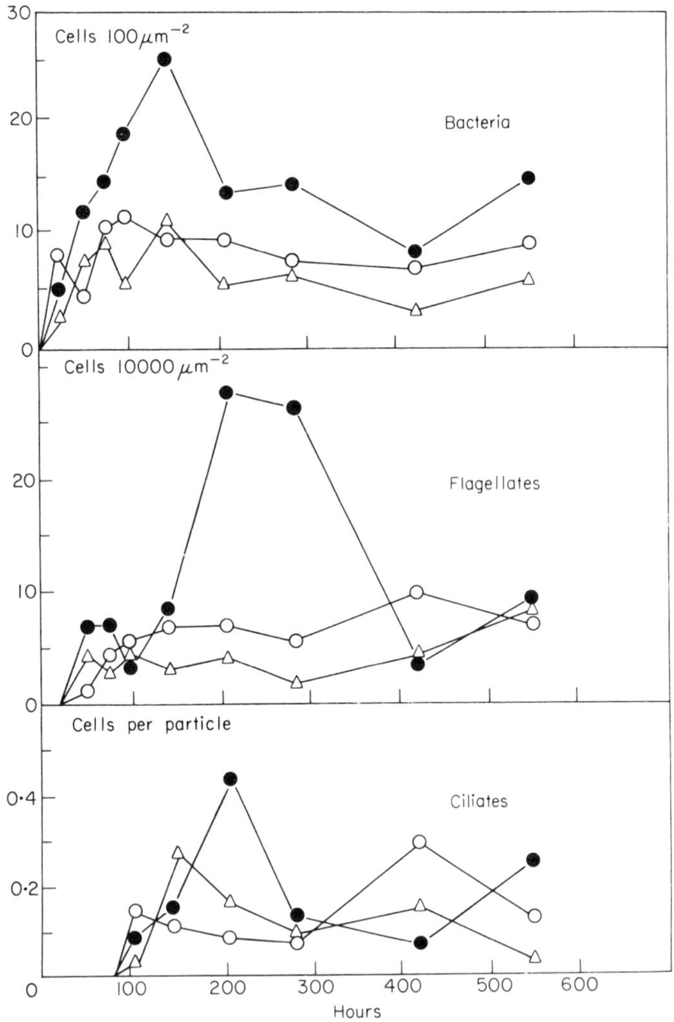

Figure 12.1. Successions of bacteria, flagellates, and ciliates during the decomposition of barley hay in seawater enriched with 5·7 mg PO_4^{3-} —P l^{-1} (△), 16·5 mg NO_3^- —N l^{-1} (○), and both PO_4^{3-} and NO_3^- (●). (Modified after Fenchel in press.)

300 hours. The other taxonomic groups, notably the micrometazoans, usually first occur at a later stage than do the flagellates and ciliates. The successional data given here are of course dependent on temperature and apply to 20–25°C. Other factors, e.g. nutrient levels, the presence or absence of turbulence, and inoculum size also affect the exact course of the succession.

The effect of grazing on bacterial numbers and turnover and on rate of decomposition

Comparisons of microcosms inoculated with a natural assemblage of bacteria and microcosms also inoculated with either one or several species of protozoa makes it possible to study the effect of protozoan grazing in decomposer systems (Barsdate *et al.* 1974; Fenchel in press; Harrison & Mann 1975b). Protozoa-free inocula were obtained by filtering a suspension of natural detritus twice through a 3 μm Nucleopore filter. The filtrates were usually rendered free of even the smallest zooflagellates but were assumed to contain all types of bacteria originally present in the sample.

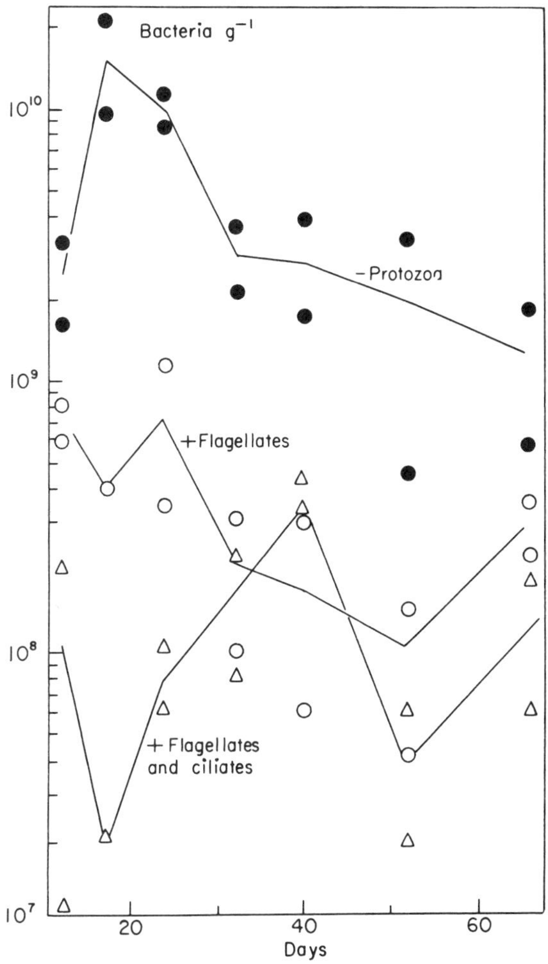

Figure 12.2. Numbers of bacteria per gram dry detritus derived from dead eelgrass leaves, to show the effect of incubation with and without various protozoa (after Harrison & Mann 1975b).

Ungrazed systems always have higher population densities of bacteria than do grazed systems. Fenchel (in press) found that ungrazed systems had a density about two times higher than a parallel system which was also inoculated with a mixed assemblage of protozoa, and about 50% higher than a system in which the only grazer was an unidentified species of choanoflagellate. In each of these systems the bacterial population levels remained relatively constant for at least the 1,000 hours during which they were followed. Barsdate *et al.* (1974) found a two- to fivefold decrease in bacterial numbers in freshwater systems grazed by the ciliate *Tetrahymena* relative to similar protozoa-free systems. Harrison & Mann (1975b) found an even greater reduction (about a factor of 10) in grazed relative to ungrazed systems based on eelgrass detritus (Fig. 12.2).

The previously mentioned oscillations in bacterial numbers characteristic for the initial stages of successions can be attributed to the grazing of protozoa which develop large populations later than do the bacteria.

It can thus be concluded that protozoan grazing is a significant factor

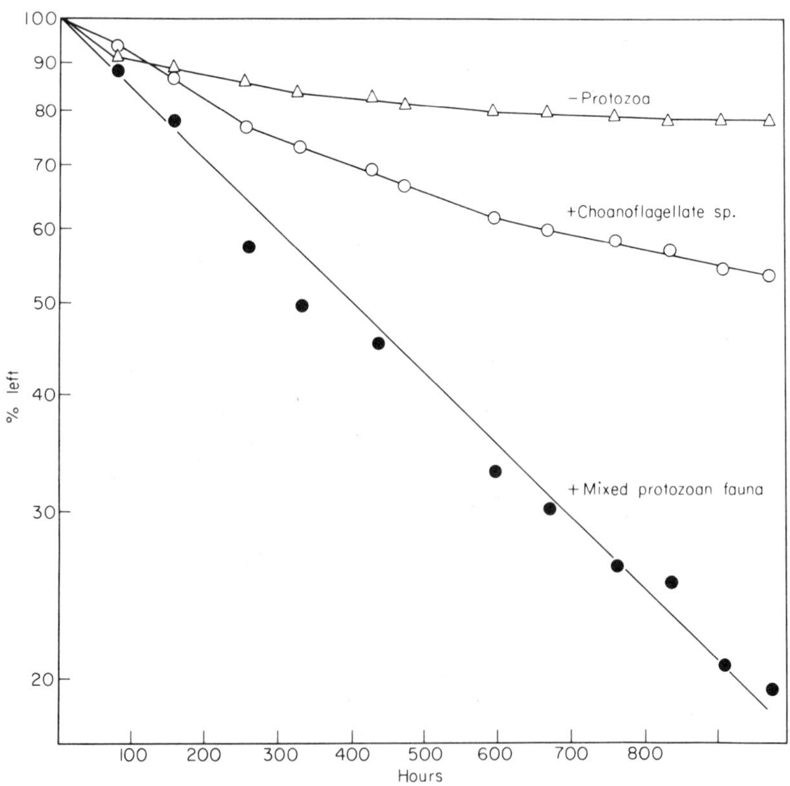

Figure 12.3. Rates of decomposition of ^{14}C labelled barley hay in seawater to show the effect of incubation with bacteria only or with bacteria plus various protozoa (Modified after Fenchel in press).

in controlling bacterial numbers in decomposer systems. Measurements on the grazing rates and bacterial turnover are difficult to carry out, however, and only a few, in part crude, estimates are available. Barsdate *et al.* (1974) found that in systems with *Tetrahymena* as the only grazer this ciliate ingested between 100 and 400 bacteria per individual per hour and would be responsible for a turnover of the bacterial population every 24–36 hours. However, it was not studied whether the ciliate had any selectivity with respect to bacterial type. Fenchel (1973) estimated that small zooflagellates consumed 4–5 bacteria per individual per hour, which meant that flagellates alone could be responsible for a turnover of the bacteria every 20 hours in the systems studied. Although these estimates are very crude the magnitudes found are of significance when discussing mineral cycling in the systems.

The stimulating effect of grazing on the rate of decomposition has been demonstrated several times. Harrison & Mann (1975b) found that in flasks containing seagrass detritus and seawater which were inoculated with bacteria, bacteria and zooflagellates, or bacteria, zooflagellates, and ciliates a loss of organic matter of 10, 16, and 20% respectively was measured 35 days after the inoculation. Fenchel (1975) found a considerable effect of protozoan grazers on the rate of decomposition of homogeneously ^{14}carbon labelled, water-extracted, barley hay (Fig. 12.3). The rate of decomposition was measured as the rate of liberation of $^{14}C\ CO_2$. Whereas the system with a natural assemblage of protozoa showed a rapid, nearly constant, rate of decomposition, the pure bacterial system had a lower and declining rate, and the system inoculated with one species of protozoa took an intermediate position.

The effect of nutrient levels on bacterial numbers and on the rate of decomposition

Whereas vascular plants have carbon/nitrogen ratios of around 15 (varying in eelgrass, for example, from 10 to 40) and carbon/phosphorus ratios near 200, the carbon/nitrogen and carbon/phosphorus ratios of bacteria are around 5·7 and 27 respectively (Harrison & Mann 1975a; Spector 1956). Bacteria, therefore, assimilate dissolved inorganic nutrients while decomposing dead plant material, and the complex consisting of detrital particles and microbial biomass becomes enriched with these essential nutrients (Fig. 12.4). The significance of this process for detritus feeding animals has been demonstrated previously (Fenchel 1972; Kofoed 1975; Newell 1965). Less is known about other aspects of the problem; for example, how nutrient levels control the rate of decomposition, how grazing on bacteria affects the nutrient–bacteria relationship, and how the fact that decomposers assimilate

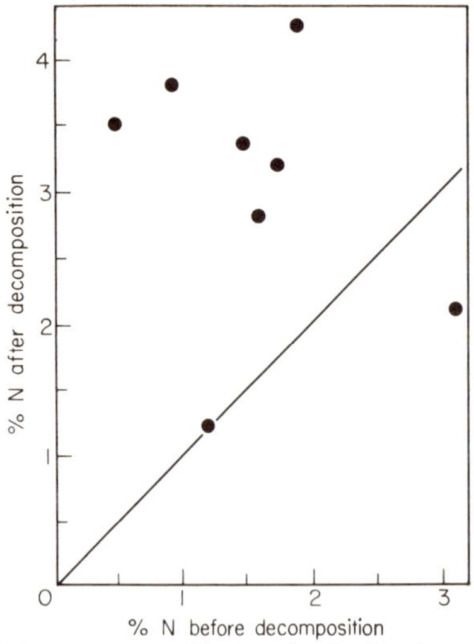

Figure 12.4. Changes in the % nitrogen (N) during decomposition of a variety of types of marine macrophyte detritus, under varying conditions and for different periods, to show the trend towards an enrichment with N. The line indicates no change in %N. (Modified after Harrison & Mann 1975b.)

inorganic nutrients accords with the general concept of mineralization. In this section we make some general observations on the affect of nutrient levels on microbial biomass and decomposition rates. In the following two sections more detailed radiotracer experiments with ^{32}P will be discussed.

The systems in which the effect of nutrients has been studied typically consist of 200 mg water-extracted plant material in 0·5 l of natural sea- or freshwater. In many experiments ^{14}C labelled hay was used in order to measure decomposition rates continuously (Barsdate *et al.* 1974; Fenchel 1973, in press; Fenchel & Rahn unpublished). In these systems the largest pool of nutrients was initially found in the organic material unless the water had been enriched. Thus, for example, the ^{14}C labelled hay contained 380 μg phosphorus per gram and if the water contained 30 μg P l^{-1} then there were initially 15 μg phosphorus in the water but 76 μg phosphorus in the organic substrate. Similar proportions held for nitrogen.

The levels of dissolved nutrients in non-enriched systems fell drastically following inoculation. In Fig. 12.5 this is shown in the case of phosphorus for two systems inoculated either with bacteria or with bacteria and a ciliate. These systems had initially been enriched with NO_3^-. The minimum levels of dissolved PO_4^{3-} occurred at the time of the maximum microbial biomass (cf. Fig. 12.1). Eventually the level of dissolved inorganic phosphorus increased again as decomposition proceeded. In these systems phosphorus is obviously limiting the standing crop of microbial biomass. Fig. 12.1 shows that enrichments with inorganic phosphorus and nitrogen increased not only the bacteria but also the higher trophic levels, i.e., the flagellates and

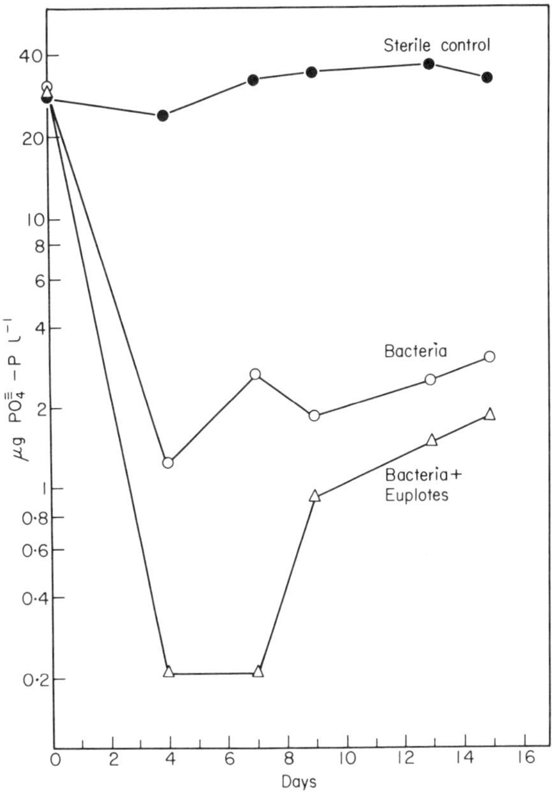

Figure 12.5. Concentrations of dissolved PO_4^{3-}—P in a seawater microcosm containing barley hay, to show differences during incubation with bacteria alone and with bacteria plus a ciliate grazer (Fenchel & Rahn unpublished results).

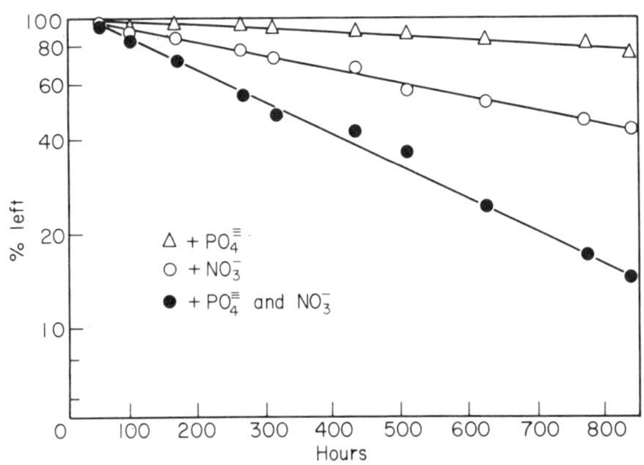

Figure 12.6. Rates of decomposition of ^{14}C labelled barley hay showing the effect of initial enrichment of the medium with phosphorus and/or nitrogen. The data are from the same experiment as shown in Fig. 12.1. (After Fenchel 1973.)

ciliates. The rate of decomposition also responded strongly to enrichment with these mineral nutrients (Fig. 12.6). Initial concentrations up to 10 mg PO_4^{3-} —P l^{-1} and 16 mg NO_3^- —N l^{-1} stimulated the breakdown rates.

Cycling of phosphorus and the effect of grazers

Barsdate *et al.* (1974) and Fenchel & Rahn (unpublished results) studied the phosphorus-cycling of laboratory microcosms by adding ^{32}P labelled PO_4^{3-} to the water. By taking samples from the system at regular intervals, fractionating these samples into dissolved and particulate material, and fractionating these in turn into organic and inorganic phosphorus it was possible to measure various pool sizes (dissolved, both inorganic and organic phosphorus; organic bacterial (+protozoan) phosphorus; and the inorganic polyphosphate pool ('surplus phosphorus') of bacteria). It was also possible to estimate transfer rates of phosphorus between these pools under 'steady state' conditions (i.e., during an interval of some hours in which no substantial change in the system could be expected). The main results of these experiments may be summarized as follows:

(1) There is a very high rate of phosphorus-cycling in these systems. At low concentrations of dissolved PO_4^{3-} (i.e., <5 µg PO_4^{3-} —P l^{-1}) turnover times of this pool were as low as 2 min. The turnover time of the total phosphorus pool of bacteria varied from 5 min. to more than 10 hours.

(2) Bacteria have a high rate of excretion of dissolved phosphorus both in organic and in inorganic forms. Grazers played a modest role compared with bacteria in the regeneration of inorganic phosphorus and pure bacterial systems showed a high rate of phosphorus-cycling. As an example, in one system consisting of bacteria and the protozoan *Tetrahymena* the latter was responsible for only 4–5% of the regeneration of phosphorus. This is consistent with the fact that the phosphorus-pool of bacteria had a turnover time of about one hour whereas the ciliate was responsible for turning over the bacterial biomass only about once every 20 hours. The bacterial uptake and excretion of phosphorus per bacterium, however, were usually higher in grazed than in ungrazed systems. This is the result, not cause, of a higher bacterial growth rate in grazed relative to ungrazed systems.

(3) There was a significant tendency for a higher absolute uptake rate of phosphorus per bacterium at lower concentrations of inorganic phosphorus. This surprising and in part contradictory result has been observed previously with algae and bacteria and has in part been attributed to systematical errors in the chemical quantitative determination of orthophosphate at very low concentrations (see references and discussion Barsdate *et al.* 1974).

These results show that there is a rapid mineral cycling, at least of phosphorus, in such model decomposer systems which is fairly independent of the mineralization of the organic substrate. This is in contrast to the results of Johannes (1965) who considered excretion of phosphorus from bacteria to be insignificant and the availability of PO_4^{3-} to be dependent on the regeneration of bacterial phosphorus through either lysis of cells or grazing by, e.g., protozoa. The conclusions of Johannes (1965) were mainly based on two experiments. In one he showed that the rate of mineralization, measured as the increase in concentration of orthophosphate, was higher in grazed than in ungrazed systems consisting of dead *Spartina* leaves, sea-water, and bacteria, with or without the bacterivorous ciliate *Euplotes*. That the mineralization proceeded faster in grazed systems is not in contrast to the present findings. Secondly, however, he showed that $^{32}P\ PO_4^{3-}$ was liberated from labelled bacteria only when a ciliate grazer was added.

The experiments were based on bacterial cultures with dissolved, easily decomposable organics (glucose, amino acids) and PO_4^{3-}. After 24 hours ciliates were added to some of the cultures and the level of dissolved phosphorus was measured in the following days in grazed as well as ungrazed cultures. It is likely that the bacteria had already utilized the organic substrates after 24 hours so there could be no compensatory growth after grazing by *Euplotes*. The regeneration of PO_4^{3-} in the grazed systems thus probably reflects the mineralization of the consumed bacteria and not a steady state condition. But it was not considered that the bacteria in the control might well have excreted and assimilated phosphorus thus keeping the dissolved inorganic phosphorus at a low level. The discrepancy between the interpretations of Johannes (1965) and Barsdate *et al.* (1974) is resolved when it is recognized that the bacteria maintain a cycle of phosphorus much more rapid than, and fairly independent of, the mineralization process and that the bacterial demand for phosphorus will keep the orthophosphate level low as long as the total pool of available phosphorus is low, i.e., until a substantial part of the organic substrate has been mineralized or if the initial PO_4^{3-} content were high.

That grazing by protozoa is not a significant factor for the regeneration of inorganic phosphorus in these systems can also be seen from other observations. Thus, if this mechanism were important the levels of dissolved orthophosphate would be expected to be higher in grazed than in ungrazed systems. This is not generally the case; frequently a somewhat higher level is found in ungrazed systems (cf. Fig. 12.5). This indicates only that the total microbial biomass is slightly higher in the grazed systems and consequently the fractional turnover rate of the dissolved pools is higher.

As previously discussed, the rates of decomposition were higher in grazed than in ungrazed systems. If the interpretation of Johannes (1965) were correct, one might expect this effect to be greatest at low concentrations of dissolved phosphorus. Fenchel (in press), however, studying rates of breakdown of a number of grazed and ungrazed systems with varying initial enrichments of phosphorus, found that the largest increase in rate in grazed compared to ungrazed systems occurred in the systems with the highest initial concentrations of phosphorus.

The rate of mineralization compared to the bacterial uptake of phosphorus

In the previous section it was shown that it is necessary to distinguish between the rates of nutrient uptake and excretion by the decomposer micro-organisms and the actual rate of mineralization, i.e., the release of nutrients from the organic substrate. This is shown schematically in Fig. 12.7. Using

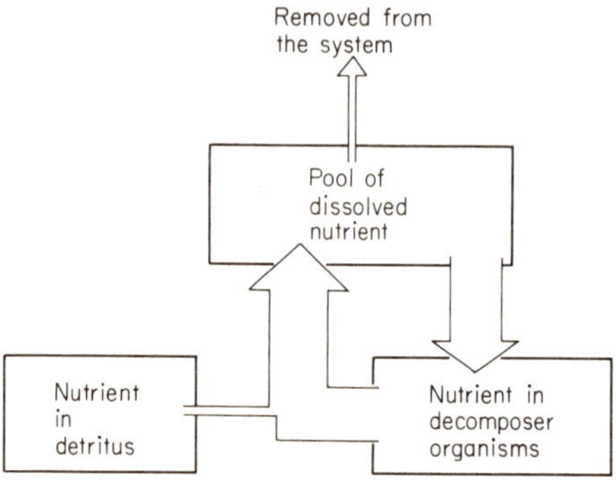

Figure 12.7. Schematical presentation of the nutrient cycling of a decomposer system showing the relation between the actual nutrient cycling and the simultaneous mineralization of the organic substrate. (After Barsdate *et al.* 1974.)

ion exchange resin which efficiently adsorb dissolved $PO_4{}^{3-}$, Barsdate *et al.* (1974) measured the rate of regeneration of phosphorus from the substrate independently from the uptake and release of phosphorus by the bacteria. First the level of dissolved $PO_4{}^{3-}$ was reduced strongly by placing one set of bags with ion exchange resin in the microcosms. These were replaced by another set of bags which were left for several hours. These in turn were removed and their contents of adsorbed $PO_4{}^{3-}$ measured; this was considered to measure the actual regeneration of phosphorus during the incubation period of the second set of bags. The uptake rate of the bacteria could thereafter be measured with the radiotracer method. For a pure bacterial system a regeneration rate of 0·028 μg P l^{-1} hr^{-1} was found. In the same system, however, bacteria took up and released 18·9 μg P l^{-1} hr^{-1}. In a parallel system with ciliate grazers the corresponding rates were 0·84 and 147 μg P l^{-1} hr^{-1}.

Possible mechanisms by which grazers stimulate bacterial activity

It is well known that, within certain limits, exploitation of a density dependent population may increase its productivity, as is the case with the decomposer bacteria associated with particulate debris. It is not precisely known, however, which mechanism is responsible for the stimulatory effect of protozoa on bacteria in our systems. In the case of larger detritus feeding invertebrates the mechanical effect of stirring and of decreasing particle size may be important (e.g. Fenchel 1970); such mechanisms are not likely to be significant in the case of protozoa.

It is clear that the availability of organic substrate is not limiting ungrazed bacteria in our systems since at most 20% of the particle surface is covered by bacteria. Mineral nutrients may be limiting, but at least with respect to phosphorus the effect of regeneration by grazers is of little significance. It is possible that dissolved nutrients or oxygen could become depleted in the immediate surroundings of the cells. Protozoa could create microturbulence around the bacterial cells and at the same time reduce the density of bacteria; both these mechanisms would lessen competition. It cannot be ruled out, of course, that grazers do excrete some growth promoting substance other than orthophosphate. The systems studied by the present authors have always been inoculated with mixed assemblages of bacterial types, but it is not unlikely that grazing selects for faster growing bacteria. Future critical experiments, which must also include systems with only one species of bacterium, may elucidate the mechanisms involved in more detail.

Significance of the results

Experiments on decomposer microcosms have given us an insight into some of the relationships between organic debris, bacteria, and protozoa as they affect rates of decomposition, mineralization, and the internal cycling of nutrients. It would be risky, however, to blindly apply such data to the situation in nature where a multitude of factors may differ from those in the microcosms. Nutrient supply, to take only one example, is less likely to be limiting in the sea where circulation of water over the detritus would bring nutrients to the bacteria. As an indication of some of the unknown factors one can compare the experiments of Zieman (1968) and Harrison & Mann (1975b) on *Thalassia* and *Zostera* respectively, two seagrasses. In the laboratory *Zostera* decomposed at a much slower rate than did *Thalassia* but one must consider not only possible differences in nutrient content of the seawater, temperature, and other physico-chemical factors, but also the chemical composition of the leaves and, for example, whether or not dead *Zostera* leaves might contain a bacteria-inhibiting substance (phenols are a plausible example although to our knowledge this problem has not been investigated).

Certain other characteristics of microcosms can be more directly related to natural detritus systems. In particular, the assimilation and internal cycling of essential mineral nutrients by micro-organisms may be of significance. Vascular plants such as marsh grasses and seagrasses produce large annual crops of organic matter of which almost all is transformed into detritus before being utilized by consumers (Mann 1972). From an evolutionary viewpoint one could propose that these plants would be competitive only if they balanced on the one hand their low rates of growth (compared with, for example, the smaller algae) and their need to provide a large amount of purely structural tissue with a means to avoid being consumed during the growing season. One mechanism by which vascular plants could achieve this compromise is by an economical use of nutrients and the build-up of tissues with a ratio of carbon to minerals too high to be nutritious for consumers (see Harrison & Mann 1975a for data on *Zostera*). When this carbon-rich, mineral-poor plant material dies, the process of decomposition is accompanied, therefore, by the assimilation of dissolved mineral nutrients by bacteria and the subsequent retention of these minerals within the microbe-detritus system. The details of these and other processes have been successfully determined using laboratory microcosms.

References

BARSDATE R.J., FENCHEL T. & PRENTKI R.T. (1974) Phosphorus cycle of model ecosystems: significance for decomposer food chains and effect of bacterial grazers. *Oikos* **25**, 239–51.

FENCHEL T. (1970) Studies on the decomposition of organic detritus derived from the turtle grass *Thalassia testudinum*. *Limnol. Oceanogr.* **15**, 14–20.

FENCHEL T. (1972) Aspects of decomposer food chains in marine benthos. *Verh. Deutsch. Zool. Ges. 65 Jahresversamml.* **14**, 14–22.

FENCHEL T. (1973) Aspects of the decomposition of seagrasses. *International Seagrass Workshop*, Leiden, Netherlands.

FENCHEL T. (in press) The significance of bacterivorous protozoa in the microbial community of detrital particles. In: *Freshwater Microbial Communities*, 2nd edn, ed. J. Cairns. Marcel Dekker, Inc. New York.

HARGRAVE B.T. (1970) The effect of a deposit-feeding amphipod on the metabolism of benthic microflora. *Limnol. Oceanogr.* **15**, 21–30.

HARRISON P.G. & MANN K.H. (1975a) Chemical changes during the seasonal cycle of growth and decay in eelgrass (*Zostera marina* L.) on the Atlantic coast of Canada. *J. Fish. Res. Board Can.* **32**, 615–21.

HARRISON P.G. & MANN K.H. (1975b) Detritus formation from eelgrass (*Zostera marina* L.): the relative effects of fragmentation, leaching, and decay. *Limnol. Oceanogr.* **20**, 924–34.

JOHANNES R.E. (1965) Influence of marine protozoa on nutrient regeneration. *Limnol Oceanogr.* **10**, 434–42.

JOHANNES R.E. (1968) Nutrient regeneration in lakes and oceans. *Adv. Microbiol. Sea* **1**, 203–13.

KOFOED L.H. (1975) The feeding biology of *Hydrobia ventrosa* II The allocation of the parameters of the carbon-budget and the significance of secretion of dissolved organics. *J. Exp. Mar. Biol. Ecol.* **19**, 243–56.

MANN K.H. (1972) Macrophyte production and detritus food chains in coastal waters. *Mem. Ist. Ital. Idrobiol. 29 Suppl.*, 353–83.

NEWELL R. (1965) The role of detritus in the nutrition of two deposit feeders, the prosobranch *Hydrobia ulvae* and the bivalve *Macoma balthica*. *Proc. Zool. Soc. Lond.* **144**, 25–45.

POMEROY L.R. (1970) The strategy of mineral cycling. *Ann. Rev. Ecol. Syst.* **1**, 171–90.

SPECTOR W.S. (ed.) (1956) *Handbook of Biological Data*. W.B. Saunders, Philadelphia.

ZIEMAN J.C. JR (1968) *A study of the growth and decomposition of the sea-grass Thalassia testundinum*. Unpublished M.Sc. Thesis. University of Miami.

13 The central role of invertebrate faeces in sediment decomposition[1]

B. T. HARGRAVE *Department of the Environment, Fisheries and Marine Service, Marine Ecology Laboratory, Bedford Institute of Oceanography, Dartmouth, Nova Scotia B2Y 4A2, Canada*

Abstract

Measurements of oxygen uptake, an index of aerobic bacterial metabolism, indicate that invertebrate faeces are foci of microbial activity. Bacterial populations on natural sediment and detritus particles, which are more sparsely populated than has been previously suggested, achieve equilibrium densities determined by nutrient availability.

Passage through an invertebrate gut completely alters bacterial populations through digestion. Little of the ingested substrate is digested, thus egestion rates are directly proportional to feeding rates. Surface area of egested material for some benthic invertebrates is logarithmically related to body weight.

Egested faecal material is rapidly colonized by successions of micro-organisms and maximum oxygen consumption is achieved within 2–3 days. Thereafter, metabolic activity falls sharply to a low and stable level corresponding to uningested substrates. Reworking of sediment and detritus by benthic invertebrates thus provides space for microbial growth. In this sense all deposit feeding invertebrates enhance their food supply and stimulate decomposition of sedimentary organic matter.

Introduction

There has been a tradition amongst ecologists to separate studies of primary and secondary production. In terrestrial systems where herbivores generally consume a small portion of plant production, primary production productivity is usually considered to be dominated by climate-controlled variables to which nutrient availability is related (Eckardt 1968). Similarly, in the past, aquatic food chain models have emphasized 'primary producer-herbivore-carnivore' links as dominant, being driven by seasonal pulses of nutrients and light (Steele 1974). It has only been during the last decade, however, that studies of grazing and decomposition have stressed interactions between producers and consumers. Predator control of herbivores (Hairston *et al.* 1960), the existence of feeding thresholds in herbivorous marine copepods (Parsons *et al.* 1967), and the stimulatory effect of grazing organisms on the metabolism of their food supply (Hargrave 1970a; Cooper 1972; Fenchel 1972) all indicate the existence of control mechanisms based on *interactions* between producers and consumers. Organizational structure and homeostasis of communities depend on such interactive mechanisms. Important questions of limits to energy flow through ecosystems can be considered by studies which focus on these processes.

THE RELATIVE IMPORTANCE OF DECOMPOSER FOOD CHAINS

Important insights concerning ecosystem functioning can be gained by identifying processes which limit the transfer of energy or nutrients. These processes may control total production and confer characteristic properties to ecosystems. For example, sedimented aquatic organic debris (detritus), like litter in terrestrial systems, represents a non-circulating nutrient supply and a potential energy source for consumer organisms. While such reserves stabilize seasonal pulses of input (Pomeroy 1970), these reservoirs of organic matter and inorganic nutrients also represent 'bottle-necks' to energy flow (Macfadyen 1961). Processes which mobilize this material are crucial ones for ecosystem functioning.

The conversion of organic matter through decomposer food chains into tissues of organisms at higher trophic levels may be of major importance in determining overall productivity in many terrestrial and aquatic ecosystems. This is particularly so in environments where nonliving organic material, either in particulate or dissolved form, accumulates. Obvious aquatic habitats where this occurs are productive marine littoral areas, salt marshes and estuaries where macrophytes dominate primary production. Little direct consumption of macrophytes occurs and much of their production enters pelagic and benthic food chains in particulate and dissolved form (Teal 1962; Khailov & Burlakova 1969; Mann 1972). Recent studies by Mann (1973) document cases of overgrazing of *Laminaria* by localized concentrations of sea urchins in a marine bay. Over a large area, however, sea urchins only utilized a small percentage of the seaweed production (Miller & Mann 1973).

Fenchel (1972) summarized previous studies which indicate that the bulk of primary production by macrophytes enters decomposer food chains which, because of shallow water, are generally localized in surface sediments. As water depth increases primary production from phytoplankton replaces macrophyte production as a source of organic matter. It is generally thought that most of this production passes through pelagic grazing food chains with only a fraction remaining as detrital input to bottom communities. An inverse relation between water depth, sedimentation and oxygen uptake by sediment cores in various aquatic systems supports this view (Hargrave 1973, 1975). The efficiency of pelagic herbivores in consuming phytoplankton production, however, may change seasonally (Poulet 1973) and inefficient feeding during bloom conditions (Cushing & Vucetic 1963) may produce nutrient rich faecal material which directly enters decomposer food chains in bottom sediments. Thus, even plankton-dominated ecosystems may possess well-developed benthic production systems fueled by processes of decomposition.

DETRITUS AS A FOOD SOURCE

Despite early emphasis that nonliving organic matter served as a food base in certain marine communities (Fenchel 1972), ecosystem production models have not been developed which adequately represent the dynamic nature of interactions between particulate substrates and attached microbial populations. The widespread consumption of detritus by aquatic vertebrates and invertebrates has been well documented (Boysen-Jensen 1914; Darnell 1967; Mann 1972). Generally, only a small amount of ingested debris is assimilated, however, and it was Newell (1965) who first demonstrated that this was exclusively microbial biomass and not the nonliving organic substrate itself. While it had been thought that bacteria were a food source for detritus feeders, the increases in nitrogen in *Hydrobia* faecal pellets kept in seawater after egestion, which decreased on subsequent re-ingestion, allowed Newell to suggest that only micro-organisms were of nutritional value and that removal stimulated cell growth on egested material. Two estuarine benthic invertebrates have been shown to assimilate detrital substrates more efficiently than associated bacteria (Adams & Angelovic 1970). The conclusions are based on indirect calculations, however, and the experiments deserve to be repeated.

Numerous studies have confirmed Newell's observations of a high assimilation efficiency for micro-organisms by various detritus feeders (Fenchel

1970, 1972; Hargrave 1970b). Direct feeding experiments, analyses of digestive enzymes and observations of reduced microbial density on detritus in the gut and on faeces, cited by these and other authors, give evidence of the thoroughness of digestion at least for certain cell types. In addition the biomass of some intertidal invertebrates correlates positively with levels of sediment carbon and nitrogen in those environments where these measures are thought to reflect microbial biomass (Newell 1970). On the other hand, Sanders & Hessler (1969) observed that organic carbon in deep sea sediments varies within an order of magnitude, far less than the corresponding variation in animal density. Deposited organic matter in the deep sea, however, is probably largely refractory in comparison to that present in intertidal sediments. In any case, correlations of microbial biomass and fauna density must be viewed with caution. Cause and effect can only be inferred since factors such as particle size markedly affect the level of organic carbon and nitrogen in sediments (Newell 1970). It cannot easily be established that distributional patterns do not reflect selection of substrates on the basis of physical properties related to grain size alone.

INTERACTIONS BETWEEN DETRITUS FEEDERS AND THEIR FOOD SUPPLY

The role of soil fauna, particularly earthworms, in decomposition processes and enhanced soil fertility has been known since Darwin's (1881) classical study. It has only been realized recently, however, that some aquatic benthic invertebrates may have similar effects. Recent studies reviewed by Satchell (1974) support ideas suggested earlier by Macfadyen (1961) concerning the catalytic effect which fauna may have on microbial growth in soil and litter. Among many possible effects, the following were suggested:

(1) distribution of cells to sites more suitable for growth.
(2) elimination of bacteriostasis through digestion.
(3) enrichment by excretions.
(4) removal of scenescent colonies.

Satchell argues that (2) and (3) may be directly related since nonspecific biostasis is easily overcome by addition of readily available energy sources. He also argues that there has been no direct evidence that browsing of scenescent bacterial colonies on natural substrates stimulates growth. In addition, he cites studies by van der Drift & Witkamp as evidence that the main effect of arthropod feeding in litter decomposition is fragmentation and exposure of fresh surfaces for microbial attack. Comminution also increases moisture content through increased water holding capacity and decreased evaporation. Stabilized and increased levels of humidity which result are major beneficial effects for microbial activity.

Aquatic deposit and detritus feeding invertebrates also enhance energy flow through sediments by their mechanical and feeding activity. Oxygen penetration into soft sediments can be increased by tube dwelling invertebrates (Edwards 1958; Hargrave 1972a). Sediment displacement by ingestion and egestion occurring at different depths has been demonstrated for bivalves (Rhoads & Young 1970; Rhoads 1973) and tubificids (Davis 1974). Mixing and displacement of sediment by burrowing and feeding increase the depth to which aerobic metabolism can occur and permit physical cycling of sedimented material which otherwise would remain buried.

Benthic deposit and detritus feeding invertebrates also produce many fragments as faeces and uningested particles which provide added or renewed surfaces for microbial colonization and growth (Hargrave 1970a; Fenchel 1970, 1972). Selective feeding on specific sizes of particles may also serve to accumulate organically rich material in faeces (Whitlatch 1974). Brundin (1949) observed that sediments in several Swedish lakes were largely composed of chironomid faecal pellets. Pellets of many species of invertebrates have a characteristic morphology and these may form a substantial fraction of some estuarine bottom deposits (Carriker 1967; Kraeuter & Haven 1970). Sediments which are intensively 'worked' by benthic organisms also tend to be physically unstable, have a higher water content and may be more readily resuspended than compact unworked sediments (Rhoads 1973). Resuspension of these sediments moves both organic and inorganic material stratified beneath the sediment surface into the water column. Such upwelling serves to increase the exposed surface of deposited material, facilitates adsorption of dissolved material on to the suspended particulate matter and thus provides additional enriched substrate for microbial growth.

Enhanced growth of micro-organisms on aquatic invertebrate faecal material was first demonstrated indirectly by Newell (1965). Large increases in nitrogen content accompanied by small decreases in carbon in *Hydrobia* faeces observed three days after feeding and similarly after re-ingestion were assumed due to bacterial growth. Light had no effect on these changes. Other studies, however, indicate that conditioning of invertebrate faeces does not always produce similar changes. *Hyalella* faeces increased organic content only in the light while a decrease occurred in darkness (Hargrave 1970a). Johannes & Satomi (1966) noted that dissolved organic matter stimulated growth of autotrophic and heterotrophic micro-organisms on *Palaemonetes* faeces during the first four days irrespective of light conditions. Colonization and subsequent changes in organic content of faeces and detritus obviously depend on the kinds of micro-organisms present, organic matter in the substrate and the nature of dissolved organic and inorganic material in solution.

Leaching of organic material is a major factor determining initial weight loss from fresh detritus (Zieman 1968; Oláh 1972). In the laboratory, *Zostera* leaves lost 35% of their dry weight in 100 days and over 80% of this loss was attributable to solubilization of organic material (Harrison 1974). Sterilized fragmented leaves lost weight more quickly, presumably due to leaching from the increased exposed surface. Balanced against these losses and those due to microbial respiration are processes which add organic material to decomposing particulate matter. *Zostera* and *Spartina* both increase in nitrogen content during ageing and disintegration (Boysen-Jensen 1914; Odum & de la Cruz 1967; Harrison 1974) presumably due to bacterial nitrogen fixation. Enhanced organic content of faeces in the presence of soluble organic matter (Johannes & Satomi 1966; Hargrave 1970a) provides additional evidence for incorporation of soluble compounds into microbial biomass. Organic matter added to invertebrate faeces as mucous or digestive fluid exudate would also form a readily available substrate for microbial attack.

All of these processes, both mechanical and chemical, serve to stimulate microbial activity. The interactions have a component of self-regulation. Micro-organisms by their ubiquitous distribution, adaptability and potential for rapid growth, appear to be held within limits by substrate availability. Space and nutrients are rapidly utilized and thus processes which provide these may regulate growth. Invertebrates, by their presence and activity, thereby ensure a continual production of microbial biomass.

Decomposition and substrate availability

SPATIAL DISTRIBUTION OF BACTERIA ON SOLID SURFACES

Colonization of solid surfaces by aquatic micro-organisms is related both to the physical characteristics of the attachment surface as well as to nutrients in the substrate and in solution. In his classical study, ZoBell (1943) considered how surfaces influence bacterial activity and increase cell numbers. In these and subsequent studies, however, enumerations by plate counts were related to sediment weight to give a false impression of high bacterial densities. It was not until the development of direct fluorescence microscopic counting techniques that the spatial distribution of micro-organisms on detritus and sediment was given perspective.

Published observations of direct counts which indicate spacing between bacterial cells and adjacent micro-colonies on well-colonized surfaces have been reviewed (Hargrave & Phillips in press). Usually between 100 and 400 μm² is available per bacterial cell. If one cell occupies 1 μm², then <1% of the surface area is occupied. Gray *et al.* (1968) have made similar calculations for sand grains in soil. Bacteria attached to surfaces covered 0·042% of the total available area. These observations tend to indicate that space *per se* is not limiting cell numbers. Certain substrates, however, such as fresh detritus and invertebrate faeces, can have much higher densities (Table 13.1). The most striking example is that of artificially fragmented *Zostera* detritus aged 4–8 days in the presence of nitrate and phosphate (Fenchel 1972). Densities of 27 cells 100 μm⁻² were observed. If the above assumption of cell area is applied, 25% of the detrital surface would be covered.

The efficiency with which invertebrates remove bacteria from ingested particles, discussed above, is also demonstrated by changes in bacterial

Table 13.1. Estimated areal coverage of bacterial populations determined on different substrates by direct microscopic counts.

Substrate	Habitat	μm² available cell⁻¹	Reference
pebbles	marine intertidal	300	Batoosingh & Anthony (1971)
sand and silt	marine intertidal sandflat	130–411	Dale (1974)
silt–clay	lake sediments	70–311	Tsernoglou & Anthony (1971)
lake sediment	Marion Lake surface	60	Hall & Hyatt (1974)
glass, plastic slides	marine	20–60	O'Neill & Wilson (1971)
Zostera detritus	artificially fragmented laboratory seawater cultures aged 4 days	4–6 (enriched) 50–100 (unenriched)	Fenchel (1972)
Thalassia detritus	artificially fragmented aged 2–4 days	110	Fenchel (1970)
	Parhyalella faeces fed *Thalassia*	50	
sediment detritus	marine intertidal	15	Fenchel (1972)
	Macoma gut contents		
	stomach	17	
	anterior gut	100	
	middle gut	200	
	posterior gut	1,000	

density as sediment passes through an organism's gut (Fenchel 1972) (Table 13.1). *Macoma* egests particles almost devoid of bacteria when fed heavily colonized detritus. Faeces can be rapidly colonized, however. *Parahyalella* produces faeces which achieve densities (100–150 μm^{-2} $cell^{-1}$) comparable to fresh detritus after five days (Fenchel 1970).

These observations suggest that characteristic stabilized 'equilibrium' densities for microbial populations on surfaces exist. The carrying capacity, quantified as available space per cell, must reflect nutrient availability in solution and in the substrate since the presence of inorganic nutrients increases densities to levels seldom observed in nature. Removal of micro-organisms from particles by invertebrate feeding provides space for colonization which might occur to cell densities determined by nutrient availability. Nutrient additions associated with faeces and excretion of soluble organic and inorganic substances by benthic invertebrates could thus significantly enhance microbial densities above levels which would occur in their absence.

OXYGEN UPTAKE AS AN INDEX OF AEROBIC DECOMPOSITION

Counts of micro-organisms, no matter how accurate or on what basis, cannot quantify their importance in decomposition processes. Measures of changes in numbers with time or related metabolic rates are more informative. It is widely accepted, however, that no single measure can describe total microbial metabolism. Measurements of substrate uptake kinetics, for example, only quantify the response of certain metabolic types to the added substrate. Similarly, anaerobic decomposition processes can only be followed by considering the dynamics of specific inorganic terminal hydrogen acceptors used by different types of micro-organisms (such as CO_2, NO_2^-, SO_4^{2-}). No one measure exists which provides a holistic view of total microbial metabolism in mixed natural populations.

Oxygen consumption reflects the total aerobic oxidative metabolism of all organisms in a community. Anaerobic metabolic processes are not measured but reduced organic and inorganic material itself will consume oxygen which may serve as an index of potential anaerobic metabolism. The method of following oxygen uptake in samples incubated in closed containers over short periods is rapid and convenient and has been applied to studies of decomposition in litter (Howard 1967), soil (Greenwood 1968) and aquatic sediments (Hargrave 1972b). Interpretation of these measurements is made difficult by possible experimental artifacts and necessary assumptions of homogeneous distribution of oxygen throughout a sample during the time period of measurement. Extrapolation of laboratory measured rates to field conditions are at best tenuous. A positive linear relation between respiration and microbial biomass during loss of weight by *Spartina* detritus decomposing under laboratory conditions (Gosselink & Kirby 1974), however, does indicate that oxygen consumption can provide a useful comparative measure of aerobic microbial activity which is related to decomposition. Carbon dioxide evolution may provide a similar and quantitatively related measure of microbial activity (Witkamp 1973).

The usefulness of measures of oxygen consumption as a comparative index of decomposition is improved if rates are expressed on a standard basis. Particle size (assumed inversely proportional to surface area) and organic content are key parameters controlling both microbial biomass and oxygen uptake in aquatic sediments and detritus (Hargrave 1972b; Dale 1974). Thus, oxygen uptake expressed unit per area or per unit organic carbon or nitrogen constitutes a standardized measure of aerobic heterotrophic activity applicable to any substrate undergoing decomposition. Comparison of oxygen uptake by algal and bacterial populations attached to sand, pebbles, rocks, and detritus on the basis of surface area and organic content, for example, demonstrated that at 10°C in oxygenated water all substrates consumed between 0·01 and 1 μg O_2 cm^{-2} hr^{-1} (Hargrave & Phillips, in press). Community oxygen consumption on a total weight basis was also directly related to the percentage of carbon and nitrogen present and there was a linear relation between oxygen uptake per unit of carbon and uptake per unit of nitrogen over a wide range of metabolic rates on all substrates. These observations suggest conservative properties for respiratory energy loss by attached microbial communities imposed by the availability of space and nutrients.

LIMITS TO DECOMPOSITION—AN HYPOTHESIS

Greenwood (1968) reviews evidence which suggests that decomposition of organic matter in soil is limited by the rate of enzymic hydrolysis. Structural stability in undisturbed soil is viewed as reducing contact between enzymes on microbial cell surfaces and organic matter. Decomposition is thus enhanced by any process which rearranges soil aggregates to facilitate exposure of enzymes to organic substrates.

Observations of bacterial biomass and metabolic rates discussed above suggest that microbial populations on aquatic sediment and detritus particles are also held at stable levels. After colonization, growth and metabolism appear to be regulated by limits of space and nutrient availability. The presence of invertebrates in sediments should considerably alter such a balance.

Previously mentioned increased oxygen availability due to mechanical stirring and removal of microbial biomass during passage of particles through the guts of organisms will increase both space and nutrients. The amount of space released will be proportional to the efficiency of cell removal during digestion. After an initially rapid growth, when space and nutrients are utilized, microbial biomass and metabolism will stabilize at levels equivalent to those in uningested sediments determined by particle surface area and nutrient supply.

The existence of stable levels of aerobic respiration by microbial communities in sediment and detritus, predictable on the basis of surface area and nutrient content, can readily be established. Similarly, exposure of such substrates to ingestion by benthic invertebrates and subsequent determinations of oxygen uptake by egested material should quantify the magnitude and duration of any stimulatory effect which ensues. These comparisons are made in the observations which follow.

Invertebrate faeces as sites of microbial activity

RELATIONS BETWEEN ANIMAL SIZE AND SURFACE OF EGESTED MATERIAL

Allometric equations based on weight or calories describe relations between ingestion and egestion for many terrestrial and aquatic detritus and deposit feeding invertebrates (Reichle 1971; Hargrave 1972c). Slope coefficients approximate to the value (0·67) expected if energy intake is proportional to body surface area when correlations are derived over a wide range (several logarithmic cycles) of animal sizes (van der Drift 1950).

Surface area of material egested by invertebrates might also be expected to show a similar proportionality to body size. Appropriate substrates were offered as food to several different freshwater and marine benthic invertebrates in excess of what could be consumed over a 6–10 hour period. Planar surface estimates were derived by measuring the diameter and length of egested pellets and faecal strings (Table 13.2). Despite the limited data, an allometric relation exists between animal dry weight and surface area of faecal material egested per unit time (Fig. 13.1). Surface area of faeces produced by the sea urchin *Strongylocentrotus* grazing on *Laminaria*, however, is an order of magnitude less than would be expected from the relationship. This might reflect the higher assimilability of this food source relative to that of sediments or detritus. Organisms which ingest material of low nutritional value must process substantially more substrate to assimilate equivalent amounts of organic matter.

Table 13.2. Calculated surface area of faecal material egested by various marine and fresh-water deposit and detritus—feeding benthic invertebrates. Estimated hourly rates of pellet production at 10°C with excess substrate available for ingestion.

Organism	Faecal pellets hr^{-1}	mm^2 $pellet^{-1}$	mm^2 an^{-1} hr^{-1}	Animal density m^{-2}	cm^2 exposed m^{-2} hr^{-1}
Hydrobia ulvae[1]	—	—	1·9	$4{\cdot}4 \times 10^4$	840
Chironomus anthracinus[2]	4	0·20	0·8	800	6
Hyalella azteca[3]	8	0·27	2·2	800	18
Littorina saxatilis[4]	4	0·60	2·4	500	12
Littorina littorea[5]	6	5·30	32	40	13
Limnaea palustris[2]	3	0·32	0·96	100	10
Yoldia limatula[5]	—	—	31	20	6
Arenicola marina[6]	—	—	126	5	6
Strongylocentrotus[5] *droehbachiensis*	10	1·50	15	37	6

[1] Data from Fenchel (1972).
[2] Collected from Lake Esrom (Hargrave 1972a,b).
[3] Data from Hargrave (1972c).
[4] Collected from *Spartina* salt marsh, Lawrencetown, N.S.
[5] Collected from St. Margaret's Bay, N.S. (Miller & Mann 1973).
[6] Collected from intertidal sediments, Petpeswick Inlet, N.S.

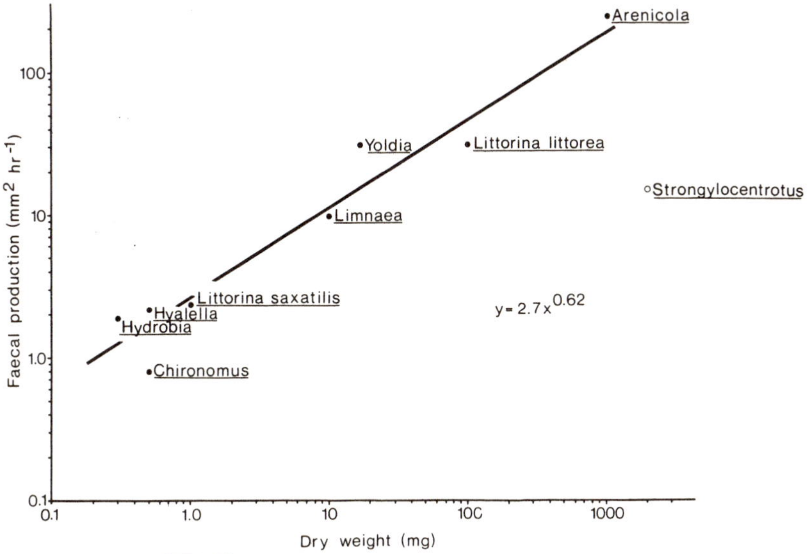

Figure 13.1. Faecal production, estimated as planar surface equivalents, compared with body dry weight for a variety of aquatic bottom living invertebrates. Data for *Hydrobia* from Fenchel (1972). Regression line calculated excluding measurements with the sea urchin *Strongylocentrotus*.

The importance of small organisms in producing surfaces through egestion is underestimated when comparison with large organisms is made on a unit organism basis. Small organisms are usually more numerous and their role is given perspective when differences in density are considered (Table 13.2). All of the species for which I have obtained data produce approximately similar amounts of egested surface when estimates are made on a population basis. Fenchel's (1972) calculations of *Hydrobia* egestion produce estimates of surface exposure two orders of magnitude higher—a direct result of the high densities which the snail populations achieve.

Fragmentation may also occur during feeding which would increase surfaces in addition to those produced as faeces. Fenchel (1970) observed that 20 *Parahyalella* feeding for 96 hours increased the amount of fine particles in *Thalassia* detritus which corresponded to an increase in external surface by a factor of 2·3. Both faeces and fragmented particles can thus represent considerable increases in substrate surface.

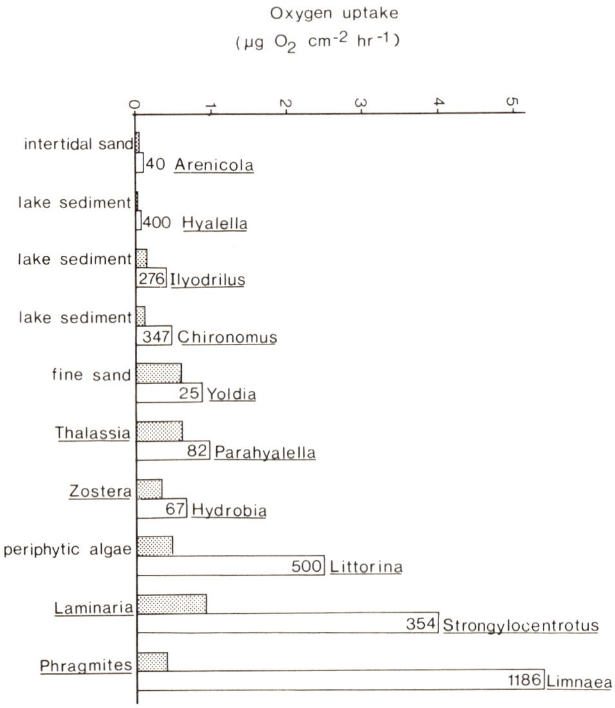

Figure 13.2. Comparison of oxygen uptake on an areal basis of different substrates offered as a food source and faeces produced after ingestion of the food by a variety of benthic invertebrates. Data for *Hydrobia* from Fenchel (1972). Each measurement represents the mean of at least three separate determinations. Numbers in histogram bars express oxygen consumption of faecal material (unshaded) as a percentage of oxygen consumed by substrates offered as food (shaded).

SUBSTRATE OXYGEN UPTAKE BEFORE AND AFTER INGESTION

Various invertebrates (Table 13.2) were held in the laboratory for 6–18 hours and offered different substrates as a food source. The substrates (sand and mud sediments, detritus and periphytic communities on pieces of macrophyte debris), collected with the organisms, were readily ingested and represented natural food material. Faeces were collected after feeding for one to two days in aerated water from the collection site. Subsamples of substrates, not offered as food, were held under similar conditions. Substrates and faeces were then incubated in the dark for short periods (2–3 hours) and decreases in dissolved oxygen measured in the presence and absence of 1% buffered formalin. The difference in the two measures was assumed to represent biological respiration of microbial populations attached to surfaces (Hargrave 1972b). Rates of oxygen uptake of both food substrates and faeces were

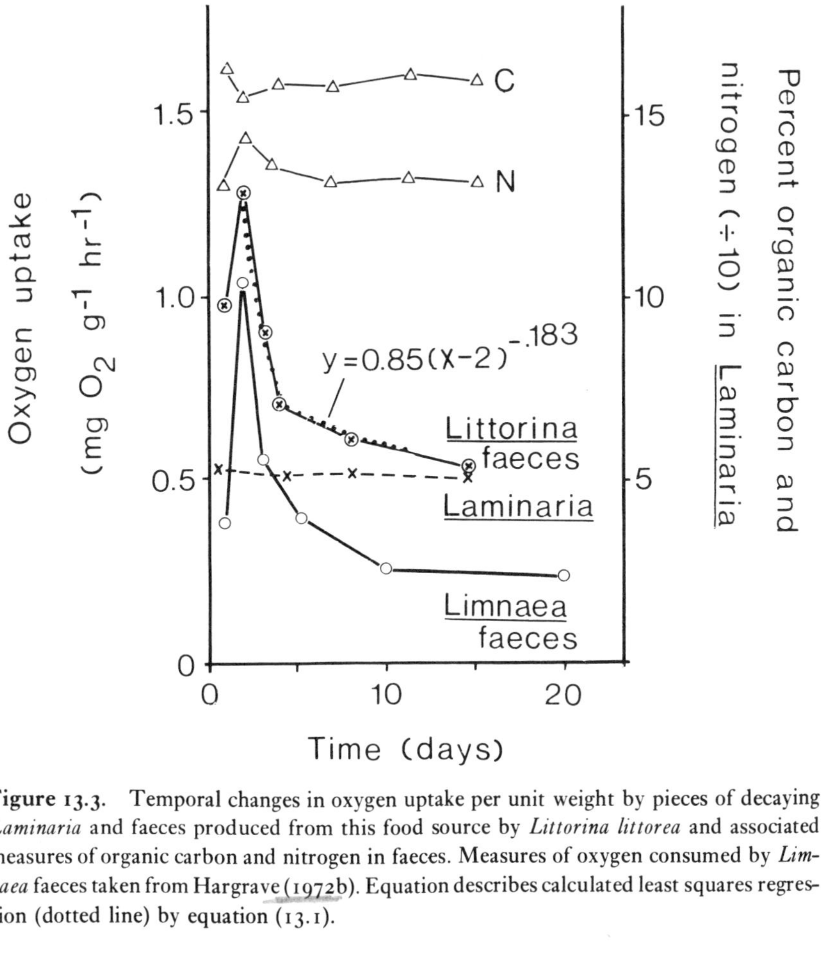

Figure 13.3. Temporal changes in oxygen uptake per unit weight by pieces of decaying *Laminaria* and faeces produced from this food source by *Littorina littorea* and associated measures of organic carbon and nitrogen in faeces. Measures of oxygen consumed by *Limnaea* faeces taken from Hargrave (1972b). Equation describes calculated least squares regression (dotted line) by equation (13.1).

expressed on a surface area and organic content basis as described by Hargrave & Phillips (in press).

Microbial communities associated with faecal material produced by all invertebrates examined consumed more oxygen on a surface area basis than substrates offered as food (Fig. 13.2). There was no consistent relation between the amount of stimulation in oxygen uptake and the type of substrate or its organic content. Respiration in sand sediments with the least organic matter (1% weight loss on ignition) was increased the least. Oxygen uptake by faeces produced from *Zostera* and *Thalassia* detritus was not enhanced to a significantly greater degree despite a much higher organic content. Faeces formed from soft mud sediments and from browsing on periphytic communities (by *Littorina* and *Limnaea*) displayed the greatest stimulation of metabolic activity.

QUANTIFYING ENHANCED METABOLIC ACTIVITY OF INVERTEBRATE FAECES

Earlier observations (Hargrave 1972b) demonstrated a marked time-dependent decrease in oxygen uptake by *Limnaea* faeces. Maximum respiration occurred one to two days after egestion and then decreased rapidly. Similar changes in oxygen uptake occurred with *Littorina* faeces which was produced from decaying *Laminaria* (Fig. 13.3). Metabolic activity of faeces was almost identical to uningested decomposed *Laminaria* tissue after 15 days.

There were only slight changes in faecal organic matter during the 15-day incubation. Nitrogen increased and carbon decreased slightly but changes were not nearly as great as those observed by Newell (1965). Expression of oxygen uptake by faeces on the basis of nitrogen content still shows stimulation in metabolic activity for a few days followed by a decrease to levels which approach those observed in decaying *Laminaria* (Fig. 13.4). Estimates of oxygen uptake by fresh *Spartina* detritus also expressed on the basis of nitrogen content, show similar time-dependent decreases in respiration comparable to those observed with *Littorina* faeces.

Time course changes in metabolic activity of microbial populations on faecal material may be described by an equation of the form:

$$R = a(t-n)^{-b} \qquad (13.1)$$

where R is oxygen consumption by the faecal material

t is time after egestion (hours or days)

n is the time taken for a maximum metabolic level to be reached where $(t-n) > 0$

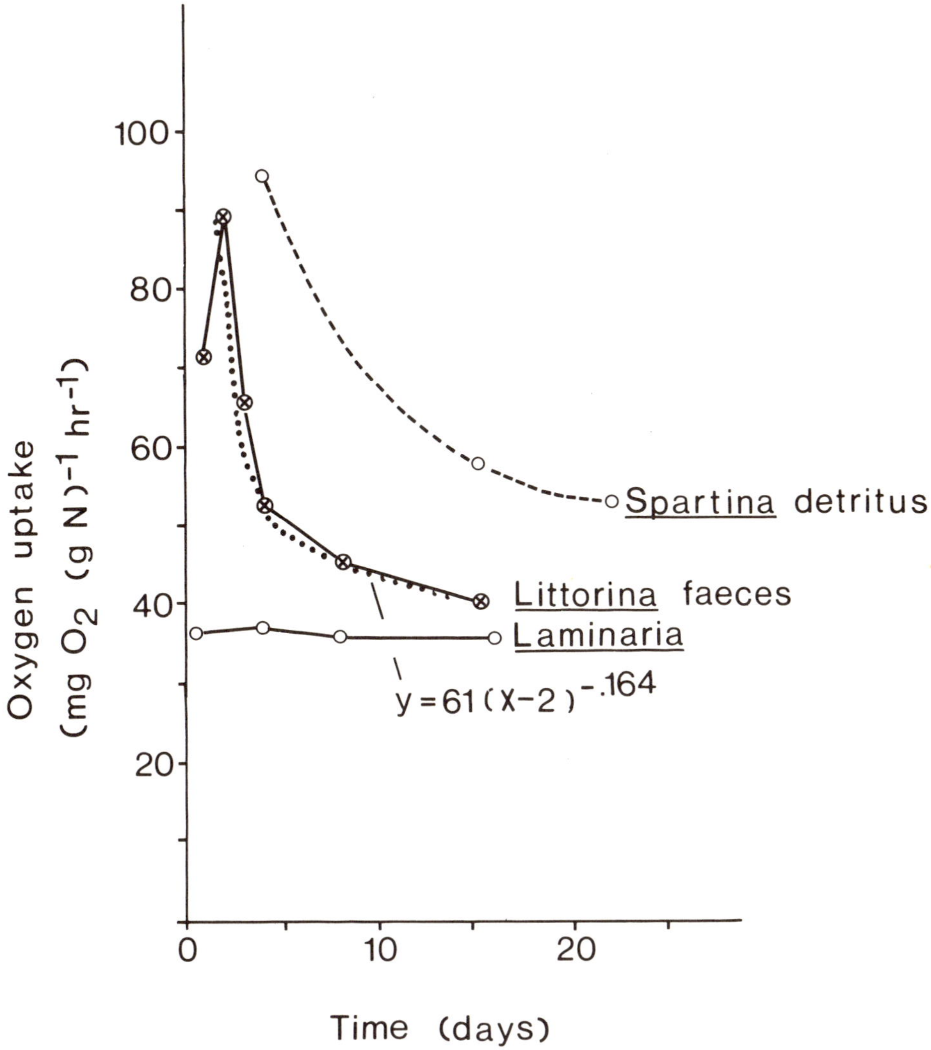

Figure 13.4. Oxygen uptake by *Littorina* faecal pellets and *Laminaria* (Fig. 13.3) expressed on the basis of nitrogen content over a period of enhanced metabolic activity. Data for *Spartina* detritus from Gosselink and Kirby (1974). Calculated least squares regression (dotted line) by equation (1).

a is a constant equivalent to R when $(t-n) = 1$

b is a negative rate constant representing the rate at which metabolic activity decreases.

Since $(t-n)$ is always positive $(n < t)$, the equation describes a logarithmic decrease in metabolic activity at rate $-b$ which begins from a value of maximum R as $(t-n)$ approaches 0. The value n represents the colonization time during which readily available substrates are metabolized, while b describes the rate of decrease in metabolism.

Presumably, biostasis, equivalent to small changes in R per unit time, results from decreased nutrient availability. Both n and b should be temperature dependent.

Data for oxygen uptake by *Littorina* faeces on the basis of total weight, carbon and nitrogen content were fitted by a least squares regression to the above equation with $n=2$. Although the slope coefficients were slightly different for each measure (total weight, $b=-0{\cdot}183$; carbon, $b=-0{\cdot}190$; nitrogen, $b=-0{\cdot}164$), the resulting negative logarithmic curves satisfactorily described the decrease in metabolic activity of microbial communities associated with faecal pellets (Figs. 13.3, 13.4).

Similar calculations of oxygen uptake by freshly wetted *Spartina* (Fig. 13.4) taken from data presented by Gosselink & Kirby (1974) indicate that an exponential decrease described by:

$$R = a \, . \, e^{-b(t-n)} \tag{13.2}$$

adequately describes the limited data (4 measurements over 22 days). Without more data the suitability of logarithmic or exponential curves to describe time-related changes in metabolic activity cannot be assessed. For example, oxygen uptake by fungal spores inoculated in sand cultures increased rapidly over two days and then decreased to lower and constant levels after four days (Witkamp 1973)—a pattern remarkably similar to that observed with faecal pellets (Fig. 13.3). Measurements at 12, 24, 48 and 96 hours, however, are insufficient to permit quantitative assessment of these changes with time.

The shape of the decay curve quantifying decreases in metabolic activity with time probably reflects the nature of substrate available for decomposition. Different organic constituents in litter are decomposed at markedly different rates and loss in weight may be proportional to the relative amounts of refractory and labile compounds remaining (Minderman 1968). If changes in metabolic rate during decomposition also indicate the loss of readily available organic matter, these substrates in faeces are consumed within a few days—the period of enhanced metabolism.

Nitrogen content in substrates is often increased during decomposition (Newell 1965; Odum & de la Cruz 1967; Harrison 1974) and it is generally observed that the higher the nitrogen content in terrestrial litter the more rapid the weight loss (Frankland 1974). Inorganic nitrogen stimulates both colonization (Fenchel 1972) and weight loss of vegetative material added to water (Hynes & Kaushik 1969). Also, nitrogen may be mineralized more rapidly than carbon or phosphorus during sediment decomposition (Seki *et al.* 1968). All of these observations suggest a central role for nitrogen in decomposition processes. Expression of measures of oxygen uptake on the basis of nitrogen content should standardize this measure of decomposition in relation to a possible limiting factor. The pronounced time-dependent changes in oxygen consumption which remain after correction for changes in nitrogen content (Fig. 13.4) indicate the importance of other factors which limit microbial metabolism on these substrates.

The relative importance of faeces in benthic communities

Macfadyen (1963) has provided an example of the usefulness of measures of oxygen uptake to assess the relative roles of various types of organisms in the metabolism of soil communities. Fenchel (1970), however, argues that such measurements of metabolic rates do not always give a complete picture of the role that organisms play. Oxygen uptake by macro-invertebrates in both soil and aquatic benthic communities form a small part of total community metabolism (Macfadyen 1963; Fenchel 1972), yet these animals promote energy flow in excess of that represented by their own metabolism.

The significance, relative to that of the total community, of stimulated metabolism due to the production of faecal material can be partially quantified by comparing the metabolic activity of organisms with that of the faeces which they produce (Fig. 13.5). Organisms with a high respiration per individual produce faeces with high metabolic activity. Either a logarithmic or linear expression describes the limited data satisfactorily which indicates a direct proportionality between oxygen uptake by an individual and the faeces which it produces. Organisms with a high metabolic level thus egest substrates which, after two days, are colonized by micro-organisms with a corresponding high metabolic rate. These maximum rates of oxygen uptake are not sustained (Fig. 13.3) but the constant production of faeces must result in a continual supply and accumulation of egested material. The relationship (Fig. 13.5) suggests that the cumulative effect of enhanced metabolism on faecal surfaces should be proportional to total fauna biomass producing the egested material.

Future studies of energy flow through sediment, detritus and litter-based communities will have to quantify the role of invertebrate faecal material more completely than has been done to date. Although large organisms produce large amounts of faeces (Fig. 13.1), the cumulative amount produced by all organisms in a benthic community, particularly the more abundant smaller species, must be considerable. Perhaps we should follow Johannes & Satomi's (1966) suggestion and treat faecal pellets as organisms. Their

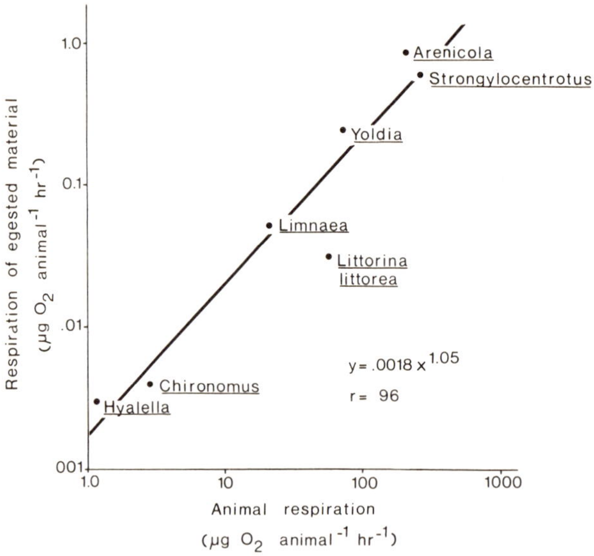

Figure 13.5. Comparison of oxygen uptake by eight species of aquatic benthic invertebrates and the faeces which they produce. Individual animal respiration measured over 3–5 hours in closed containers without substrate. Oxygen consumption of egested material, aged two days in aerated water, as described in Hargrave & Phillips (in press) calculated $mm^{-2}\times$hourly rate of faecal pellet production (Table 13.2). All measurements between 10–20°C.

presence constitutes a critical interface for cell growth, the consumption and release of nutrients and organic matter, and a renewable food supply. These surfaces may also provide sites where microbial biostasis, which limits decomposition, is continually broken. Such processes are undoubtedly central both for decomposition and overall energy flow through these communities.

Acknowledgements

Dr. T. Hughes kindly read a draft of the manuscript. Drs. K. H. Mann, P. Harrison, and J. Anderson offered useful comment.

References

Adams S.M. & Angelovic J.W. (1970) Assimilation of detritus and its associated bacteria by three species of estuarine animals. *Chesap. Sci.* 11, 249–54.

Batoosingh E. & Anthony E.H. (1971) Direct and indirect observations of bacteria on marine pebbles. *Can. J. Zool.* 17, 655–64.

BOYSEN-JENSEN P. (1914) Studies concerning the organic matter of the sea bottom. *Rep. Danske biol. Stat.* **22**, 1–39.

BRUNDIN L. (1949) Chironomiden und andere Bodentiere der Sudschwedischen Urgebirgseen. *Inst. Freshwater Res., Drottningholm. Rep.* **30**, pp. 914.

CARRIKER M.R. (1967) Ecology of estuarine benthic invertebrates: a perspective. In: *Estuaries*, ed. G.H. Lauff, pp. 442–87. Publ. Am. Assoc. Advan. Sci. **83**.

COOPER D.C. (1972) Enhancement of net primary productivity by herbivore grazing in aquatic laboratory microcosms. *Limnol. Oceanogr.* **18**, 31–7.

CUSHING D.H. & VUCETIC T. (1963) Studies on a *Calanus* patch. III. The quantity of food eaten by *Calanus finmarchicus*. *J. mar. Biol. Ass. UK.* **43**, 349–71.

DALE N.G. (1974) Bacteria in intertidal sediments: factors related to their distribution. *Limnol Oceanogr.* **19**, 509–18.

DARNELL R.M. (1967) Organic detritus in relation to the estuarine ecosystem. In: *Estuaries*, ed. G.H. Lauff, pp. 376–82. Publ. Amer. Assoc. Advan. Sci. **83**.

DARWIN C. (1881) *The Formation of Vegetable Mould through the Action of Worms.* John Murray, London.

DAVIS, R.B. (1974) Stratigraphic effects of tubificids in profundal lake sediments. *Limnol. Oceanogr.* **19**, 466–88.

DRIFT J. VAN DER (1950) Analysis of the animal community in a beech forest floor. *Tijdschr. Ent.* **94**, 1–168.

ECKARDT F.E. (1968) Functioning of terrestrial ecosystems at the primary production level. *Proc. Copenh. Symp.* UNESCO. Paris.

EDWARDS R.W. (1958) The effect of larvae of *Chironomus riparius* Meigen on the redox potentials of settled activated sludge. *Ann. Appl. Biol.* **46**, 457–64.

FENCHEL T. (1970) Studies on the decomposition of organic detritus derived from the turtle grass *Thalassia testudinum*. *Limnol. Oceanogr.* **15**, 14–20.

FENCHEL T. (1972) Aspects of decomposer food chains in marine benthos. *Verh. Deutsch. Zool. Gesell. 65. Jahresversamml.* **14**, 14–22.

FRANKLAND J.C. (1974) Decomposition of lower plants. In: *Biology of plant litter decomposition*, eds. C.H. Dickinson & G.J.F. Pugh, pp. 3–36. Academic Press, London.

GOSSELINK J.G. & KIRBY C.J. (1974) Decomposition of salt marsh grass, *Spartina alterniflora* Loisel. *Limnol. Oceanogr.* **19**, 825–32.

GRAY T.R.G., BAXBY P., HILL I.R. & GOODFELLOW M. (1968) Direct observation of bacteria in soil. In: *The Ecology of Soil Bacteria*, eds. T.R.G. Gray & D. Parkinson, pp. 173–92. University of Toronto Press.

GREENWOOD D.J. (1968) Measurement of microbial metabolism in soil. In: *The Ecology of Soil Bacteria*, eds. T.R.G. Gray & D. Parkinson, pp. 138–57. University of Toronto Press.

HAIRSTON N.G., SMITH F.E. & SLOBODKIN L.B. (1960) Community structure, population control and competition. *Am. Nat.* **94**, 421–5.

HALL K.J. & HYATT K.D. (1974) Marion Lake IBP—from bacteria to fish. *J. Fish. Res. Board Can.* **31**, 893–911.

HARGRAVE B.T. (1970a) The effect of a deposit-feeding amphipod on the metabolism of benthic microflora. *Limnol. Oceanogr.* **15**, 21–30.

HARGRAVE B.T. (1970b) The utilization of benthic microflora by *Hyalella azteca*. *J. Anim. Ecol.* **39**, 427–37.

HARGRAVE B.T. (1972a) Oxidation-reduction potentials, oxygen concentration and oxygen uptake of profundal sediments in a eutrophic lake. *Oikos* **23**, 167–77.

HARGRAVE B.T. (1972b) Aerobic decomposition of sediment and detritus as a function of particle surface area and organic content. *Limnol. Oceanogr.* **17**, 583–96.

HARGRAVE B.T. (1972c) Prediction of egestion by the deposit-feeding amphipod *Hyalella azteca*. *Oikos* **23**, 116–24.

HARGRAVE B.T. (1973) Coupling carbon flow through some pelagic and benthic communities. *J. Fish. Res. Board Can.* **30**, 1317–26.

HARGRAVE B.T. (1975) The importance of total and mixed-layer depth in the supply of organic material to bottom communities. *Symp. Biol. Hung.* **15**, 157–65.

HARGRAVE B.T. & PHILLIPS G.A. (in press) Oxygen uptake by microbial communities on solid surfaces. In: *The Aquatic Microbial Communities*, ed. J. Cairns, Marcel Dekker, New York.

HARRISON P.G. (1974) *Growth and detritus formation in a temperate seagrass*, Zostera marina. Unpublished Ph.D. Thesis, Dalhousie University. 109 pp.

HOWARD P.J.A. (1967) A method for studying the respiration and decomposition of litter. In: *Progress in Soil Biology*, eds. O. Graff & J.F. Satchell, pp. 464–72. Vieweg and North-Holland, Braunschweig and Amsterdam.

HYNES, H.B.N. & KAUSHIK N.K. (1969) The relationships between dissolved nutrient salts and protein production in submerged autumnal leaves. *Verh. Internat. Verein. Limnol.* **17**, 95–103.

JOHANNES R.E. & SATOMI M. (1966) Composition and nutritive value of faecal pellets of a marine crustacean. *Limnol. Oceanogr.* **11**, 191–7.

KHAILOV K.M. & BURLAKOVA Z.P. (1969) Release of dissolved organic matter by marine seaweeds and distribution of their total organic production to inshore communities. *Limnol. Oceanogr.* **14**, 521–7.

KRAEUTER J. & HAVEN D.S. (1970) Fecal pellets of common invertebrates of lower York River and lower Chesapeake Bay, Virginia. *Chesapeake Sci.* **11**, 159–73.

MACFADYEN A. (1961) Metabolism of soil invertebrates in relation to soil fertility. *Ann. Appl. Biol.* **49**, 215–18.

MACFADYEN A. (1963) The contribution of the microfauna to total soil metabolism. In: *Progress in Soil Biology*, eds. O. Graff & J.F. Satchell, pp. 3–17. Vieweg and North-Holland, Braunschweig and Amsterdam.

MANN K.H. (1972) The ecological energetics of the seaweed zone in a marine bay on the Atlantic coast of Canada. II. Productivity of the seaweeds. *Mar. Bio.* **14**, 199–209.

MANN K.H. (1973) Seaweeds: their productivity and strategy for growth. *Science* **182**, 975–81.

MILLER R.J. & MANN K.H. (1973) Ecological energetics of the seaweed zone in a marine bay on the Atlantic coast of Canada. III. Energy transformation by sea urchins. *Mar. Biol.* **18**, 99–114.

MINDERMAN G. (1968) Addition, decomposition and accumulation of organic matter in forests. *J. Ecol.* **56**, 355–62.

NEWELL R. (1965) The role of detritus in the nutrition of two marine deposit feeders, the prosobranch *Hydrobia ulvae* and the bivalve *Macoma balthica*. *Proc. Zool. Soc. Lond.* **144**, 25–45.

NEWELL R.C. (1970) *Biology of Intertidal Animals*. Logos Press, London, 555 pp.

ODUM E.P. & DE LA CRUZ A.A. (1967) Particulate organic detritus in a Georgia salt marsh-estuarine ecosystem. In: *Estuaries*, ed. G.H. Lauff, pp. 383–8. Publ. Amer. Assoc. Advan. Sci. **83**.

OLAH J. (1972) Leaching, colonization and stabilization during detritus formation. *Mem. Ist. Ital. Idrobiol.* 29 Suppl., 105–27.

O'NEILL T.B. & WILCOX G.L. (1971) The formation of a 'primary film' on materials submerged in the sea at Port Hueneme, California. *Pac. Sci.* **25**, 1–12.

PARSONS T.R., LEBRASSEUR R.J. & FULTON J.D. (1967) Some observations on the dependence of zooplankton grazing on the cell size and concentration of phytoplankton blooms. *J. Oceanogr. Soc. Jap.* **23**, 10–17.

POMEROY L.R. (1970) The strategy of mineral cycling. *Ann. Rev. Ecol. and Syst.* **1**, 171–90.

POULET S.A. (1973) Grazing of *Pseudocalanus minutus* on naturally occurring particulate matter. *Limnol. Oceanogr.* **18**, 564–73.

REICHLE D.E. (1971) Energy and nutrient metabolism of soil and litter, pp. 465–77. Productivity of forest ecosystems. *Proc. UNESCO Brussels Symp. 1969.*

RHOADS D.C. (1973) The influence of deposit-feeding benthos on water turbidity and nutrient recycling. *Amer. J. Sci.* **273**, 1–22.

RHOADS D.C. & YOUNG D.K. (1970) The influence of deposit-feeding organisms on sediment stability and community trophic structure. *J. Mar. Res.* **28**, 150–78.

SANDERS H.L. & HESSLER R.R. (1969) Ecology of the deep-sea benthos. *Science* **163**, 1419–24.

SATCHELL J.E. (1974) Litter-interface of animate/inanimate matter. In: *Biology of Plant Litter Decomposition*, eds. C.H. Dickinson & G.J.F. Pugh, pp. xiii–xliv. Academic Press, London.

SEKI H., SKELDING J. & PARSONS T.R. (1968) Observations on the decomposition of a marine sediment. *Limnol. Oceanogr.* **13**, 440–9.

STEELE J.H. (1974) *The structure of marine ecosystems.* Harvard University Press, Cambridge, Massachusetts.

TEAL J.M. (1962) Energy flow in the salt marsh ecosystem of Georgia. *Ecology* **43**, 614–24.

TSERNOGLOU D. & ANTHONY E.H. (1971) Particle size, water-stable aggregates, and bacterial populations in lake sediments. *Can. J. Microbiol.* **17**, 217–27.

WHITLATCH R.B. (1974) Food-resource partitioning in the deposit feeding polychaete *Pectinaria gouldii. Biol. Bull.* **147**, 227–35.

WITKAMP M. (1973) Compatibility of microbial measurements. In: Modern Methods in the Study of Microbial Ecology, ed. Th. Rosswall, pp. 179–88. *Bull. Ecol. Res. Comm. (Stockholm) 17.*

ZIEMAN J.C. (1968) *A study of the growth and decomposition of the sea-grass* Thalassia testudium. Unpublished M.Sc. Thesis, Inst. Marine Sci., University Miami (Fla.).

ZOBELL C.E. (1943) The effect of solid surfaces upon bacterial activity. *J. Bacteriol.* **46**, 39–56.

14 Detritus, micro-organisms and animals in fresh water

A. D. BERRIE *Freshwater Biological Association, River Laboratory, Wareham, Dorset, U.K.*

Abstract

Detritus is a major source of food for micro-organisms and animals in freshwater ecosystems. It appears to be a low quality food for animals. They pass it rapidly through their guts and are only able to assimilate a small proportion of the material ingested. There is evidence that they utilize the micro-organisms attached to the detritus rather than the detritus itself and these are a high quality food. The micro-organisms are also important in making the detritus attractive to animals and less refractory to digestion.

A high proportion of the energy entering freshwater ecosystems is in the form of detritus from allochthonous sources and much of the autochthonous primary production passes to decomposers rather than grazers. Micro-organisms constitute a trophic level intermediate between autotrophic plants and animals. Plants are quantitatively dominant in ecosystem metabolism but animals are capable of stimulating increased activity among autotrophs and heterotrophs at lower trophic levels.

Introduction

This contribution has to cover a wide field which has attracted considerable attention among freshwater biologists in recent years. Organic detritus includes all forms of nonliving organic material, ranging from large fragments to dissolved compounds. Detritus contains energy and nutrients which can be exploited particularly by micro-organisms. Microbial colonization usually results in a complex association of bacteria, fungi and algae and this colonized detritus forms a food source for animals.

The fact that detritus is an important component of freshwater ecosystems has been stated by various authors since at least the 1920s but studies of detritus have proved difficult due to its complex nature and the lack of suitable analytical techniques. Similarly, the potential of bacteria as food for invertebrates was recognized at least by the 1930s. This relationship has been examined further (Rodina 1963; Marzolf 1964) although most work seems to have been done in the marine field. Detailed investigations have faced the problem of distinguishing between organic material assimilated from detritus and that from the associated micro-organisms. Radioactive labelling techniques have allowed further progress but leave the problem of labelling the nonliving detritus. Recent advances include the use of two tracers for improved estimates of assimilation processes (Calow & Fletcher 1972) and the production of artificial detritus incorporating a radioactive label and acceptable to microbes and invertebrates (Saunders 1972).

It is now clear that studies of the detritus food chain are of major importance in understanding aquatic ecosystems (Melchiorri-Santolini & Hopton 1972). The subject has many aspects and it cannot be reviewed in detail here. The paper will concentrate on particular aspects of the role of freshwater organisms in decomposition processes which are relevant to an understanding of the trophic economy of freshwater ecosystems. The viewpoint is somewhat biased towards the author's personal interest in river systems but appears also to have relevance to lakes.

The food of freshwater animals

Freshwater invertebrates can be classified in many ways on the basis of their feeding habits (e.g. Cummins 1973). For the present purpose it is convenient to divide them into five categories.

1 Suspension feeders which have some means of filtering suspended particles or organisms from the water
2 Shredders and grazers which prefer to bite pieces off plants or from large particles such as leaf litter
3 Gatherers and scrapers which move around collecting food particles from the substratum
4 Deposit feeders which live in the bottom sediments and ingest this material
5 Predators which consume other animals.

It is conventional to think of most animals as herbivores feeding directly on autotrophic plants. In fresh water this plant material comprises phytoplankton, attached algae and higher aquatic plants. Phytoplankton is most important in the larger lakes and rivers. Vascular plants are confined to the shallow margins of lakes and to suitable sections of rivers. Phytoplankton may be extensively grazed by zooplankton but few freshwater invertebrates appear to graze directly on living tissue of higher plants. Autotrophic plants can only grow in the euphotic zone and large parts of many lakes and the beds of turbid rivers are unsuitable for these plants.

In a study of the gut contents of 43 taxa of invertebrates in the River Lambourn, Scorgie (1974) found that 4 had 100% animal food, 9 had over 95% detritus and 7 had over 75% fresh plant material (mostly diatoms). On the basis of the predominant food 6 taxa were carnivores, 22 were detritivores and 12 were herbivores with 3 evenly divided between fresh plant material and detritus. This qualitative emphasis on detritus is reinforced by evidence from quantitative invertebrate samples which showed that the numerically dominant taxa in the river include a high proportion of detritus in their food. These results agree broadly with existing information from other localities.

Detritus is also important as a food for fish. In the River Thames it forms over half the volume of the stomach contents in gudgeon and in roach over 50 mm length; and over one third of the contents in dace (Berrie 1972a and b).

Within the scope of the feeding methods listed above it is important to realize that most freshwater invertebrates tend to eat what is available. They are seldom specific in their feeding habits and, consequently, the results of food studies are very variable. It is difficult to depict generalized feeding relationships in freshwater ecosystems because the interconnections in the food webs tend to be very complex. Even predators may take substantial quantities of plant or detrital material when circumstances make this necessary. Also, many predators only become carnivorous in the later stages of their life and may be herbivores or detritivores in the early stages.

In spite of these complexities the available evidence shows that detritus is an important food source for animals. Information about its utilization and nutritional characteristics is essential to an understanding of aquatic ecosystems.

Nature and origin of detritus

Particulate detritus may be suspended in the water column, deposited on the bed of the lake or stream or on plants and other surfaces. There is continuous interchange between these positions depending on the water currents and turbulence, the quantities and nature of the material present, the amount of macrophyte present and various other factors. Benthic deposits of particulate material generally contain a lower proportion by weight of organic matter than suspended particles. In the River Thames the suspended tripton has 20–80% organic matter whereas epiphytic detritus has about 40% organic matter and bottom sediments about 15% (Berrie 1972b). In the Lambourn the seston has 30–90% organic matter and the easily suspended fraction of the bottom deposits only 11–12% organic matter (Scorgie 1974).

Dissolved organic matter is produced by bacteria, plants and animals living in the water and leached from decomposing tissues both in the water and from the surrounding land. Spring water contains small quantities and rain water reaching the water via tree canopies can collect quantities as high as 800 mg C l^{-1} although values of 20–40 mg C l^{-1} are more usual (Farr personal communication). Many effluents contain high levels of dissolved organic matter. Wetzel & Manny (1972) have shown that leaf material in an artificial, recirculating stream increased the level of dissolved organic carbon by an order of magnitude. Bacterial populations built up rapidly utilizing the labile organic carbon and nitrogen compounds within 72 hours. This indicated that woodland streams have a high capacity for processing natural dissolved organic compounds.

Petersen & Cummins (1974) have made a quantitative assessment of the allochthonous input to streams and have shown that the dry weight ranges from 1–5 g m^{-2} day^{-1}. It is interesting to compare this with the figure of 2·7 g m^{-2} day^{-1} given by Macfadyen (1971) for the normal autotrophic production of dry matter by terrestrial and freshwater communities. This suggests that the contribution made by allochthonous material to the energy budget of streams is approximately the same as that made by primary production within the streams.

The main autochthonous source of detritus is provided by aquatic macrophytes decomposing *in situ*. Most available studies of streams have shown that terrestrial leaf litter contributes much more material than the aquatic macrophytes but in chalk streams production by *Ranunculus* can exceed 400 g m^{-2} yr^{-1} and that of *Rorippa* can reach 600 g m^{-2} yr^{-1} (Dawson 1973).

Decomposition of leaf litter in streams has received considerable attention in recent years and has been reviewed recently by Bärlocher & Kendrick (1974), Petersen & Cummins (1974) and Willoughby (1974). When the leaves enter the water they undergo an initial period of leaching which is fairly rapid. During this they lose some 15% of their organic matter. They then undergo a sequence of events which has been called processing. The first part of this is biological and consists of the leaf material being colonized by micro-organisms, both bacteria and fungi. These start the decomposition process and render the leaf material attractive to invertebrates. The second stage of the processing is when the large invertebrates begin to attack the leaf material and shred it into smaller particles. The rate at which this process occurs depends on the composition of the invertebrate community. Where few suitable invertebrates are present the process is slow and the leaf material may last for up to a year or more (Mathews & Kowalczewski 1969). Usually the leaves are attacked rapidly and may be reduced to the veins only within a matter of weeks. A sequence of colonization can be shown in which the shredding invertebrates build up in numbers on the leaf material during this phase and the numbers then decline as the material is reduced to fine detritus. The invertebrates which actively break down leaf material make it available to other animals in two distinct ways. First they produce particles of waste material which they do not ingest and these may be small enough to be ingested by microphagous invertebrates which cannot ingest the larger pieces. Secondly, a great deal of the material passes through their guts and emerges as only partially digested faeces. These are extensively utilized by other animals in the system and this aspect is discussed by Hargrave (Chapter 13, this volume). A difference between freshwater and terrestrial ecosystems is that leaf decomposition in the latter is dependent on adequate moisture whereas this is not a problem in water and decomposition may proceed more rapidly.

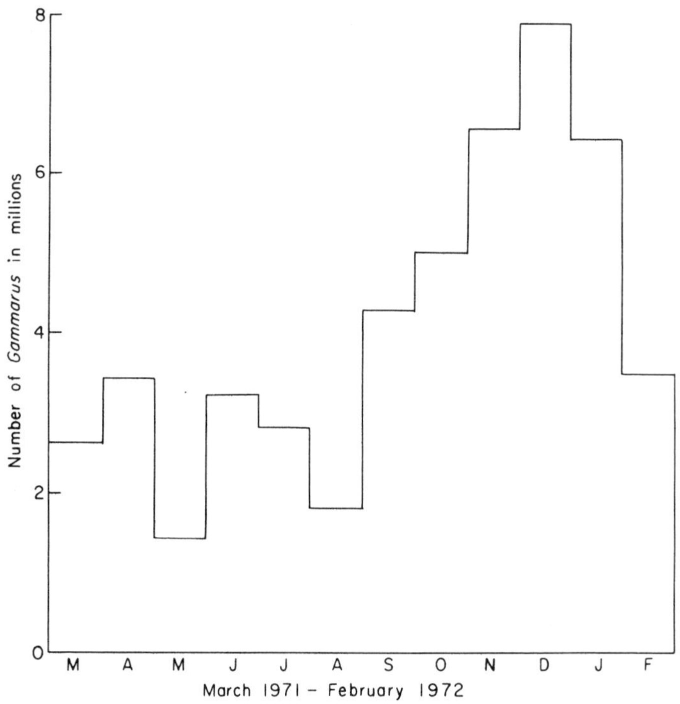

Figure 14.1. Seasonal changes in the estimated total number of *Gammarus pulex* in two 50 metre lengths of the River Lambourn, Berkshire.

Most terrestrial leaf litter enters temperate freshwater systems during a short period in the autumn and a rich food supply becomes available for detritivores at that time. In the River Lambourn *Gammarus pulex* is the most abundant invertebrate species. Large *Gammarus* are the animals mainly responsible for shredding leaf material in the river and smaller ones feed on smaller fragments. The changes in population density of *Gammarus* were recorded in two 50 m lengths of the river, one partially shaded by trees and the other unshaded. The combined data (Fig. 14.1) show a substantial and progressive increase in density during the autumn months with the highest density in December. Biomass figures are not yet available for the *Gammarus* population in each month but the total dry weight of invertebrates at the sites is highest during October–December. It seems likely that the *Gammarus* population will be large in terms of biomass as well as numbers during the autumn and this corresponds with the period in which terrestrial leaf litter becomes available. These observations are in general agreement with previous studies which have been discussed by Hynes (1970). Although Scorgie (1974) has calculated that the leaf litter is not sufficient to support *Gammarus* throughout the year, it does make a significant contribution to the trophic economy of the river.

Utilization of detritus by animals

In the River Thames, the leaves of the yellow water lily, *Nuphar lutea*, become covered with a coating of attached algae and detritus. This reaches quantities of about 1 g dry weight per leaf or 0·1 g dm^{-2} (Berrie, Bottrell & Mackey, unpublished). A community of invertebrates utilizes this source of food during the summer months. Investigation showed that the most important groups were snails and cladocerans. The snails increased rapidly in numbers during May and decreased after a period of high discharge in June. During July and August *Sida crystallina*, a detritus-feeding cladoceran, reached very high population densities and the quantity of detritus on the leaves was greatly reduced. This was almost certainly because most of the material was passing through the guts of these animals at any point in time. After the *Sida* population declined during late August, the snails increased in numbers again. Such communities seem likely to produce competition for the available resources and this may have considerable effects on the population dynamics of the component species.

The question arises of how animals actually utilize detritus as food. Do they digest part of the detritus particles or do they strip off the bacteria coating the particles and digest these leaving the detritus virtually unaltered? The possibilities are not mutually exclusive and both processes could be involved. Investigation of this problem poses a number of technical difficulties but there is evidence that tubificid worms (Brinkhurst & Chua 1969; Wavre & Brinkhurst 1971), snails (Calow 1974), amphipods (Hargrave 1970b) and cladocerans (Bottrell, unpublished) can ingest bacterial tissue from detritus and that this may well be the main nutritional component in their food.

Many animals are unable to digest structural substances such as cellulose and lignin from plants, and other animals have difficulty doing so. *Hyalella azteca* can assimilate detritus but it is not able to assimilate cellulose and lignin-like substances although these compounds may account for 50% of the organic matter in the subsurface sediments on which it lives (Hargrave 1970b). However, there is some evidence of cellulose digestion by *Gammarus pulex* (Bjarnov 1972). Micro-organisms can fulfil a useful intermediate role by breaking down these refractory compounds and using some of the carbon content as their energy source. Nitrogen will be incorporated into the microbial tissues which then constitute a food source of higher nutritional value for the animals.

Gammarus pseudolimnaeus ingests sterile maple leaves which have been treated with hot HCl in preference to ones treated with cold HCl or with hot or cold water (Bärlocher & Kendrick 1975b). The same response was shown using treated filter paper, indicating a preference for partly hydrolysed tissues. Cellobiose occurs as an intermediate product of cellulose degradation and enzymes active on cellobiose occur in several freshwater invertebrates (Bjarnov 1972). Bärlocher & Kendrick (1975b) conclude that micro-organisms attacking leaf litter may fulfil a dual role by converting the material into microbial tissue and by catalysing the partial degradation of refractory material into substances that can be digested by detritus feeders.

Many animals have an ability to select the best material which is available. Odum (1968) states that striped mullet prefer very fine particles which may be richer in absorbed organic matter and adsorbed bacteria. These were present in higher values in the stomachs of the fish than in the sediments. Britton (1968) made similar observations on roach in the Thames (Berrie 1972a) and Hargrave (1970b) showed that the same was true of the amphipod *Hyalella*.

A number of studies are now available on the assimilation efficiencies of various invertebrates feeding on detritus and on isolated micro-organisms. Some of the data are given in Table 14.1. It should be stressed that the rates are affected by a number of factors such as body size, water temperature and food concentration and the original sources should be examined for full details.

In general the assimilation efficiencies for detritus are low with means ranging from 6 to 35%. Only a few animals have been shown to achieve

Table 14.1. Assimilation efficiencies of some freshwater invertebrates.

Species	Assimilation efficiency % Mean	Range	Source
Detrital Food			
Hyalella azteca	15		Hargrave 1971
Gammarus pulex		30–40	Nilsson 1974
Gammarus pulex	20	14–26	Scorgie 1974
Gammarus pseudolimnaeus	10		Bärlocher & Kendrick 1975a
Asellus aquaticus	30	26–44	Prus 1971
Eurycercus lamellatus	20		Smirnov 1962
Eurycercus lamellatus	34·6		Kryuchkova *et al.* 1971
Cladocera 6 spp.		10–30	Bottrell unpublished
Daphnia sp.	7	2–18	Saunders 1972
Pteronarcys scotti	10·6	8·5–15·9	McDiffett 1970
Potamopyrgus jenkinsi	6·9	2·8–14·5	Heywood & Edwards 1962
Rutilus rutilus	7		Britton 1968
Bacterial Food			
Hyalella azteca		60–83	Hargrave 1970b
Planorbis contortus		81–97	Calow & Fletcher 1972
Fungal Food			
Gammarus pseudolimnaeus		42–76	Bärlocher & Kendrick 1975a

rates higher than 20%. The figures for bacterial and fungal food are of particular interest. These indicate that the animals can utilize the microbial tissue much more efficiently than detritus, and it seems likely that they will continue to do so when living on detritus with an associated microflora. Hargrave (1970b) showed that *Hyalella azteca* digested various bacteria and algae (except blue-green algae) and epiphytic material from *Chara* with efficiencies of over 50% and for some bacteria and epiphytes up to 92%.

The gut clearance times of several species of invertebrates are given in Table 14.2. These data show that the food is passed through the gut quite rapidly. The material being assimilated must be in a readily available form as it is unlikely that refractory substances could be digested in such times. There is evidence that the retention time of material in the gut of some

Table 14.2. Gut clearance times for some freshwater invertebrates.

Species	Approximate gut clearance time	Source
Hyalella azteca	30 min	Hargrave 1970b
Gammarus pulex	30–40 min	Scorgie 1974
Simocephalus vetulus	20–35 min	Bottrell unpublished
Eurycercus lamellatus	7 min	Smirnov 1962
Baetis rhodani	30 min	Brown 1961
Simulium spp.	20–30 min	Ladle *et al.* 1972

species is influenced by the availability of food. Richman (1958) found an inverse correlation between the assimilation efficiency in *Daphnia pulex* and the concentration of algal food. This could result from a physical problem or a physiological adaptation but it should be advantageous to the animals.

Appleby & Brinkhurst (1970) calculated that the community of tubificid worms living in the bottom sediments of Toronto Harbour, Lake Ontario, would ingest the top 20 mm layer 4–12 times per year. Similarly, Hargrave (1972) states that the population of the deposit-feeding amphipod *Hyalella azteca* will pass the top 1 mm of sediment in Marion Lake, British Columbia, through its guts 1·58 times per year. This supports the evidence for re-utilization of detritus discussed above.

Energy and nutrients in freshwater ecosystems

A common feature of both rivers and lakes is the fact that the water in them is moving progressively downstream. There is a continuous range of variation from short, swift-flowing rivers where the water may pass through the whole system in a matter of a few hours to systems containing large lakes where the water may remain for many years. A number of budgets have been produced for rivers and streams which show a high input of allochthonous material to the systems. The proportion of allochthonous to autochthonous input tends to depend on the degree of afforestation in the catchment. The presence of large numbers of trees along the banks of a stream leads to the input of leaf litter and the trees also shade the stream and thus reduce the production of algae and macrophytes in the water. There

Table 14.3. Speculative organic budget for Bere Stream in tonnes of organic matter per year (Data from Westlake *et al.* 1972 using revised figures by these authors).

Input		Output	
Net primary production		Secondary respiration	30?
Vascular plants	13		
Algae	6		
Trees	2	Losses downstream	
Marginal plants	1	Dissolved organic	51
Commercial cress beds	38	Suspended organic	72
Spring water	17	Large drift	5
Balance (probably allochthonous groundwater, drainage effluents, etc.)	81?		
Total	158		158

is some evidence quoted above that the two inputs may be similar quantitatively in certain circumstances.

Westlake *et al.* (1972) produced an energy budget for the Bere Stream in Dorset which flows through arable and pasture land with some tree cover and has a flourishing growth of vascular plants and algae. An updated version of their data is presented in Table 14.3. Even in these favourable conditions autochthonous primary production only accounted for some 12% of the energy input to the stream. A remarkably large quantity of energy was present in the form of detritus. Its organic weight totalled 40 kg m^{-2} yr^{-1} and it had an energy content equivalent to 57% of the useful annual solar energy input. This is probably at least an order of magnitude larger than the food requirements of the invertebrates in the stream.

In a study of Bear Brook, New Hampshire, Fisher & Likens (1973) showed that over 99% of the energy input was allochthonous in origin (Table 14.4). In this case woodland forms a continuous canopy over the brook and

Table 14.4. Annual energy budget for Bear Brook in kcal m^{-2} yr^{-1} (Data from Fisher & Likens 1973).

Input		Output	
Moss production	10	Secondary respiration	2,035
Leaf litter	1,750	Outlet streams	
Other tree fragments	921	Dissolved organic	2,800
Inlet streams		Particulate organic	1,204
Dissolved organic	2,800		
Particulate organic	558		
Total	6,039		6,039

a considerable proportion of the energy came from leaf litter. The canopy also prevented the growth of algae and vascular plants in the brook.

An organic carbon budget for Lawrence Lake, Michigan, has been constructed by Wetzel *et al.* (1972). This is a small marl lake, 4·96 ha in area, with a relatively small drainage basin which is 7·1 times the area of the lake. Table 14.5 shows that the annual carbon input is almost as great as the net production within the lake. Although the balance may be different in large lakes, it is clear that allochthonous detritus can be an important energy input to lakes.

In addition to energy, allochthonous sources provide a supply of inorganic nutrients to freshwater ecosystems. This has recently been elegantly demonstrated by Lund and his colleagues in a series of experiments in Blelham Tarn, Cumbria, a small eutrophic lake 10·2 ha in area. Two rubber tubes 45·5 m in diameter were used which each isolated over 18,000 tonnes of water (c. 2·6% of the total volume) from the rest of the water in the tarn

Table 14.5. Net ecosystem budget for Lawrence Lake in g C m^{-2} yr^{-1} (Data from Wetzel *et al.* 1972).

Input		Output	
Net autochthonous ecosystem production	28·3		
Inlet streams		Outlet streams	
Dissolved organic	14·9	Dissolved organic	35·8
Particulate organic	3·0	Particulate organic	2·8
Groundwater		Sedimentation	14·8
Dissolved organic	6·0		
Particulate organic	1·2		
Total	53·4		53·4

(Lund 1972; Lack & Lund 1974). This showed that the nutrients within the tubes were rapidly used up and carried to the bottom. The water within the tubes became oligotrophic compared with the water outside which was receiving a continuous input of nutrients from the in-flow stream.

Thus rivers and lakes are open ecosystems which continuously receive inputs of nutrients and organic material from their catchments. The material that is entering them is more or less balanced by material that is leaving them and ultimately entering the sea. This leads to conceptual difficulties which are particularly acute when studying a section of a river. Much of the material passes straight through and is not utilized within the system. If it was heavily utilized this could dramatically alter the conditions lower down the system. Ladle *et al.* (1972) calculated that when populations of larval Simuliidae in the Bere Stream reach high densities and the throughput of suspended solids is low, it appears theoretically possible for the larvae to filter out all the particles from the water in about 0·6 km. In practice, significant clearance of a component of the suspended solids by filter-feeding animals has only been recorded immediately below the outfall of lakes where substantial quantities of planktonic material are discharged which are readily distinguishable from other suspended solids.

The importance of decomposer organisms in freshwater ecosystems

The information now available can be used to construct a simple conceptual model of the food resources and their utilization in a freshwater ecosystem (Fig. 14.2). A network of minor connections could be superimposed but these might confuse the main lines. The relative importance of the components will vary between different systems and, although a large lake will

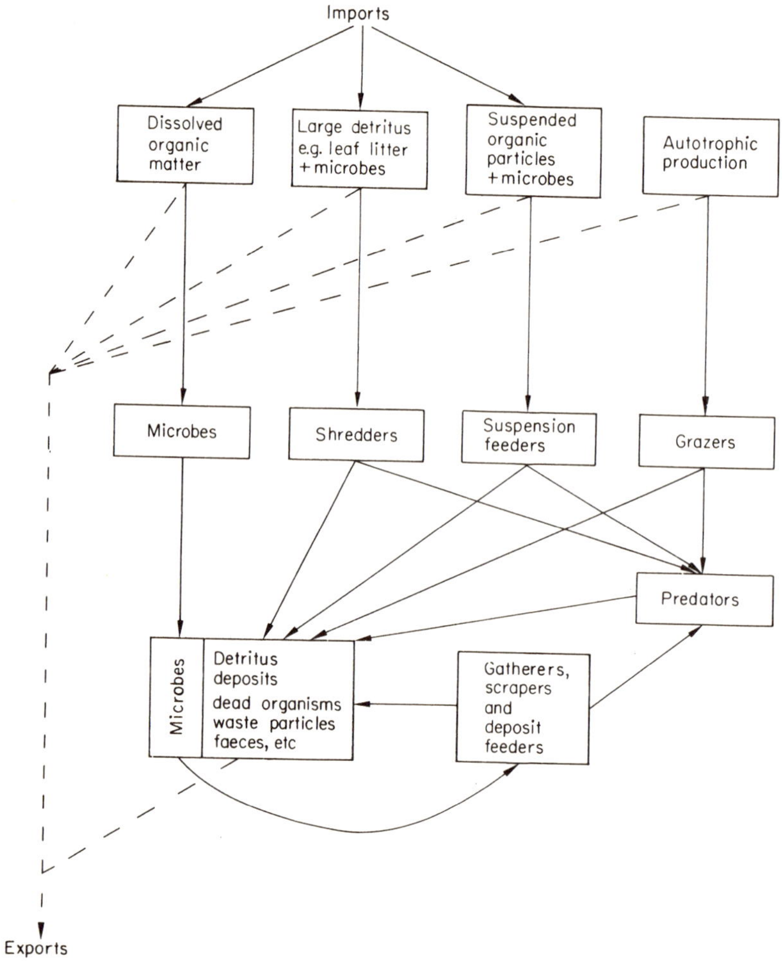

Figure 14.2. Food resources and their utilization in freshwater ecosystems.

function quite differently from a small woodland stream, the basic principles are similar. The invertebrate community has been divided into the functional groups indicated above except that grazers on live plant material have been separated from shredders of large detritus such as leaf litter. The detritus pool is a major component of the system. It is exploited by micro-organisms and, through their activity, by the detritus-feeding invertebrates which represent a large proportion of the benthic and epiphytic fauna. The shredders, grazers, suspension feeders and predators all contribute to this pool by their own activities. Particles which they do not ingest and partially digested faeces probably add much more material to the pool than that ultimately coming from their dead bodies or exuviae.

A simple diagram of the cycle of materials in ecosystems is shown in

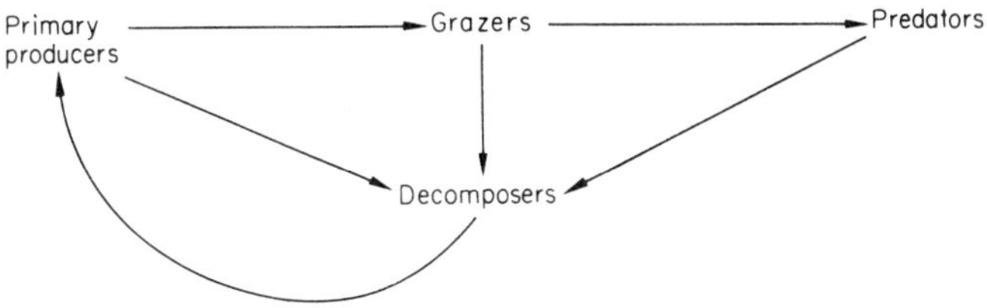

Figure 14.3. Main pathways usually described for the transfer of materials in ecosystems.

Fig. 14.3 following the pattern usually described in basic texts. Organic matter is built up by the plants which form the food of the grazing animals at the next trophic level. These in turn are eaten by predators. At each level organisms die and are broken down by decomposers, releasing some of the nutrients to be taken up again by plants.

Some modification of this concept seems necessary to describe freshwater ecosystems (Fig. 14.4). A large food source for secondary production is provided by allochthonous material, and much of the autochthonous primary production may pass to decomposers rather than grazers. Microbes constitute a trophic level intermediate between autotrophic plants and animals, as well as being the decomposers which are responsible for recycling nutrients. The lines indicating transfers from animals to decomposers and the recycling of nutrients have been omitted from this figure for clarity.

When production and energy flow are measured in ecosystems it is clear that plants are quantitatively dominant. Autotrophic plants must produce the food supply for the secondary production, and most of the heterotrophic metabolism of communities is usually attributable to microbes. In aquatic ecosystems, heterotrophic production by both microbes and animals may be increased by substantial inputs of allochthonous material which supplement the autochthonous production of food for them. The role of animals in the ecosystem is not adequately described by a quantitative assessment of their metabolic activity. There is growing evidence that their activity can affect the functioning of the whole system. The role of shredders in breaking down large detritus has been discussed and this process proceeds more slowly when only microbial activity is involved. The shredding process,

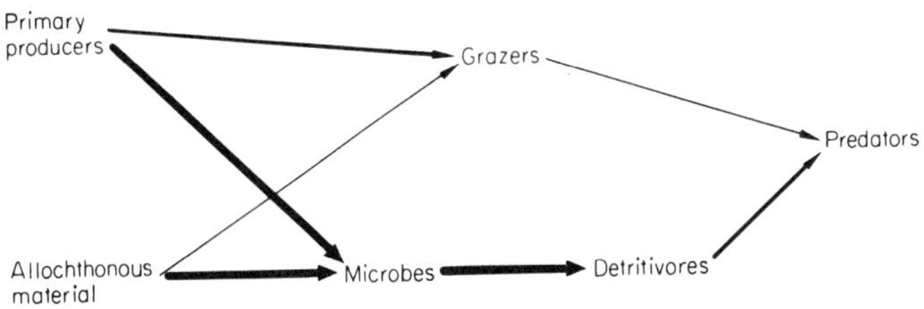

Figure 14.4. Main trophic pathways in freshwater ecosystems.

together with grazing and suspension feeding, makes material available to detritus-feeding animals which might otherwise be utilized more exclusively by microbes or lost from the system. Hargrave (1970a) found that benthic community respiration increased in a non-linear manner with increasing density of the deposit-feeding amphipod *Hyalella azteca*. He showed that epibenthic algal production and bacterial respiration were both stimulated with increasing density of *Hyalella* up to and above natural maximum densities. He also showed that dissolved organic compounds excreted by *Hyalella* significantly increased the rate of recolonization of faecal material by micro-organisms.

Animals appear to be capable of stimulating increased activity among autotrophs and heterotrophs at lower trophic levels thus increasing the rate of decomposition processes and the metabolic activity of the system as a whole. The mechanisms involved include creating space for the production of additional organisms to replace those that are eaten and secretion of chemical substances which stimulate the other organisms.

Acknowledgements

I am grateful to Mr. H.H. Bottrell, Dr. H.R.A. Scorgie, Dr. J.F. Wright and other colleagues in my research group at the University of Reading who have provided unpublished information from their studies on the Rivers Thames and Lambourn. These studies were financed by the Natural Environment Research Council, the Thames Water Authority and the Central Water Planning Unit. Dr. M. Ladle and Mr. I.S. Farr kindly read the manuscript and provided several helpful comments.

References

Appleby A.G. & Brinkhurst R.O. (1970) Defecation rate of three tubificid oligochaetes found in the sediment of Toronto Harbour, Ontario. *J. Fish. Res. Bd Can.* **27**, 1971–82.

Bärlocher F. & Kendrick B. (1974) Dynamics of the fungal population on leaves in a stream. *J. Ecol.* **62**, 761–91.

Bärlocher F. & Kendrick B. (1975a) Assimilation efficiency of *Gammarus pseudolimnaeus* (Amphipoda) feeding on fungal mycelium or autumn-shed leaves. *Oikos* **26**, 55–9.

Bärlocher F. & Kendrick B. (1975b) Leaf-conditioning by microorganisms. *Oecologia* (Berl.) **20**, 359–62.

Berrie A.D. (1972a) Productivity of the River Thames at Reading. In: *Symposium on Conservation and Productivity of Natural Waters*, eds. R.W. Edwards & D.J. Garrod. *Symp. zool. Soc. Lond.* **29**, 69–86.

Berrie A.D. (1972b) The occurrence and composition of seston in the River Thames and the role of detritus as an energy source for secondary production in the river. *Mem. Ist. Ital. Idrobiol.* 29 Suppl., 473–83.

BJARNOV N. (1972) Carbohydrases in *Chironomus*, *Gammarus* and some Trichoptera larvae. *Oikos* **23**, 261–3.

BRINKHURST R.O. & CHUA K.E. (1969) Preliminary investigation of the exploitation of some potential nutritional sources by three tubificid oligochaetes. *J. Fish. Res. Bd Can.* **26**, 2659–68.

BRITTON R.H. (1968) The diet of roach and bleak in the River Thames in relation to their growth. *J. Anim. Ecol.* **37**, (3), 28P–29P.

BROWN D.S. (1961) The food of the larvae of *Chloëon dipterum* L. and *Baetis rhodani* (Picter) (Insecta, Ephemeroptera). *J. Anim. Ecol.* **30**, 55–75.

CALOW P. (1974) Evidence for bacterial feeding in *Planorbis contortus* Linn. (Gastropoda: Pulmonata). *Proc. malac. Soc. Lond.* **41**, 145–56.

CALOW P. & FLETCHER C.R. (1972) A new radiotracer technique involving ^{14}C and ^{51}Cr for estimating the assimilation efficiencies of aquatic, primary consumers. *Oecologia* **9**, 155–70.

CUMMINS K. W. (1973) Trophic relations of aquatic insects. *A. Rev. Ent.* **18**, 183–206.

DAWSON F.H. (1973) *The production ecology of* Ranunculus penicillatus *var.* calcareus *in relation to the organic input into a chalk stream.* Unpublished Ph.D. Thesis, University of Aston in Birmingham.

FISHER S.G. & LIKENS G.E. (1973) Energy flow in Bear Brook, New Hampshire, an integrative approach to stream ecosystem metabolism. *Ecol. Monogr.* **43**, 421–39.

HARGRAVE B.T. (1970a) The effect of a deposit-feeding amphipod on the metabolism of benthic microflora. *Limnol. Oceanogr.* **15**, 21–30.

HARGRAVE B.T. (1970b) The utilization of benthic microflora by *Hyalella azteca* (Amphipoda). *J. Anim. Ecol.* **39**, 427–37.

HARGRAVE B.T. (1971) An energy budget for a deposit-feeding amphipod. *Limnol. Oceanogr.* **16**, 99–103.

HARGRAVE B.T. (1972) Prediction of egestion by the deposit-feeding amphipod *Hyalella azteca*. *Oikos* **23**, 116–24.

HEYWOOD J. & EDWARDS R.W. (1962) Some aspects of the ecology of *Potamopyrgus jenkinsi* Smith. *J. Anim. Ecol.* **31**, 239–50.

HYNES H.B.N. (1970) *The ecology of running waters.* Liverpool University Press.

KRYUCHKOVA N.M., OSTAPENYA A.P., PAVLYUTIN A.P. & RUBINOVA L.A. (1971) The balance of energy of *Eurycercus lamellatus* (O.F. Müller). In: *Bioproductivity of lakes in Belorussia*, ed. P.G. Petrovich, pp. 104–14. Minsk.

LACK T.J. & LUND J.W.G. (1974) Observations and experiments on the phytoplankton of Blelham Tarn, English Lake District. I. The experimental tubes. *Freshwat. Biol.* **4**, 399–415.

LADLE M., BASS J.A.B. & JENKINS W.R. (1972) Studies on production and food consumption by the larval Simuliidae (Diptera) of a chalk stream. *Hydrobiologia* **39**, 429–38.

LUND J.W.G. (1972) Preliminary observations on the use of large experimental tubes in lakes. *Verh. int. Verein. theor. angew. Limnol.* **18**, 71–7.

MCDIFFETT W.F. (1970) The transformation of energy by a stream detritivore, *Pteronarcys scotti* (Plecoptera). *Ecology* **51**, 975–88.

MACFADYEN A. (1971) The soil and its total metabolism. In: *Methods of Study in Quantitative Soil Ecology: population, production and energy flow*, ed. J. Phillipson, pp. 1–13. IBP Handbook No. 18, Blackwell Scientific Publications, Oxford.

MARZOLF G.R. (1966) The trophic position of bacteria and their relation to the distribution of invertebrates. In: *Organism-substrate relationships in streams*, eds. K.W. Cummins, C.A. Tyron & R.T. Hartman, *Spec. Publs Pymatuning Lab. Fld Biol.* **4**, 131–5.

MATHEWS C.P. & KOWALCZEWSKI A. (1969) The disappearance of leaf litter and its contribution to production in the River Thames. *J. Ecol.* **57**, 543–52.

MELCHIORRI-SANTOLINI U. & HOPTON J.W. (eds.) (1972) Detritus and its role in aquatic ecosystems. *Mem. Ist Ital. Idrobiol.* **29** Suppl.

NILSSON L.M. (1974) Energy budget of a laboratory population of *Gammarus pulex* (Amphipoda). *Oikos* **25**, 35–42.

ODUM W.E. (1968) The ecological significance of fine particle selection by the striped mullet *Mugil cephalus*. *Limnol. Oceanogr.* **13**, 92–8.

PETERSEN R.C. & CUMMINS K.W. (1974) Leaf processing in a woodland stream. *Freshwat. Biol.* **4**, 343–68.

PRUS T. (1971) The assimilation efficiency of *Asellus aquaticus* L. (Crustacea: Isopoda). *Freshwat. Biol.* **1**, 287–305.

RICHMAN S. (1958) The transformation of energy by *Daphnia pulex*. *Ecol. Monogr.* **28**, 273–91.

RODINA A.G. (1963) Microbiology of detritus of lakes. *Limnol. Oceanogr.* **8**, 388–93.

SAUNDERS G.W. (1972) The transformation of artificial detritus in lake water. *Mem. Ist. Ital. Idrobiol.* **29** Suppl., 261–88.

SCORGIE H.R.A. (1974) *Studies on the food of invertebrates in the River Lambourn, with particular reference to the role of detritus in the trophic economy of the community.* Unpublished Ph.D. Thesis, University of Reading.

SMIRNOV N.N. (1962) *Eurycercus lamellatus* (Chydoridae: Cladocera): field observations and nutrition. *Hydrobiologia* **20**, 280–94.

WAVRE M. & BRINKHURST R.O. (1971) Interactions between some tubificid oligochaetes and bacteria found in the sediments of Toronto Harbour, Ontario. *J. Fish. Res. Bd Can.* **28**, 335–41.

WESTLAKE D.F., CASEY H., DAWSON H., LADLE M., MANN, R.H.K. & MARKER A.F.H. (1972) The chalk-stream ecosystem. In: *Productivity Problems of Freshwaters*, eds. Z. Kajak & A. Hillbricht-Ilkowska, Warszawa-Krakow, 615–38.

WETZEL R.G., RICH P.H., MILLER M.C. & ALLEN H.L. (1972) Metabolism of dissolved and particulate detrital carbon in a temperate hard-water lake. *Mem. Ist. Ital. Idrobiol.* 29 Suppl., 185–243.

WETZEL R.G. & MANNY B.A. (1972) Decomposition of dissolved organic carbon and nitrogen compounds from leaves in an experimental hard-water stream. *Limnol. Oceanogr.* **17**, 927–31.

WILLOUGHBY L.G. (1974) Decomposition of litter in fresh water. In: *Biology of plant litter decomposition*, eds. C.H. Dickinson & G.J.F. Pugh, **2**, 659–81. Academic Press, London.

Section Three
Decomposition processes as components of whole ecosystems

15 Decomposition in freshwater

G. W. SAUNDERS *Division of Biomedical and Environmental Research, U.S. Energy Research and Development Administration, Washington, D.C., U.S.A.*

Introduction

Decomposition is the major process involved in the dissipation of energy stored in organic matter in aquatic systems. A part of decomposition may be accomplished through strictly physical-chemical mechanisms such as photo-oxidation, but by far the dominant pathway of decomposition is mediated through biological mechanisms.

It is this potential energy, that is dissipated over a longer time period than the energy storing processes, and that permits the system to function when the light is cut off or where there is no light.

The general question that students of decomposition are addressing concerns which general pathway of decomposition predominates in a particular aquatic ecosystem. Is most organic matter decomposed by microbes with the whole animal kingdom functioning as a small spinning satellite which siphons off only a very small amount of the total organic matter potentially available to it? Or does most of the organic matter present in lakes and streams pass into the guts of animals, to be digested there, assimilated across the gut wall, and ultimately degraded via the intermediary metabolism of the animal community? If the animal pathway is dominant, is this represented by a so-called grazing food chain or by a detrital food chain? Is the organic matter ingested by animals digested by their own intestinal enzymes, or is an indigenous intestinal microflora required to first hydrolyse refractory food material into a dissolved form, or must that ingested organic matter be first transformed into bacterial protoplasm which is subsequently stripped from particles in the gut and then digested and assimilated?

There are estimates from lakes that suggest that zooplankton activities provide the dominant mechanism of decomposition although this pathway is not correlated with any trophic state for these lakes (cf. IBP/PF Synthesis

Volume on Freshwater Production). It has been shown that zooplankton can clear more than 300% of the water volume per day in an eutrophic lake (Haney 1973). It is believed that in the pelagic ocean the major pathway of energy flow is through the grazing food chain (Steele 1974) although this notion has recently been questioned by Pomeroy (1974) who suggests that a major component of microbial communities has not been observed in the ocean until recently. For a heterotrophic small woodland stream it has been estimated that 30% of the particulate organic input is processed by animals (Cummins, personal communication). However, recent comprehensive material budgets for a lake (Table 15.1) and for a stream (Table 15.2) suggest that animals are insignificant in freshwater decomposition process.

Most of the nonliving organic matter in waters is considered to be refractory to decomposition. If this is true, nonliving organic matter must on the average accumulate relative to living organic matter. This means that inorganic nutrients will be trapped in very large metabolically immobile pools. If animals are to be the major pathway in decomposition processing, it means that by virtue of their collecting capabilities and their enzymatic capabilities they must be better able to decompose organic matter than bacteria. It is not clear that this is so.

Table 15.1. Annual budget for Lawrence Lake, Michigan, in gm C m^{-2} yr^{-1} (derived from Wetzel *et al.* 1972).

Inputs		
Macrophytes		87·9
Epiphytic algae		37·9
Epipelic algae		2·0
Heterotrophy		2·8
Algal secretion and autolysis		14·7
Phytoplankton		43·4
Allochthonous particulate organic		4·1
Allochthonous dissolved organic		21·0
Littoral plant secretion		5·5
Bacterial chemosynthesis		7·1
	Total	226·4
Outputs		
Benthic respiration		117·5
Outflow dissolved organic		35·8
Outflow particulate organic		2·8
Permanent particulate sedimentation		14·8
CO_2 precipitation as $CaCO_3$		2·0
Bacterial respiration dissolved organic matter		20·6
Bacterial respiration particulate organic matter		8·6
Algal respiration		13·0
	Total	215·1

Table 15.2. Annual energy budget for Bear Brook, New Hampshire (Fisher & Likens 1973), kcal m^{-2} yr^{-1}.

Inputs	
Litter fall	
Leaf	1,370
Branch	520
Miscellaneous	370
Wind transport	
Autumn	290
Spring	90
Throughfall	31
Fluvial transport	
Coarse particulate organic	430
Fine particulate organic	128
Dissolved organic, surface	1,300
Dissolved organic, subsurface	1,500
Moss production	10
Total	6,039
Outputs	
Fluvial transport	
Coarse particulate organic	930
Fine particulate organic	274
Dissolved organic	2,800
Respiration	
Macroconsumers	9
Microconsumers	2,026
Total	6,039

The objective here is to provide some quantitative perspective relative to microbial decomposition processes in freshwaters. Plant litter decomposition has recently been reviewed (Dickinson & Pugh 1974). The presentation is mainly qualitative and descriptive. Detritus decomposition in freshwater and marine systems has been treated in a recent symposium (Melchiori-Santolini & Hopton 1972). However, there has been no recent review concerning decomposition of dissolved organic matter in freshwaters, although Hood's (1970) review of Concepts and Data up to 1968, with its emphasis on marine waters, can probably be extrapolated to freshwater.

Until the conditions that determine whether decomposition proceeds through animal or microbial activities and until studies of primary production and decomposition are brought into better balance, we can not hope to understand very much concerning the structure and functioning of freshwater ecosystems.

Organic matter in fresh water

It is well known that in lakes, streams, and the upper oceans that the concentration of dissolved organic matter is approximately an order of magnitude greater than the particulate matter. Recent investigations indicate that in the pelagic zone of lakes and in the upper ocean, organic detritus usually is greater than living material by a factor of 5–15 times (Duursma 1961; Finenko & Zaika 1970; Jorgensen 1966; Nauwerck 1963; Poltz 1972; Saunders 1972a; Wetzel *et al.* 1972). During times of phytoplankton blooms the above ratio may approach unity and the amount of seston may even exceed the dissolved organic matter (Weinmann 1970). If organic detritus is decomposed at a rate relatively slower than the rate at which phytoplankton is removed by death, grazing, and settling, organic detritus would tend to accumulate in the open water and the above ratio is a logical consequence.

The dissolved organic matter contains only a small fraction of simple organic molecules as amino acids, sugars, and carboxylic acids (Brehm 1967; Gocke 1970; Poltz 1972; Weinmann 1970). These simple molecules amount to about 5–10% of the total dissolved organic matter. The remainder of the total dissolved organic matter has traditionally been considered as highly refractory, i.e., resistant to decomposition (Wetzel *et al.* 1972; Wetzel & Otsuki 1974), and composed mainly of humic substances. However, it has been demonstrated that there are peptides, and polysaccharides, as well as UV-absorbing and fluorescing compounds in the higher molecular weight phase of the dissolved organic pool although not enough work has yet been done to demonstrate that the proportion of nonhumic to humic substances varies in any systematic manner in classes of lakes. It is known that a significant fraction of the dissolved organic matter in the ocean (Sharp 1973) and in lakes (Brehm 1967; Schindler & Alberts 1974) is actually colloidal in size. Gocke (1970) has demonstrated that colloidal substances released by *Scenedesmus quadricauda* in culture are not readily decomposed by bacteria from lake water. This suggests that if dissolved organic substances are resistant to decomposition by bacteria it might be due to their physical-chemical state as well as to their molecular structure. At any rate it seems to be important to begin to partition what has been operationally defined as dissolved and that which is colloidal. Perhaps greater differentiation according to molecular size ultimately will be necessary.

The particulate organic matter in lakes and streams has been little characterized chemically. In larger lakes most of the particulate matter would be planktogenic. For smaller lakes a greater proportion of particles are derived from the littoral zone and macrophytes (Wetzel *et al.* 1972). In forested areas there may be significant inputs of dead leaves to the littoral zone. However, most of the dead particulate organic matter suspended in the water appears to be very small in size, some of it a few micrometres in dimension but a large proportion at the lower end of the frequency distribution, a micrometre or less in dimension. Processing of larger particles from leaves and macrophytes would normally occur in the surficial sediments and in the littoral or near-littoral zone. Particles derived from trees and shrubs and macrophytes will tend to have more structural components than those derived from phytoplankton and thus will tend to contain more pectin, cellulose, and lignin-like material. This will have bearing on the specific manner and the rates at which decomposition occurs for these two broad classes of particulate organic matter.

Studies of decomposition in streams have mainly emphasized those in woodlands (Cummins 1974) in which most particulate organic matter comes from leaf and grass litter in the autumn (fall)–winter period and dissolved organic matter from residues, percolated through the soil, which are derived from microbial decomposition. Coarse particulate matter (larger than 1 mm diameter) is associated with sediments during periods of normal or low flow (Cummins 1974), while fine particulate organic matter suspended in the water is a direct product of decomposition (Ladle 1972) and the physical flocculation of dissolved materials (Lush & Hynes 1973). Manny & Wetzel (1973) have shown that dissolved organic carbon and dissolved organic nitrogen concentration were nearly constant throughout a diel period in a hard water stream, whereas particulate organic carbon varied considerably. The latter may have correlated with moving and feeding activities of immature insects.

Convention suggests that almost all of the material produced in lakes during a year is decomposed. Only a small fraction of the annual production actually enters the sediments. This latter has been estimated to vary from 1 to 15% for lakes ranging from oligotrophic to eutrophic (Ohle 1956), 25% for eutrophic Linsley Pond (Deevey & Stuiver 1964) but curiously up to 50% for an oligotrophic marl lake (Wetzel *et al.* 1972). Wetzel & Hough (1973) and Wetzel & Otsuki (1974) point out that a large majority of the lakes of the world have a significant macrophytic flora and an accompanying epiphytic microflora. It may well be that lack of attention to this component of lakes has resulted in too high an estimate of relative decomposition of annual production in the total basin volume.

Earlier work suggested that most of the decomposition of plankton occurs in the water column, i.e., the plankton are produced in the upper photic zone and is almost completely decomposed before it reaches the sediments (Kleerekoper 1952). Ohle (1956) has suggested that the organic matter produced in the photic zone is decomposed and re-photosynthesized 2–3 times before it settles into the hypolimnion. As the concentration of nonliving particulate and dissolved organic matter is one to two orders of magnitude greater than the living matter, most of it must be resistant to decomposition. Or is it?

Whether or not the organic detritus and the dissolved organic matter that is found in lakes and streams is resistant to decomposition or refractory is relative. From more recent work it is clear that major fractions of the total dissolved organic pool are susceptible to decomposition. The need is to place into perspective the meanings of relatively rapid, relatively slow, and resistant to decomposition in the function of any aquatic system.

I have pointed out (Saunders 1971) that the concentrations of dissolved carbohydrate, dissolved organic nitrogen, and dissolved organic carbon fluctuate annually in a small productive lake and that the patterns of fluctuation are different and out of phase with one another. The concentrations observed at any instant in time are the result of cumulative differences between the production and the utilization or removal of the material under consideration. The change in concentration is therefore a net change and does not reflect the true rate of production or utilization. The fact that the annual cycles of these three classes of dissolved organic matter are out of phase states that they are not ever in constant relationship to one another. Such changes in relationship presumably reflect changes in the quality and quantity of micro-organisms as well as their metabolic activities. This suggests that an aquatic system is totally dynamic in its operation. With regard to decomposition this must be true on very short timescales as well as longer ones and must apply to simple molecules as well as to complex ones.

In Table 15.3 data is presented concerning the maximum and minimum concentrations of total dissolved organic carbon, dissolved carbohydrate, and dissolved organic nitrogen or protein in lakes of different sizes and degrees of trophy. All of these generalized organic substances exhibit the potential to be decomposed to a significant degree, in general from 50% to near 100%. This does not occur in the whole lake volume but only at certain depths or over certain depth intervals. However, the very shallow productive lakes reported by Barica (1974) exhibit the largest variations in concentrations of dissolved organic substances. This is all the more remarkable as these ponds are more contiguous to the soil and might be expected to contain relatively high concentrations of humic substances. If the latter is true then in these cases humic substances are subject to considerable degradation and are not stable.

The simple organic molecules, i.e., sugars, amino acids, and carboxylic acids, occur at low concentrations in freshwaters, the order of 10^{-6} to 10^{-8} M from direct analysis or estimated as (K_t+S_n) using bioassay. Little is known concerning temporal variations in concentrations. However, for pro-

Table 15.3. Percentage difference between the maximum and minimum concentrations of proximate dissolved organic substrates in lakes ($mg\,l^{-1}$).

Lake	Max.	Min.	% difference	Reference
Dissolved Organic Carbon				
Frains 1970	14·3	10·1	29	Storch 1971
	15·8	12·8	20	Storch 1971
1960	15	6·5	57	Saunders unpublished
Section Four 1970	6	2	67	Fast *et al.* 1973
Hemlock 1970	10	6	40	Fast *et al.* 1973
Lawrence 1967	5	2	60	Miller 1972
	4	1	75	Miller 1972
1968	9	2	78	Wetzel *et al.* 1972
1969	9	2	78	Wetzel *et al.* 1972
1970	8	3	63	Wetzel *et al.* 1972
1971	6	3	50	Wetzel *et al.* 1972
Duck 1968	16	7	66	Miller 1972
Lötsjön 1965	30	5	73	Allen 1969
Lake 885 1971–2	62·4	0·4	99	Barica 1974
Lake 154 1971–2	51·6	1·4	98	Barica 1974
Lake 302 1971–2	84·7	14·4	83	Barica 1974
Lake 255 1971–2	32·4	4·8	85	Barica 1974
Dissolved Carbohydrate				
Frains 1960	5·2	1·3	75	Saunders unpublished
Lac Leman 1973	0·4	0·1	75	Lugrin 1974
Dissolved Protein				
Frains 1960	5	1·2	75	Saunders unpublished
	21	3	86	Saunders unpublished
Lac Leman 1973	0·8	0·05	90	Lugrin 1974
Dissolved Organic Nitrogen				
Lawrence 1970	0·23	0·065	72	Manny 1972
Wintergreen 1970	1·7	0·8	53	Manny 1972
Lake 885 1971–2	11·51	0	100	Barica 1974
Lake 154 1971–2	8·9	0·54	94	Barica 1974
Lake 302 1971–2	8·1	0·2	98	Barica 1974
Lake 255 1971–2	2·99	0·208	93	Barica 1974

ductive Plüss See (Brehm 1967) free amino acids fluctuated in concentration during the year but the proportion of the amino acids held a remarkable constancy. Brehm (1967) suggests that their concentration might be in some sort of equilibrium with the biota. There is a tendency for most of the free amino acids to have higher concentrations in the photic zone than in the hypolimnion. The peptide amino acids also varied considerably during the year in the surface water.

If the largest pool of dead organic matter in lake waters is potentially susceptible to relatively short-term and near total degradation, why is this not apparent in many lakes? Either the imports of dissolved organic matter must be essentially balanced by the removal processes or factors other than the chemical nature of the organic molecules must control their decomposition.

Russell (1968) has pointed to the corollary question in soils. There is some evidence that humus is fairly readily decomposable by soil organisms but that is may be protected from decomposition by being absorbed by clay. What is it that protects humic substances from being decomposed in fresh waters?

The decomposition reaction

The measured concentration of any substance in an aquatic system is the balance between the input rates and the removal rates. Removal may be due to settling or lateral transport by physical mechanisms, or by feeding, or by decomposition. Although decomposition can be effected by nonbiological reactions, most of decomposition as we understand it is mediated by living organisms. The decomposition reaction is in fact a reaction between a substrate and an enzyme or an enzyme system. Therefore the decomposition reaction may be expressed as a bimolecular second order reaction (Saunders 1972c):

$$\frac{-d[S]}{dt} = c.[S].[E] \tag{15.1}$$

where [S] = substrate concentration
[E] = effective enzyme concentration
c = empirical coefficient which expresses the degree to which environmental variables control the reaction rate.

This expression will apply over all water types and all concentration levels. If the enzyme concentration remains constant then the equation will reduce to the expression for an apparent first order reaction:

$$\frac{-d[S]}{dt} = c_1.[S] \tag{15.2}$$

Both total organic matter and numbers of bacteria vary by one to two orders of magnitude from oligotrophic to eutrophic systems (Kuznetsow 1959). Therefore in productive lakes and streams we would expect decomposition to proceed disproportionately faster than in unproductive systems. This certainly turns out to be the case for small dissolved organic molecules.

Usually the material being decomposed is not a single chemical material but a heterogeneous set of organic substrates. This is certainly true of particulate matter such as plankton, macrophytes and leaf litter and of dissolved organic matter excreted or hydrolysed from particulate organic matter. It is also true of the total dissolved organic pool which must contain a large diversity of organic chemical substrates.

The usual technique of determining decomposition in fresh waters (and in other systems) is to determine the weight loss in time of packets of leaves, or macrophytes, or suspended particles in a unit volume of contained water sample. Another technique is to determine the remaining radioactivity in labelled particulate or dissolved organic matter. Either technique may be only approximate because micro-organisms will develop on particulate material and be included in its weight determination or they may metabolize labelled dissolved organic material and release it in another dissolved form.

The kinetics of decomposition are described generally in Fig. 15.1.

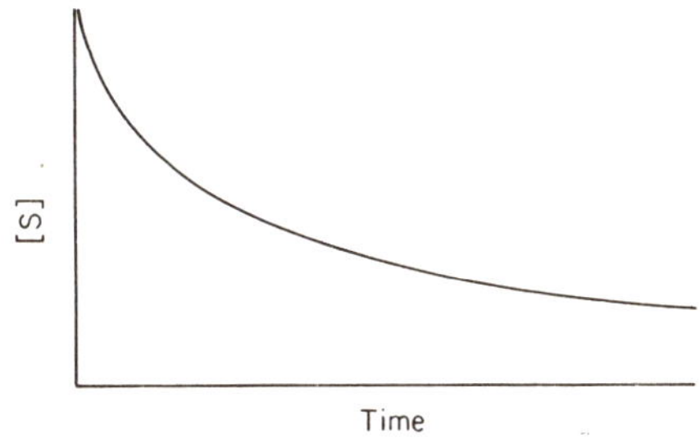

Figure 15.1. A generalized curve for decomposition.

Whether the substrate is simple or heterogeneous the curve can usually be analysed into two components having different reaction rate constants, assuming that the reaction rates follow first order kinetics. As discussed above this assumption is not strictly true. The curve tends to decrease approaching some constant value asymptotically. The percentage of material remaining after some long period of time varies widely, depending on the nature of the initial substrate and the conditions of the experiment, but usually is the order of 20–30% of the initial material. The remaining material is considered to be resistant to decomposition. However, there are at least two cases, to be discussed below, in which that conclusion is only apparent, suggesting that it is an artefact of the experimental procedure.

DECOMPOSITION OF PARTICULATE ORGANIC MATTER — process how happens

The first step in the decomposition of particulate organic matter is lysis. Apparently endogenous enzymes cause higher molecular material to hydrolyse. If the dead particulate organic matter is in contact with water the soluble organic matter will be leached. Leaching is essentially complete within the first twenty-four hours. It has been shown that leaching is the order of 20–40% of the initial biomass. Part of the organic matter leached must be derived from dissolved cytoplasmic substrates. It is difficult to differentiate between this organic material and that derived by autolysis. However, estimates suggest that autolysis is significant; algae, 20–30% (Golterman 1964, 1972); aquatic macrophytes, 30–40% (Otsuki & Wetzel 1974); seaweeds, 58% (Khailov & Burlakova 1969); zooplankton, 20–30% (Krause 1959); leaf litter, 5–30% (Cummins 1974). In the case of terrestrial plant litter, lysis undoubtedly occurs shortly after death. Extraction by leaching depends upon contact and wetting with water. Particles that fall directly into a water body will contribute all the products of lysis to the system; the contribution of these materials from plant parts that first fall on to land will depend upon the weather and how much leaching occurs before these parts enter the water.

Streams

The decomposition of leaf litter in streams is the most studied aspect of decomposition of particulate organic matter (Boling *et al.* 1975; Cummins 1974; Willoughby 1974). The general picture concerning leaf litter decomposition is that following the brief period of leaching there is a period of conditioning during which fungi develop on and in the leaf material. In streams the hyphomycetous fungi are the principle colonizers and conditioning agents. They transform leaf material into fungal protoplasm and increase in biomass until a maximum is reached. The leaf material is considered to be fully conditioned at the time of maximal fungal biomass development. Subsequent to this maximal colonization there is a rapid decrease in microbial activity on the leaf, probably associated with exhaustion of easily metabolized substrate in the leaf. Invertebrates normally do not feed on the leaves until they have become conditioned and it is presumed this is because they feed on the fungi rather than the leaf material *per se*. During conditioning by fungi and feeding by invertebrates there must be some release of dissolved organic matter to the surrounding water, although this has not been studied. Nor has the role of bacteria in decomposition of litter been studied, so that there are no data available on the relative quantitative roles of bacteria and fungi in the decomposition of leaf litter.

lakes

Leaf litter decomposition in lakes has not yet been quantitatively studied. Willoughby (1974) implicates Actinomycetales in the shallow littoral zone of lakes. It may well be that the agents of microbial decomposition of leaf litter in lakes and streams are quite different. The decomposition of hickory (*Carya glabra*) leaf litter packs in the littoral zone of three lakes was very much slower than in three small streams (Cummins, unpublished). This may reflect differences in the role of fungi in leaf decomposition in lakes and streams. Pieczynska (1972) investigated the decomposition of *Potamoge-*

ton spp., leaves of *Phragmites communis*, and leaves of *Salix* sp. in the littoral zone of a productive lake. The average rate of decomposition ranged from 0·4 to 2·5% per day. This is comparable to the rates for leaf decomposition in streams (Table 15.4) although the temperature was undoubtedly somewhat warmer than in the streams.

Table 15.4. Percentage decomposition of leaf litter in stream water.

Substrate	%/day	Author
Slow Leaves		
Acer macrophyllum	0·25–0·70	Sedell *et al.* 1974
Populus spp.	0·30–0·46	Petersen & Cummins 1974
Carya glabra	0·3	Petersen & Cummins 1974
Fagus grandifolia	0·25	Petersen & Cummins 1974
Quercus borealis	0·27	Petersen & Cummins 1974
Ulmus americana	0·05–0·15	Petersen & Cummins 1974
Pseudotsuga menziesii	0·20–0·75	Sedell *et al.* 1974
Tsuga heterophylla	0·20–0·75	Sedell *et al.* 1974
Medium Leaves		
Salix lucida	0·62–0·78	Petersen & Cummins 1974
Juglans nigra	0·70	Petersen & Cummins 1974
Betula lutea	0·5–1·2	Petersen & Cummins 1974
Carpinus caroliniana	0·83	Petersen & Cummins 1974
Alnus gentinosa	0·75	Petersen & Cummins 1974
Acer saccharum	0·33–1·07	Petersen & Cummins 1974
Acer rubrum	0·41–0·62	Petersen & Cummins 1974
Acer platinoides	0·76	Petersen & Cummins 1974
Acer circinatum	0·5–1·9	Sedell *et al.* 1974
Fast Leaves		
Tilia americana	1·75	Petersen & Cummins 1974
Cornus amomum	1·15	Petersen & Cummins 1974
Decodon verticillatus	1·01	Petersen & Cummins 1974
Fraxinus americana	1·20	Petersen & Cummins 1974
Alnus rubra	1·2–1·6	Sedell *et al.* 1974

Leaf decomposition in streams is species dependent. Rates of decomposition have been classified as rapid, intermediate, and slow (Kaushik & Hynes 1971; Petersen & Cummins 1974; Triska 1970). The differences in relative processing rates probably reflect differences in the biochemical nature of the different species of leaves. Relative decomposition rates of leaf litter in streams are presented in Table 15.4.

Whether or not fungi are involved in decomposition of plant materials is probably a matter of particle size. Olah (1972) has observed leaching of ground up *Phragmites* sp. particles followed by a succession of bacterial types during decomposition. Fungi were not implicated in the decomposition of these particles. As the organic particles became smaller the numbers of bacteria inhabiting them became larger until the particles became so small that they could no longer provide surface for bacterial adherence. Measurements of microbial metabolism in aquatic systems (Hargrave 1972; Petersen & Cummins 1974) have shown a general inverse relation with particle size.

Decomposition of algae in lake water has not received much attention quantitatively. In two cases it has been estimated to be very rapid (Frankland 1974). Saunders (1972a) followed the decomposition of *in situ* particulate organic matter labelled with ^{14}C for one day. Presumably this was mostly killed phytoplankton. The initial rate of loss of ^{14}C was 14·1% per day. Three-day-old detritus prepared in the same manner exhibited initial loss rates of 0·3% per day and 5·6% per day. The temperature of the lake water in which the analyses were conducted was 25, 1, and 10°C, respectively. The initial decomposition rates of phytoplankton in this productive lake are the order of 10% per day. It remains to be determined if such a rate is representative for unproductive lakes also or whether much slower rates would occur in such lakes. It appears that decomposition is initiated immediately so that hydrolytic exoenzymes must be present in the lake water. Saunders and Cummins (unpublished) have independent evidence to support the notion of hydrolytic exoenzymes in lake water and that there may be differences between lake water and streams. Finely ground and washed ^{14}C labelled tobacco leaf cellulose was placed in 3·8 litre vessels of surface lake water and stream water from a productive woodland stream. Two subsamples of stream water were used; one with conditioned leaf material added and one without leaf material. All vessels were stirred slowly with a magnetic stirrer. Radioactive carbon dioxide was recovered immediately and continuously in the lake water sample. There was a four-hour lag period before significant quantities of radioactive carbon dioxide were recovered from the stream water with added leaf material and an eight-hour lag period in the stream water with no additions. This suggests that a cellulase was initially present in the lake water whereas it took some time to build up a comparable activity of cellulase in the stream water and that a significant fraction of cellulolytic activity in the stream water was associated with the leaf inoculum.

Saunders (1972a) has shown that the dissolved organic ^{14}C released from radioactive detritus averaged little more than 1% per day. In the winter this was approximately 50% of the total loss from detritus per day; in the summer it was approximately 20%. Only three sets of analysis were conducted so it is difficult to generalize. However, he did calculate that release of dissolved organic matter from detritus would amount to 2–6 times that secreted by algae in the upper photic zone. He also suggested that organic detritus represents a larger reservoir of dissolved organic matter than phytoplankton in the photic zone and an even relatively larger reservoir in the aphotic zone whether the release is due to hydrolysis by enzymes dispersed in the lake water or is due to extracellular release by aggregates of bacteria surrounding the detrital particles.

There are no comparable data concerning decomposition of particulate organic matter in unproductive aquatic systems.

A general conception of decomposition of particulate organic matter in productive water bodies is summarized in Fig. 15.2. All particulate matter

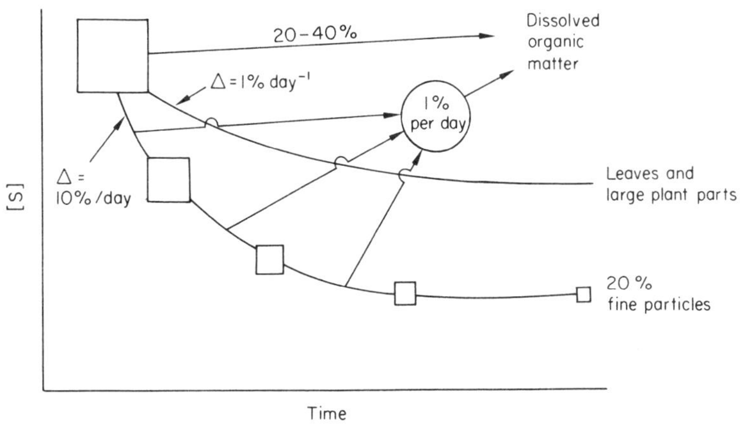

Figure 15.2. A general scheme for decomposition of particulate organic matter in productive water bodies.

partially autolyses upon death. When the particles are wetted by water this lysed material is extracted and amounts to 20–40% of the initial dry weight of material. The remaining organic matter decomposes more slowly. Large particles, particularly leaf litter in streams and possibly other large plant parts, are colonized and conditioned by fungi. The initial decomposition rate is the order of 1% per day. A residuum of decomposition occurs which may decompose at a rate of only a few percent per year. There must be classes of organic matter that decompose at intermediate rates but because of the very slow rates and thus the long times involved it is difficult to determine these accurately from kinetic studies. Where animals feed on conditioned leaves, decomposition rates are increased (Sedell *et al.* 1974). The increased loss rate must be due in part to decomposition and in part to mechanical loss of small particles during feeding.

Fine particles, phytoplankton and zooplankton, exhibit initial decomposition rates the order of 10% per day. There is a residuum that decomposes at a rate of approximately 1% per day or less. There may be a fraction that decomposes at a rate of only a few percent per year (Frankland 1974). In the case of both large and small particles about 20% of the initial material appears to be resistant to decomposition, although this percentage may vary over a wide range depending upon conditions during the analysis.

DECOMPOSITION OF DISSOLVED ORGANIC MATTER

Decomposition of dissolved organic matter also has not been studied directly. It has not been a simple matter to analyse for substrate concentration of simple organic molecules, primarily because of their extreme dilution which is the order of 10^{-6} to 10^{-8} M. The higher molecular weight substances such as polysaccharides, peptides and proteins, and lipids are difficult to isolate and identify, let alone quantitatively estimate. The total dissolved organic pool may be estimated relatively easily and with reasonable accuracy and precision. However, changes in the pool are ordinarily relatively small, a few percent per day, and no measure of input rate is made. Therefore, measured changes have a high variance and usually reflect net rates of change rather than true rates of decomposition of the total dissolved organic pool.

Investigations of decomposition have therefore focused on the decomposition of radioactive sugars, amino acids, and acetate estimated as turnover time of the substrate. The method is essentially a bioassay utilizing Michaelis–Menten kinetics as suggested by Parsons & Strickland (1962) and applied in fresh water by Hobbie & Wright (1965a). The method assumes that input and output balance. A second approach is to determine the rate of decrease of radioactive substrate added to water samples in presumed tracer quantities. This method estimates the true relative rate of decomposition. Measures of substrate concentration usually are not made.

The turnover times for glucose in a wide spectrum of lake types are represented in Fig. 15.3. There is an obvious general relation between glucose

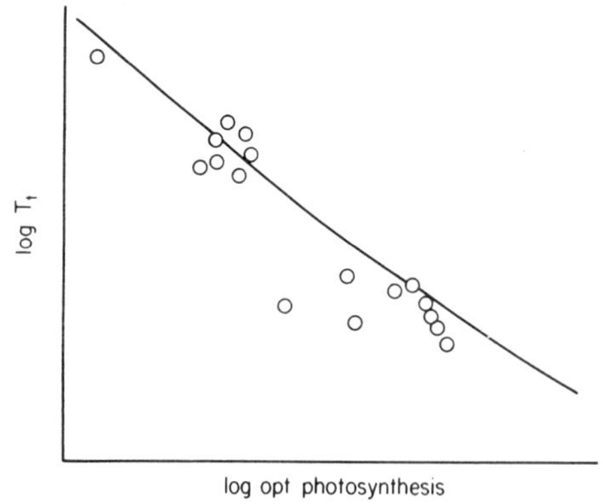

Figure 15.3. The log–log relationship between turnover time of glucose in water bodies of different trophic state and optimal photosynthesis. T_t in hours; photosynthesis in g Cl^{-1} d^{-1}. (Allen 1969; Hobbie & Wright 1965b; Hobbie *et al.* 1972; Miller 1972; Rodhe 1965; Seki *et al.* 1974; Saunders unpublished; Saunders *et al.* unpublished; Vaccaro *et al.* 1968; Wetzel 1967.)

turnover time and primary production, with glucose turnover time ranging from 0·5 hours in very productive water to several thousand hours in the unproductive oceans. The turnover times and optimal rates of photosynthesis are crudely averaged from minimal data. The ranges may vary from one to two orders of magnitude for both turnover time and optimum photosynthesis. While this general relationship is interesting it is in a sense more interesting to question why the turnover time for glucose varies over 1–2 orders of magnitude within a single water body. What is it that controls the rate of decomposition of glucose since, as far as we know, the concentrations of glucose in seawater, sediments, and lake water are not fundamentally different? Glucose in unproductive waters has a turnover time that is longer than the organic substances considered to be resistant to decomposition in productive lakes. It is not its concentration and certainly it is not its molecular structure that makes glucose appear to be resistant to decomposition in unproductive waters. Other factors must control this.

The turnover time for amino acids and acetate are roughly comparable to that of glucose in those systems where they have been studied. Therefore one would predict the general relationships given in Fig. 15.3 for all simple organic metabolites.

Decomposition of dissolved organic matter released by phytoplankton has been studied in Frains Lake, Michigan, by Storch (1971). From February through October the turnover time for this uncharacterized substrate ranged from 48 to 240 hours. Although algae may release simple organic substrates and higher molecular weight materials, kinetic studies of phytoplankton extracellular release suggest that phytoplankton release predominantly high molecular weight organic matter (Saunders 1972b; Saunders, unpublished). Dissolved organic matter released extracellularly by bacteria decomposes at the same rate as algal extracellular organic matter (Saunders, unpublished). The turnover time for glucose in Frains Lake, winter and summer, ranged from 1 to 5 hours (Saunders, unpublished). Therefore, the organic matter released by algae and bacteria in Frains Lake is kinetically different from glucose and other simple substrates and is probably higher molecular weight material.

Dissolved organic matter leached from leaves in a productive artificial stream contained two components that disappeared at rates of approximately 25% per day and 1% per day (Cummins *et al.* 1972; Wetzel & Manny 1972). These rates are comparable to those observed for decomposition of organic compounds released by algae in productive lakes.

The net rate of loss of dissolved carbohydrate and of dissolved organic nitrogen in Frains Lake during the spring of 1960 approximates 0·4% per day, which provides a turnover time of 6,000 hours. These values represent a minimum decomposition rate or a maximum turnover time because the supply rates were not known. Where decrease in the total dissolved organic carbon pool could be observed in Frains Lake, the daily decomposition ranged from 2·7 to 6·9% providing turnover times of 900–2,300 hours.

Therefore in productive lake systems there are classes of dissolved organic matter that are decomposed at relatively characteristic rates. Simple organic molecules decompose at rates that range from 500 to 5,000% per day. Higher molecular weight organic substances released by phytoplankton and bacteria decompose in the range of 10–50% per day. Other higher molecular weight dissolved substances decompose on the order of 1% per day and it is assumed there is at least one additional class that decomposes at a rate very much less than 1% per day.

There are no comparable data available for unproductive aquatic systems.

The fact that there are several pools of particulate and dissolved organic matter that decompose at relatively rapid and slower rates provides a certain amount of flexibility in the operation of the whole system. The smaller rapidly metabolized pools sustain a vital activity within the system. However, as they are utilized and their concentrations decrease so must the metabolic activity of the micro-organisms consuming them. The larger more slowly metabolized organic pools may also serve other functions than being directly involved in the energetics of the system. For example they may contain a necessary and sufficient amount of extracellular hydrolytic enzymes that keep the system running or they may be necessary to chelate metal ions and maintain a balance appropriate to the function of the system. Of course one major problem of analysis is to determine if such functions do indeed occur in fresh waters, whether they are significant in the metabolism of the system, and what controls the functions.

The dynamics of decomposition

Since rate of decomposition is a function of both substrate concentration and effective enzyme concentration the pattern in the kinetics of the absolute rate of decomposition will vary according to the form of the input rate of the substrate to the system being studied and according to any changes in the effective enzyme available to react. Three basic patterns of supply and removal rates can occur.

The first relationship is that of a constant rate of input of substrate balanced by an equal and constant rate of decomposition. This is a steady state system and is not likely to occur very frequently in any natural system. The concentrations of amino acids in Plüss See (Brehm 1967) exhibit major seasonal variations and also differences in vertical distributions. In the photic zone there are daily fluctuations in primary production, the direct source of organic substrates. Therefore a constant rate of supply and rate of decomposition are likely to be very rare and to occur for only very short periods of time.

A second possible relationship is to have an instantaneous input of substrate coupled with an instantaneous increase in the decomposition rate. As the concentration of substrate decreases the absolute decomposition rate would decrease in an exponential form. The actual pattern will depend on whether or not enzyme saturation is approached or exceeded. However, it is difficult to imagine that an instantaneous input of substrate is anything but a relatively rare event.

The third pattern is to have a fluctuating input rate and a fluctuating decomposition rate with a lag in the fluctuating decomposition rate and thus a time difference in the maxima of the two rate fluctuations. This pattern is well known in simple chemical reaction sequences, where A→B→C→D→E. If A decomposes into B the concentration of A will decrease and that of B will fluctuate through a maximum as B decomposes into C, and so on through C to D and E. The extent of the phase difference in the oscillations depends on whether the individual reaction rates are fast or slow. However, it is a physical law that this sequence of fluctuations must occur. This must be the usual case in natural systems and it is the common pattern that must be dealt with if decomposition is to be analysed in natural systems. The fluctuations may be periodic or aperiodic. The fluctuations may have different periods. There are diel periods, periods of a few days, weekly periods, seasonal periods, and longer term periods. There are many known and obvious examples from natural waters. Extracellular release of dissolved organic matter by phytoplankton oscillates on a 24-hour basis (Saunders 1971, 1972b; Saunders & Storch 1971; Saunders *et al.* 1975) and the dissolved material may have a longer period component, leaf leachate in an artificial stream fluctuates in concentration over a few days and has two decomposition rate components (Cummins *et al.* 1972; Wetzel & Manny 1972), the concentration of dissolved organic matter released by macrophytes oscillates within a few days (Otsuki & Wetzel 1974), the input of leaf litter to a stream oscillates over several months in the autumn (Boling *et al.* 1975), the death of macrophytes fluctuates over a few months period, and there are other examples. The coupled oscillatory reaction sequence (Saunders & Storch 1971) must be the ordinary case in the sequence of decomposition reactions that dissipates the energy stored through photosynthetic processes, whether that sequence occurs in the photic zone or whether it is redistributed vertically out of the photic zone by sedimentation of plant material or by turbulent mixing.

DYNAMICS OF DISSOLVED ORGANIC MATTER DECOMPOSITION

A simulation experiment was conducted in a hypereutrophic lake during a bloom of *Microcystis* sp. Two days prior to the simulation, the oscillation in the rate of release of dissolved organic carbon was determined. During the simulation glucose-carbon was added to a sample of lake water throughout the day at the same average hourly rates estimated from the previous extracellular release analysis. Throughout the day radioactive glucose was used in tracer quantities to estimate the relative rate of assimilation of glucose-carbon by the bacterial community. Knowing the relative rate of assimilation and the rate of glucose additions to the experimental vessel, the absolute rate of assimilation of nonradioactive glucose was calculated. The results are presented in Fig. 15.4 where assimilation rate is plotted versus substrate concentration. The arrows on the curve indicate the sequence in time but linear distance along the curve is not proportional to time. This is a modified Michaelis–Menten curve for the relation between uptake velocity and substrate concentration. It is modified because the uptake velocity does not approach a limiting value asymptotically but with time begins to increase and as the substrate concentration decreases there is an apparent hysteresis in the uptake curve.

This simulation indicates that if phytoplankton releases glucose extracellularly in an oscillatory manner and if the microbial enzyme system is not

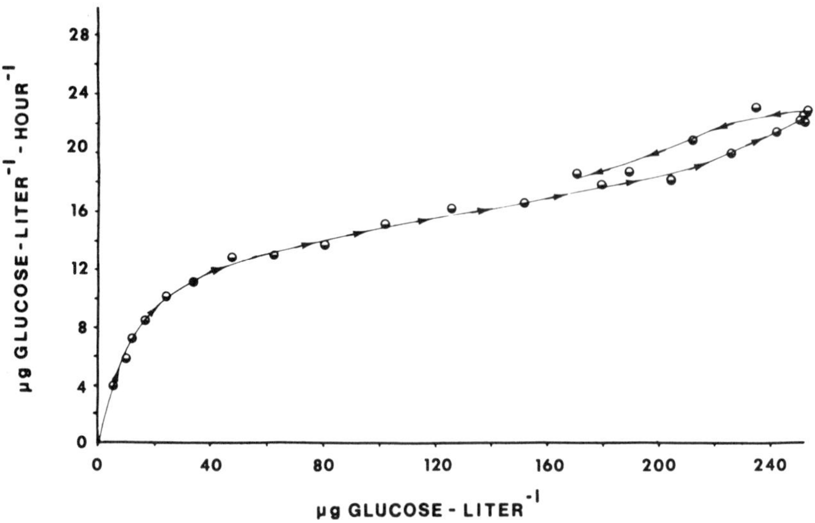

Figure 15.4. Modified Michaelis–Menten curve of velocity of glucose uptake versus glucose concentration. From a simulation experiment in surface water sample from Sanctuary Lake, Pennsylvania, June 1970. Time sequence 0600 to 0600 indicated by arrows.

saturated for the substrate, the assimilation rate and thus the decomposition rate will oscillate merely because the uptake velocity moves along the Michaelis–Menten curve. The upward deviation of the curve at higher substrate concentrations indicates an effective increase in enzyme concentration. I interpret the hysteresis to mean that the microbial community changed its internal enzyme configuration to increase its capacity to assimilate glucose, in response to the signal that higher input rates of glucose would be sustained. This does not require growth of the microbial community.

The modified Michaelis–Menten curve can be used to model the kinetics of substrate concentration, assuming that the oscillatory input from the phytoplankton and the capacity of the microbial community to utilize substrate for the next several days remained constant. The result would be that each day the glucose concentration would fluctuate, the average glucose concentration would increase until after the third day the concentration would be in steady state oscillation. However, in this model the average glucose concentration would be unrealistically high and it would be expected that growth of bacteria would occur.

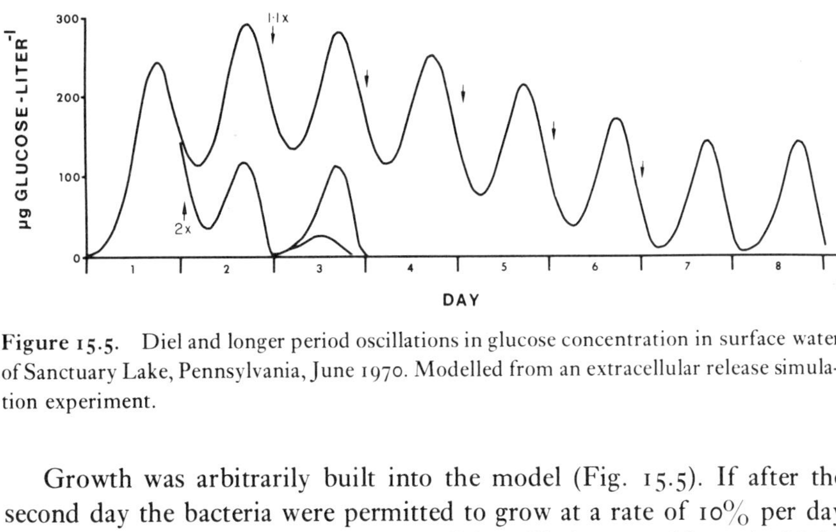

Figure 15.5. Diel and longer period oscillations in glucose concentration in surface water of Sanctuary Lake, Pennsylvania, June 1970. Modelled from an extracellular release simulation experiment.

Growth was arbitrarily built into the model (Fig. 15.5). If after the second day the bacteria were permitted to grow at a rate of 10% per day the longer term oscillation containing diel oscillations would result. If the bacteria doubled the first day but did not increase subsequently the intermediate oscillation would occur. If the bacteria doubled the first day and doubled again the second day the short-term oscillation with the small amplitude diel oscillation on the third day would occur.

These models are suggestive of reality in that one conceives of the phytoplankton and bacterial communities fluctuating about one another in longer term oscillations. The longer term oscillations are constituted of short-term more tightly coupled diel oscillations. This simple model would apply to all substrates that would be metabolized at comparable rates to glucose. This would mean generally the sugars, amino acids, and carboxylic acids. The glucose concentrations model unrealistically high; this suggests that glucose is not the substrate being released extracellularly by the phytoplankton.

It is also clear that the bacterial community cannot keep up with the glucose being released unless population growth occurs.

Fogg (1966) has pointed out that there are two broad classes of organic substrates that may be released by algae. The first consists of small molecules that are released from intermediary metabolic pools. It is possible to estimate accurately the amount of organic carbon released extracellularly, using radioactive carbon as a tracer, provided that the intermediary pool is in isotopic equilibrium with the inorganic pool. This is the usual assumption when extracellular release is quantitatively estimated. The second class of substrate is higher molecular weight material. Normally the pool from which this material is derived would not be in isotopic equilibrium with the inorganic carbon pool. This is because the photosynthetic increment is small compared to the initial biomass of algae. Therefore it is not possible to estimate accurately release of this material by algae. Storch (1971) has attempted to estimate upper and lower boundaries for phytoplankton extracellular release.

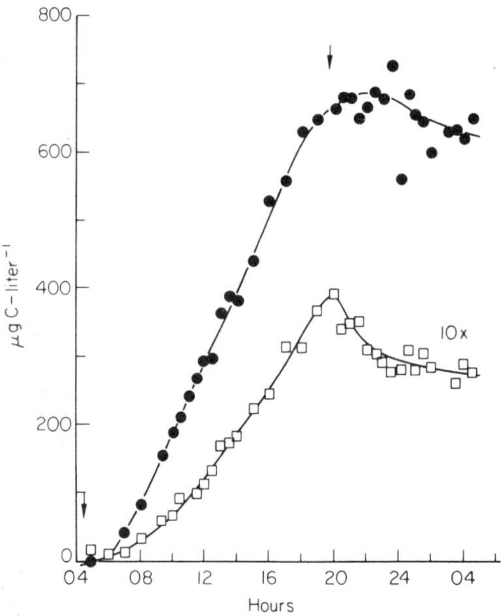

Figure 15.6. Cumulative ^{14}C photosynthesis and phytoplankton extracellular release in surface sample from Wintergreen Lake, Michigan, 2 July 1969. Arrows denote sunrise and sunset. Photosynthesis, ●; extracellular release, □. Note extracellular release is ten times actual absolute values.

Saunders (1972b) has demonstrated three patterns in diel extracellular release of dissolved organic matter by phytoplankton. In one pattern (Case 3) the substrate released decreases in concentration after sunset (Fig. 15.6). The concentration of radioactive substrate after sunset may be described by an equation that has two exponential terms. An analysis of the decrease in extracellular dissolved organic matter assuming first order kinetics for the two apparent substrates is presented in Fig. 15.7 for a *Microcystis* community in an eutrophic lake in Michigan. The two substrates are decomposed at relative rates of 44·5% per hour and 1·7% per hour, respectively. In this case it is assumed that extracellular release stops at sunset. Therefore

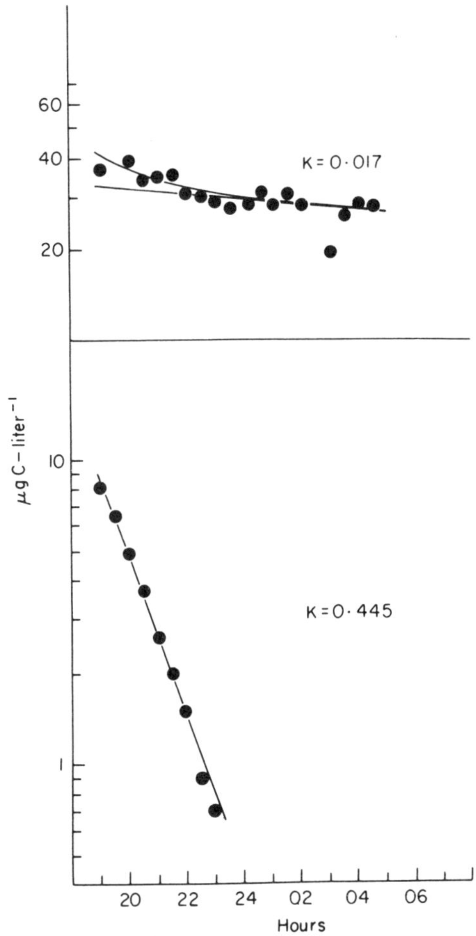

Figure 15.7. Analysis of decrease of extracellular radioactive organic matter after sunset, Wintergreen Lake, Michigan, 2 July 1969. Upper figure: total extracellular release and fitted linear extrapolation of slowly metabolized component. Lower figure: fitted decrease of rapidly metabolized component.

the decrease in concentration after sunset is a measure of the true rate of decomposition. The relative rate of decomposition of 44·5% per hour is comparable to that of glucose (Fig. 15.3) and other simple organic molecules in productive lakes. The relative rate of 1·7% per hour falls within the range found by Storch (1971) for the decomposition rate of phytoplankton extracellular organic matter in a productive lake. It can be shown that it is highly improbable that the slowly decomposing substrate is derived by rapid bacterial transformation of the simple substrate.

Saunders (1972a) has argued that a lag between fixation of carbon in phytoplankton and extracellular release of carbon is indicative that the released material is derived from a pool far along in the biosynthetic sequence or that it is higher molecular weight organic matter. There is a significant lag in this example (Fig. 15.6). If the two die away curves (Fig. 15.7) are extrapolated to sunset and if it is assumed that both source pools were at isotopic equilibrium with the inorganic carbon pool, the rapidly decomposing substrate has an initial concentration of approximately 8 μg C per litre and the more slowly decomposing substrate a concentration of approximately 33 μg C per litre at sunset. This is four times the smaller pool. However, it is not likely that the source pool was at isotopic equilibrium with the inorganic carbon pool. Therefore the second organic substrate would be some unknown and indeterminate increment greater than 33 μg C per litre. It could be from 2 to 7 times greater (Storch 1971).

Therefore this is an example in which two substrates, identified kinetically but not chemically, are released by phytoplankton. The higher molecular weight substrate is released at a rate that oscillates during the daylight. The rapidly metabolized substrate probably has an oscillatory input rate also. A rough calculation will show that within a factor of two the two substrate pools are energetically equivalent in microbial community metabolism during this 24-hour period. However, this particular pattern of extracellular release, Case 3, has been observed only infrequently. In most cases a die away of substrate is not observed after sunset although the rate of extracellular release may decrease and a lag between particulate fixation and extracellular release is the usual situation. This suggests that in general it is higher molecular weight organic matter that is released extracellularly by phytoplankton in lakes.

If phytoplankton release simple organic molecules at relatively low absolute rates these molecules may be assimilated almost as rapidly as they are released (cf. Fig. 15.3). However, when this is the case the small molecules may not be energetically very important to the microbial community. An example of the kinetics of decomposition of radioactive glucose added to surface water from a eutrophic lake in tracer quantities is presented in Fig. 15.8. The radioglucose decomposed at a rate of 96% per hour, so that this

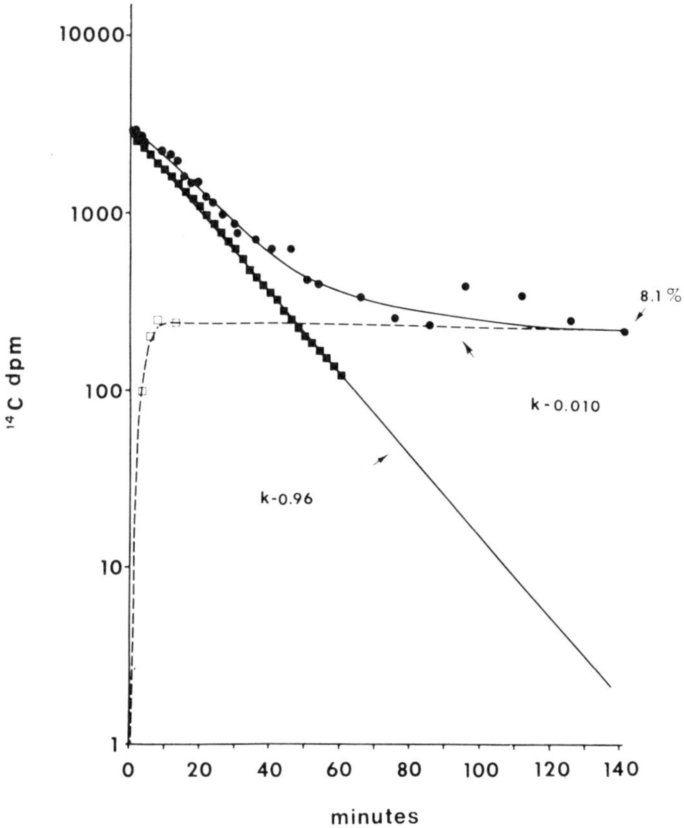

Figure 15.8. Kinetics of tracer ^{14}C glucose decomposition and bacterial extracellular release expressed as equivalent glucose per litre.

reaction was essentially complete within one hour, and by two hours 8·1% of the original radioactivity was in the dissolved phase. I interpret this to mean that the radioactive glucose was transformed into another organic substance which was released extracellularly by the bacterial community. Most of the daughter radioactive organic matter was formed and released within the first 20 minutes of the experiment, as deduced by extrapolation. Extracellular release by bacteria, in the several analyses I have conducted, ranges from 3 to 15% of the initial radioactive substrate. Glucose, amino acids, and acetate fall within this range. The relative rate of decomposition of the extracellular material is 1% per hour. The kinetics were followed for several hours beyond that indicated in Fig. 15.8. One percent per hour is approximately the same rate of decomposition as for organic matter released by phytoplankton in productive lakes. The chemical structure of the organic matter released extracellularly by bacteria and phytoplankton probably is not the same but kinetically the released substances appear to behave similarly.

In the case where bacteria assimilate algal extracellular release of higher molecular weight, do they transform this substrate, release extracellularly approximately 5% of the assimilated material, and subsequently decompose this daughter substrate at the same relative rate as the organic matter released by the algae? Or is the initial substrate transformed to a metabolite that is decomposed at a much slower rate?

Storch (1971) has shown that phytoplankton extracellular release in Frains Lake, Michigan, was decomposed at an initial relative rate that varied from 10 to 50% per day. This is much slower than the rate of decomposition of small organic molecules. Saunders (1972b, unpublished) believes that in general phytoplankton communities release only higher molecular weight dissolved organic matter. A model of the dynamics of this type of more slowly decomposing organic matter is presented in Fig. 15.9. The model is based on the simulation model for glucose release by phytoplankton (Fig. 15.5). Three assumptions are involved. The first is that the polymer carbon is released in an oscillatory manner and at the same hourly rates as estimated for the glucose simulation experiment. The second is that the relative rate of decomposition of this polymer is 1% per hour which is a representative

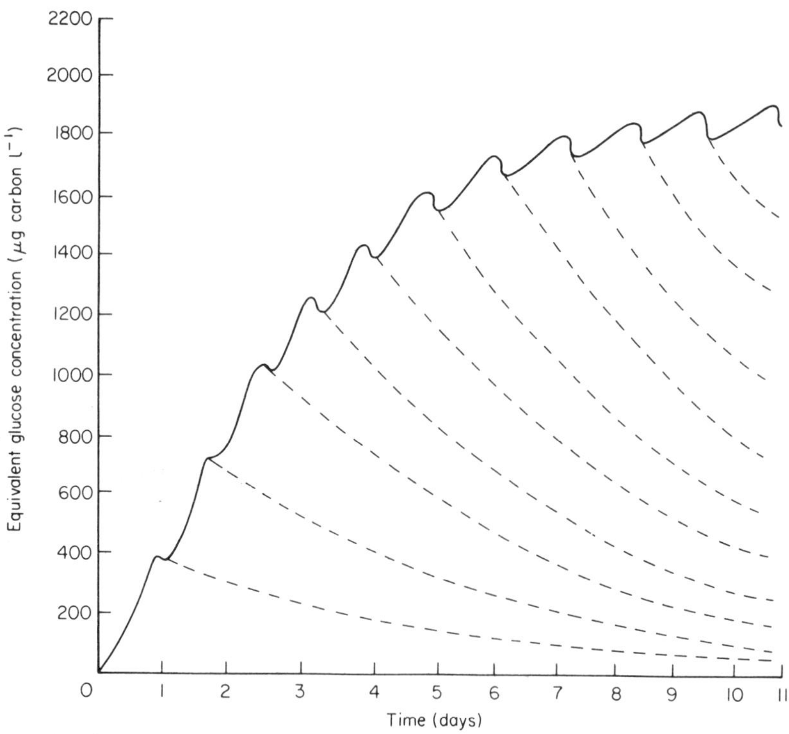

Figure 15.9. A model of the dynamics of higher molecular weight dissolved organic extracellular release expressed as equivalent glucose per litre.

rate for phytoplankton extracellular release in productive lakes. The third is that of no growth in the microbial community. Each daily increment is summed and decomposition of the daily input and the residuum from previous daily inputs is followed separately. The polymer concentration oscillates on a diel basis. This is because the absolute rate of input oscillates as must the absolute rate of decomposition. The concentration of equivalent glucose accumulates until by day ten it is approaching 2,000 μg per litre and is near steady state oscillation. This concentration is approximately ten times that of the glucose simulation model (Fig. 15.5). If the microbial decomposers are allowed to grow or die, there would be longer term fluctuations as in the glucose model. However, the relative amplitude of any oscillation in the polymer model would be less than in the glucose model. In this sense the polymer system tends to be more stable and it also has a relatively larger reserve. If the polymeric material has some function other than as a substrate in the energetics and growth of microbes, that function would tend to be more stable also.

In reality this model would not normally be observed in nature. As lakes normally are permanent and have been in operation for longer than a few days, the pools of higher molecular weight organic matter will have already been accumulated. Therefore, fluctuations in the model after ten days would be more representative of the operation of a natural system than the model as it is presented in Fig. 15.9.

Storch (1971) determined the extracellular release rates by phytoplankton from February to November in Frains Lake, Michigan. He also estimated the relative rate of assimilation of this pool by the bacterial community. His calculations show that the dissolved organic matter released by the phytoplankton is quantitatively utilized and that the release rates and decomposition rates tend to oscillate about one another and may be out of phase by as much as two weeks, although in general the phase differences are shorter than that.

If algae, bacteria, and macrophytes tend to release higher molecular weight organic substrates and if upon death hydrolysis of protoplasm is not complete during lysis, then inputs of polymers will tend to dominate over small molecules and higher molecular weight dissolved organic pools will accumulate.

If polymers are released that decompose at very slow relative rates, they will tend to accumulate to a greater degree and to be relatively more stable. However, the accumulation will depend not only on the relative decomposition rate but the absolute rate of input. If this is small then the pool size will tend to be small.

Humic substances are known to be resistant to decomposition. This may be because of their molecular structure or because of their physical form, i.e., they may be colloidal. However, data presented earlier (Table 15.1) suggests that humic substances are susceptible to relatively significant degrees of decomposition. Bacteria isolated from water do not grow or maintain themselves on humic substances isolated from water. However, it has been shown that a bacterium isolated from lake water can grow in media in which humic substances are the only source of carbon and energy (De Haan 1972). It has also been shown that a *Pseudomonas* sp. has a greater growth yield in a medium containing fulvic acid and lactate than on lactate alone (De Haan 1974). This may in part explain how humic substances at times appear to be readily decomposed. This process of co-metabolism (Horvath 1972) in fresh water is worthy of more study.

DYNAMICS OF PARTICULATE ORGANIC MATTER DECOMPOSITION

It is difficult to develop any specific statement concerning decomposition of particulate organic matter in lakes because very few studies have been conducted. The mortality rates for phytoplankton, macrophytes, bacteria, and zooplankton are not known. However, Saunders (1963) has estimated very high concentrations of bacteria in Frains Lake during September through November when littoral macrophytes were dying. Other smaller fluctuations must occur when phytoplankton populations die. The relative rate of decomposition of phytoplankton in a productive lake has been given earlier (Fig. 15.3) and that for macrophytes by Pieczynska (1972), which is comparable to decomposition rates of leaves in a productive stream. Fine particles appear to decompose relatively ten times faster than large organic particles. The organisms that decompose leaves and macrophytes in lakes may be qualitatively very different from those in streams. However, the end point must be fundamentally the same.

Leaf decomposition has been studied mostly in streams and has been described briefly in this paper. Greater detail and scope for detritus processing in streams have been reviewed by Boling *et al.* (1975) and Cummins (1974). The dynamics of large particle processing in streams is undoubtedly analogous to that in lakes. Boling *et al.* (1975) provide an input schedule for groups of leaf matter, categorized as slow, medium, and fast decomposing, into a small productive temperate woodland stream. The input of leaves occurs mainly during the autumn and of course fluctuates through a maximum. The dynamics of the substrate (leaf) concentration in an aquatic system would be analogous to the dynamics presented in Fig. 15.9 for phytoplankton extracellular release. The time scale is different because leaf material decomposes roughly ten times more slowly than algal extracellular material. The shape of the dynamics would also be different because for leaves the rate of input increases rapidly each day for two to six weeks, depending on the class of leaf. By early December leaf input is near zero per day. The leaf substrate would be broken down very slowly and in principle it is possible to describe the decomposition of each daily input. In general each class of leaf material would increase rapidly in concentration during early autumn. Then there would be a long slow decrease in concentration through the year until the following autumn. The class of leaf material decomposing at 1·5% per day would have a small residuum of material available after 3–4 months. The leaf material decomposing at 0·5% per day would still have some residuum present after one year.

In the case of the woodland stream there is a smaller amplitude input of leaves in the late spring. This should also occur in lakes. Although there were no data taken, Boling *et al.* (1975) modelled the input of fine particulate matter to the stream. The input was small in autumn and greater in the winter and spring months. Fine particulate organic matter was dominant in this model. The same pattern of input should occur in lakes. In addition, during the spring and autumn mixing periods there will be a significant redistribution of fine particulate matter within a lake (Wetzel *et al.* 1972).

DYNAMICS OF DECOMPOSITION PROCESSES IN FRESH WATER

If the evidence from lakes and streams is combined, there is a general picture that can be developed concerning decomposition in fresh water:

1 The total rate of decomposition is a general function of the nutrient level in the system, i.e., productive as compared to unproductive systems (cf. Fig. 15.3).

2 In any system decomposition rates will oscillate around the general level of activity set by nutrient input.

3 The biota that perform decomposition processes cope with increased inputs depending upon the magnitudes of increase. Decomposition activity may increase merely because the activity follows along the hyperbola for uptake velocity described by Michaelis–Menten. If there is a somewhat greater increase in substrate, the biota may increase the amount of enzyme per cell, and lastly the amount of enzyme may be increased by individual or population growth. The reverse sequence may occur when substrate levels decrease.

4 There are a larger number of particulate and dissolved organic pools of different metabolic mobility in any system. There is a small pool of readily available organic matter and larger pools of less readily available substrate.

The inputs to these pools are not in phase. The less mobile pools contain a large reserve of essential nutrients. The effect of such a concentration structure and metabolic structure is to provide both flexibility and stability within the system.

5 The molecular structure and concentration of an organic substrate do not totally control its rate of decomposition. The turnover time for glucose ranges over four orders of magnitude for the water bodies investigated but it also varies by 2–3 orders of magnitude within a single aquatic system. Given the assumption of infinite metabolic flexibility in microbes, there should be little dead organic matter in any water body unless it is metabolically intractable. It was shown that the large pools of dead organic matter are susceptible to nearly total decomposition. In general they are not totally decomposed. Therefore, other variables must also control decomposition processes and it should be a major goal in studying decomposition to determine what these control variables are and the dominance with which they operate in different systems.

Conclusion

Decomposition is a subtle process and involves a complex set of interactions and a variety of time scales. Decomposition intensities fluctuate in response to production of organic matter and also feed back on that production. It is obvious that there is only a minimal body of data available and that we have only a partial perspective concerning decomposition in a wide spectrum of types of water bodies.

The intensity of decomposition in waters of different trophic state has been examined only for glucose (Fig. 15.3). However turnover time cannot provide any information concerning the absolute rates of decomposition and of microbial community metabolism. It is therefore imperative to refine chemical analytical methods and to be able to apply these methods quickly and accurately to estimate substrate concentration. In general polymers must be hydrolysed into simple molecules before they can be assimilated by microbes. The simple organic molecules then may act as the valve between the preponderant concentration of higher molecular weight organic matter and microbial metabolism. If the absolute assimilation of the total spectrum of simple organic substrates in a water sample is measured this would at least be a measure of microbial decomposition of the total dissolved organic pool. In a fluctuating system the frequency of estimating assimilation rates would be high and for simple substrates in productive systems the sampling would be done within a 24-hour period. In unproductive systems the period of sampling might be longer. There would be no need to estimate the nonsimple substrates because the valve of simple substrates would act as a filter or channel for all microbial decomposition of dissolved material.

Some comparative studies of decomposition of larger molecules in different trophic systems would demonstrate whether these molecules were decomposed at a slower relative rate than glucose. The turnover time for this class of organic matter would then be the order of tens of thousand of hours in very unproductive systems (Fig. 15.3).

On the other hand a large particle of plant material such as a macrophyte or a leaf may decompose at the same rate in eutrophic and oligotrophic waters primarily because such material is in a very concentrated form and only needs to be populated with microbes in order to be decomposed. If such were the case this would reflect a fundamental difference in the function of productive and unproductive water bodies. In the latter, transport of energy through the particulate pathway would be relatively more important than in productive systems.

The manner in which organic matter entering a water body or produced in a water body is redistributed in that system is an important mechanism to analyse because it dictates where decomposition will take place and how organisms will be distributed. Estimates of settling and sedimentation rate of organic particles are critical to the understanding of how decomposition processes are distributed.

Quantitative estimation of decomposition in sediments will place in perspective the relation of littoral and pelagic subsystems in lake metabolism. Perhaps our present conception of their relative importance will be revised as suggested by Wetzel *et al.* (1972) and Table 15.1.

An analysis of mortality will describe the timing and the rate of production of different forms of dead organic matter. This may be used, in part, to describe the dynamics of organic pools and how decomposition processes are coupled to population oscillations.

Usually detritus and microbes are intimately combined and it has proven difficult to isolate detritus, bacteria, algae, and other particles in water samples. However, metabolic activities can be isolated kinetically using appropriate substrates (Saunders 1971, 1972a) and it may be feasible to separate particles physically by density gradient centrifugation. While the methods may be tedious this kind of detail will undoubtedly be necessary to any ultimate understanding of decomposition processes.

Descriptions of distributions or of processes are necessary but not sufficient to the understanding of decomposition. Ultimately the question must be asked as to how the distributions came about, what controls the pattern and the intensity of the process. This must lead to perturbation analysis. It is only through such analysis that decomposition processes can be accurately understood. The crude models presented above suggest the patterns and time scale for certain decomposition events. They also suggest that these events can be analysed *in situ* and that the pattern unique to each system can be resolved at a level that is nontrivial.

References

Allen H.L. (1969) Chemo-organotrophic utilization of dissolved organic compounds by planktic algae and bacteria in a pond. *Int. Revue ges. Hydrobiol.* **54**, 1–33.

Barica J. (1974) Some observations on internal cycling, regeneration, and oscillation of dissolved nitrogen and phosphorus in shallow self-contained lakes. *Arch. Hydrobiol.* **73**, 334–60.

Boling R.H., Goodman E.D., Zimmer J.V., Cummins K.W., Reice S.R., Petersen R.C. & Van Stickle J.A. (1975) Toward a model of detritus processing in a woodland stream. *Ecology* **56**, 141–51.

Brehm J. (1967) Untersuchungen uber den Aminosäure-Haushalt holsteinischer Gewässer, insbesondere des Plüss-Sees. *Arch. Hydrobiol.*/Suppl. 32, 313–435.

Cummins K.W. (1974) Structure and function in stream ecosystems. *Bioscience* **24**, 631–41.

Cummins K.W. *et al.* (1972) Organic enrichment with leaf leachate in experimental lotic ecosystems. *Bioscience* **22**, 719–22.

Deevey E.S. & Stuiver M. (1964) Distribution of natural isotopes of carbon in Linsley Pond and other New England lakes. *Limnol. Oceanogr.* **9**, 1–11.

De Haan H. (1972) Some structural and ecological studies on soluble humic compounds from Tjeukemeer. *Verh. Internat. Verein. Limnol.* **18**, 685–95.

De Haan H. (1974) Effect of a fulvic acid fraction on the growth of a Pseudomonas from Tjeukemeer (The Netherlands). *Freshwat. Biol.* **4**, 301–10.

Dickinson C.H. & Pugh G.J.F. (eds.) (1974) *Biology of plant litter decomposition*. Vols. 1 & 2. Academic Press, London.

Duursma E.K. (1961) Dissolved organic carbon nitrogen, and phosphorous in the sea. *Neth. J. Sea Res.* **1**, 1–148.

Fast A.W. *et al.* (1973) Effects of artificial aeration on the chemistry and algae of two Michigan lakes. *Water Resources Res.* **9**, 624–47.

Finenko Z.Z. & Zaika V.E. (1970) Particulate organic matter and its role in the productivity of the sea. In: *Marine food chains*, ed. J.H. Steele, pp. 32–44. University of California Press, Berkeley.

Fisher S.G. & Likens G.E. (1973) Energy flow in Bear Brook, New Hampshire: an integrative approach to stream ecosystem metabolism. *Ecol. Monogr.* **43**, 421–39.

Fogg G.E. (1966) The extracellular products of algae. *Oceanogr. Mar. Biol. Ann. Rev.* **4**, 195–212.

Frankland J.C. (1974) Decomposition of lower plants. In: *Biology of plant litter decomposition*, eds. C.H. Dickinson & G.J.F. Pugh, Vol. 1, pp. 3–36. Academic Press, London.

Gocke K. (1970) Untersuchungen uber Abgabe und Aufnahme von Aminosauren und Polypeptiden durch Planktonorganismen. *Arch. Hydrobiol.* **67**, 285–367.

Golterman H.L. (1964) Mineralization of algae under sterile conditions or by bacterial breakdown. *Verh. Internat. Verein. Limnol.* 15-544-548.

Golterman H.L. (1972) The role of phytoplankton in detritus formation. *Mem. Ist. Ital. Idrobiol.* 29 Suppl., 89–103.

HANEY J.F. (1973) An *in situ* examination of grazing activities of natural zooplankton communities. *Arch. Hydrobiol.* **72**, 82–132.

HARGRAVE B.T. (1972) Aerobic decomposition of sediment and detritus as a function of particle surface area and organic content. *Limnol. Oceanogr.* **17**, 583–96.

HOBBIE J.E. & WRIGHT R.T. (1965a) Bioassay with bacterial uptake kinetics: glucose in freshwater. *Limnol. Oceanogr.* **10**, 471–4.

HOBBIE J.E. & WRIGHT R.T. (1965b) Competition between planktonic bacteria and algae for organic solutes. In: *Primary productivity in aquatic environments*, ed. C.R. Goldman, pp. 175–85. *Mem. Ist. Ital. Idrobiol.* 18 Suppl., University of California Press.

HOBBIE J.E. *et al.* (1972) Carbon flux through a tundra pond ecosystem at Barrow, Alaska. *U.S. Tundra Biome Report* **72–1**, 1–29.

HOOD D.W. (ed.) (1970) Symposium on organic matter in natural waters. *Inst. Marine Science, Univ. Alaska, Occasional Publ.* **1**, 1–265.

HORVATH R.S. (1972) Microbial co-metabolism and the degradation of organic compounds in nature. *Bacteriol. Rev.* **36**, 146–55.

JORGENSEN C.B. (1966) *Biology of suspension feeding.* Pergamon Press, Oxford.

KAUSHIK N.K. & HYNES H.N.B. (1971) The fate of dead leaves that fall into streams. *Arch. Hydrobiol.* **68**, 465–515.

KHAILOV K.M. & BURLAKOVA Z.P. (1969) Release of dissolved organic matter by marine seaweeds and distribution of their total organic production to inshore communities. *Limnol. Oceanogr.* **14**, 521–37.

KLEEREKOPER H. (1952) The mineralization of plankton. *J. Fish. Res. Bd. Can.* **10**, 284–91.

KRAUSE H.R. (1959) Biochemische Untersuchungen über den postmortalen Abbau von totem Plankton unter aeroben und anaeroben Bedingungen. *Arch. Hydrobiol.*/Suppl. **24**, 297–337.

KUZNETSOW S.I. (1959) Die Rolle der Mikroorganismen im Stoffkreislauf der Seen. VEB Deutscher Verlag der Wissenschaften, Berlin.

LADLE M. (1972) Larval Simuliidae as detritus feeders in chalk streams. *Mem. Ist. Ital. Idrobiol.* 29 Suppl., 429–39.

LUGRIN M. (1974) *Premieres donnees sur les composes organic dissous dans le basin du Lac Leman.* Unpublished Ph.D. Thesis, Univ. Paris.

LUSH D.L. & HYNES H.B.N. (1973) The formation of particles in freshwater leachate of dead leaves. *Limnol. Oceanogr.* **18**, 968–77.

MANNY B. (1972) Seasonal changes in dissolved organic nitrogen in six Michigan lakes. *Verh. Internat. Verein. Limnol.* **18**, 147–56.

MANNY B. & WETZEL R.G. (1973) Diurnal changes in dissolved organic and inorganic carbon and nitrogen in a hardwater stream. *Freshwater Biol.* **3**, 31–43.

MELCHIORI-SANTOLINI U. & HOPTON J.W. (ed.) (1972) Detritus and its role in aquatic ecosystems. *IBP/UNESCO Symposium. Mem. Ist. Ital. Idrobiol.* 29 Suppl., 1–540.

MILLER M.C. (1972) *The carbon cycle in the epilimnion of two Michigan Lakes.* Unpublished Ph.D. Thesis, Michigan State University.

NAUWERCK A. (1963) Die Beziehungen zwischen Zooplankton and Phytoplankton im See Erken. *Symb. Bot. Upsaliensis* **17**, 1–163.

OHLE W. (1956) Bioactivity, production, and energy utilization in lakes. *Limnol. Oceanogr.* **1**, 139–49.

OLAH J. (1972) Leaching, colonization and stabilization during detritus formation. *Mem. Ist. Ital. Idrobiol.* 29 Suppl., 105–27.

OTSUKI A. & WETZEL R.G. (1974) Release of dissolved organic matter by autolysis of a submerged macrophyta, Scirpus subterminalis. *Limnol. Oceanogr.* **19**, 842–5.

PARSONS T.R. & STRICKLAND J.D.H. (1962) On the production of particulate organic carbon by heterotrophic processes in sea water. *Deep-sea Res.* **9**, 211–22.

PETERSEN R.C. & CUMMINS K.W. (1974) Leaf processing in a woodland stream. *Freshwat. Biol.* **4**, 343–68.

PIECZYNSKA E. (1972) Ecology of the eulittoral zone of lakes. *Ekologia Polska* **20**, 637–732.

POLTZ J. (1972) Untersuchungen uber des Vorkommen und den Abbau von Fetten und Fettsäuren in Seen. *Arch. Hydrobiol. Suppl.* **40**, 315–99.

POMEROY L.R. (1974) The ocean's food web, a changing paradigm. *Bioscience* **24**, 499–732.

RODHE W. (1965) Standard correlations between pelagic photosynthesis and light. In: *Primary productivity in aquatic environments*, ed. C.R. Goldman, pp. 365–81. *Mem. Ist. Ital. Idrobiol. 18 Suppl.* University of California Press, Berkeley.

RUSSELL E.W. (1968) The agricultural environment of soil bacteria. In: *The ecology of soil bacteria*, eds. T.R. Gray & D. Parkinson, pp. 77–89. University of Toronto Press.

SAUNDERS G.W. (1963) The biological characteristics of fresh water. *Proc. Great Lakes Res. Conf.* Great Lakes Res. Division Pub. **10**, 245–57.

SAUNDERS G.W. (1971) Carbon flow in the aquatic system. In: *The structure and function of fresh-water microbial communities*, ed. J. Cairns, pp. 31–45, Res. Div. Monogr. 3, Blacksburg, Virginia.

SAUNDERS G.W. (1972a) The transformation of artificial detritus in lake water. *Mem. Ist. Ital. Idrobiol.* 29 Suppl., 261–88.

SAUNDERS G.W. (1972b) The kinetics of extracellular release of soluble organic matter by plankton. *Verh. Internat. Verein. Limnol.* **18**, 140–6.

SAUNDERS G.W. (1972c) Summary of the general conclusions of the symposium. *Mem. Ist. Ital. Idrobiol.* 29 Suppl., 533–40.

SAUNDERS G.W., COFFMAN W.P., MICHAEL R.G. & KRISHNASWAMY S. (1975) Photosynthesis and extracellular release in ponds of South India. *Verh. Internat. Verein. Limnol.* **19** (in press).

SAUNDERS G.W. & STORCH T.A. (1971) Coupled oscillatory control mechanism in a planktonic system. *Nature, Lond.* **230**, 58–60.

SCHINDLER J.E. & ALBERTS J.J. (1974) Analysis of organic-inorganic associations of four Georgia reservoirs. *Arch. Hydrobiol.* **74**, 429–40.

SEDELL, J.R., TRISKA F.J., HALL J.D., ANDERSON N.H. & LYFORD J.H. (1974) Sources and fates of organic inputs in coniferous streams. In: *Integrated research in the coniferous forest biome*, eds. R.H. Waring & R.L. Edmonds, pp. 57–69, Bull. No. 5 Coniferous Forest Biome US/IBP. University of Washington, Seattle.

SEDELL J.R., TRISKA F.J. & TRISKA N.S. (1975) The processing of conifer and hardwood leaves in two coniferous forest streams: I. Weight loss and associated invertebrates. *Verh. Internat. Verein. Limnol.* **19** (in press).

SEKI H., NAKAI T. & OTOBE H. (1974) Turnover rate of dissolved materials in Philippine Sea at winter of 1973. *Arch. Hydrobiol.* **73**, 238–44.

SHARP J. (1973) Size classes of organic carbon in seawater. *Limnol. Oceanogr.* **18**, 441–7.

STEELE J.H. (1974) *The structure of marine ecosystems.* Harvard University Press, Cambridge, Mass.

STORCH T.A. (1971) *Production of extracellular dissolved organic carbon by phytoplankton and its utilization by bacteria.* Ph.D. Thesis. The University of Michigan.

TRISKA F.J. (1970) *Seasonal distribution of aquatic hyphomycetes in relation to the disappearance of leaf litter from a woodland stream.* Unpublished Ph.D. Thesis. Univ. Pittsburgh, Pennsylvania.

VACCARO R.F., HICKS S.E., JANNOSCH N.W. & CASEY F.G. (1968) The occurrence and role of glucose in sea water. *Limnol. Oceanogr.* **13**, 356–60.

WEINMANN G. (1970) Gelöste Kohlenhydrate und andere organische Stoffe in naturlichen Gewässern und in Kulturen von *Scenedesmus quadricauda*. *Arch. Hydrobiol.*/Suppl. **37**, 164–242.

WETZEL R.G. (1967) Dissolved organic compounds and their utilization in two marl lakes. *Hidrologiai Kòzlòny* **47**, 298–303.

WETZEL R.G. & HOUGH R.A. (1973) Productivity and role of aquatic macrophytes in lakes. An assessment. *Polskia Archiwum Hydrobiologii* **20**, 9–19.

WETZEL R.G. & MANNY B.A. (1972) Decomposition of dissolved organic carbon and nitrogen compounds from leaves in an experimental hardwater stream. *Limnol. Oceanogr.* **17**, 927–31.

WETZEL R.G. & OTSUKI A. (1974) Allochthonous organic carbon of a marl lake. *Arch. Hydrobiol.* **73**, 31–56.

WETZEL R.G., RICH P.H., MILLER M.C. & ALLEN H.L. (1972) Metabolism of dissolved and particulate detrital carbon in a temperate hard-water lake. *Mem. Ist. Ital. Idrobiol.* **29** Suppl., 185–243.

WILLOUGHBY L.G. (1974) Decomposition of litter in Freshwater. In: *Biology of plant litter decomposition*, eds. C.H. Dickinson & G.J.F. Pugh, Vol. II, pp. 659–81. Academic Press, London.

16 Processes in decomposition and nutrient transfer in forest systems[1,2]

M. WITKAMP and B. S. AUSMUS *Oak Ridge National Laboratory, Oak Ridge, Tennessee, 37830 U.S.A.*

Introduction

Forest ecosystems cover most of the vegetated land area on earth (Whittaker & Likens 1973) and have a higher sustained productivity than other natural terrestrial ecosystems. This high productivity is achieved and maintained through effective recycling of nutrient elements by allocating a significant part of the energy captured by primary producers to heterotrophic organisms.

Basically, forest ecosystems are nutrient element accumulating systems in which the elements are continuously recombined in a variety of organic and inorganic compounds. Solar energy captured by autotrophs is used to drive these chemical transformations through a system of consumers, predators, decomposers and nonbiotic components. The result is almost complete reuse, preservation, and accumulation of essential elements which cannot be sustained by annual increments from the mineral substrate and precipitation at rates required to maintain productivity.

We postulate that within a given climate and water regimen, neither of which can be greatly modified by forest development, the success in competition between ecosystems depends on their ability to recycle and accumulate nutrients which limit productivity. In general these will be nitrogen (N) and phosphorus (P) and potassium (K). Annual inputs of elements to detritus are primarily within roots and leaf detritus but 50 to 75% of the above-ground input of K and cesium (Cs) and about half of manganese (Mn) and magnesium (Mg) are leached from foliage by rain (Nye 1961; Wither-

[1] Research supported by the Eastern Deciduous Forest Biome, US–IBP, founded by the National Science Foundation under Interagency Agreement AG–199, BMS69—01147 AO9 with the U.S. Energy Research and Development Administration, ORNL.

[2] Contribution No. 229 from the Eastern Deciduous Forest Biome US–IBP. Publication No. 829, Environmental Sciences Division, ORNL.

spoon 1964; Duvigneaud & Denayer-DeSmet 1969; Best & Monk 1975; Bernhard-Reversat 1975; Ausmus *et al.* 1976a).

The inputs of organic matter containing the elements consists mainly of carbohydrates and lignin. Decomposition of these compounds is of little direct import to the nutrient requirements of trees as it yields primarily carbon dioxide, which is rapidly lost from the system and is usually not limiting to primary production, and little water. Complete oxidation of 300 g m^{-2} dry litter or roots yields less water than 0·2 mm of precipitation.

In this paper we will outline some of the major processes in mineral transfer from detritus to roots. We will give examples of mineral losses and subsequent losses in productivity that occur when the forest ecosystem is disrupted. Finally, we will discuss how the various mineral transfer processes prevent nutrient losses from the ecosystem, and illustrate how these processes operate differently among various types of forests in different climates.

Major processes of nutrient element transfer from surface litter to roots include leaching by rain; faunal comminution and digestion of detritus; microbial immobilization and release of mineral elements; recycling of nutrient elements through microflora and fauna; and biotic translocation of nutrient elements in rhizomorphs, roots, and soil fauna.

Decomposition and mineral transfer in the soil

LEACHING

Leaching processes affect element pathways through foliar leaching and detrital leaching. Foliar leaching inputs increase towards leaf senescence and culminate for most elements in abscission. Many monovalent cations from the canopy quickly move through the litter layer into the top soil where they may be retained by the adsorption complex (Gonzalo de las Salas 1973; Riekerk 1971) or can be directly taken up by roots in the absence of a litter layer (Went & Stark 1968).

Leaching of detritus in the field is always concurrent with biological decay (Witkamp & Frank 1969a). Evidence of mineral transfer through physical leaching alone is primarily experimental (Witkamp & Frank 1970). We carried out microcosm experiments using local litter of *Liriodendron tulipifera* L. and local forest soil. The turnover time for *Liriodendron* leaf litter in the field is 1·14 years (Ausmus & Witkamp 1974). The microcosms were leached at the local rate of precipitation of about 2·5 cm per week and maintained at 2°C to keep microbial decomposition at a minimum (Fig. 16.1). Leaching of ^{137}Cs from *L. tulipifera* L. litter microcosms increased with increasing rates of simulated rainfall (Witkamp & Frank 1970). In the

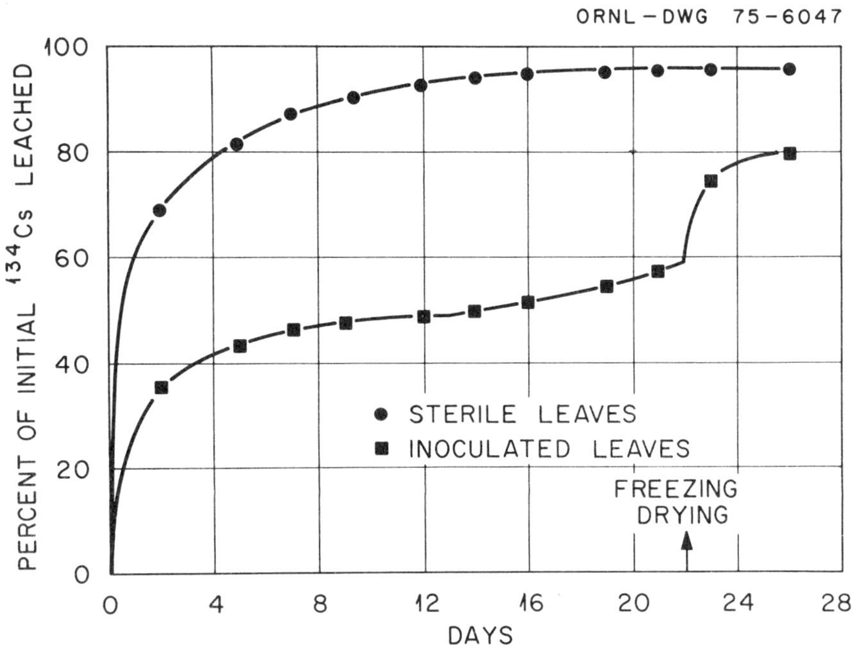

Figure 16.1. Exponential loss of ^{134}Cs from steam sterilized fragmented *Liriodendron* leaves (1 g dry wt.) upon nine successive leachings of 2·5 ml water each, and the greatly reduced losses of ^{134}Cs from identical leaves inoculated with *Penicillium* sp. at day 0. Net immobilization of ^{134}Cs in developing mycelium is indicated by the area between the two curves and increases to 43% of the initial ^{134}Cs of the leaves until day 12 after which ^{134}Cs release from senescent mycelium and breakdown of leaf tissue becomes greater than gross immobilization by the fungus. Acute killing of the mycelium on day 22 by 2 hr of freezing at −10°C and air drying releases about half of the immobilized ^{134}Cs in two subsequent leachings.

same experiment losses of ^{137}Cs and K from the continuously moist litter were significantly (P = 10%) affected by precipitation rates while losses of less soluble Mg and of C were not.

Other experiments have shown that leaf litter fragmentation increases leaching of ^{137}Cs 3·2 fold and of ^{60}Co 2·5 times over the amounts leached from entire leaves of five local tree species (Witkamp 1969). Initial leaching of ^{137}Cs from *Liriodendron* litter in microcosms was asymptotic with loss coefficients between 10 and 20% per day for about 5 to 10% of the total ^{137}Cs in the litter. After this initial pulse of easily leachable ^{137}Cs, slower release by microflora for the remaining ^{137}Cs prevailed with loss coefficients of 0·25 to 0·5% per day (Witkamp & Frank 1970).

In field experiments with three species, decomposing litter loss coefficients increased with increasing rates of leaching. For example, loss coefficients for ^{137}Cs from oak litter exposed on the forest floor, in a forest pond, and a brook were 29, 71, and 137% per month respectively. For less soluble ^{60}Co corresponding loss coefficients were smaller and increased less sharply viz. from 16 to 36 to 49% per month (Witkamp & Frank 1969a). Similar relationships were found for alder and sycamore litters.

Thus, in general, leaching from foliage will depend on the solubility of the ion and vigour of the foliage. Ions readily leached from foliage appear to pass quickly through litter into soil where they may be adsorbed into the soil exchange complex or taken up directly by roots (Gonzalo de las Salas 1973). Upon leaf fall elements remaining in cell contents are released rather quickly from the litter. Elements in structural material are released more gradually from detritus depending on the rate of decay.

FAUNAL BREAKDOWN

Decomposer fauna affects element cycling through mechanical comminution of the detritus, enzymatic breakdown of ingested matter, channelization of woody substrates, and inoculation of detritus and living roots with microbial propagules. The mechanical fragmentation of leaf litter destroys the protective cuticle; exposes cell contents; and increases surface/volume ratios, water-holding capacity, aeration, and downward mobility of particulates and solubles. Subsequent ingestion and excretion of detritus by fauna raises pH and reduces contents of fat and easily decomposable carbohydrates of the litter (van der Drift & Witkamp 1960; Nicholson *et al.* 1966; McBrayer *et al.* 1974). Nutrient element concentration of litter before and after intestinal passage is generally unchanged depending on the stage of development of the animal, but molecular forms may be different. In addition the nitrogen in enzymatic and structural protein, which may make up 60% of the plant cells (Parsons & Tinsley 1975), may be incorporated in chitin, uric acid, and animal protein. Calcium pectate of plant membranes may be transferred to oxalate and carbonate in millepede exoskeletons or snail shells.

Comminution of oak leaves by decomposer fauna in a microcosm increased CO_2 evolution from the leaf fragments seven fold over that from entire control leaves during three weeks (van der Drift & Witkamp 1960). Since the release of many elements is proportional to CO_2 evolution and loss of litter weight (Witkamp 1969), nutrient releases from faeces can be expected to increase over release from entire leaves.

Field experiments comparing litter breakdown in the presence and absence of the consumer fauna show increases in litter weight loss by one-third while half as much ^{134}Cs was retained under the effect of litter fauna (Crossley & Witkamp 1966). Energy expenditure by consumers in the same forest accounted for less than 1% of the total decomposer effort (Reichle *et al.* 1975). Thus, a small initial energy expenditure by the decomposer fauna feeding upon the detritus yielded a considerable increase in release of mineral element from that detritus.

The effect of litter consumers on subsequent decomposition is greatest for recalcitrant substrates with high C/N by facilitating microbial attack. Less recalcitrant substrate with lower C/N such as herbaceous litter is suitable for direct microbial attack, in particular by bacteria, and absence of consumers does not decrease the rate of decomposition (Malone & Reichle 1973).

Litter and soil fauna respires between 7% of the litter input in tropical forest, 4% and 6% in deciduous temperate and 4% in boreal forest (Whittaker 1974; Petersen 1974; de Boois 1974; van der Drift 1971; Kitazawa 1971). Faunal biomass declines from tropical rain forest to boreal forest by a factor of six. In some tropical rain forest there may be a rather small mass of decomposer fauna because of lack of litter caused by rapid decay (Kitazawa 1971; Lee 1972). Faunal respiration in North American temperate forest may be less than 1% of the detrital input (McBrayer *et al.* 1974) primarily because of the absence of leaf litter consuming earthworms which make up the bulk of the biomass in European temperate forest. Introduction of *Lumbricus terrestris* in a black locust (*Robinia pseudoacacia* L.) stand in Ohio lead to greater mobility of P and cations (Vimmerstedt & Finney 1973).

Faunal decomposer activity is relatively high and diverse in temperate forest with sufficient litter providing a stable environment with relatively good quality food which improves following microbial colonization. Fungivorous fauna in a *Liriodendron* forest consumes 86% of fungal net production and 99% of the fungivores in turn fell as prey to the predator fauna (McBrayer *et al.* 1974). In particular, K and P assimilation is high in the fungivores. Sodium assimilation is lower but Na has already been concentrated in fungi (McBrayer *et al.* 1974; Ausmus & Witkamp 1974). Density of both fungivorous arthropods and nematodes is positively correlated with fungal density (McBrayer *et al.* 1974). Faunal bodies as well as faeces add compounds of greater N, P, and Na concentration, and availability to microbial attack and thus add to the rate of decomposition of detritus, in particular that of high C/N such as wood (Bachelier 1973).

In contrast to the limited direct above-ground consumption of detritus, below-ground consumption may be appreciable. In a *Liriodendron* forest nematodes were calculated to consume 8·5% of the annual production of roots utilizing their high soluble carbohydrate content (Ausmus *et al.* 1976b). This mechanism shortcuts utilization of the detrital pool by adding energy and nutrients directly from living plants to the decomposer web. Above-ground, in a similar shortcut, aphids add smaller amounts of plant sap, high in sugars, directly to the litter layer.

In climates where moisture is limiting decomposition of surface litter, the soil fauna may effectively transport all the litter underground where conditions for decay are more favourable. For example, the earthworms remove leaf litter from mediterranean forests (Bouché 1972), and termites may consume close to 100% of the detrital input in savanna (Wood Chapter 6, this volume).

It thus appears that faunal decomposer activity based on detrital energy increases and diversifies mineral releases from detritus in time and space to an extent far beyond that suggested by faunal elemental or energy pools.

MICROBIAL DECOMPOSITION

Microbial attack on tree parts begins even before senescence. Leaves are invaded by microfloras corresponding to tree species and their position in the crown (Cowley 1970). Branches and standing dead may show considerable decay before falling (Chesters 1950) and roots are invaded by fungi and bacteria well before they slough (Bowen & Rovira 1968).

Once fallen, leaf litter decays at rates averaging from 485% per year in tropical deciduous forest (Malaisse *et al.* 1975) to less than 20% in tundra (Heal & French 1974). In the early stages of the microbial succession associated with decay, bacteria, yeasts and mucoraceous fungi using easily decomposable and soluble sugars, amino and other organic acids immobilize much of the soluble cations thereby causing a drop in pH. This pH drop, together with the fact that only the more resistant substrates remain, favours fungi for continued colonization. With increasing narrowing of C/N and C/P ratios, increasing amounts of senescent fungal hyphae, and improved moisture conditions in the decaying litter results in increasing bacterial densities (Witkamp 1974; Ausmus & Witkamp 1974). Greatest microbial mass in surface litter of our *Liriodendron* forest is as much as 100 g m^{-2}, 80% of which is from fungi. In general, decomposer cells may immobilize considerable amounts of N, P, K (Witkamp & Frank 1970; Ausmus *et al.* 1976a). The mineral elements retained in the microbial cells are protected from leaching (Witkamp 1969). Nitrogen that is not microbially immobilized may be rapidly lost from the system (Nömmik & Popovic 1971).

Potential element immobilization on leaf litter and soil substrates is best known from microcosm studies with *Liriodendron* litter. Microbial immobilization of mineral elements in microcosms may account for 40 to 80% of the retention of easily leachable elements such as Cs (Fig. 16.1) and K. Immobilization of N and P may be of the same order. The high percentages of elements immobilized are a result of a microbial bloom in the newly established microcosms. In deciduous forest the percentage of element immobilization in litter is in general lower than that measured under the optimized conditions of the microcosm. Lower values in the field during summer may be caused by interrupted microbial activity following drying, and in particular by faunal consumption of microbial standing crop (Witkamp & Barzansky 1968).

The effect of fauna on element immobilization became evident when addition of millipedes to *Liriodendron* litter microcosms lowered microbial ^{137}Cs immobilization from 36·0 to 10·8% of the original litter content (Witkamp & Frank 1969b). Addition of snails further decreased ^{137}Cs immobilization even though corresponding respiration rates measured as CO_2 evolution increased respectively from 300 to 400 to 660 μl CO_2 hr^{-1}. Apparently the consumers stimulated microbial activity but decreased microbial standing crop and immobilization. Since half times (T/2=the time required for loss of half of the element from a compartment) were 2·5 months for microflora alone and only 6·6 days for microflora and millipedes combined, faunal consumption of microflora speeds up element turnover.

Freezing and thawing, wetting and drying, and faunal consumption all speed the release of immobilized elements (Witkamp 1969). In the field, net release of several elements from microbially decomposed matter is proportional to loss of substrate (Witkamp 1969). The initially constant release coefficient may later decrease when only the most recalcitrant litter components and breakdown products remain (Minderman 1968).

The balance of the initial immobilization and subsequent release is a net release of mineral elements in forms suitable for root uptake. Some of this release will be taken up by plant roots and the soil exchange complex, while the rest may be recycled within the decomposer compartment. Effectively, these processes minimize losses from the ecosystem of elements whose absence would severely limit primary production.

RECYCLING OF NUTRIENT ELEMENTS WITHIN THE DECOMPOSER COMPARTMENT

Neither litter comminution nor enzymatic breakdown affect the transfer of nutrient elements independently. Comminution stimulates microbial growth and microbial growth and decay stimulate faunal consumption. There are a variety of interactions among decomposer organisms and the substrate that tend to conserve nutrient elements within the decomposer compartment and diversify their release in time.

Mineral transfers between the micro-organisms have not been measured directly in the field but there is descriptive and indirect analytical evidence of such transfer among litter micro-organisms. The decline of initial bacterial populations on soluble and easily decomposable derivatives of freshly fallen litter may be accompanied by an increase in predaceous protozoa

(Habte & Alexander 1975) which in turn may be preyed upon by nematodes which may fall prey to nematode trapping fungi, gamasid mites, or bacteria. Actinomycetes and bacteria decompose exoskeleton and tissues of mites; bacteria and actinomycetes decompose senescent hyphae (Novogroudsky 1948), and bacteria cause final decomposition of most biota and their decomposition products.

Indirect analytical evidence of mineral recycling among microbiota is found in the half-time for ^{137}Cs in the microbial pool in forest microcosms. Half-times are of the order of 2·5 months which, under the conditions of the rapidly developing microflora of the microcosm, far exceed the life span of the individual microbes. Further evidence of nutrient cycling within microbiota is found in the continued reduction of C/N and C/P ratios of a decomposing substrate with its decomposer population. C/N ratios in wood may be initially as high as 1,000 and decline to around 10 in humus, reflecting repeated recycling and preservation of better than 99% of N while C is lost from the system in respiration. Assuming one third utilization of C (Waksman 1952) and complete reuse of catabolized N, then one population turnover would reduce a C/N of 100 to 33; the next turnover would yield a C/N of 11. In fact there are no complete turnovers of entire populations and no 100% N recoveries but the example shows that population half-times in the field must be much longer than the half-hour generation times of the laboratory. These longer half-times in the field agree independently with direct microbial observations and numerical population growth estimates (Clarholm 1973; Parinkina 1974) and substrate potential for population turnover (Clark 1967).

Recycling of nutrient elements within the decomposer compartment also takes place between fauna and microflora. Highly developed forms are the fungal gardens of termites and reingestion of faeces by coprophages. Coprophagous fauna recycles their food by utilizing gut flora as well as microflora that develops outside the body on the faeces. Rabbits produce soft faeces with 25% protein and high P levels in addition to the hard pellets. Nitrogen assimilation by the animals may be reduced by 50% when reingestion of the soft faeces is prevented (McBee 1971). Seemingly, coprophagy is an efficient mechanism to preserve nutrient elements and to increase digestibility of poor food (McBee 1971). In our *Liriodendron* forest *Apheloria montana*, a litter-feeding millipede, is coprophagous. The animals lack cellulases and gain little weight (1·1% in 30 days) when fed leaf litter directly even though they increase their ingestion rate by 60% over that of controls whose diet included faeces and who gained 16% weight over the same period. Apparently microbial conversion of unavailable substrate carbohydrates to available microbial organics and transfer of nutrient elements from unavailable to available food sources enabled the animals to maintain energetic and elemental balance (Na, Ca, K, P) by ingesting faeces and associated microflora (McBrayer 1973).

Effects of decomposer invertebrates on woody substrate decomposition rates may be the most significant of invertebrate microflora interactions. The tunnelling activities of termites and passalid beetles in fallen boles and the subsequent microbial succession greatly enhances wood incorporation into soil organic matter. The rate at which wood decays may depend on the microbial-faunal interactions which concentrate nitrogen in the wood by translocation from litter and soil via fungal mycelia and by nitrogen fixation.

The various interactions among decomposer organisms and their recycling of mineral elements is an ecosystem process that prevents sudden losses and diversifies release time and release rates of these elements for subsequent root uptake or preservation by the soil–humus adsorption complex. The fuel for this function is that part of solar energy captured by the system that becomes available to decomposers as detritus.

MINERAL MOVEMENT THROUGH SOIL

The movement of minerals through the soil profile is usually pictured as downward transport of mineral elements by water. In fact in most forests there is little downward water transport during the growing season when temperatures and evapotranspiration are high. Even in the tropical rain forest with abundant precipitation, only 1% of Strontium (Sr) and Cs moved from the litter layer to a depth greater than 12·5 cm in six months (Jordan 1970) indicating little water transport. Transport of nutrients in soil may rather result from water uptake by roots and diffusion following mineral uptake by the roots. In nontropical forest during autumn and winter there will be net downward movement of nutrient elements along with water particularly along channels made by animals and roots.

In a series of microcosm experiments using field materials and realistic temperature and moisture conditions we investigated the effects of litter-leached microbial products and other organic compounds on element transfer. When ^{137}Cs, ^{60}Co (Cobalt) and ^{54}Mn cations are added to local soil only about 1% leached to a depth of 2·5 cm during 50 days of simulated rain. When the water had first percolated through slowly decomposing litter from three to six times more of the ions moved down to 2·5 cm most likely complexed with soluble organic materials and thus protected against adsorption on to the soil's cation exchange complex. This assumption is further supported by the greater effect of leaf leachate on ^{60}Co and ^{54}Mn which are less soluble and more effectively complexed than on ^{137}Cs which is highly soluble and mostly in ionic state (Table 16.1).

As a result of their increased transfer these complexed elements were taken up by grass rooting 'below' the 2·5 cm zone in three to fifteen times greater amounts. Again the effect was greater for ^{60}Co and ^{54}Mn than for ^{137}Cs. From these experiments it would appear that the organics released from surface litter aid in effective transfer of elements from surface detritus to primary producers.

Cation mobilization by *Alnus*, *Carpinus* and *Quercus* extracts was reported by Lossaint (1959). Migration of metals complexed with organics appears to be faster at temperatures around 0°C than at higher temperatures when soil microflora may attack the complexes and immobilize the metal ions (Dommergues & Mangenot 1970).

Table 16.1. Percent of nuclide ions from tree litter (average of three species) or from aqueous solution that during seven weeks of leaching (2·5 cm per week) drained through 2·5 cm of forest soil (average of two) and was taken up in grass on the same soils.

	Through soil		In grass	
	Litter	Water	Litter	Water
^{137}Cs	2·5	0·7	0·6	0·17
^{60}Co	6·0	1·1	4·6	0·30
^{54}Mn	7·0	1·5	1·5	0·27

Increased cation mobility through acidification of the soil solution is well documented (Boquel & Suavin 1974; Dommergues & Mangenot 1970). Special cases of this phenomenon in forest soil are the reduction of pH by leachate from acid litter and by dissolved CO_2 from detritus decomposers and root respiration. After 15 weeks of simulated rain, pine litter at 2°C loses 25% of its ^{54}Mn content in acid leachate (pH 3·7) while basic leachate (pH 8·4) of black locust litter removes only 2·5% of the ^{54}Mn of the locust leaves (Witkamp & Ausmus 1975). Total cation (Na+K+Mn+Ca) mobility in the B horizon of a Douglas Fir plantation was highly correlated ($r=0{\cdot}97$) with dissolved CO_2, apparently because hydrogen ions of the bicarbonate replace cations from the soil exchange complex (McColl & Cole 1968). There is also evidence that the bicarbonate ion may be taken up by roots to compensate for exudation of organic acids (Jacoby & Laties 1971).

It appears from these field observations and microcosm experiments that bicarbonate ions and organic decomposition products of the detritus expedite effective translocation of mineral elements from the detritus to roots for renewed uptake by the primary producers.

UPWARD MOVEMENT

The general concepts of mineral transfer are a net downward movement under the effect of mass transport in water, particle transport in small organic fragments of faeces by gravity, or incorporation in deeper layers by fauna. These concepts have hindered the idea that minerals may also move upward. There is much evidence that this is the case. First there is a large underground input from roots. During growth roots exude amino acids providing nitrogen to the soil pool. Upon sloughing the annual mineral input of roots should equal or exceed input from litter. Annual production and turnover of roots $<0{\cdot}5$ cm diameter in *Liriodendron* forest was estimated as $9\,t\,ha^{-1}$ (Harris *et al.* 1973). This root contribution to detrital mass is one-and-three-quarter times as large as all above-ground inputs and roots provide 2·3 times more N to the forest detrital pool than all other inputs. This large input and subsequent movement to the surface by aqueous or gaseous diffusion in tissues and casts of burrowing soil fauna, or in fungal mycelium and rhizomorphs may help explain numerous reports on absolute N increases in litter, in particular those poor in nitrogen, and in wood at the litter surface (Anderson 1973; Wood 1974; Remacle & Vanderhoven 1972). In the past much of this nitrogen increase in leaf litter has been ascribed to N fixation even though pH levels made this unlikely. Our unpublished data indicate considerable initial N fixation in woody substrate. This may provide initial N input to decomposers while subsequent development of mycelium and rhizomorphs and earthworm invasion may transport N upward from soil and leaf litter to surface wood.

More direct evidence of mineral movement from soil to surface litter comes from our simulation models of data from microcosm experiments on leaf decomposition. Cesium—137 which usually is strongly adsorbed on to soil, had a monthly gross transfer coefficient from soil to litter of 17·2%. The addition of millipedes to the microcosm increased the coefficient to 20·8% per month and with millipedes and snails the coefficient was 22·4% per month. Apparently this fauna was importing ^{137}Cs from soil through either direct contact or via faeces (Witkamp & Frank 1969b).

Other evidence of upward movement of ^{137}Cs from soil to litter was found in the field when untagged litter was exposed on soil in a *Liriodendron* dominated forest tagged with ^{137}Cs. In 6 weeks leaves with little or no microbial activity accumulated some ^{137}Cs ($3{\cdot}6\,\mu pm\,mg^{-1}$ dry wt^{-1}) but accumulation by identical leaves with active microflora ($22\times$ more active) was 2·4 times greater, suggesting active transport of ^{137}Cs from soil to litter (Witkamp & Barzansky 1968).

In our *Liriodendron* forest movement of nitrogen from the large input of root detritus ($900\,g\,m^{-2}\,yr^{-1}$ at 0·74% N) to the considerable annual input of wood detritus (230 $g\,m^{-2}$ at 0·25% N) with N (C/N>25) limiting for microbial attack may occur via rhizomorphs, decomposer fauna and possibly diffusion, gaseous transfer, and splashing. Apparently a variety of biotic and abiotic mechanisms translocate elements along gradients of decreasing element concentration and expedite decomposer activity where concentrations of these elements may otherwise be limiting to decomposition processes.

The upward transport of nutrient elements from roots and humus soil pools can be an important mechanism for continuous litter decomposition in particular at times of nutrient deficiency or in surface detritus lacking sufficient nutrients.

NUTRIENT ELEMENT UPTAKE BY ROOTS

Selective uptake of nutrients by roots, especially of P, appears to be highly efficient. The effectiveness of anion transfer from detritus to plants was well illustrated in the earlier mentioned microcosm experiment on the transfer of ^{32}P from litter to grass. Within six weeks of incubation only 9% of the P was released from the litter and its microflora. Half of this was taken up by the grass. This in contrast to ^{54}Mn, 22% of which was released from identical litter and only 1·5% of which was taken up grass. Uptake efficiency of P, possibly also in organic form, is further increased by the mycorrhizal associations between fungi and tree roots.

More effective uptake of nutrient elements by trees with mycorrhiza than by trees without is well known from the literature (Skinner & Bowen 1974; Mosse 1972; Trappe & Fogel in prep.; Bowen 1973). We found in a microcosm experiment that pine seedlings (*Pinus taeda* L.) with mycorrhiza took up 12 times more ^{54}Mn from leachate of decomposing pine litter than seedlings without mycorrhiza (Witkamp & Ausmus 1975). Greater absorptive area and ability to adsorb organic compounds containing nutrient element of mycorrhiza roots over non-mycorrhizal roots are postulated to reduce nutrient losses from the system.

Mycelia of many of the mycorrhizal fungi are also active in the decomposition of detritus, thus providing a direct link between detrital nutrient elements and roots via hyphae, hyphal strands, and rhizomorphs (Miller & Laursen 1974). This link is most evident in the tropical rain forests, where moisture and temperature are sufficiently favourable for fungal activity to decompose litter at rates that prevent accumulation of surface litter and soil organic matter. As a result, a dense root mat with associated mycorrhizal fungi may lie on top of the mineral soil with only a thin discontinuous layer off leaf litter over the roots. The input of high energy materials along with the nutrient elements of detritus supports survival and function of mycorrhizal fungi. Mycorrhiza increase uptake and consequently decrease loss of nutrient elements from the system.

Nutrient cycling in ecosystems

BENEFITS TO FOREST ECOSYSTEMS DERIVED FROM NUTRIENT RECYCLING

The benefits of nutrient recycling are obviously derived from a reduced need to explore a large soil area for freshly weathered minerals. At the same time detrital energy permits decomposer fungi to explore sites of primary mineralization that are inaccessible to roots. Experience with vegetation on virgin soil, that is soil without organic matter or that has been deprived of humus and nutrients such as under the kudzu vine (Witkamp *et al.* 1966), or the irreversibly damaged tropical forest stands (Went & Stark 1968) illustrate the inability of such soil to maintain forest type productivity.

Additional benefit of the multitude of biotic processes transferring nutrient elements at different rates and along different pathways are the more gradual nutrient release (Ausmus *et al.* 1976a) and increased system stability (De Angelis 1975). Our previous examples have identified diversifications of nutrient element release rates and pathways through faunal comminution of detritus, microbial and faunal immobilization, release from decomposer biomass by senescence, death or predation, coprophagy, complexing with soluble organics, adsorption or incorporation into soil humus and transfer in fauna and fungi. Decomposer activity and resultant releases of nutrient elements have a biotic basis and, at least at the same latitudes and altitudes, are similarly related to 'favourable' conditions of moisture and, in particular, temperature as is primary production and its requirement for these nutrient elements.

External nutrient cycles, which use the energy content of plant litter to preserve and resupply nutrients to primary producers via heterotrophic organisms, thus appear to be effective. A complementary internal recycling processes in plants is the withdrawal of nutrients and energy resources from moribund leaves, that do no longer capture sufficient solar radiation to make up for their maintenance cost and the redistribution of these resources to tissues with a positive energy balance. These processes are most evident in conifers which generally translocate a greater percentage of their nutrient elements from needles to wood and bark than deciduous trees withdraw from senescent leaves (Duvigneaud & Denayer-DeSmet 1969; Nelson *et al.* 1970). This results in higher C/N and C/P ratios in coniferous detritus which, coupled with the high lignin and resin content of needles, results in slow decomposition and low rates of nutrient release. In some areas this can lead to chronic litter accumulation, bog formation and eventually the loss of the coniferous forest ecosystem.

The evolved periodic senescence of plant leaves and roots, in particular those of deciduous trees, appears to be an effective way to maintain system productivity levels well above those that could be achieved without external recycling. A key to the sustained high primary productivity of forest ecosystems is that this production is based on accumulated nutrient capital rather than on annual renewable resources supplied by wet and dry fallout and weathering of parent material and soil.

NUTRIENT LOSSES FROM DISRUPTED ECOSYSTEMS

Another way of illustrating the effectiveness of external nutrient recycling as a maintenance property of the forest ecosystem is by observing what happens when it is disrupted. Invariably disruption of the closed cycling system leads to nutrient losses from the ecosystem. The deforestation of an experimental watershed in New England increased N export in streams from 0·9 mg l^{-1} before to 53 mg l^{-1} after cutting (Likens *et al.* 1969), or changed an organic matter increment of 4·5 kg ha^{-1} yr^{-1} to a loss of 57·0 kg ha^{-1} yr^{-1} (Likens *et al.* 1969, 1970; Marks & Bormann 1972). The cutting of tropical forest also led to increased litter breakdown, mineralization and subsequent losses of nutrient elements (Edmiston 1970; Witkamp 1970; Gonzalo de las Salas 1973). In other Brazilian and African studies increased N loss was recorded following the clearance of tropical rain forest (Bernhard-Reversat 1974). The clearing of Amazon rain forest may lead to irreversible loss of nutrients and humus because of increased mineralization and absence of concomitant root uptake. The minerals weathered and accumulated for thousands of years of ecosystem development are lost in a few years and scrubland or other degraded ecosystems can become established. Revegetation may be difficult even when nutrients are artifically replaced by fertilizer treatment (Went & Stark 1968).

Destruction of Appalachian forest by sulphur dioxide fumes from copper smelters led to the erosion of an entire forest profile and the loss of all the accumulated nutrients. Attempts to revegetate this area with kudzu vine (*Pueraria lobata*, an oriental nitrogen fixing legume), accompanied by liberal addition of fertilizer (NPK) failed after initial success because the nutrients were not recycled. The harsh desert like climate at the surface prevented microbial activity because of lack of moisture. Fauna was slow to follow because of lack of inoculum. As a result within ten years 80% of the biologically available P and N and about as much K had accumulated in litter (Witkamp *et al.* 1966). In contrast virgin forest even in tropical high rainfall areas may lose only in the order of 1% of their annual nutrient input (Turvey 1974).

Apparently the evolved mechanisms for preservation and secular accumulation of nutrient elements based on detrital energy and breakdown products result in an essentially closed system for these elements in undisturbed forest ecosystems. Disruption of the system can destroy element preservation and accumulation mechanisms resulting in nutrient losses, subsequent decreased primary productivity, and in extreme cases degradation or total loss of the system.

Concepts of nutrient cycling in forests

Diversification of the pathways along which nutrients lost from primary producers become available for renewed root uptake is presumed to stabilize the ecosystem as failure of a single pathway becomes less critical to sustained mineral recycling. In general, primary production (and consequent nutrient requirements by the roots of primary producers) and decomposer activity (and subsequent release of nutrients from turnover of the microbial population) are similarly affected by temperature and moisture. The resulting annual synchrony in nutrient release from detritus and uptake by plants for a variety of ecosystems shows consistency along altitudinal and latitudinal gradients. There is less release of nutrient elements in the cold or dry climates when roots are generally less active than in warmer climates with more rapid primary production and root uptake. The synchrony between release of nutrient elements from detritus and their uptake by roots is not directly proportional. Decay rates decrease more rapidly than rates of litter input with increasing latitude (Heal & Chapman 1972). As a result there is a decreasing ratio of litter input to standing crop of litter with increasing latitude (Kononova 1975). Decay rates and consequent release rates of nutrient elements may decrease twenty-fivefold from tropical rain forest to boreal forest while primary productivity and associated root uptake and detrital inputs differ by a factor of 2·5 (Whittaker & Likens 1973) to 3 (Bray & Gorham 1964).

One probable reason for the smaller difference in primary productivity than in decay rates is that the relationship between primary production and temperature appears to be linear (Whittaker & Woodwell 1971) or asymptotic (Schultze & Koch 1971) as contrasted with the $Q_{10} \simeq 2$ (de Boois 1974) or greater (Flanagan & Veum 1974) exponential relationship that is usually found for decomposer activity along the prevailing range of temperatures. In temperate forest with litter decay completed in about one year nutrient release from detritus will equal root uptake requirements. In colder climates primary production and detrital input will exceed decay. Here litter and humus accumulate to levels at which a new equilibrium is established between rates of detrital input and decay. In warmer temperate climates and in the tropics the potential for detrital decay exceeds the annual input. By recycling mineral elements more than once a year, productivity can be increased to rates well above those in boreal and cool temperate zones (Klinge *et al.* 1975) even on nutrient deficient soils (Gonzalo de las Salas 1973). Thus, rates of litter decay appear to control rates of nutrient release and, since nutrients are probably always limiting, of primary productivity of forests in different climates.

Optimal release of nutrient minerals from detritus in boreal forest is somewhere between litter and humus layers, and here the majority of root mycorrhiza and fungal mycelium is located. The inherent acidity of the coniferous litter, possibly reinforced by the predominant ammonia—nitrogen economy which may cause roots to excrete H^+ to maintain their electrical balance (Flint & Gesper 1974), and the high lignin and wax content of the litter led to podzolization and accumulation of a large pool of slowly decomposing humus. This pool contains over 90% of the systems nutrient element content (Rodin & Bazilevich 1967) either as part of the organic compounds, in particular N, or adsorbed on to the greatly increased cation exchange capacity of the humus over that of the soil and fresh litter. Thus, deep leaching of nutrient elements is prevented and stability of the system is improved by the presence of a large slowly exchanging pool (Ovington 1965; Jordan *et al.* 1972; O'Neill *et al.* 1975).

The rapid decay of detritus in tropical rain forests, on the other hand, leads to almost complete oxidation of the organic compounds and uptake of the nutrient elements by primary producers. This results in nutrient element storage in the biomass of primary producers in particular in trunks and boles (Jordan *et al.* 1972; Longman & Jenik 1974). The tropical forest therefore contains a large and slowly recycling pool of nutrient elements which contributes to the systems stability (Jordan & Kline 1972; Child & Shugart 1972). Removal of this above-ground pool by human activity is more likely than removal of the soil nutrient pool from boreal forest. Moreover the stability of rapidly recycled elements, in particular the essential elements, was found in model simulations to be less than that of non-essential elements (Jordan *et al.* 1972). The tropical forest is therefore more vulnerable to human interference, in particular to modern agriculture practice, than other forest types and needs careful protection against over-exploitation.

Mineral transfer mechanisms in temperate forest take an intermediate position with, in general, smaller nutrient element pools in biomass and soil organic matter than either tropical rain forest or boreal forest respectively. The stability of temperate forest against peturbation stems from both these pools as well as from greater diversification of functions and pathways than in either tropical or boreal forest. In particular, the regulating and stabilizing role of decomposer fauna (O'Neill *et al.* 1975) is optimal as a result of continuous availability of palatable leaf litter. This is in contrast to lack of litter in tropical rain forest limiting litter feeding populations (Kitazawa 1971) and the low nutritional value, high resin content and low temperatures that limit litter consumption by fauna in boreal forest.

The decreasing ratios of primary productivity over decomposition with increasing temperatures are confounded by differences in litter decomposability. In warmer climates C/N values and lignin concentration of litter and root detritus tend to be lower than in cold climates (Jenny 1950). This trend may well be a result of the availability of much of the systems nitrogen in the short detritus-to-primary producer cycle in tropical forest, while in boreal forest most of the nitrogen has been accumulated in the slow cycle of soil humus.

The differences in cold and warm climate nutrient cycles may also have led to the greater development in tropical forests of external recycling via decomposers while internal recycling by nutrient translocation may prevail in conifer dominated boreal forests (Duvigneaud & Denayer-DeSmet 1969).

Conclusions

In conclusion, the decomposer organisms appear to control nutrient element supply and, since this supply is limiting, primary productivity. Decomposer organisms at the same time prevent mineral losses and thus allow secular nutrient accumulation which led to the dominant position of forest ecosystems on the earths vegetated land areas. This position is maintained by the decomposer-controlled release of nutrients. Forest stability, at least in temperate and boreal forests, is also to a large extend decomposer-controlled (O'Neill *et al.* 1975). Since decomposer activity is fuelled by a part of the solar energy captured by the trees the allocation of energy to detritus appears to benefit forest stability as well as productivity (Reichle *et al.* 1976).

The preceding concepts are based on oversimplifications of the real world. The authors are well aware of this but believe that the indicated trends at this time provide a unifying working hypothesis. In this hypothesis forests are viewed as nutrient accumulating systems. Consequently major competition is for nutrients and major utilization of captured solar energy in the form of respiration and detrital inputs are underground where the competition occurs.

The mechanisms evolved to preserve and recycle nutrient elements are primarily driven by detrital energy. Their multiplicity helps to stabilize the system. Over different climates decreasing ratios of detrital input/decay with increasing temperatures are a result of lesser temperature response by primary producers than by decomposers. As a result there is decreasing underground storage of the accumulated nutrient inventory with increasing temperature and a parallel shift in the site of ecosystem stability.

Within a climate the difference in temperature response by primary producers and decomposers may cause a periodicity shift between root and decomposer utilization of the detrital nutrient pools (Ausmus *et al.* 1976a).

References

ANDERSON J.M. (1973) The breakdown and decomposition of sweet chestnut (*Castanea sativa* Mill) and beech (*Fagus silvatica* L.) leaf litter in two deciduous woodland soils. II. Changes in the carbon, hydrogen, nitrogen and polyphenol content. *Oecologia* **12**, 275–88.

AUSMUS B.S., EDWARDS N.T. & WITKAMP M. (1976a) Microbial immobilization of carbon, nitrogen, phosphorus and potassium: implications on forest ecosystem processes. This volume, Chapter 17.

AUSMUS B.S., FERRIGINI R. & MCBRAYER J.F. (1976b) Microcosm assessment of the role of litter feeding fauna on decomposition processes. *Soil Biol. Biochem.*

AUSMUS B.S. & WITKAMP M. (1974) *Litter and soil microbial dynamics in a deciduous forest stand.* Eastern Deciduous Forest Biome–IBP–73–10.

BACHELIER G. (1973) Action de la faune du sol sur l'humification des materiaux végétaux. *Rev. Écol. Biol. Sol* **10**, 453–73.

BERNHARD-REVERSAT F. (1975) Nutrients in throughfall and their quantitative importance in rain forest mineral cycles. In: *Tropical Ecological Systems*, eds. F.B. Golley & E. Medina, pp. 137–59. Springer, Heidelberg.

BEST G.R. & MONK C.D. (1975) Cation flux in hardwood and white pine watersheds. In: *Proc. Symp. Mineral Cycling in Southeastern Ecosystems.* eds F.G. Howell *et al.* pp. 847–61. CONF. 740513, U.S. Dept. Commerce, Springfield, Virginia 22151.

BOOIS H.M. DE (1974) Measurement of seasonal variation in the oxygen uptake of various litter layers of an oak forest. *Pl. and Soil* **40**, 545–55.

BOQUEL G. & SUAVIN L. (1974) Solubilisation du fer par deux souches bactériennes en présence de litiére de Teck. *Rev. Écol. Biol. Sol* **11**, 187–95.

BOUCHÉ M.B. (1972) *Lombricides de France-Ecologie et Systematique.* Institut National de la Recherche Agronomique 149 rue de Grevelle 75–Paris-7e.

BOWEN G.D. (1973) Mineral nutrition in ectomycorrhizae. In: *Ectomycorrhizae—their ecology and physiology*, eds. G.G. Marks & T.T. Kozlowski, pp. 151–205. Academic Press, New York.

BOWEN G.D. & ROVIRA A.D. (1968) The influence of micro-organisms on root growth and metabolism. In: *Root growth*, ed. W.J. Whittington, pp. 170–201. Butterworths, London.

BRAY J.R. & GORHAM (1964) Litter production in forests of the world. *Advan. Ecol. Res.* **2**, 101–57.

CHESTERS C.G.C. (1950) On the succession of microfungi associated with the decay of logs and branches. *Trans. Lincolnshire Naturalist Union*, 129–35.

CHILD G.I. & SHUGART H.H., JR. (1972) Frequency response analysis of magnesium cycling in a tropical forest ecosystem. In: *Systems analysis and simulation in ecology*, ed. B.C. Patten, pp. 103–35. Academic Press, New York.

CLARHOLM M. (1973) Direct counts of bacteria in tundra peat for estimating generation time and biomass production. *IBP Swedish Tundra Biome Project Techn. Rep.* **16**, 43–55.

CLARK F.E. (1967) Bacteria in Soil. In: *Soil Biology*, eds. A. Burges & F. Raw, pp. 15–49. Academic Press, London.

COWLEY G.T. (1970) Vertical study of microfungal populations on leaves of *Dacroydes excelsa* and *Manilkara bidentata*. In: *A tropical rain forest*, eds. H.T. Odum & R.F. Pigeon, pp. F 41–2. TID–24270. U.S. Dept. Commerce, Springfield, Virginia 22151.

CROSSLEY D.A., JR. & Witkamp M. (1966) The role of arthropods and microflora in breakdown of white oak litter. *Pedobiologia* **6**, 293–303.

DE ANGELIS D.L. (1975) Stability and connectance in food web models. *Ecology* **56**, 238–43.

DOMMERGUES Y. & MANGENOT F. (1970) *Ecologie Microbienne du Sol.* Masson, Paris, 769 pp.

DRIFT J. VAN DER (1971) Production and decomposition of organic matter in an oakwood in the Netherlands. *Proc. Symp. Productivity forest ecosystems of the world. IBP-UNESCO.*

DRIFT J. VAN DER & WITKAMP M. (1960) The significance of the breakdown of oak litter by *Enoicyla pusilla* Burm. *Arch. Neerland. Zool.* **3**, 489–92.

DUVIGNEAUD P. & DENAYER-DESMET S. (1969) Biological cycling of minerals in temperate deciduous forest. In: *Analysis of temporate forest ecosystems*, ed. D.E. Reichle, pp. 199–225. Springer, New York.

EDMISTON J. (1970) Soil studies in El Verde rain forest. In: *A tropical rain forest*, eds. H.T. Odum & R.F. Pigeon, pp. H79–87 TID 24270 U.S. Dept. Commerce, Springfield, Virginia 22151.

FLANAGAN P.W. & VEUM A.K. (1974) Relationships between respiration, weight loss, temperature and moisture in organic residues in tundra. In: *Soil organisms and decomposition in tundra*, eds. A.J. Holding *et al.*, pp. 249–77. Swedish IBP Committee, Stockholm.

FLINT P.S. & GERSPER P.L. (1974) Nitrogen nutrient levels in arctic tundra soils. In: *Soil organisms and decomposition in tundra*, eds. A.J. Holding *et al.*, pp. 375–87. Swedish IBP Committee, Stockholm.

GONZALO DE LAS SALAS F. (1973) Eigenschaften und Dynamik eines Waldstandortes im Grenzbereich des immergrunen tropischen Regenwaldes im mittleren Magdalenatal (Kolumbien). Göttinger Bodenk. Ber. **27**, 1–106.

HABTE M. & ALEXANDER M. (1975) Protozoa as agents responsible for the decline of *Xanthomonas campestris* in soil. *Appl. Microbiol.* 159–64.

HARRIS W.F., GOLDSTEIN R.A. & HENDERSON G.S. (1973) Analysis of forest biomass pools: Annual primary production and turnover of biomass for a mixed deciduous forest watershed. In: *Proceedings working party forest biomass IUFRO*, ed. H. Young, pp. 41–64. University of Maine Press, Orono.

HEAL O.W. & CHAPMAN S.B. (1972) Soil fauna and decomposition. In: *Soil fauna and decomposition processes*, ed. D. Parkinson, pp. 13–40. IBP/PT theme 8 Louvain, Belgium. University of Calgary Press, Calgary, Alberta.

HEAL O.W. & FRENCH D.D. (1974) Decomposition of organic matter in tundra. In: *Soil organisms and decomposition in tundra*, eds. A.J. Holding *et al.*, pp. 279–309. Swedish IBP Committee, Stockholm.

JACOBY B. & LATIES G.G. (1971) Bicarbonate fixation and malate compartmentalization in relation to salt induced stoichiometric synthesis of organic acid. *Plant Physiol.* **47**, 525–31.

JENNY H. (1950) Causes of the high nitrogen and organic matter in temperature and tropical regions. *Soil Sci.* **69**, 63–9.

JORDAN C.F. (1970) Movement of ^{85}Sr and ^{134}Cs by the soil water of a tropical rain forest. In: *A tropical rain forest*, eds. H.T. Odum & R.F. Pigeon, pp. 201–4. TID-24270. U.S. Dept. Commerce, Springfield, Virginia 22151.

JORDAN C.F. & KLINE J.R. (1972) Mineral cycling: some basic concepts and their application in tropical rain forest. *Ann. Rev. Ecol. System.* **3**, 33–50.

JORDAN C.F., KLINE J.R. & SASSCER D.S. (1972) Relative stability of mineral cycles in forest ecosystems. *Am. Nat.* **106**, 237–53.

KITAZAWA Y. (1971) Biological regionality of soil fauna and its function in forest ecosystem types. In: *Proc. Symp. Productivity forest ecosystems of the world.* IBP-UNESCO.

KLINGE H., RODRIGUES W.A., BRUNING E. & FITTKAU E.J. (1975) Biomass and structure in a central Amazonian rain forest. In: *Tropical Ecological Systems*, eds. F.B. Golley & E. Medina, pp. 115–22. Springer, Heidelberg.

KONONOVA M.M. (1975) Humus of virgin and cultivated soils. In: *Soil components. II. Organic components*, eds. J.E. Gieseking, pp. 475–526. Springer, N.Y.

LEE K.E. (1972) Soil fauna energetics in tropical forest. In: *Soil fauna and decomposition processes*, ed. D. Parkinson, pp. 54–68. IBP/PT theme 8, Louvain. University of Calgary Press, Calgary, Alberta.

LONGMAN K.A. & JENIK J. (1974) *Tropical forests and its environments.* Longman Inc., 72 Fifth ave, New York, New York 10011.

LIKENS G.E., BORMANN F.H. & JOHNSON N.M. (1969) Nitrification importance to nutrient losses from cutover forest ecosystem. *Science* **163**, 1205–8.

LIKENS G.E., BORMANN F.H., JOHNSON N.M., FISHER D.W. & PIERCE R.S. (1970) Effects of forest cutting and herbicide treatment on nutrient budgets in the Hubbard Brook Watershed Ecosystem. *Ecol. Monog.* **40**, 23–47.

LOSSAINT P. (1959) Etude expérimentale de la mobilisation du fer des sols sous l'influence des litières forestières. *Ann. Agron* **10**, 369–542.

MALAISSE F., FRESON R., GOFFINET G. & MALAISSE-MOUSSET M. (1975) Litterfall and litter breakdown in Miombo. In: *Tropical Ecological Systems*, eds. F.B. Golley & E. Medina, pp. 137–52. Springer, N.Y.

MALONE C.R. & REICHLE D.E. (1973) Chemical manipulation of soil biota in fescue meadow. *Soil Biol. Biochem.* **5**, 629–39.

MARKS P.L. & BORMANN F.H. (1972) Revegetation following forest cutting: mechanisms for return to steady state nutrient cycling. *Science* **176**, 914–15.

MCBEE R.H. (1971) Significance of intestinal microflora in herbivory. *Ann. Rev. Ecology Systematics* **2**, 165–76.

MCBRAYER J.F. (1973) Exploitation of deciduous leaf litter by *Apheloria montana* (Diplopoda: Eurydesmidae). *Pedobiologia* **13**, 90–8.

MCBRAYER J.F., REICHLE D.E. & WITKAMP M. (1974) Energy flow and nutrient cycling in a cryptozoan food-web. Eastern Deciduous Forest Biome–IBP–73–8.

MCCOLL J.G.M. & COLE D.W. (1968) A mechanism of cation transport in a forest soil. *Northwest Science* **42**, 134–40.

MILLER O.K. JR. & LAURSEN G.A. (1974) Below-ground fungal biomass on U.S. tundra biome sites at Barrow Alaska. In: *Soil organisms and decomposition in tundra*, eds. A.J. Holding *et al.*, pp. 151–8. Swedish IBP Committee, Stockholm.

MINDERMAN G. (1968) Addition, decomposition and accumulation of organic matter in forests. *J. Ecol.* **56**, 355–62.

MOSSE B. (1972) Effects of different endogone strains on the growth of *Paspalum notatum*. *Nature* **239**, 222–3.

NELSON L.E., SWITZER G.L. & SMITH W.H. (1970) Dry matter accumulation in young loblolly pine (*Pinus taeda* L.) In: *Tree growth and forest soils*, eds. C.T. Youngberg & C.B. Davey, pp. 261–74. Corvallis, Oregon State University Press.

NICHOLSON P.B., BOCOCK K.L. & HEAL O.W. (1966) Studies on the decomposition of the faecal pellets of a millipede (*Glomeris marginata* (Villers)). *J. Ecol.* **54**, 755–66.

NÖMMIK H. & POPOVIC B. (1971) Recovery and vertical distribution of ^{15}N labelled fertilizer nitrogen in forest soil. *Studia Forestalia Suecica* **92**, 1–20.

NOVOGROUDSKY D.M. (1948) The colonization of soil bacteria on fungal hyphae. *Microbiologiya* **17**, 28–35.

NYE P.H. (1961) Organic matter and nutrient cycles under moist tropical forest. *Pl. and Soil.* **13**, 333–46.

O'NEILL R.V., HARRIS W.F., AUSMUS B.S. & REICHLE D.E. (1975) A theoretical basis for ecosystem analysis with particular reference to element cycling. In: *Proc. Symp. Mineral cycling in southeastern ecosystems.* eds. F.G. Howell *et al.*, pp. 28–40. CONF. 740513. U.S. Dept. Commerce, Springfield, Virginia 22151.

OVINGTON J.D. (1965) Nutrient cycling in woodlands. In: *Experimental pedology*, eds. E.G. Hallsworth & D.V. Crawford, pp. 208–18.

PARSONS J.W. & TINSLEY J. (1975) Nitrogenous substances. In: *Soil components. I. Organic-components*, ed. J.E. Gieseking, pp. 263–304. Springer, N.Y., 534 pp.

PARINKINA O.M. (1974) Bacterial production in tundra soil. In: *Soil organisms and decomposition in tundra*, eds. A.J. Holding *et al.*, pp. 65–77. Swedish IBP Committee (Stockholm).

PETERSEN H. (1974) A biomass model from the Danish Beech Forest Project. *Göttinger Bodenk. Ber.* **30**, 170–6.

REICHLE D.E., MCBRAYER J.F. & AUSMUS B.S. (1973) Ecological energetics of decomposer invertebrates in a deciduous forest. In: *Progress in soil zoology*, ed. J. Vanec, pp. 283–91. Academia, Prague.

REICHLE D.E., O'NEILL R.V. & HARRIS W.F. (1976) Principles of energy and material exchange in ecosystems. In: *Proc. First International Congress of Ecology*. The Hague, The Netherlands.

REMACLE J. & VANDERHOVEN Ch. (1973) Evolution of carbon and nitrogen contents in incubated litters. *Pl. and Soil.* **39**, 201–3.

RODIN L.E. & BAZILEVICH N.I. (1967) *Production and mineral cycling in terrestrial vegetation.* Oliver and Boyd, London, 288 pp.

RIEKERK H. (1971) The mobility of phosphorus, potassium and calcium in a forest soil. *Soil. Sci. Soc. Am. Proc.* **35**, 350–6.

SCHUBE E.D. & KOCH W. (1971) Measurement of primary production with cuvettes. In: *Productivity of forest ecosystems*, ed. P. Duvigneaud, pp. 141–57. UNESCO, Paris.

SKINNER M.F. & BOWEN G.D. (1974) The uptake and translocation of phosphate by mycelial strands of pine mycorrhiza. *Soil. Biol. Biochem.* **6**, 62–8.

TURVEY N.D. (1974) Water in the nutrient cycle of a Papuan rain forest. *Nature* **251**, 414–15.

TRAPPE J.M. & FOGEL R.D. (in prep.) Ecosystematic functions of mycorrhiza. In: *The belowground ecosystem: A synthesis of plant-associated processes*, ed. J.K. Marshall US/IBP Interbrome Symposium, Colorado State University, Fort Collins, 5–7 Sept. 1973.

VIMMERSTEDT J.P. & FINNEY J.H. (1973) Impact of earthworm introduction on litter burial and nutrient distribution in Ohio strip-mine spoil banks. *Soil. Sci. Soc. American Proc.* 37, 388–91.

WAKSMAN S.A. (1952) *Soil microbiology*. J. Wiley, New York.

WENT F.W. & STARK N. (1968) Mycorrhiza. *BioScience* 18, 1035–9.

WHITTAKER J.B. (1974) Interactions between fauna and microflora at tundra sites. In: *Soil organisms and decomposition in tundra*, eds. A.J. Holding *et al.* Swedish IBP Committee, Stockholm, 1974.

WHITTAKER R.H. & LIKENS G.E. (1973) Carbon in the biota. In: *Carbon in the biosphere*, eds. G.M. Woodwell & E.V. Pecan, pp. 281–302. CONF. 720510. U.S. Dept. Commerce, Springfield, Virginia 22151.

WHITTAKER R.H. & WOODWELL G.M. (1971) Measurement of net primary production of forest. In: *Productivity of forest ecosystems*, ed. P. Duvigneaud, pp. 159–75. UNESCO, Paris.

WITHERSPOON J.P. JR. (1964) Cycling of ^{134}Cs in white oak trees. *Ecol. Monographs* 34, 403–20.

WITKAMP M. (1969) Environmental effects on microbial turnover of some mineral elements. *Soil Biol. Biochem.* 1, 167–84.

WITKAMP M. (1970) Aspects of soil microflora in a gamma-irradiated rain forest. In: *A tropical rain forest*, eds. H.T. Odum & R.F. Pigeon, pp. F 29–33. TID-24270. U.S. Dept. Commerce, Springfield, Virginia 22151.

WITKAMP M. (1974) Direct and indirect counts of fungi and bacteria as indexes of microbial mass and productivity. *Soil Sci.* 118, 150–5.

WITKAMP M. & AUSMUS B.S. (1975) Effect of tree species, temperature and soil on transfer of ^{54}Mn from litter to roots in a microcosm. In: *Mineral cycling in southeastern ecosystems*. eds. F.G. Howell *et al.*, pp. 694–9. CONF. 740513. U.S. Dept. Commerce, Springfield, Virginia 22151.

WITKAMP M. & BARZANSKY B. (1968) Microbial immobilization of ^{137}Cs in forest litter. *Oikos* 19, 392–5.

WITKAMP M. & FRANK M.L. (1969a) Loss of weight, ^{60}Co, and ^{137}Cs from tree litter in three subsystems of a watershed. *Environmental Sci. Technol.* 3, 1195–8.

WITKAMP M. & FRANK M.L. (1969b) Cesium-137 kinetics in terrestrial microcosms. In: *Proc. 2nd National Symp. radioecology*, eds. D.J. Nelson & F.C. Evans, pp. 635–43. US AEC CONF-670503. U.S. Dept. Commerce, Springfield, Virginia 22151.

WITKAMP M. & FRANK M.L. (1970) Effects of temperature, rainfall, and fauna on transfer of ^{137}Cs, K, Mg, and mass in consumer-decomposer microcosms. *Ecology* 51, 465–74.

WITKAMP M., FRANK M.L. & SHOOPMAN J.L. (1966) Accumulation and biota in a pioneer ecosystem of kudzu vine at Copperhill, Tennessee. *J. Appl. Ecol.* 3, 383–91.

WOOD T.G. (1974) Field investigations on the decomposition of leaves of *Eucalyptus delegatensis* in relation to environmental factors. *Pedobiologia* 14, 343–71.

17 Microbial immobilization of carbon, nitrogen, phosphorus and potassium: implications for forest ecosystem processes[1,2]

B. S. AUSMUS, N. T. EDWARDS, and M. WITKAMP *Environmental Sciences Division, Oak Ridge National Laboratory,*[3] *Oak Ridge, Tennessee 37830, U.S.A.*

Abstract

Carbon, nitrogen, phosphorus, and potassium immobilization by microbial populations were estimated in 01 litter, 02 litter, and soil (0–10 cm) monthly through one year in a second-growth, mesic hardwood forest, dominated by *Liriodendron tulipifera*. Elemental concentrations of litterfall, roots, and other inputs (dryfall, stemflow and throughfall), of elemental export in groundwater and of litter and soil strata were available from related studies. Elemental concentrations of microbes (bacteria, actinomycetes, fungal mycelia, fungal rhizomorphs, and fungal fruiting bodies) were determined. Using these concentrations and monthly estimates of field microbial populations, elemental pools immobilized in microbes were calculated for the forest floor strata. Highest microbial immobilization of carbon, nitrogen, phosphorus and potassium occurred in 01 litter during May, in 02 litter during July, and in soil (0–10 cm) during September and October. The period of maximum root growth (January–March) coincided with minimal microbial elemental immobilization. Seasonal microbial immobilization patterns suggest (1) advantages to canopy species for winter root growth, and (2) low elemental export by groundwater due to high microbial immobilization during periods of low root uptake.

[1] Research supported by the Eastern Deciduous Forest Biome, US–IBP, funded by the National Science Foundation under Interagency Agreement AG–199, BMS 69–01147 A09 with Energy Research and Development Administration—Oak Ridge National Laboratory.
[2] Contribution No. 255. Eastern Deciduous Forest Biome.
[3] Operated for the Energy Research and Development Administration by the Union Carbide Corporation. Publication No. 826, Environmental Sciences Division, Oak Ridge National Laboratory.

Introduction

Forest ecosystems exhibit mechanisms for capture, distribution, and conservation of elements essential to system persistence (Witkamp 1971; Szabo 1974; Dommerques & Mangenot 1971; Gosz *et al.* 1972; Likens *et al.* 1972; Reiners 1968, 1973; Satchell 1974). The mechanisms of nutrient immobilization, mineralization, release, and redistribution (e.g., root uptake) have been postulated to be controlled by belowground decomposer-plant interactions (Parkinson 1967; Witkamp 1971; Henderson & Harris 1974; Harris *et al.* 1975). If we are to investigate interactions of plants and decomposers, procedures are necessary to quantify microbial immobilization and release of mineral elements in forest floor strata through time.

Indirect measurements of microbial immobilization and release of elements have shown microbial influence on element leaching and root uptake (Witkamp 1971; Witkamp & Frank 1969a,b; Witkamp & Frank 1970). Determinations of microbial elemental composition, biomass, and activity are difficult due to rapid metabolic responses to environmental changes, small organism sizes, and inadequate techniques (Parkinson 1970, in prep.; Ausmus & Witkamp 1973; Szabo 1974). In addition, corollary information on inputs (litterfall, root sloughing), exports (CO_2 efflux, soil water percolate), substrates (litter and detrital root pools), and other decomposer organisms (arthropods, nematodes, earthworms) is essential to understanding forest elemental conservation mechanisms (Reichle *et al.* 1975; Mitchell 1974; Frankland 1974; Bell 1974; Jensen 1974; Kaarik 1974; Waid 1974; Lodha 1974).

Our objectives in this paper are to: (1) estimate the microbial immobilization of carbon, nitrogen, phosphorus and potassium monthly in 01 litter (L layer), 02 litter (F and H layers),[1] and mineral soil (0–10 cm); (2) compare microbially immobilized element pools with total element pools in litter and soil strata, (3) contrast microbial immobilization of these four elements with decomposer invertebrate immobilization, and (4) identify possible mechanisms controlling nutrient conservation in a second-growth, mesic hardwood forest.

Materials and methods

The study area is located near Oak Ridge, Tennessee, USA, within a 30 hectare, second-growth mesic hardwood forest dominated by *Liriodendron tulipifera* L. (Edwards & Sollins 1973). Mean canopy height was 30 m; mean

[1] H layers are inseparable from F or mineral soil due to the minimal litter accumulation at the study site caused by high decomposition rates.

stand age was 48 years. Soils within the area are alluvial, cherty silt loams. The climatic regime is humid-mesothermal (Thornthwaite 1948). Mean annual soil temperature is 13°C (Reichle *et al.* 1973b). Soil moisture was rarely limiting, although short periods of drought in the litter layers were detected in late summer (Edwards 1975a).

Litterfall was collected in 25 plastic tubs (0·25 m^2 area) located randomly within the site. Collections were made monthly except during autumnal litterfall when more frequent samples were necessary and during minimal litterfall when sampling frequencies were decreased.

Wong Samplers below the forest canopy were used to collect throughfall (collectors were closed except during precipitation). Troughs around five trees (three tulip poplars, an oak, a dogwood) were used to collect stemflow. Elemental inputs of canopy leaching were calculated by subtracting elemental input to the canopy by precipitation from element input to the forest floor by throughfall and stemflow (Edwards 1975a).

Three tube lysimeters inserted to approximately 60 cm were used to collect soil water. Water was removed biweekly except during periods of high rainfall. Water fluxes used to estimate elemental losses per unit area were estimated using the PROSPER model of water flow through atmosphere-soil-plants (Goldstein & Mankin 1974). We have made the assumption that element export below 60 cm represents losses from the forest since occurrence of feeding roots is infrequent beyond this soil depth.

Elemental concentrations were determined for detrital sources, precipitation, groundwater, and for micro-organisms. Microbial elemental concentrations were determined using forest substrates and microbial populations which were grown in the laboratory; microbial samples were separated from substrates by scraping or flotation (Ausmus & Witkamp 1973). Mean litter elemental pools were calculated from samples taken at two sites within the study area (N.T. Edwards, unpublished data; McBrayer *et al.* 1973). Elemental concentrations were determined monthly at one site (N.T. Edwards, unpublished data) and seasonally (Jan., Feb., Apr., July, Sept., Nov.) at the other (McBrayer *et al.* 1973). Root elemental concentrations were determined monthly (Cox *et al.* in prep.; W.F. Harris and D.E. Todd, unpublished data).

Three elements were detected using a Lira Model 300 gas chromatograph: carbon by flame ionization, phosphorus by flame photometry, and nitrogen by Coulson conductivity (Horton *et al.* 1971, 1973). Analysis of potassium was performed using a Perkin–Elmer Model 403 atomic spectrophotometer.

ATP determinations of samples from 01 litter, 02 litter, and soil (0–10 cm) were the basis for estimating microbially immobilized elemental pools. On eighteen dates during the year samples were extracted using the cold sulphuric acid technique (Lee *et al.* 1971a,b) and analysed for ATP content using the luciferin–luciferase enzyme assay (McElroy 1947) measured with a GeMSAEC centrifugal photometric analyser (Anderson 1969; Coleman *et al.* 1971; Ausmus 1973). Replicate samples were analysed after removal of roots, or roots and invertebrates to provide a correction for non-microbial ATP within each forest floor stratum. Bacterial ATP was estimated as a fraction of each sample based on microbial respiration rates with and without antibiotic amendments determined on a Gilson respirometer (Anderson & Domsch 1973). Using McElroy's (1947) ATP : element content conversion ratios, we established ATP to element conversion ratios for

Table 17.1. ATP concentration and carbon, nitrogen, phosphorus, and potassium concentrations of bacteria, actinomycetes, and fungi in logarithmic growth phases and of fungal fruiting bodies and rhizomorphs.

Microbial group	% dry weight C ±2·5*	N ±0·22*	P ±0·05*	K ±0·25*	ATP concentration mg g⁻¹ dry wt
Bacteria	50·0	4·00	0·91	1·50	1·0±0·2
Actinomycetes	44·0	4·20	1·00	4·00	2·3±0·1
Fungal mycelia	45·0	2·80	0·24	0·12	2·3±0·1
Fungal fruiting bodies	37·7	4·40	1·00	4·90	2·3±0·6
Fungal rhizomorphs	36·2	4·13	0·93	3·16	2·1±0·7

ATP : element conversions calculated as mg m^{-2} ATP extracted from a stratum (corrected for extraction efficiencies) and separated into bacterial and non-bacterial (all other microbes) using respiration rates with and without antibiotics (Anderson & Domsch 1973). ATP values were multiplied by mean elemental concentrations to estimate mean elemental pool immobilized within microbes of that forest floor stratum.

* Estimates of analytical precision, standard errors of the mean.

microbial groups grown on litter substrates in the laboratory or scraped from field samples (Table 17.1). The use of the ATP assay to estimate elemental pools of micro-organisms is an extension of the phenomenon noted by McElroy (1947) and many other research workers that although the absolute values of ATP concentration and elemental concentrations change through the microbe life cycle, the ratio of ATP to at least C, N, P, K, Na are quite constant. Therefore, by measuring ATP content of detrital substrates we can accurately estimate microbial elemental pools.

Differences in elemental concentrations among life forms of the same microbial population, especially fungi, are real and must be kept in mind when estimating microbially immobilized elemental pools. However, since, at any point in time, the portion of microbial mass which are spores and fruiting bodies is small (Witkamp 1966 estimated 10% of total fungal mass as spores), these concentration differences have been ignored for this first approximation of microbial immobilization.

Results and discussion

FOREST LITTER NUTRIENT STATUS

Examination of differences in patterns of element accumulation in litter among several forest types yield indirect evidence of microbial immobilization of selective nutrients within the upper litter layer. Using estimates of litterfall and litter turnover obtained from decomposition process level studies, estimates of annual deposition and minimal litter nutrient pools were made on the U.S. Forest Service inventory plots within the southern Appalachians. Under high decomposition rates (due to forest litter characteristics and meterological conditions) greater amounts of carbon may be lost annually from 01 litter than litter input. This condition exists in our study area due to favourable substrate quality (low lignin concentration; high nitrogen concentration), litter temperatures rarely below freezing, and rare (late summer) litter moisture deficit (Edwards 1975a). As shown in Fig. 17.1, carbon is accumulated above annual deposition only in pine and pine-hardwood stands. Forests dominated by hardwoods typically retained only the annual carbon input as litterfall. The pattern of accumulation of phosphorus and loss of potassium from litter of several forest types suggested selective immobilization of nutrient elements by litter microbial populations (Fig. 17.1).

The *Liriodendron* forest site shows little nitrogen loss or gain during litter decomposition and retains approximately 82% of the litterfall nitrogen in summer 01 litter (Ausmus & Witkamp 1973; McBrayer *et al.* 1973). This is probably not true of woody litter at the site which has been found to gain nitrogen through time (D.E. Todd, unpublished data). However, nitrogen is preferentially accumulated over carbon (Fig. 17.1). This preferential conservation of nitrogen is even greater in the litter layer of oak stands (Cromack 1972; Kelly 1973) which has a higher initial C/N.

Phosphorus accumulation in litter is affected by forest type (Fig. 17.1). Hardwood stands preferentially accumulate phosphorus over carbon to a greater degree than pine and pine-hardwood stands. The *Liriodendron* stand preferentially accumulates phosphorus above even that of cove hardwood stands.

Potassium seems to be lost from the litter layer equally across forest types (Fig. 17.1). The *Liriodendron* stand retains approximately 38% of the annual potassium input as litterfall compared to approximately 20% in other stands

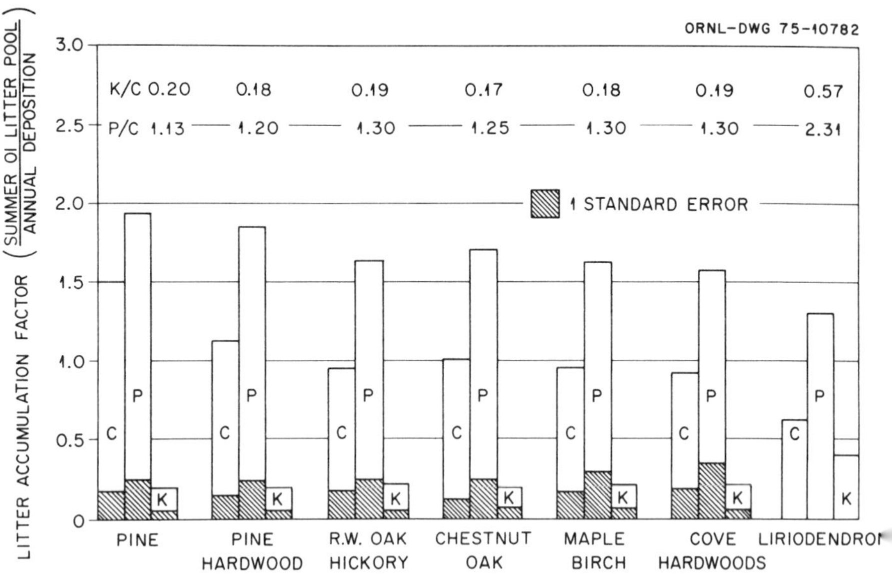

Figure 17.1. Accumulation and loss of elements in the o1 litter layer of forest types occurring within the Appalachian region of the U.S. The litter dynamics of the study site are indicated.

examined. These losses of potassium through the upper litter layer suggest limited control of potassium losses by the litter decomposer community. This hypothesis seems plausible since (1) forest types differ little in litter potassium content, (2) potassium is readily leached, and (3) microbial populations do not concentrate potassium above substrate concentrations (Cromack *et al.* 1975).

Thus, it seems that litter nutrient status is directly dependent upon microbial immobilization of the elements against leaching processes. Elements preferentially immobilized by the microbial community are retained and accumulated in the litter while those not preferentially immobilized are leached. It is necessary to know soil microbial immobilization patterns to determine whether nutrients leached from the litter are lost from the system or available for root uptake.

Microbial immobilization

In o1 litter, maximum microbial immobilization of carbon, nitrogen, phosphorus and potassium occurred during May (Figs 17.2, 17.3, 17.4, 17.5); minimum microbial immobilization occurred during February and March. Carbon immobilization ranged from $1{\cdot}9\,g\,m^{-2}$ to $45{\cdot}2\,g\,m^{-2}$ (Fig. 17.2) which is equivalent to a maximum of 37% of the o1 litter carbon pool. Nitro-

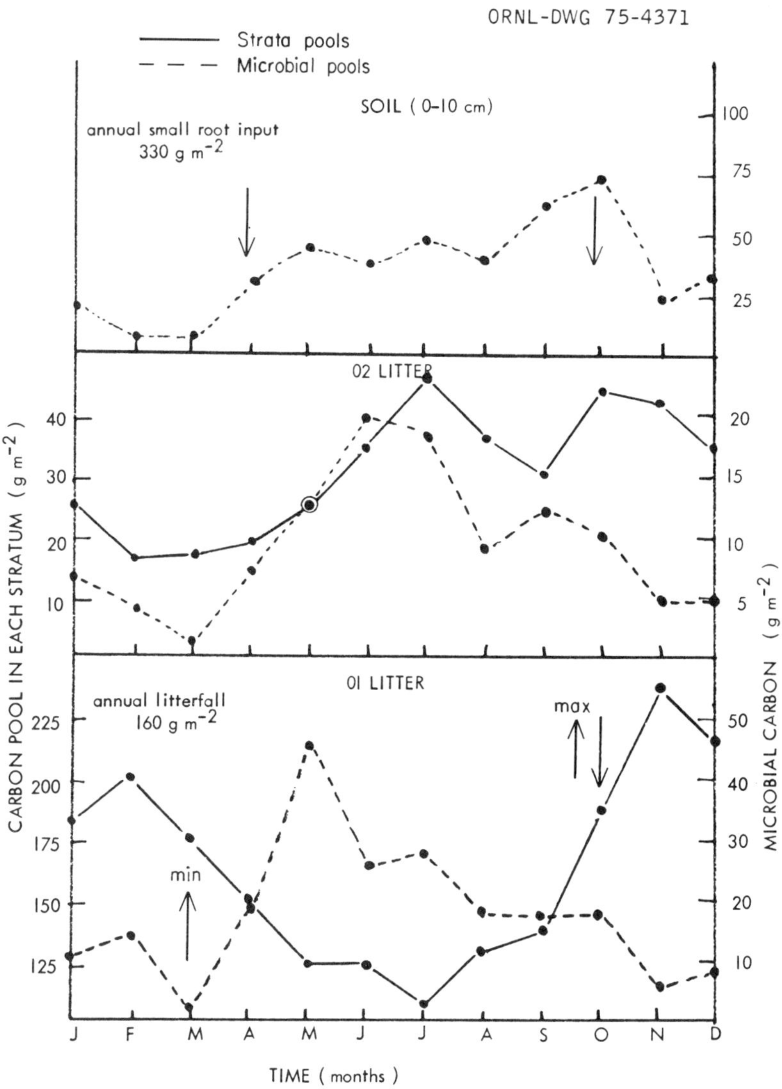

Figure 17.2. Microbial immobilization of carbon in 01 litter, 02 litter, and soil calculated monthly through the year. Carbon pools of 01 and 02 litter are indicated monthly. Periods of maximum litterfall and root sloughing (↓) and maximum and minimum CO_2 efflux (↑) are noted.

gen immobilization ranged from 0·09 g m^{-2} to 3·15 g m^{-2}, equivalent to 2 to 99% of the 01 litter nitrogen pool (Fig. 17.3). Phosphorus immobilization ranged from 0·02 g m^{-2} to 0·58 g m^{-2}, equivalent to 6 to 200% of the 01 litter phosphorus pool (Fig. 17.4). Potassium immobilization ranged from 0·06 g m^{-2} to 1·6 g m^{-2}, equivalent to 11 to 400% if only the 01 litter

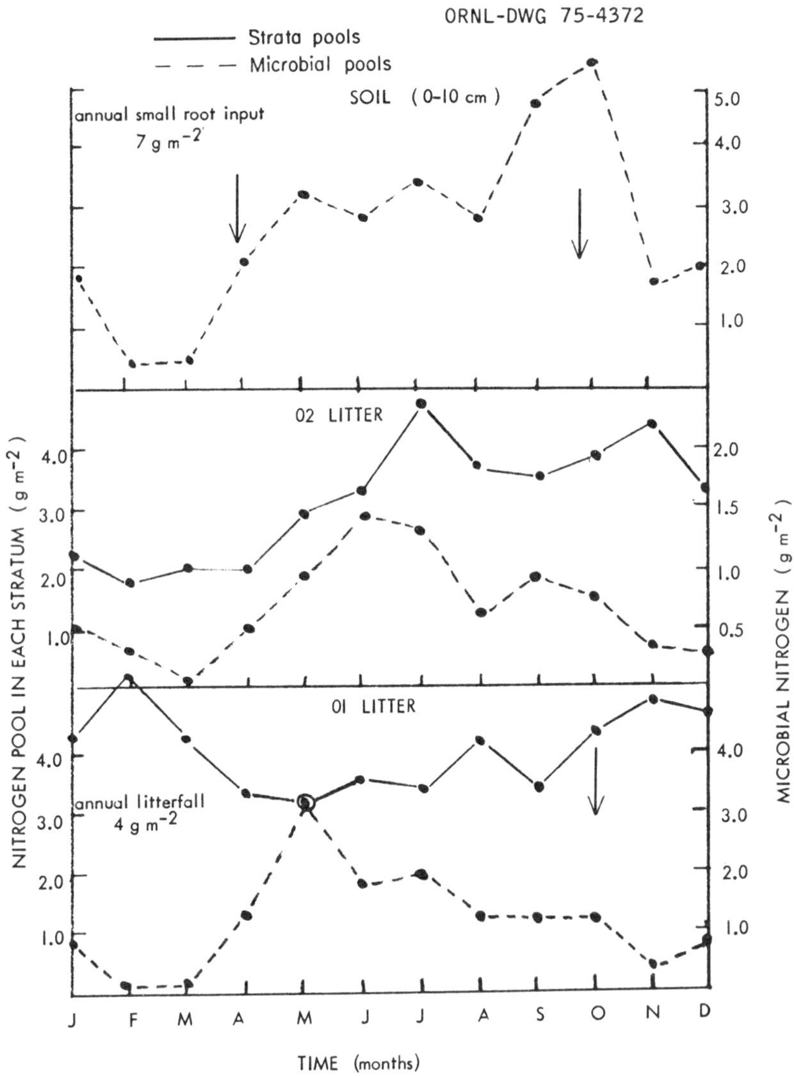

Figure 17.3. Microbial immobilization of nitrogen in 01 litter, 02 litter, and soil calculated monthly through the year. Nitrogen pools of 01 and 02 litter and periods of maximum litterfall and root sloughing are shown with arrows.

potassium pool is considered (Fig. 17.5). It should be remembered that comparison of estimated microbial elemental pools to that of 01 litter layer is simplistic, not considering soluble elemental pools which might be immobilized, nitrogen fixation processes, or translocation of elements from 02 litter and/or soil via fungal mycelial growth and faunal activities. The influence of these processes will be discussed later.

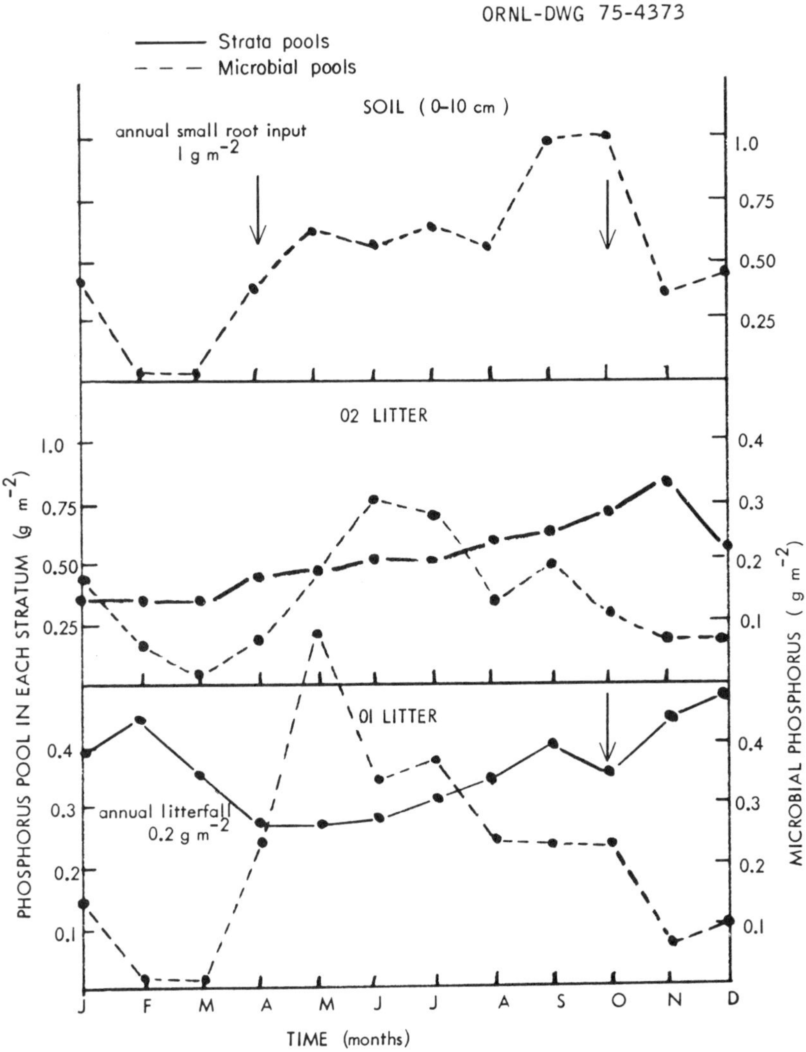

Figure 17.4. Microbial immobilization of phosphorus in 01 litter, 02 litter, and soil calculated monthly through the year. Phosphorus pools of 01 and 02 litter and periods of maximum litterfall and root sloughing are shown with arrows.

In 02 litter, maximum microbial immobilization of the four elements occurred during July, the minimum occurred during March (Figs 17.2, 17.3, 17.4, 17.5). Carbon immobilization ranged from 1·1 g m^{-2} to 19·3 g m^{-2}, equivalent to 3 to 25% of the 02 litter carbon pool (Fig. 17.2). Nitrogen immobilization ranged from 0·08 g m^{-2} to 1·45 g m^{-2} which is equivalent to 4 to 44% of the 02 litter nitrogen pool (Fig. 17.3). Microbial immobilization of phosphorus ranged from 0·01 g m^{-2}, equivalent to 3 to 59% of the 02 litter phosphorus pool (Fig. 17.4). Potassium immobilization ranged

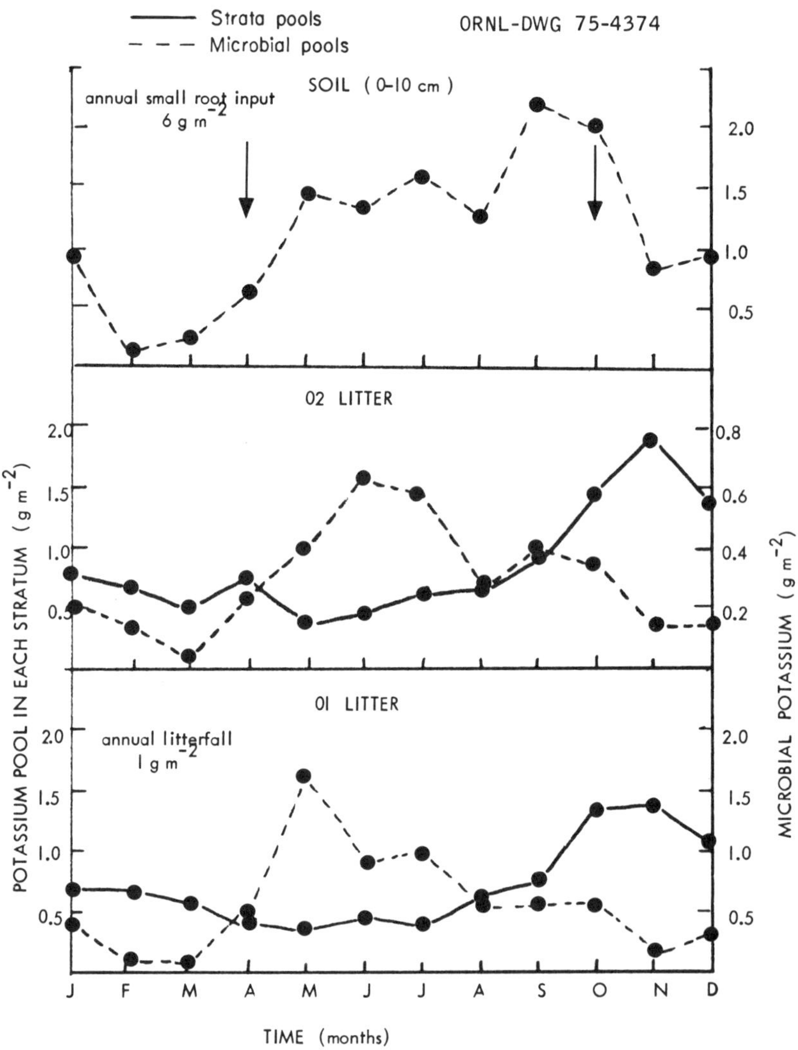

Figure 17.5. Microbial immobilization of potassium in 01 litter, 02 litter, and soil calculated monthly through the year. Potassium pools of 01 and 02 litter and periods of maximum litterfall and root sloughing are shown with arrows.

from $0{\cdot}04\,g\,m^{-2}$ to $0{\cdot}58\,g\,m^{-2}$, equivalent to 8 to 91% of the 02 litter potassium pool (Fig. 17.5).

Soil (0–10 cm) microbial dynamics are more stable and immobilize greater amounts of nutrients than either litter layer (Figs 17.2–17.5). Maximum microbial immobilization occurred during September–October with a lesser peak during May. Minimum microbial immobilization in soil, as in the litter layers occurred during February and March. Microbial carbon immobilization ranged from $5{\cdot}2\,g\,m^{-2}$ to $73{\cdot}8\,g\,m^{-2}$, much less than the estimated soil organic carbon in these horizons. Nitrogen immobilization ranged from 0·39 g m^{-2} to 5·4 g^{-2}. Phosphorus immobilization ranged from 0·08 g m^{-2} to 1·03 g m^{-2}. Potassium immobilization ranged from 0·18 g m^{-2} to 2·17 g m^{-2}.

Relationship of microbial immobilization to detrital sources and pools

Introduction of carbon, nitrogen, phosphorus and potassium into the detrital food web is dominated by belowground processes. Belowground to aboveground input ratios were 2 : 1 for carbon, 1·5 : 1 for nitrogen, 3·6 : 1 for phosphorus, and 1·4 : 1 for potassium. These estimates include roots <0·5 cm in diameter as the only belowground detrital source, while litterfall, meterological inputs, and canopy leaching were included as aboveground detrital sources (Table 17.2). Aboveground detrital inputs are primarily in the form

Table 17.2. Detrital and solution sources of carbon, nitrogen, phosphorus and potassium and their export from the *Liriodendron* forest ecosystem.

	g m^{-2} yr^{-1}					
	Detrital Inputs				Exports	
Element	Below-ground small roots	litterfall	Above-ground ppt	canopy leaching	gaseous	ground-water
Carbon	330	160	—	—	1065	1·3
Nitrogen	7	4	0·7	0·2	?	0·35
Phosphorus	1	0·2	0·06	0·02	—	0·004
Potassium	6	1	0·32	2·9	—	0·9

of litterfall for carbon, nitrogen, and phosphorus (Fig. 17.6). Precipitation inputs were significant for nitrogen, phosphorus, and potassium (8–22%). Canopy leaching accounted for 63% of the potassium input aboveground. Greatest potassium leaching occurred during September and October (Fig. 17.7).

The seasonality of microbial immobilization among the three forest floor strata reflects the periodicity of detrital inputs aboveground and belowground. The rate of increasing immobilization of nutrients in 01 litter in May is greater than that of other forest floor strata caused by increasing spring temperatures (rate of increase approximately twice that of 02 litter and soil). The timing of this burst of microbial activity approximates the period of significant detrital input of *Liriodendron* flower parts and other reproductive structures (McBrayer *et al.* 1973). Summer estimates of micro-

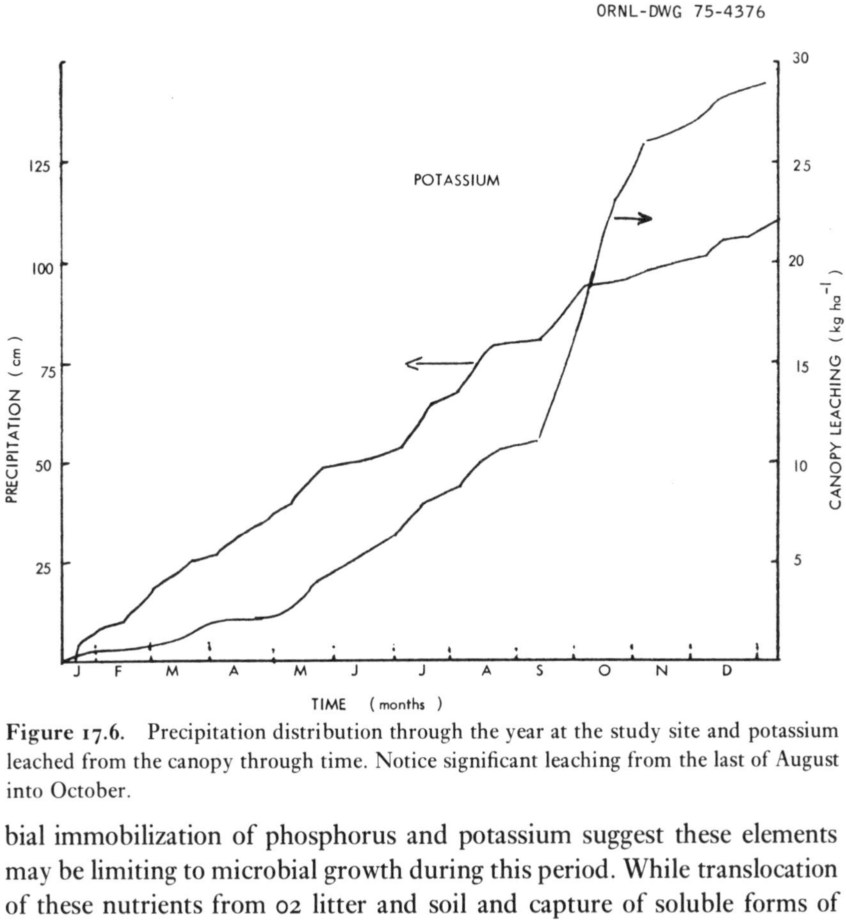

Figure 17.6. Precipitation distribution through the year at the study site and potassium leached from the canopy through time. Notice significant leaching from the last of August into October.

bial immobilization of phosphorus and potassium suggest these elements may be limiting to microbial growth during this period. While translocation of these nutrients from O2 litter and soil and capture of soluble forms of these ions could partially compensate for this implied stress, it is likely that microbial mass is dominated by inactive protoplasm with reduced elemental concentrations (Dowding, this volume, Chapter 7) during this period. A large amount of inactive microbial protoplasm would cause an overestimate of phosphorus and potassium immobilization when based on ATP to elemental conversion factors calculated using data on active growth forms (Cromack *et al.* 1975; Stark 1973).

Microbial immobilization in O2 litter is greatest during the summer. Leaching from the O1 litter layer following short dry periods and the consistently high moisture of soil layers allows high summer microbial activity in the O2 litter layer. Further, there was a high correlation between maximum nutrient elemental pools in O2 litter and maximum microbial immobilization (Figs 17.2, 17.3, 17.4, 17.5).

Soil microbial immobilization peaked twice during the year. Spring increase in microbial activity corresponds to increased soil temperatures,

ORNL-DWG 75-4375

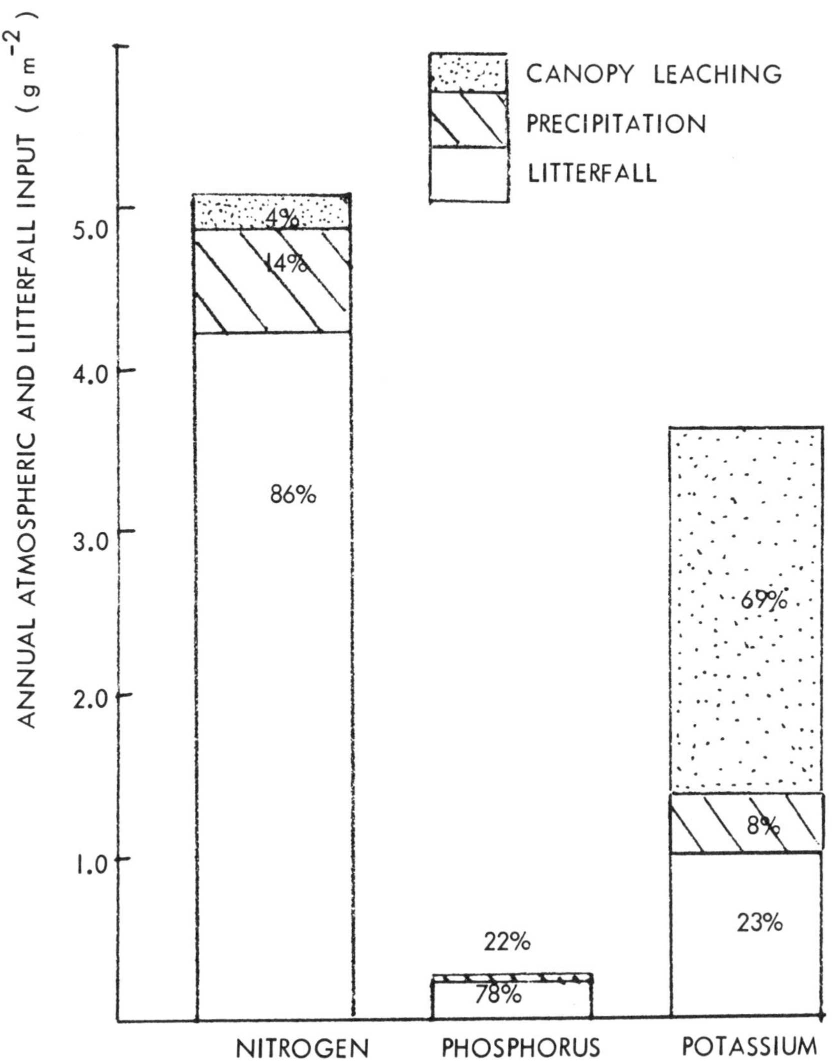

Figure 17.7. Sources of nitrogen, phosphorus, and potassium input to the forest floor from aboveground. Only potassium is significantly leached from the canopy. Significant amounts of the three elements were input as precipitation.

spring *Liriodendron* root sloughing, and input of reproductive parts to the 01 litter layer. Spring microbial activity was expected to be even higher since this is the period of largest root sloughing during the year. It is possible that these sloughed roots are catabolized by the mycorrhizal fungi. If these fungi are at a competitive advantage for these root substrate due to their prior association, their respiration and growth rates would be less than those of fast responding soil microbial species. If the 'sugar microbes' were catabolizing these spring sloughed roots, one would expect to see a period of rapid microbial activity followed by a significant microbial decrease as the populations exhausted the substrates (Fig. 17.2). Summer microbial populations were stable indicating a minimum soil moisture stress during high summer temperatures of the study site. Fall increases in microbial immobilization were correlated with more favourable temperature–moisture regimes, continued catabolism of root detritus, and maximum canopy leaching of potassium. The fall immobilization maximum, greater than the spring peak, may be produced by microbial populations capable of rapid response to changes in solution concentrations of potassium, or other nutrients, within favourable temperature–moisture regimes. The rapid decrease in microbial immobilization by November without drastic temperature reduction supports this hypothesis.

Comparison of microbial and invertebrate elemental immobilization

Using studies of decomposer invertebrates made within the study area, nutrient immobilization by microbial and faunal population were compared (McBrayer *et al.* 1973; Reichle *et al.* 1973, 1975; Ausmus *et al.*, in prep.). Since patterns of immobilization were similar among the four elements examined, nitrogen immobilization will serve as a basis for comparison among decomposer groups (Fig. 17.8).

Arthropod immobilization was weakly correlated to microbial immobilization through the year (McBrayer *et al.* 1973). While arthropod dynamics mimmicked microbial dynamics due to the dominance of the fungus–fungivore trophic pathway, arthropod immobilization averaged sevenfold less than microbial immobilization of nitrogen (McBrayer *et al.* 1973). Arthropods averaged higher nitrogen immobilization than other faunal groups, but during winter earthworms immobilized as much nitrogen as arthropods. Nematode immobilization of nitrogen was an order of magnitude less than that of earthworms and arthropods (Fig. 17.8).

During February and March, arthropods and earthworms immobilize as much as nitrogen as microbial populations which are at an annual minimum. Significant amounts of nitrogen, or other elements were not leached out of the decomposition strata into groundwater. We believe this to be the result of significant root uptake activity during this period. Nematodes, dominated by phytophages, probably have little effect on nutrient dynamics directly by immobilization, but may greatly affect nutrient dynamics increasing decomposer community by forming the lowest link in the soil

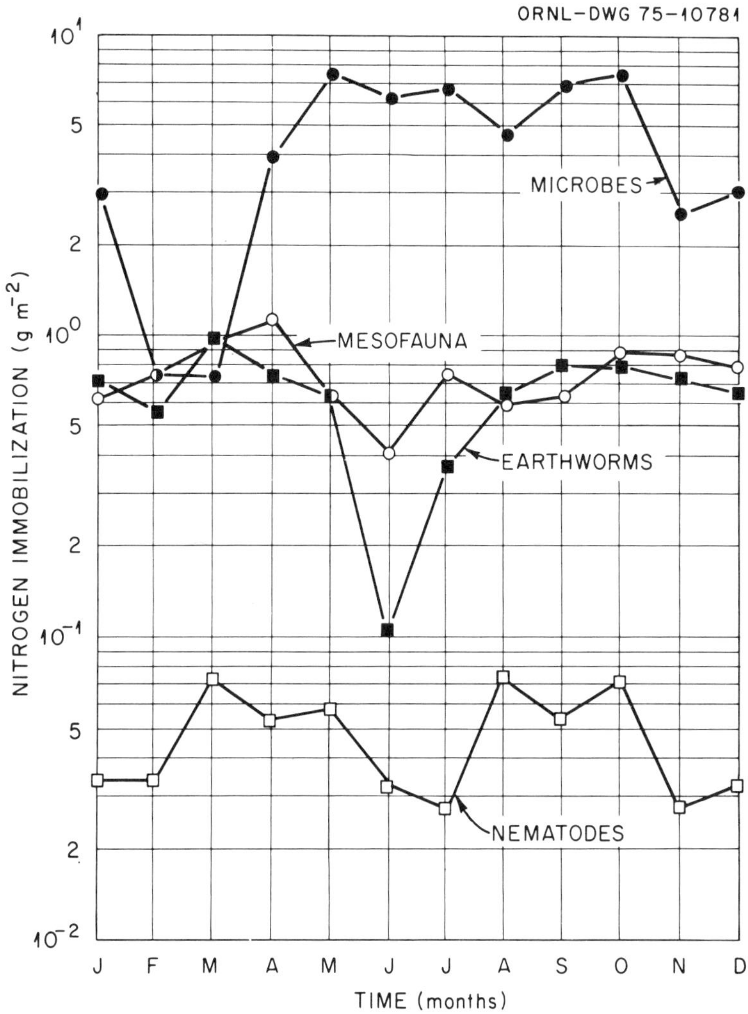

Figure 17.8. Comparison of microbial nitrogen immobilization with that of mesofaunal arthropods, earthworms, and nematodes. While microbial immobilization is much greater than that of faunal, faunal immobilization is much more constant through the year.

decomposer food web and by affecting live root processes (mycorrhizal induction and root harvest).

Implications of microbial immobilization patterns to ecosystem processes

Within this *Liriodendron* forest the gross primary production was 2,162 g carbon $m^{-2} yr^{-1}$. Autotrophic respiration (including the entire root community), was 66% of the gross primary production (1,436 g cm^{-2} yr^{-1}). Of the net primary production, 42% was expended for decomposer respiration (670 g cm^{-2} yr^{-1}) leaving a net ecosystem production of 56 g cm^{-2} yr^{-1}. The forest floor (litter and soil) of this forest annually has a CO_2 efflux of 1,065 g cm^{-2} yr^{-1} of which 20–35% evolves from root activity (20% if rhizosphere microbial respiration is separated from root respiration, Reichle *et al.* 1975). The currency for translocation and retention of nutrient elements is carbon allocated for maintenance of the primary producers and decomposers. Over long time periods carbon is accumulated, primarily as soil organic matter, which serves as the biochemical framework for element immobilization by decomposers and redistribution to primary producers via root uptake. While the biochemical form of carbon (e.g., sucrose vs cellulose) limits the rate of microbial and faunal catabolism, the total availability of carbon is not limiting to decomposers during any period of the year.

Nitrogen is preferentially accumulated over carbon in all forest floor strata. Availability of nitrogen is extremely limited in 01 litter foliage and especially woody substrates. While calculations of microbial immobilization of nitrogen suggest almost all 01 litter nitrogen could be within microbial cells during May, translocation of nitrogen from 02 litter and from soil and nitrogen fixation processes in woody litter (Ausmus, unpublished data from *Liriodendron* site and J. Roskoowski, unpublished data, Hubbard Brook) could supplement nitrogen availability. It may be hypothesized that translocation of nitrogen and nitrogen fixation rates in substrates are controlled by substrate nitrogen concentrations.

Minimal decomposer nitrogen immobilization occurs at the season (February–March) of maximum root growth in the *Liriodendron* stand (Harris *et al.*, in prep.). This is also the season that nitrate and ammonia are readily available from soil solution for root uptake due to near saturated soil conditions (G.H. Henderson, unpublished data). Even though considerable soil water percolation occurs during this period (February–March) less than 3% of the annual detrital nitrogen input is lost to groundwater. The phase shift between high microbial immobilization and maximum assimilation of nitrogen in new root tissue is strong evidence that the interaction of autotroph and decomposer populations within this forest ecosystem represents a system-level mechanism for effective nitrogen retention, controlled remineralization and redistribution to support sustained autotroph productivity.

Phosphorus is preferentially accumulated over carbon in 01 litter. This element is probably limiting to litter microbial populations from March through July. We have no direct evidence that this limitation is met by translocation from 02 litter and soil or by decreased microbial phosphorus concentrations.

By contrast, potassium inputs are dominantly soluble, and not retained in 01 litter. While this forest retains 38% of the annual potassium input in 01 litter, other forest types retain an average of 20%. This suggests limited biological control of potassium movement in litter. Further, potassium is limiting to 01 litter populations from April through July (immobilization estimates > available pool). However, groundwater losses of potassium are less than 10% of the annual detrital input. Therefore, it would seem that control of potassium loss is retained in soil but not litter. Well-known soil chemical reactions involving ionic equilibria of soil solution and the organic–inorganic exchange complex could account for conservation of potassium in the soil.

The strategy of elemental retention and redistribution within the forest system, we hypothesize, is dependent upon maintenance of the forest floor (litter and soil) detrital pools and its decomposer community. The cost to the forest is primarily carbon lost in heterotrophic respiration. Seasonality of microbial immobilization of carbon, nitrogen, phosphorus, and potassium within forest floor strata with maximum immobilization occurring at different times among strata and minimum immobilization synchronized with likely periods of maximum root uptake resulted in low amounts of nutrient export from the forest. Decomposer invertebrates, while possessing smaller immobilized pools of these elements may regulate microbial immobilization and release rates through harvest (Mitchell 1974) or indirectly via substrate influences (Van der Drift & Witkamp 1960; Witkamp & Crossley 1966; Patten & Witkamp 1967; McBrayer 1973). Maximum growth of *Liriodendron* roots (defined to include the accompanying root community of mycorrhizal fungi and root surface microbes) occurs at the time of minimum decomposer nutrient immobilization. This harmony between plants and decomposers, we believe, is the major mechanism underlying forest nutrient conservation to support sustained productivity.

References

ANDERSON N.G. (1969) Analytical techniques for cell fractions. XII. A multiple-cuvet rotor for a new microanalytical system. *Anal. Biochem.* **28**, 545–64.

ANDERSON J.P.E. & DOMSCH K.H. (1973) Quantification of bacterial and fungal contributions to soil respiration. *Arch Mikrobiol.* **93**, 113–27.

AUSMUS B.S. (1973) The use of the ATP assay in terrestrial decomposition studies. IBP International Symposium: Modern Methods in the Study of Microbial Ecology, Uppsala, Sweden, June 1972. *Bull. Ecol. Res. Comm.* (Stockholm) **17**, 223–34.

AUSMUS B.S., FERRIS J.M., REICHLE D.E. & WILLIAMS E.C. (in prep.) The role of primary consumers in forest root dynamics. In: *The belowground ecosystem: A synthesis of plant associated processes*, ed. J.K. Marshall. US/IBP Interbiome Symposium, Colorado State University, Fort Collins, 5–7 Sept. 1973.

AUSMUS B.S. & WITKAMP M.W. (1973) *Litter and soil microbial dynamics in a deciduous forest stand.* Oak Ridge National Laboratory, Oak Ridge, Tennessee. EDFB–IBP–73–10.

BELL M.K. (1974) Decomposition of herbaceous litter. In: *Biology of plant litter decomposition*, Vol. I, pp. 37–67, eds. C.H. Dickinson & G.J.F. Pugh. Academic Press, New York.

COLEMAN R.L., SHULTS W.D., KELLEY M.T. & DEAN J.A. (1971) Environmental applications of a centrifugal photometric analyser. *Anal. Lab.* **1**, 26–32.

COX T.L., AUSMUS B.S., HARRIS W.F. & TODD R.L. (in prep.) The role of roots in biochemical cycles and detritus-based food chains in the eastern deciduous forests. In: *The belowground ecosystems: A synthesis of plant associated processes*, ed. J.K. Marshall. US/IBP Interbiome Symposium, Colorado State University, Fort Collins, 5–7 Sept. 1973.

CROMACK K., JR (1972) *Litter production and decomposition in a mixed hardwood watershed at Coweeta Hydrologic Station, North Carolina.* Unpublished Ph.D. Dissertation, The University of Georgia, Athens.

CROMACK K.D., TODD R.L. & MONK C.D. (1975) Elemental concentration by microorganisms. *Soil Biol. Biochem.* **7**, 1–4.

DOMMERGUES Y. & MANGENOT J. (1971) Interactions de la microflore et de la faune du sol. In: *Ecologie microbienne du sol.* Massonet Cie, Paris.

DRIFT J. VAN DER & WITKAMP M. (1960) The significance of the breakdown of oak litter by *Enoicyla pusilla. Burm. Arch. Neerl Zool.* **13**, 486–92.

EDWARDS N.T. (1975a) Seasonality of nutrient cycling in a southern *Liriodendron* forest. *Bull. Ecol. Soc. Amer.*

EDWARDS N.T. (1975b) Effects of temperature and moisture on CO_2 evolution in a mixed deciduous forest floor. *Soil Sci. Soc. Amer. Proc.* **39**, 361–5.

EDWARDS N.T. & SOLLINS P. (1973) Continuous measurement of CO_2 evolution from partitioned forest floor components. *Ecology* **54 (2)**, 406–12.

FRANKLAND J.C. (1974) Decomposition of lower plants. In: *Biology of plant litter decomposition*, Vol. I, pp. 3–36, eds. C.H. Dickinson & G.J.F. Pugh. Academic Press, New York.

GOLDSTEIN R.A. & MANKIN J.B. (1974) Documentation of PROSPER: A model of atmosphere–soil–plant water flow. EDFB–IBP–73–9.

GOSZ J.R., LIKENS G.E. & BORMANN F.H. (1972) Nutrient content of litterfall on the Hubbard Brook Experimental Forest, New Hampshire. *Ecology* **53**, 769–84.

HARRIS W.F., KINERSON R.S. & EDWARDS N.T. (in prep.) Comparison of belowground biomass of natural deciduous forests and loblolly pine plantations. In: *The Belowground Ecosystems: A synthesis of plant-associated processes*, ed. J.K. Marshall. US/IBP Interbiome Symposium, Colorado State University, Fort Collins, 5–7 Sept. 1973.

HARRIS W.F., SOLLINS P., EDWARDS N.T., DINGER B.E. & SHUGART H.H. (1975) Analysis of carbon flow and productivity in a temperate deciduous forest ecosystem. Productivity of World Ecosystems. *Proc. Symp. V General Assembly of the Special Committee for the IBP*, Seattle, Wash. Natl. Acad. Sci., Washington, D.C., 31 Aug.–1 Sept. 1972, pp. 116–22.

HENDERSON G.S. & HARRIS W.F. (1974) An ecosystem approach to characterization of the nitrogen cycle in a deciduous forest watershed. In: *Forest soils and forest land management*, eds. B. Bernier & C.H. Winget, pp. 179–93. Proc. Fourth Amer. Forest Soils Conf., Larval University, Quebec, August 1973.

HORTON A.D., SHULTS W.D. & MEYER A.S. (1971) Determination of nitrogen, sulfur, phosphorus, and carbon in solid ecological materials via hydrogenation and element-selective detection. *Analytical Letters* **4**, 613–21.

HORTON A.D., SHULTS W.D., MEYER A.S. & MATHEWS D.R. (1973) Determination of nitrogen, sulfur, phosphorus, and carbon in solid biological materials via hydrogenation and element selective detection. *Environ. Sci. Technol.* 7, 449–51.

JENSEN V. (1974) Decomposition of Angiosperm tree leaf litter. In: *Biology of plant litter decomposition*, Vol. 1, pp. 69–104, eds. C.H. Dickinson & G.J.F. Pugh. Academic Press, New York.

KAARIK A.A. (1974) Decomposition of Wood. In: *Biology of plant litter decomposition.* Vol. 1, pp. 129–74, eds. C.H. Dickinson & G.J.F. Pugh. Academic Press, New York.

KELLY J.M. (1973) *Dynamics of litter decomposition, microbiota, populations, and nutrient movement following nitrogen and phosphorus additions to a deciduous forest stand.* Unpublished Ph.D. Dissertation, the University of Tennessee, Knoxville.

LEE C.C., HARRIS R.F., WILLIAMS J.D.H., ARMSTRONG D. & SYERS J.K. (1971a) Adenosine triphosphate in lake sediments II. *Determination Soil Sci. Soc. Amer. Proc.* 35, 82–6.

LEE C.C., HARRIS R.F., WILLIAMS J.D.H., ARMSTRONG D.E. & SYERS J.K. (1971b) Adenosine triphosphate in lake sediments II. Origin and significance. *Soil Sci. Soc. Amer. Proc.* 35, 86–91.

LIKENS G.E. & BORMANN F.H. (1972) Nutrient cycling in ecosystems. In: *Ecosystem Structure and Function*, pp. 25–67, ed. J.A. Wiens (Proceed. 31st Annual Biology Colloquium). Oregon State University Press.

LODHA B.C. (1967) Decomposition of digested litter. In: *Biology of plant litter decomposition*, Vol. I, pp. 213–41, eds C.H. Dickinson & G.J.F. Pugh. Academic Press, New York.

MCBRAYER J.F. (1973) Exploitation of deciduous leaf litter by *Apheloria montana* (Diplopoda: Eurydesmidae). *Pedobiol.* 13, 90–8.

MCBRAYER J.F., REICHLE D.E. & WITKAMP M. (1973) *Energy flow and nutrient cycling in a cryptozoan food web.* EDFB–IBP–73–8.

MCELROY W.D. (1947) The energy source for bioluminescence in an isolated system. *Proc. Natl. Acad. Sci.* 33, 342–5.

MITCHELL M.J. (1974) Interactions of oribatid mites (Acari: Crytostigmata) and fungi in an aspen woodland (*Populus tremuloides* Mich.) soil. *Bull. Ecol. Soc. Amer.* (Abstract) p. 51.

PARKINSON D. (1967) Soil micro-organisms and plant roots. In: *Soil Biology*, pp. 449–78, eds. A. Burges & F. Raw. Academic Press, New York.

PARKINSON D. (1970) Methods for the quantitative study of heterotrophic soil microorganisms. In: *Methods of Study in Soil Ecology*, pp. 101–5, ed. J. Phillipson. UNESCO, Paris.

PARKINSON D. (in prep.) Organic matter turnover in soil. In: *The Belowground Ecosystem: A Synthesis of Plant-Associated Processes*, ed. J.K. Marshall. US/IBP Interbiome Symposium, Colorado State University, Fort Collins, 5–7 Sept. 1973.

PATTEN B.C. & WITKAMP M. (1967) Systems analysis of 134 cesium kinetics in terrestrial microcosms. *Ecology* 48, 813–24.

REICHLE D.E., DINGER B.E., EDWARDS N.T., HARRIS W.F. & SOLLINS P. (1973) Carbon flow and storage in a forest ecosystem. In: *Carbon and the Biosphere*, pp. 345–65, eds. G.M. Woodwell & E.V. Pecan. Proc. 24th Brookhaven Symposium in Biology, AEC Symposium. Series No. 30, TIC, USAEC.

REICHLE D.E., MCBRAYER J.F. & AUSMUS B.S. (1975) Ecological energetics of decomposers in a deciduous forest. In: *Progress in Soil Zoology*, ed. J. Vaněk, pp. 283–92. Academia, Prague.

REINERS W.A. (1968) Carbon dioxide evolution from the floor of three Minnesota forests. *Ecology* 44, 471–83.

REINERS W.A. (1973) Terrestrial detritus and the carbon cycle. In: *Carbon and the Biosphere*, eds. G.M. Woodwell & E.V. Pecan, pp. 303–27. Proc. 24th Brookhaven Symposium in Biology, Series No. 30. Tech. Info. Cent. USAEC.

RICHARDSON C.J. (1971) Development of a 'physiological site index', an indirect method of primary production estimation via the holocoenotic approach. US–IBP Memo Report 71–69.

SATCHELL J. (1974) Introduction. In: *Biology of plant litter decomposition*, eds. C.J. Dickinson & G.J.F. Pugh. Vol. 1, pp. i–xxxvi, Academic Press, New York.

STARK N. (1973) *Nutrient cycling in a Jeffrey pine ecosystem*. Institute of Microbiology, University of Montana, Missoula, Montana.

SZABO I.M. (1974) *Microbial communities in a forest-rendzina ecosystem: the pattern of microbial communities*. Akademia Kiado, Publisher. Budapest, Hungary, 415 pp.

THORNTHWAITE C.W. (1948) An approach toward a rational classification of climate. *Geograp. Rev.* 38, 85–106.

WAID J.S. (1974) Decomposition of roots. In: *Biology of plant litter decomposition*. eds. C.H. Dickinson & G.J.F. Pugh, Vol. 1, pp. 197–211. Academic Press, New York.

WITKAMP M. (1966) Biological concentration of ^{137}Cs, and ^{90}Sr in arctic food chains. *Nucl. Saf.* **18**, 58–62.

WITKAMP M. (1971) Soils as components of ecosystems. In: *Annual Review of Ecology and Systematics*, eds. R.F. Johnston, P.W. Frank & C.D. Michner, **2**, 85–110.

WITKAMP M. & CROSSLEY D.A. (1966) The role of arthropods and microflora in breakdown of white oak litter. *Pedobiol.* **6**, 293–303.

WITKAMP M. & FRANK M. (1969a) Environmental effects on microbial turnover of some mineral elements. Part 1: Abiotic factors. *Soil. Biol. Biochem.* **1**, 167–76.

WITKAMP M. & FRANK M. (1969b) Environmental effects on microbial turnover of some mineral elements. Part II. Biotic factors. *Soil Biol. Biochem.* **1**, 177–84.

WITKAMP M. & FRANK M. (1970) Effects of temperature, rainfall, and fauna on transfer of ^{137}Cs, K, Mg, and mass in consumer–decomposer microcosms. *Ecology* **51**, 465–74.

18 A review of root production processes and their influence on soil biota in terrestrial ecosystems

D. C. COLEMAN
Natural Resource Ecology Laboratory,
Colorado State University,
Fort Collins, Colorado 80523, U.S.A.

Introduction

One of the principal outcomes of international productivity studies over the last ten years has been the interest in, and the necessity for, interdisciplinary approaches to decomposition processes. The need to consider many trophic levels, virtually the entire assemblage of catabions and anabions (Lotka 1924) in the 'belowground ecosystem', was made evident in Phillipson (1970), and, more recently, in Marshall (in prep.). Particularly in the latter volume, the accurate and complete assessment of root production and root activity has emerged as a major focal point for much new work in the post-IBP era.

Although perhaps controversial, it seems appropriate to question the paradigm under which many total-system terrestrial ecology studies have operated. Thus, for want of funds, labour, or other limiting resources, they

Table 18.1. Distribution of organic matter production ($g\,m^{-2}\,yr^{-1}$) in four terrestrial ecosystems.

Production	Deciduous forest (Tennessee)*	Lightly-grazed shortgrass prairie (Colorado)†	Cool desert shrubs, *Atriplex* (Utah)‡	Arctic tundra (Barrow, Alaska)¶
Above ground	700 (370)	118	120	150
Below ground	900	576	270	100
Below ground/Total	0·56 (0·71)	0·83	0·69	0·4

* W.F. Harris & D.C. Coleman, unpublished manuscript. Parentheses indicate increments other than woody tissue.
† D.C. Coleman, R. Andrews, J.E. Ellis & J.S. Singh (unpublished data).
‡ Fernandez & Caldwell 1975.
¶ Shaver & Billings 1975.

have assessed structural components (litter, roots), measured inputs via standing crop estimates over various time spans, and in turn calculated inputs from maxima–minima. In fact, inputs from root structural material are a majority of the total organic inputs from many biomes (Table 18.1). There are several reasons why these estimates may be considerably in error (Iwaki & Midorikawa 1968; Milner & Hughes 1968): (1) root harvest studies measure only net increments over time; (2) labile or very small particulate materials are missed in sieving and sorting, etc. However, these studies have the advantage of comparability with pre-IBP studies, such as the massive compendium of Rodin & Bazilevich (1967).

Drawing extensively on recent data from several US/IBP biome programs, and other pertinent studies, three key aspects of root production and root activity are considered.

(1) The harvest method. As usually employed, it is inappropriately used, but can be substantially improved by:
 (a) considering age and size-classes of roots and
 (b) examining seasonal root growth dynamics.

(2) The sources of materials as input to the saprophagic (Wiegert & Owen 1971) food chains. These sources may be comprised of considerable amounts of sloughed root tissues, mucilaginous material 'mucigel', etc. (Jenny & Grossenbacher 1963) or soluble materials, in addition to root hairs.

(3) The impact of these labile materials on rhizosphere and mycorrhizal associates is certainly important and is just beginning to be elucidated.

Production of structural roots

HARVEST STUDIES

Woodlands

Recent reviews have documented extensive studies of tree root production (Hermann, in prep.; Lyr & Hoffmann 1967). Much of the extant literature (Hermann, in prep.) notes a six to ten or more fold difference between above-ground and below-ground productivity (Table 18.1). Because these estimates are based on infrequent measurements and not year-round, the root production estimates may be significantly low. The history of seasonality of root growth for a few species of forest trees is long, but not well noted. Thus Resa (1877) observed new root growth occurs in autumn for deciduous trees and in autumn and spring for coniferous trees. Recently, Harris *et al.* (in prep.) have examined the production and turnover of large (>0·5 cm diam.)

Table 18.2. Annual above-ground and below-ground productivity of various forest types (kg × 10^3 ha^{-1} yr^{-1}).

Type	Age (yr)	Above ground	Below ground	Total
Beech*	100	7·9	0·9	8·8
Oak*	48	8·3	2·4	10·7
Oak*	222	8·2	0·7	8·9
Spruce*				
Site I	100	11·9	2·1	14·0
Site III	100	6·5	1·3	7·8
Radiata pine*	29	20·0	3·0	23·0
Loblolly pine†	14	7·0	9·0	16·0
Tulip poplar†	80	7·0	9·0	16·0

* From Hermann, in press.
† From Harris *et al.*, in press.

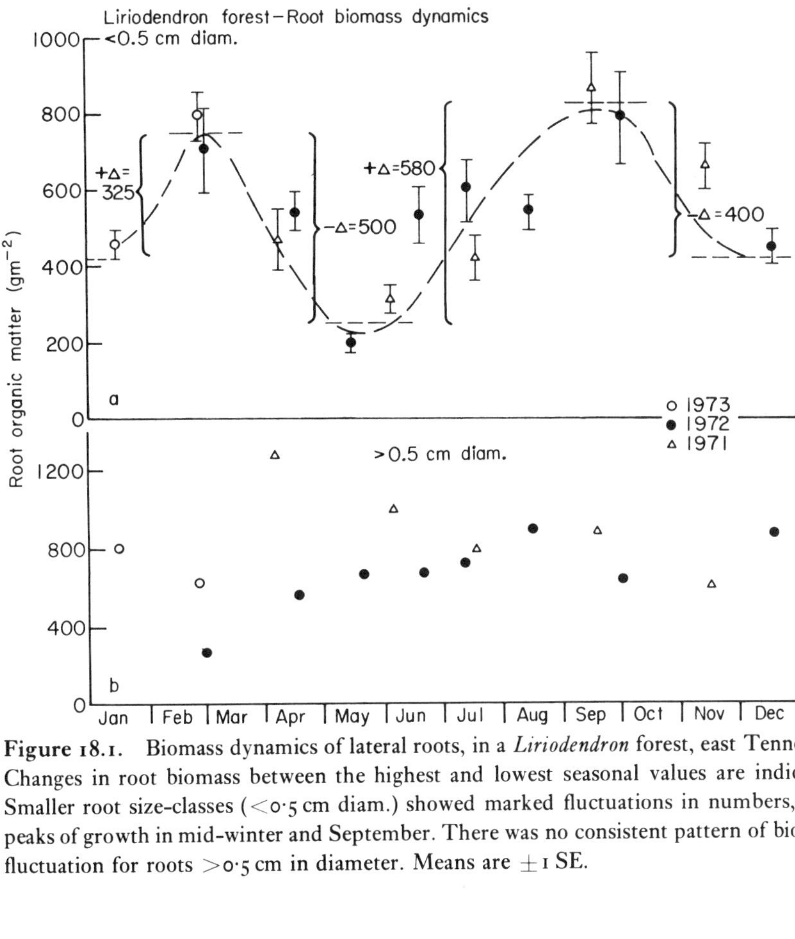

Figure 18.1. Biomass dynamics of lateral roots, in a *Liriodendron* forest, east Tennessee. Changes in root biomass between the highest and lowest seasonal values are indicated. Smaller root size-classes (<0·5 cm diam.) showed marked fluctuations in numbers, with peaks of growth in mid-winter and September. There was no consistent pattern of biomass fluctuation for roots >0·5 cm in diameter. Means are ±1 SE.

vs. small (<0.5 cm diam.) size-classes of roots (Table 18.2). Interestingly, the smaller root size-classes showed marked fluctuations in number, with peaks of growth in mid-winter and apparent rapid die-offs in spring (Fig. 18.1). Additionally, the life-span of these small roots was estimated as ≤ 1 year. Further evidence of this rapid turnover is shown in European beech, where fine roots (<2 mm diam.) dropped from 2,458 kg ha^{-1} in May to 189 kg ha^{-1} in August, a 92% drop. One year later (1970), however, the fine roots decreased from 1,169 kg ha^{-1} in May to 235 kg ha^{-1} in July, a lower loss, but still to only 20% of the original value (Hermann, in prep.). One difficulty in forests, as in other biomes, is the extremely large variability in root initiation, growth, and die-off. Thus Reynolds (1970) suggested small cells or volumes of soil as small as 30 cm on a side underwent cyclical initiation, extention, and death of fine roots, independently of environmental variation.

Tundra

Extensive studies at Point Barrow, Alaska, have uncovered a variety of adaptive or growth strategies by tundra plants. Using a 'tiller system analysis', Shaver & Billings (1975) were able to differentiate the growth habits of roots of *Carex aquatilis* Wahl., *Dupontia fischeri* R. Br., and *Eriophorum angustifolium* Honck. The turnover rates varied markedly, with *E. angustifolium* being annual to five to eight or more years for *C. aquatilis* (Table 18.3). Recent studies of translocation in *D. fischeri* found marked amounts moving to the roots, particularly in mid-summer, occasionally reaching 56% of the total (Allessio & Tieszen 1973). However, no information was given on exudates or sloughed materials. General estimates of root turnover were 100 g m^{-2} yr^{-1} or about 25% of the total mass per year.

Table 18.3. Summary of some characteristics of roots in three Arctic plant species (from Shaver & Billings 1975).

Characteristics	*Eriophorum angustifolium*	*Carex aquatilis*	*Dupontia fischeri*
Above-ground longevity	5 to 7 (8) yr	4 to 7 yr	3 to 4 (5) yr
Root longevity	Annual	5 to 8 (10) yr	4 to 6 yr
Elongation ability	1 yr	2 to 3 yr	1 yr
Seasonality of elongation	Continuous	Continuous	Early-mid season
Lateral production	None	2 to 4 yr	90% first yr
Seasonality of lateral production	—	Continuous	Mid-late season
Root origins	Stem base	Stem base	Rhizome nodes

Desert shrubland (cool semi-desert)

In studies of growth rates of three prominent shrub species of northern Utah (*Atriplex confertifolia* (Torr. & Frem.) S. Wats., *Ceratoides lanata* Nevski, and *Artemisia tridentata* subsp. *wyomingensis* Beetle) the seasonal root dynamics of these plants were followed. Fernandez & Caldwell (in press) studied growth rates of roots in the upper metre of soil in monospecific stands for each shrub, following growth over nearly eight months, from April–November 1973. The root growth was followed in insulated, Perspex-window chambers. Soil temperatures were measured by appropriate thermocouples and soil water potentials by thermocouple psychrometers. Root growth and size-class data were recorded during the season.

Table 18.4. Turnover coefficients, biomass, and productivity of root systems for shrub communities (from Caldwell & Camp 1974).

Communities	Root biomass ($g\,m^{-2}$) Tap roots	Root biomass ($g\,m^{-2}$) Lateral roots	Turnover coefficient Tap roots	Turnover coefficient Lateral roots	Root productivity ($g\,m^{-2}\,yr^{-1}$)	Shoot productivity ($g\,m^{-2}\,yr^{-1}$)
Ceratoides lanata	67	1,541	0·02	0·12	186	64
Atriplex confertifolia	48	1,823	0·12	0·24	443	154

Several major observations of interest resulted. Young roots elongated usually for only two to three weeks for any given individual meristem, with maximal rates approaching 10–14 mm day^{-1} (Fig. 18.2). Within a few weeks more, the roots underwent suberization, and gave little or no evidence of decomposition during the observation period (M.M. Caldwell, personal communication) as the upper layers dried out.

Related studies by Caldwell & Camp (1974) determined that root production was three times above-ground productivity for *Atriplex* and *Ceratoides*. In addition to measuring root production by careful harvest techniques, Caldwell & Camp (1974) also measured $^{14}C/^{12}C$ ratios in root structural carbohydrates over the time period April–September. The 'dilution effect' of gradually increasing amounts of ^{12}C in new roots vs. old roots gave an alleged independent determination of root turnover. Interestingly, the turnover times and the amounts (Table 18.4) closely paralleled one another.

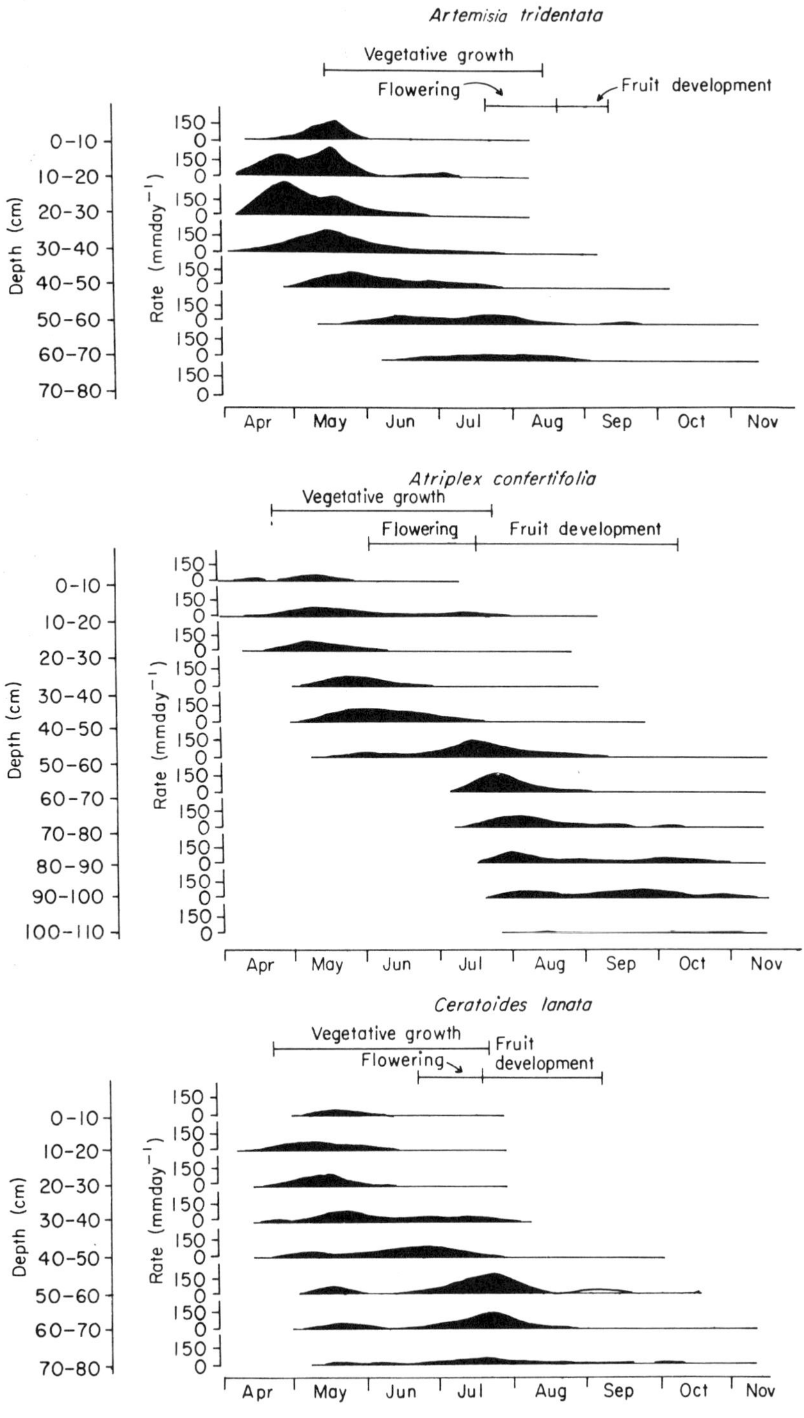

Figure 18.2. Root elongation rates (mean daily, for 10 cm increments in the soil). Rates are mm × day^{-1} for the visible root system in a 50 cm wide observation window. Phenological phases are also indicated (from Fernandez & Caldwell 1975).

Grasslands and meadows

Perhaps the greatest contrasts of all biomes occur in grassland ecosystems. Thus, going from cool, mesic through warm-arid, desert habitats, a marked shift in the root : shoot ratio takes place, ranging from 1 : 1 to 8 : 1 or more (Sims & Singh 1971). Estimates of annual root turnover have been made in several of these habitats, and range from 25% per year (Dahlman & Kucera 1967) (in more mesic) to 50% or more in the more xeric systems (Sims & Singh 1971). However, site factors may have an overriding influence, viz., 50% turnover in a hay-meadow (Nilsson 1970) vs. about 14% in a seashore-meadow (Tyler 1971).

Several workers have observed marked seasonality in root growth, e.g., between temperate (spring maxima) and tropical grasses (summer maxima) (Ueno & Yoshihara 1967) and within temperate grasses (Garwood 1967). Many pertinent observations of earlier literature on this and other aspects of herbaceous root growth are summarized by Troughton (1957) and in Whittington (1969) and are not extensively covered here. Root longevity has been studied in many areas, with widely varying conclusions drawn from them. Much earlier work (Tatarinova 1961, reviewed by Speidel & Weiss 1974) has noted the longevous nature of roots, whereas, in contrast, more recent work (Speidel & Weiss 1972, 1974) emphasizes the more transient nature of roots, particularly when observed from initiation onward. Speidel & Weiss (1974) noted that 54% of observed root tips survived ≤1 month while only 10–15% survived for ≥4 months (Table 18.5). In a fashion

Table 18.5. Demographic data on young roots in grasslands and desert.

Species	Increments (mm) Maximum	Increments (mm) Average	Elongation time (days)	Longevity (mo.)	Source
Lolium perenne	10	5			Garwood 1967
S. 23			24–35	0·2–0·5 (spring)	
			65–83	5·0–6·0 (autumn)	
Paspalum notatum	50	25	25–30	?	Ueno & Yoshihara 1967
Festuca rubra	5	3	21–36	0·5–1·0	Speidel & Weiss 1974
Bouteloua gracilis	60–100	5–10	14–21	1·0–2·0?	Ares 1976
Artemisia tridentata	14	~4	14–21	10–12?	Fernandez & Caldwell 1975
Atriplex confertifolia	7	~3	14–21	10–12?	Fernandez & Caldwell 1975

similar to the cool shrubs, 65–70% ceased elongation in less than two weeks.

Recent work of Ares (1976) in the Colorado shortgrass prairie shows trends similar to this. He observed (using insulated glass-panelled observation chambers similar to those used by Caldwell and colleagues) large amounts of root growth in April and May, some of it even before any apparent leaf growth. Ares also recorded the condition of individual roots for several months. He noted that 30–60% of the young, unsuberized roots decomposed within a few weeks after cessation of elongation in the spurt of spring growth (percent varying depending on the depth increments observed). The differences in timing of root death, as observed in the grassland studies, contrasted with the halophytic shrub roots is intriguing and remains unexplained.

In addition to root size-class observations, Ares (1976) also followed the biochemical status of the roots, noting that young, unsuberized roots underwent initiation and most of their growth occurred at times when available soil water was high (from −1 to −2 bars), even late in the season (early September) and that massive die-offs and suberization of the remainder occurred only later. Young, unsuberized roots constitute a significant portion of the total biomass at times (Table 18.6), and may be an important part of root production which is missed by occasional root sampling (every four to six weeks).

At the opposite end of the root longevity scale, recent work of Clark

Table 18.6. Ash-free dry weight ($g\,m^{-2}$, cumulative up to depth indicated) of root material at different depths at the Pawnee Site (Ares 1976).

Type of root	Depth −0·05 m	−0·15 m	−0·45 m
	15 May 1973		
Juvenile	0·5	0·5	0·8
Nonsuberized	29·5	179·5	224·5
Nonsuberized friable	17·0	26·0	41·0
Suberized	39·0	77·0	104·0
Suberized friable	31·5	46·5	105·0
	9 August 1973		
Juvenile	0·22	0·34	0·34
Juvenile friable	0·01	0·01	0·01
Nonsuberized	112·08	437·28	524·52
Nonsuberized friable	57·09	59·85	90·27
Suberized	180·00	205·46	210·50
Suberized friable	54·30	127·19	146·96

Table 18.7. Biomass carbon standing crops (g m^{-2} per 36 cm depth) in roots of blue grama shortgrass prairie, 1972 (from Clark, unpublished ms.).

Biomass carbon	29 March	10 May	12 June	8 August	3 October	$\bar{X}$	$\bar{X}$ (%)
Attached roots	130·1	133·3	112·2	126·2	95·5	119·5	17
Loose roots	425·3	339·4	371·6	301·4	399·7	368·5	54
Root debris	275·5	223·0	123·4	190·7	172·0	196·9	29
Total	830·9	695·7	607·2	618·3	667·2	684·9	100

and colleagues (unpublished) has demonstrated the existence of a large pool of 'detrital' roots (size range <2 mm to $>0{\cdot}3$ mm diam.) amounting to some 20–30% of total root carbon in shortgrass prairie ecosystems (Table 18.7). These are long-lived fragments, which have an as-yet-unknown turnover time. This further emphasizes the need to consider roots in age and size categories as shown in Fig. 18.3.

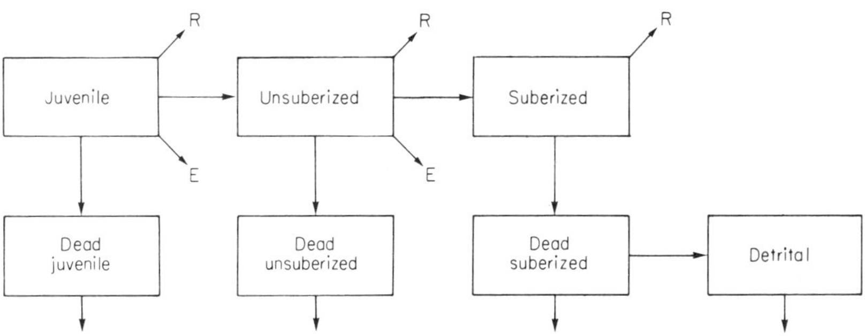

Figure 18.3. Schematic representation of root physiological classes for the shortgrass prairie. E = exudation, R = respiration activity. Detrital roots are probably several years old, having narrower C/N ratios than the younger roots.

Production of sloughed materials, exudates

Loss of materials from healthy roots have been much studied during the last decade, mostly under laboratory conditions (Head 1970, 1973). Details of several studies of loss of root cap cells, root hairs and secretions, and cortex and bark are given by Head (1973). Additionally, the studies of 'rhizosphere effects', thoroughly reviewed by Rovira (1973) and Rovira & Davey (1974), have given further scope to our understanding of the complex phenomena occurring in the volume of soil 1–2 mm from the root surface. Short-term exudation studies (McDougall 1968; McDougall & Rovira 1970) give considerable information on different categories of substances (diffusible, non-diffusible) and loci of release. The consequent microbial activity on soluble substances (Kunc & Macura 1966) and microbial and protozoan activity on soluble substances and mucigel (Jenny & Grossenbacher 1963; Greaves & Darbyshire 1972; Darbyshire & Greaves 1973) is considerable and well documented.

When one considers, in addition, the further effects of the freezing–thawing or wetting–drying phenomena (Reid 1974), the complexities multiply to near staggering proportions. Fortunately, however, a very recent comprehensive review of root exudate phenomena is available (Hale *et al.*, in press); the reader is referred to them for details on sources, mechanisms, and effects of environmental factors on root exudates. This ensuing discussion attempts to supplement these reviews and to place them in a 'whole-system' context. We will first examine agroecosystems and then 'natural' ecosystems.

AGROECOSYSTEMS

In what is now regarded as a classical study, Shamoot *et al.* (1968) grew 11 annual and perennial species (mostly crop plants) in airtight chambers, allowing the plants to photosynthetically fix ^{14}C from $^{14}CO_2$. After a growing period (usually of 212 days), plants were harvested, roots removed, and the remaining root-derived soil organic matter analysed for ^{14}C. Shamoot *et al.* (1968) found an additional 25–49 g organic debris per 100 g (dry wt.) of roots retrieved. Because this did not include more transient compounds which may have been exuded and rapidly metabolized by micro-organisms, this is probably an underestimate of organic carbon from the 'labile' root fraction.

In a more recent study, Johnen (1974) examined the growth of mustard and wheat plants in chambers which received $^{14}CO_2$ on the tops only. He measured several decomposition-soil phenomena of interest, including root respiration, decomposition, and addition to the soil organic matter in a fashion similar to Shamoot *et al.* (1968). He noted that an additional 0·3–0·6 times the root carbon was present in the soil from mustard, while only 0·2 times more carbon was in the soil from the wheat cultures. Because the wheat roots averaged 2·3 tons carbon ha^{-1} or 230 $g\,m^{-2}$, this would equal about 46 g carbon of sloughed materials and exudates in the soil. These figures are comparable to the range of the study of Shamoot *et al.* (1968).

An often-cited laboratory study (Samtsevich 1965) of the formation of mucilaginous caps on maize and wheat roots is interesting. Growing roots from seed, axenically, in moist air chambers and on filter paper, Samtsevich (1965) observed mucilaginous cylinders on most of the roots. Extrapolating from cylinder volume, assuming 1% dry matter in the material, he calculated 350 $g\,m^{-2}$ would be formed in a wheat field and 625 $g\,m^{-2}$ in a maize field. Assuming carbon is 45% of root organic matter, these values are three to six times greater than the maximal estimates of the workers who used radio-tracer-carbon and more normal (typical) growth conditions.

'NATURAL' ECOSYSTEMS

Little quantitative information on root sloughings and exudates in natural ecosystems is available. Two independent efforts have been in progress recently and will be reviewed.

Smith (in prep. and unpublished ms.) and colleagues have studied exudation of organic and inorganic compounds from tree roots for several years. Working at the Hubbard Brook Site, West Thornton, New Hampshire, they have laboriously isolated new woody roots of three species (*Betula alleghaniensis* Britton, *Fagus grandifolia* Ehrh., and *Acer saccharum* Marsh.) for study. After isolation and surface-sterilization of the new woody root tips, small air-layering chambers (Lyford & Wilson 1966) were used to collect the exudates. By calculating the number of new woody root tips per tree

Table 18.8. Root exudates from north-eastern hardwoods, New Hampshire, USA (from Smith, unpublished ms.). All values are in kg C ha^{-1} 150 $days^{-1}$.

	Tree species		
Compounds	*Betula alleghaniensis*	*Fagus grandifolia*	*Acer saccharum*
Carbohydrates	0·66	0·50	0·10
Amino acids–amides	0·08	0·11	0·02
Organic acids	0·90	1·44	0·23
Total: 4·04 *or* 0·4 g m^{-2} $season^{-1}$			

and per hectare, Smith obtained values of carbon in three moieties (carbohydrates, amino acids and amides, and organic acids) (Table 18.8). This value, 4·04 kg C ha^{-1} or only 0·4 g m^{-2}, is low relative to estimated fibrous root production and turnover. However, other sources remain to be studied before a full carbon budget can be calculated.

First, these exudates were accumulated only over the 150-day growing season, and the amounts exuded in the over 200 days not accounted for remain unknown. Second, many new fine roots are undoubtedly formed per year, perhaps in winter, as in *Liriodendron tulipifera* L. (Harris *et al.* in prep.). The organic input from these may be substantially greater than that of the exudates alone.

Recent work by Singh & Coleman (1974, in prep.) has examined the problem of 'functional' vs. 'nonfunctional' roots in the shortgrass prairie. As an ancillary study, we have done further tagging of blue grama sods with

^{14}C to determine the extent of movement of ^{14}C from the roots into the soil. Three 0·5 m^2 plots were labelled sequentially, for five consecutive days. Each plot received 0·4 mCi of $^{14}CO_2$ for 3·5–4 h per day. Three days after the final labelling period, duplicate soil cores were taken from each plot, roots were removed by hand, rinsed, autoradiographed, and combusted, and ^{14}C was counted (Singh & Coleman, in prep.). Soil samples were hand-

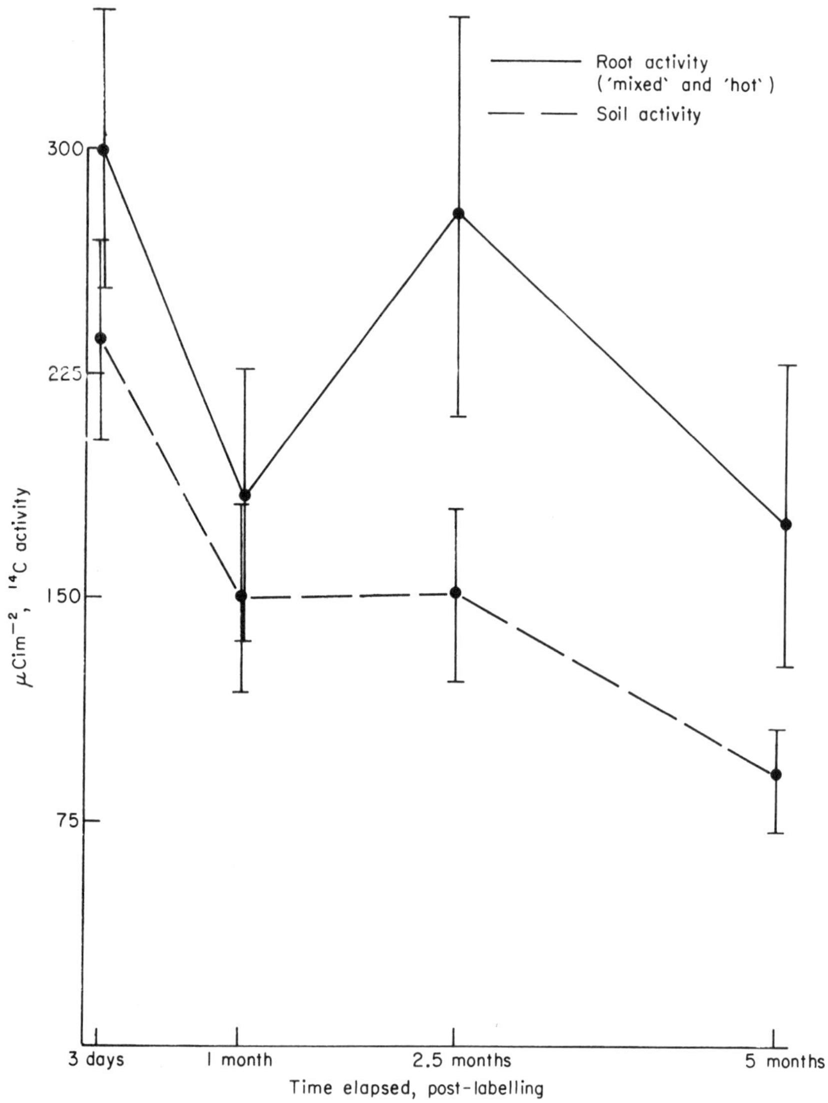

Figure 18.4. ^{14}C activity of root biomass and soil, per m^2, Pawnee Site, 1973. Means are ±1 SE. Vegetation was labelled in late June, then sampled 3 days, 1 month, 2·5 months, and 5 months later. Initial decrease is probably rapid decomposition of transient organic exudates, sloughings.

sorted to remove all visible debris, combusted in a modified Dumas nitrogen apparatus, and flushed with O_2, and evolved CO_2 was trapped in ethanolamine for counting by liquid scintillation.

Three days after labelling, root activity was 300 μCi m^{-2}, dropping to 186 at one month, increasing to 280 at 2·5 months, and dropping to 175 at five months (Fig. 18.4). Soil activity was 235 μCi m^{-2} (to 60 cm depth), 150 at one month, 157 at 2·5 months, and 91 μCi m^{-2} at five months. A comparison of soil : root activities is informative. For the four sampling dates the ratios were 0·78, 0·81, 0·56, and 0·52, respectively. The ratio of soil to root activity was high at first, then dropped to about 50% of the root activity, which approaches the upper level of activity reported by Shamoot *et al.* (1968). However, the difference between the initial activity level of 235 μCi and 150 μCi one month later, or 85 μCi, probably represents a considerable input of high-quality organic materials to the system. However, we do not as yet know the quantity of material represented by the ^{14}C-active substances in the soil organic matter fraction.

Dahlman (1967) measured ^{14}C in soil organic matter, as well as in roots. His samplings began ten months after the tops were labelled; thus no inferences about short-term transients from shoots–roots–soil are possible.

Root-mycorrhizal associations

Another area of root-rhizosphere interaction is the complex phenomenon of mycorrhizal associations. Recent references (Gerdemann 1968; Mosse 1973) document the role which mycorrhizae have in facilitating phosphorus uptake, particularly in P-deficient soils. The energy costs to the plant remain largely unknown, although some preliminary estimates have been made by Bevege *et al.* (in press). They observed that vesicular-arbuscular forms (endomycorrhizas) were much smaller 'sinks' for photosynthate than ectomycorrhizas (by ^{14}C tracer experiments). For example, Bevege *et al.* (in press) found that extramatrical *Endogone* (now *Glomus* spp.) hyphae represented only 1% of the plant weight, yet, via increased phosphate absorption, increased fresh plant (clover) weight by 150%.

Root contributions to the below-ground subsystem

Consideration should be given to a more elaborate 'word model' of the below-ground subsystem, which can be studied in a modular fashion, but one which is interlinked in many ways (Fig. 18.5). As an example of saprophagic feedback mechanisms, the enhanced shoot growth of unfertilized

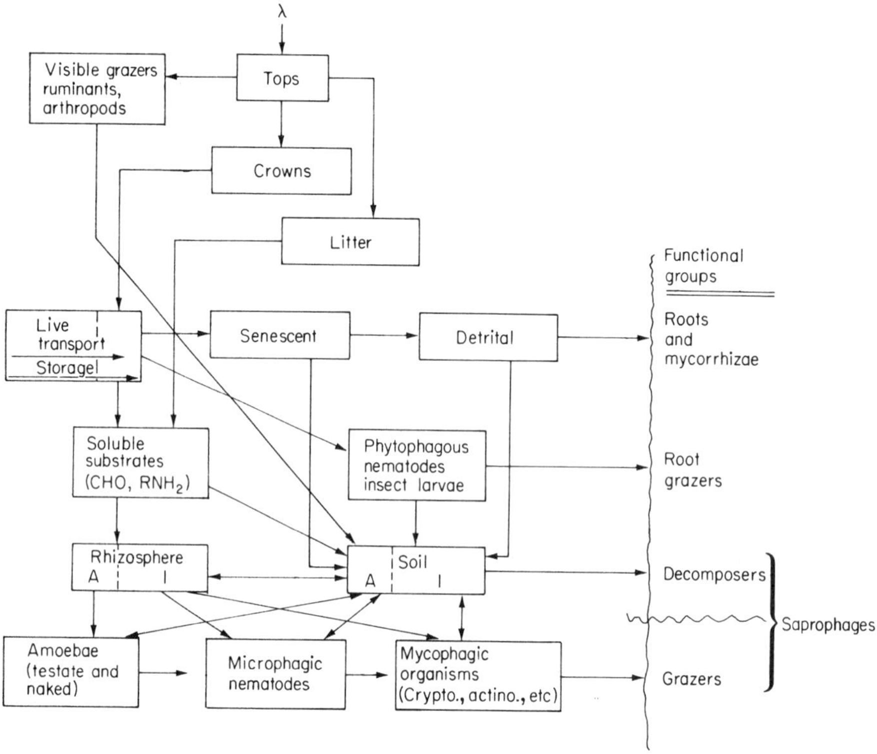

Figure 18.5. A 'word model' of the below-ground subsystem, specifically describing a shortgrass prairie. Physiological and functional groups are important in saprophagic processes. A=active, I=inactive microflora. Arrows to the right indicate respiratory energy losses. Recycling of mineral nutrients from saprophages to roots, while not represented in the diagram, is implicit.

plants with amoebae feeding in the rhizosphere (Nikulyuk & Gel'tzer 1972) is intriguing and deserves further study. Thus for a grassland ecosystem, one may view biotic control processes operating at several levels: root rhizosphere, root grazing, saprophagic activity, and saprophagic grazers. In many cases, I suggest that the transient small 'pulses' of organic materials serve as primers of microbial activity and may markedly influence the microbial activity and attendant nutrient cycling.

Several authors have recently speculated on the functional role of exudates, sloughing, organic acid release, etc. I suggest that several functions, namely facilitation of system-level retention of mineral nutrients (Harris & Coleman, unpublished ms.), exploitation of deeper areas of soil water with increasing dryness (Fernandez & Caldwell 1975), and, perhaps, phasing of root growth during nematode quiescence (e.g., Smolik 1974) may all play a part. There may be trade-offs between new root growth, with its 100 times greater water absorption than suberized roots (Ares 1976), and the simultaneous susceptibility of succulent tissues to invertebrate grazers. The opportunistic behaviour exhibited by tree and herbaceous roots is most impressive, and can be advantageously viewed in terms of organismal (root) response to occasional favourable events in an otherwise unfavourable environment.

Summary

In forest, shrub, and herbaceous ecosystems, a majority of the organic input comes from root production and decomposition phenomena. As studies continue and become more intensive, two aspects of root production are receiving greater emphasis: (1) the extent and timing of new roots' growth and die-off, and (2) the extent and quantity of organic exudates and sloughed root materials. The former probably contributes more carbon per m^2 than the latter, but data remains sparse.

Studies of exudates and sloughed materials have primarily utilized annual crop plants, but ^{14}C rhizo-deposition gave a similar 'pulse' to the saprophagic organisms after tagging perrenial shortgrass prairie. The roles of this high quality material, which at times may approach 200 g dry matter m^{-2} yr^{-1}, are important and many-faceted. These resources may have significant feedbacks on nutrient immobilization and availabilities in soil microsites via the grazing of soil animals on roots and microbial populations.

The population and possible system-level implications of these massive energy costs to plant growth are briefly examined and commented upon.

Acknowledgements

I wish to thank several colleagues in the US biome programmes who generously contributed data and/or criticisms to the paper: J. Ares, M.M. Caldwell, F.E. Clark, C.V. Cole, W.F. Harris, R.K. Hermann, and G.R. Shaver. Comments and data from M.G. Hale and W.H. Smith are also appreciated.

This paper reports on work supported in part by National Science Foundation Grants GB–31862X2, GB–41233X, and BMS73–02027 A02 to the Grassland Biome, US International Biological Program, for *Analysis of Structure, Function, and Utilization of Grassland Ecosystems.*

References

ALLESSIO M.L. & TIESZEN L.L. (1973) Patterns of translocation and allocation of ^{14}C-photoassimilate *in situ* studies with *Dupontia fischeri* R. Br. Barrow, Alaska. In: *Primary Production and Production Processes, Tundra Biome*, eds. L.C. Bliss & F.E. Wielgolaski, pp. 219–29. University of Alberta, Edmonton.

ARES J. (1976) Dynamics of the root system of blue grama (*Bouteloua gracilis* (H.B.K.) Lag.). *J. Range Manage.* **29** (3).

BEVEGE D.I., BOWEN G.D. & SKINNER M.F. (In press) *Comparative carbohydrate physiology of ecto- and endo-mycorrhizas.* Symposium on Endotrophic Mycorrhizas, Leeds, July 1974.

CALDWELL M.M. & CAMP L.B. (1974) Below ground productivity of two cool desert communities. *Oecologia (Berl.)* **17**, 123–30.

DAHLMAN R.C. (1967) *Carbon turnover and transfer studies of ^{14}C labelled roots in a prairie ecosystem.* Ph.D. dissertation, University of Missouri, Columbia.

DAHLMAN R.C. & KUCERA C.L. (1967) Carbon-14 cycling in the root and soil components of a prairie ecosystem. In: *Proceedings of the 2nd National Symposium on Radioecology*, eds. F.C. Evans & D.J. Nelson, pp. 652–60. AEC–TID 4500, Conf. 670503. Clearing House of Fed. Sci. and Tech. Info., Springfield, Va.

DARBYSHIRE J.F. & GREAVES M.P. (1973) Bacteria and protozoa in the rhizosphere. *Pest. Sci.* **4**, 349–60.

FERNANDEZ O.A. & CALDWELL M.M. (1975) Phenology and dynamics of root growth of three cool semi-desert shrubs under field conditions. *J. Ecol.* **63**, 703–14.

GARWOOD E.A. (1967) Seasonal variation in appearance and growth of grass roots. *J. Brit. Grassl. Soc.* **22**, 121–30.

GERDEMANN J.W. (1968) Vesicular-arbuscular mycorrhiza and plant growth. *Annu. Rev. Phytopathol.* **6**, 397–418.

GREAVES M.P. & DARBYSHIRE J.F. (1972) The ultrastructure of the mucilaginous layer on plant roots. *Soil Biol. Biochem.* **4**, 443–9.

HALE M.G., MOORE L.D. & GRIFFIN G.J. (In press) Root exudates and exudation. In: *Plant-Microorganisms Interactions*, eds. Y.R. Dommergues & J.P. Balandreau,

HARRIS W.F., KINERSON R.S. JR & EDWARDS N.T. (In prep.) Comparison of belowground biomass of natural deciduous forests and loblolly pine plantations. In: *The Belowground Ecosystem: A Synthesis of Plant-Associated Processes*, ed. J.K. Marshall. US/IBP Interbiome Symposium, Colorado State University, Fort Collins, 5–7 Sept. 1973.

HEAD G.C. (1970) Methods for the study of production in root systems. In: *Methods of Study in Soil Ecology*, ed. J. Phillipson, pp. 151–7. IBP/UNESCO, Paris.

HEAD G.C. (1973) Shedding of roots. In: *The Shedding of Plant Parts*, ed. T.T. Kozlowski, pp. 237–93. Academic Press, New York.

HERMANN R.K. (In prep.) Growth and production of tree roots: A review. In: *The Belowground Ecosystem: A Synthesis of Plant-Associated Processes*, ed. J.K. Marshall. US/IBP Interbiome Symposium, Colorado State University, Fort Collins, 5–7 Sept. 1973.

IWAKI H. & MIDORIKAWA B. (1968) Principles for estimating root production in herbaceous perennials. In: *Methods of Productivity Studies in Root Systems and Rhizosphere Organisms*, eds. M.S. Ghilarov, V.A. Kovda, L.N. Novichova-Ivanova, L.E. Rodin & V.M. Sveshnikova, pp. 72–8. Nauka, Leningrad.

JENNY H. & GROSSENBACHER K. (1963) Root-soil boundary zones as seen in the electron microscope. *Soil Sci. Soc. Am., Proc.* **27**, 273–7.

JOHNEN B.G. (1974) *Bildung, Menge, und Umsetzung von pflanzenwurzeln im Boden.* Ph.D. dissertation, Rhein. Fried. Wilh. Univ., Bonn.

KUNC F. & MACURA J. (1966) Decomposition of root exudates in soil. *Folia Microbiol.* **11**, 239–47.

LOTKA A.J. (1924) The energy transformers of nature. In: *Elements of Mathematical Biology*, pp. 325–35. Dover Publications, Inc. (1956 reprint), New York.

LYFORD W.H. & WILSON B.F. (1966) Controlled growth of forest tree roots: Technique and application. *Harvard For. Pap. 16*, Petersham, Mass.

LYR H. & HOFFMANN G. (1967) Growth rates and growth periodicity of tree roots. *Int. Rev. For. Res.* **2**, 181–236.

MARSHALL J.K. (ed.) (In prep.) *The Belowground Ecosystem: A Synthesis of Plant-Associated Processes.* US/IBP Interbiome Symposium. Colorado State University, Fort Collins, 5–7 Sept. 1973.

McDOUGALL B.M. (1968) The exudation of ^{14}C-labelled substances from roots of wheat seedlings. *9th Int. Congr. Soil Sci., Trans.* **3**, 647–55.

McDOUGALL B.M. & ROVIRA A.D. (1970) Movement of C^{14}-photosynthate into roots of wheat seedlings and exudation of C^{14} from intact roots. *New Phytol.* **69**, 37–46.

MILNER C. & HUGHES R.E. (1968) *Methods for the Measurement of the Primary Production of Grassland: IBP Handbook No. 6.* Blackwell Scientific Publications, Oxford, xii+70 pp.

MOOSE B. (1973) Advances in the study of vesicular-arbuscular mycorrhiza. *Annu. Rev. Phytopathol.* **11**, 171–96.

NIKULYUK B. & GEL'TZER J. (1972) *Soil Protozoa of the U.S.S.R.* Acad. Sci., Tashkent, Uzbek S.S.R., U.S.S.R. (in Russian).

NILSSON J. (1970) Notes on the biomass and productivity of belowground organs of south-Swedish hay meadow. *Bot. Not.* **123**, 183–94.

PHILLIPSON J. (ed.) (1970) *Methods for the Study of Soil Ecology.* IBP/UNESCO, Paris, 303 pp.

REID C.P.P. (1974) Assimilation, distribution, and root exudation of ^{14}C by ponderosa pine seedlings under induced water stress. *Plant Physiol.* **54**, 44–9.

RESA F. (1877) Über die Periode der Wurzelbildung. Inaug. diss., Bonn (as cited by: McDougall W.B. (1916) The growth of forest tree roots. *Am. J. Bot.* **3**, 384–92).

REYNOLDS E.R.C. (1970) Root distribution and the cause of its spatial variability in *Pseudotsuga taxifolia* (Poir.) Britt. *Plant Soil* **32**, 501–17.

RODIN L.E. & BAZILEVICH M.I. (1967) *Production and Mineral Cycling in Terrestrial Vegetation.* Oliver and Boyd, Edinburgh and London.

ROVIRA A.D. (1973) Zones of exudation along plant roots and spatial distribution of microorganisms in the rhizosphere. *Pest. Sci.* **4**, 361–6.

ROVIRA A.D. & DAVEY C.B. (1974) Biology of the rhizosphere. In: *The Plant Root and its Environment*, ed. E.W. Carson, pp. 153–204. Univ. Press of Virginia, Charlottesville.

SAMTSEVICH S.A. (1965) Active excretions of plant roots and their significance. *Fiziol. Rastenii* **12**, 731–40. (Engl. transl.)

SHAMOOT S., McDONALD L. & BARTHOLOMEW W.V. (1968) Rhizo-deposition of organic debris in soil. *Soil. Sci. Soc. Am., Proc.* **32**, 817–20.

SHAVER G.R. & BILLINGS W.D. (1975) Root production and root turnover in a wet tundra ecosystem, Barrow, Alaska. *Ecology* **56**, 401–9.

SIMS P.L. & SINGH J.S. (1971) Herbage dynamics and net primary production in certain ungrazed and grazed grasslands in North America. In: *Preliminary Analysis of Structure and Function in Grasslands*, ed. N.R. French, pp. 59–124. Range Sci. Dep. Sci. Ser. No. 10. Colorado State University, Fort Collins.

SINGH J.S. & COLEMAN D.C. (1974) Distribution of photoassimilated ^{14}carbon in the root system of a shortgrass prairie. *J. Ecol.* **62 (2)**, 359–65.

SINGH J.S. & COLEMAN D.C. (In press) Evaluation of functional root biomass and translocation of photoassimilated carbon-14 in a shortgrass prairie ecosystem. In: *The Belowground Ecosystem: A Synthesis of Plant-Associated Processes*, ed. J.K. Marshall. US/IBP Interbiome Symposium, Colorado State University, Fort Collins, 5–7 Sept. 1973.

SMITH W.H. (In prep.) Tree root exudates and the forest soil ecosystem: Exudate chemistry, biological significance and alteration by stress. In: *The Belowground Ecosystem: A Synthesis of Plant-Associated Processes*, ed. J.K. Marshall. US/IBP Interbiome Symposium, Colorado State University, Fort Collins, 5–7 Sept. 1973.

SMOLIK J.D. (1974) *Nematode Studies at the Cottonwood Site.* US/IBP Grassland Biome Tech. Rep. No. 251. Colorado State University, Fort Collins.

SPEIDEL B. & WEISS A. (1972) Zur ober- und unterirdischen Stoff-produktion einer Goldhaferwiese bei verschiedener Düngung. *Angew. Botanik* **46**, 75–93.

SPEIDEL B. & WEISS A. (1974) Untersuchungen zur Wurzelaktivität unter einer Goldhaferwiese. *Angew. Botanik* **48**, 137–54.

TATARINOVA N.K. (1961) On the longevity of the roots of meadow grasses. *Bot. Zhurn.* **46**, 925–36 (in Russ., Engl. summ.).

TROUGHTON A. (1957) *The Underground Organs of Herbage Grasses.* Commonwealth Bur. Pastures Field Crops Bull. No. 44, Hurley, Berkshire.

TYLER G. (1971) Distribution and turnover of organic matter and minerals in a shore meadow ecosystem. IV. Studies in the ecology of Baltic Sea-shore meadows. *Oikos* **22**, 265–91.

UENO M. & YOSHIHARA K. (1967) Spring and summer root growth of some temperate-region grasses and summer root growth of tropical grasses. *J. Brit. Grassl. Soc.* **22**, 148–52.

WHITTINGTON W.J. (ed.) (1969) *Root Growth.* Butterworths, London.

WIEGERT R.G. & OWEN D.F. (1971) Trophic structure, available resources, and population density in terrestrial vs. aquatic ecosystems. *J. Theor. Biol.* **30**, 69–81.

Section Four
Modelling of decomposition systems

19 Decomposition models based on climatic variables, substrate variables, microbial respiration and production

P. W. FLANAGAN *University of Alaska, Fairbanks, Alaska 99701, U.S.A.*
F. L. BUNNELL *Faculty of Forestry, University of British Columbia, Vancouver, B.C., Canada*

Abstract

Results of our efforts to model decomposition in tundra are presented. Models were developed from field studies and specific laboratory analyses. Later, model predictions were evaluated against further field and laboratory data. Respirometric measures of decomposition and micro-organism metabolism are emphasized.

Introduction

The recognition that word models are an important initial step in the development of more rigorous mathematical models was explicit in the early stages of the International Biological Programme (Coloumbe & Brown 1970). The U.S. Tundra Biome's Decomposition Group and members of the International Tundra Working Groups on Decomposition and Microbiology initiated development of mathematical models in June 1972 (see Bunnell & Tait 1974 and Bunnell & Scoullar 1975 for a discussion of historical development). The objectives of these modelling efforts were essentially threefold:

(i) to explain seasonal and annual variations in decomposition rates;
(ii) to evaluate the influences on decomposition of moisture, temperature and substrate chemistry;
(iii) to incorporate explicitly microbiological data and relationships into a model which would facilitate intersite comparisons within tundra and allow extrapolation beyond the tundra ecosystem.

It was apparent that studies in weight loss were inadequate to address these three objectives.

Measures of weight loss integrate many simultaneously occurring phenomena and a large amount of variability within and between years is expected. The initial attempts at synthesis across a wide variety of tundra sites (Heal & French 1974) documented this tremendous variability. In developing the simulation models discussed here the approach was to relate specific processes to broad integrative measures within a biologically sound and mathematically rigorous framework. Thus, laboratory measures of specific processes, such as respiration and substrate utilization (Flanagan & Veum 1974; Flanagan & Scarborough 1974) could be related to field measures of weight loss (Bunnell & Scoullar 1975; Bunnell *et al.* in press and Bunnell *et al.* in press). Our premise was that the explicit recognition of component processes would account for the great variability in integrative measures (Bunnell & Tait 1974). In developing the models we have recognized two integrative approaches to decomposition studies. One holds that the total 'organic matter-microbial' complex respires measurable CO_2 in relation to climatic and state variables; the evolved CO_2 closely reflects the decomposition process and rate. The second approach states that the microbial biomass, annual productivity and associated respiration reflect the turnover rate of organic matter and when assessed in relation to annual primary productivity, inform us of the status of organic accumulation. Each broad approach can be stated as a complex hypothesis. Models incorporating either of these hypotheses can be tested for a variety of ecosystems. We have explored the hypotheses in relation to the tundra ecosystem particularly at Barrow, Alaska.

It is not the purpose of this paper to describe in detail the models built by the decomposition group of the U.S. IBP Tundra Biome Programme or the International Tundra Biome's Decomposition Working Group. The paper presents precis of two models, GRESP and DCOMP, which constitute a still evolving framework within which we are exploring decomposition processes and the possibility of developing an understanding of their role in ecosystems. The models were presented by Bunnell & Tait (1974), evaluated by Bunnell *et al.* (in press a and b) and examined within an ecosystem perspective by Bunnell & Scoullar (1975). The approach is strictly synecological and differs clearly from the approach of Marr *et al.* (1963). We also have employed data on growth kinetics of individual micro-organisms in the model of Marr *et al.* (1963) to simulate organism turnover rates in the tundra phyllosphere. This more autoecological approach has been used by several investigators in other ecosystems (Babiuk & Paul 1970; Gray & Williams 1971).

The general paradigms involved in most earlier models of decomposition (e.g. Jenny *et al.* 1949; Olson 1963; Minderman 1968; Jones & Gore, in press) were reviewed and compared to modelling within the Tundra Biome of IBP by Bunnell & Tait (1974). The models discussed in the present paper differ from earlier works in that the variations in decomposition rates are over short periods (daily) and explicitly acknowledge the influences of changing moisture, temperature and substrate quality. Further, they explicitly relate microbial processes such as respiration to broad patterns of weight loss. Bunnell & Tait (1974) discuss the merits of explicit versus implicit treatments of theory.

However, as have most earlier studies, the models GRESP and DCOMP ignore microfloral biomass. Other workers (e.g. Gray & Williams 1971) have brought attention to the difficulties encountered in relating field measures of annual microbial biomass transients and chemostat assessments of efficiency of microbial productivity and maintenance demands to field measures of primary productivity. Annual microbial maintenance demands calculated from chemostat determination of maintenance requirements may preclude growth of field microbial populations (Gray & Williams 1971). However, microbial growth is known to occur in the sites in question. Explanations offered for this paradoxical situation include: (i) underestimated productivity; (ii) conservative or inflated estimates of yield coefficient and maintenance demand respectively; (iii) no real accumulation of organic matter in the system, but a decrease; and (iv) inflated estimates of microbial biomass.

Here we test the model of Marr *et al.* (1963) using fungal measures from Barrow.

Laboratory, field and modelling methods

Microbial ecologists must brave a wavering episteme whenever laboratory studies of processes are related to field studies of the results of processes. Epistemological questions loom especially large when the relations are to be expressed unambiguously in mathematical models (Bunnell 1973). Gross as most field data are, they are 'real', integrated measures of decomposer activities and the influences of rate regulators upon them. We assumed respirometric measures of microbial activity to be especially 'real' and selected moisture, temperature and substrate quality as the most influential regulators of both respiration and weight loss. Using Gilson respirometers we examined influences of moisture, temperature and substrate quality on microbial respiration in above-ground substrates (Flanagan & Veum 1974; Bunnell *et al.* in press). These measures were incorporated in the models GRESP; DCOMP; as described by Bunnell *et al.* (in press, a and b).

Standing dead tissues were calculated in the manner described below. Before presenting the calculations for substrate availability it must be emphasized that the figures presented are subject to spatial and temporal variations within and between specific subsites of the Barrow intensive study site (Bunnell *et al.* 1975c). The study area referred to here is the same subsite of the Barrow intensive site referred to in Flanagan & Veum (1974) and Flanagan & Scarborough (1974). This subsite is an almost pure stand of *Eriophorum angustifolium* with randomly dispersed small (25–100 cm^2) patches of *Carex aquatilis*.

Approximately 100 g m^{-2} of new leaf tissue was produced annually on the site. The average standing crop of standing dead leaf tissue over four seasons was also about 100 g m^{-2} and on average was composed of 60, 32 and 11 g m^{-2} of leaves that were one, two and three years old respectively. Four-year-old standing dead leaves were rarely encountered. The between-year variation in amount of standing dead m^{-2} was relatively small (90 g m^{-2}—118 g m^{-2}).

In the absence of lemming influences, or production of new leaf materials, approximately 30% of the weight was removed as leachate from the standing dead zone by autumn (fall) rains and the freeze–thaw cycle leaving approximately 70 g m^{-2} yr^{-1} to become new standing dead tissue (all figures given are averages derived from four years of observation). Because standing dead vegetation did not accumulate from year to year we have assumed that on average the decomposition losses from standing dead approximate the rate of production, about 70 g m^{-2}. Further, there is a loss of 70—60 = 10 g m^{-2} of standing dead vegetation during the first growing season. The effects of intensive lemming grazing greatly alter the relationships outlined above (Bunnell *et al.* 1975c). However, during the period 1972 to 1974 inclusive lemming numbers were generally much lower than the pronounced cyclic highs noted by Pitelka (1973).

Annual and instantaneous moisture and temperature dependent weight losses in terms of CO_2 loss in respiration were determined earlier for standing dead leaves (Flanagan & Veum 1974; Bunnell *et al.* in press, a). In this study these values are compared with calculated respiration for fungal maintenance and production in standing dead leaf remains. The approach allows an evaluation of the model of Marr *et al.* (1963) for predicting organic turnover rates in a relatively simple natural system using empirical data concerning specific decomposer micro-organisms and their respiration in a native substrate.

Results and discussion

A. THE CHEMOSTAT MODEL

By examining growth, respiration, production efficiency and maintenance demands of individual micro-organisms in dead leaves and incorporating these data in a model we were able to predict annual weight losses in the leaf tissue. The model of Marr *et al.* (1963) states:

$$\frac{dx}{dt}+ax=y\frac{ds}{dt} \tag{19.1}$$

where $\frac{dx}{dt}$=instantaneous growth rate

a=maintenance demand ($g\,g^{-1}\,hr^{-1}$),
x=biomass of micro-organisms,
y=production efficiency or yield coefficient ($g\,g^{-1}$ substrate),
and $\frac{ds}{dt}$=instantaneous change in substrate.

If growth ceases then

$$\frac{dx}{dt}=0$$

and

$$\frac{ax}{y}=\frac{ds}{dt} \tag{19.2}$$

From growth and respiration measures of batch cultures (e.g. Fig. 19.1a and 19.1b) the values of 'a' and 'y' were calculated as $0{\cdot}32\times10^{-3}\,g\,g^{-1}\,hr^{-1}$ and $0{\cdot}35\,g\,g^{-1}$ substrate respectively. The average standing crop of fungi (x in equation (19.1) and (19.2)) determined by the Jones & Mollison (1948) technique was 0·0045 g fungal tissue per gram standing dead leaf tissue.

Using equation (19.2) we can calculate the rate of substrate utilization for maintenance of the microfungi per square metre of standing dead:

$$\frac{ds}{dt}=\frac{0{\cdot}32\times10^{-3}\,g\,g^{-1}\,hr^{-1}\times0{\cdot}0045\,g\,g^{-1}\times100\,g\,m^{-2}}{0{\cdot}35\,g\,g^{-1}}$$

$$=4{\cdot}1\times10^{-4}\,g\,hr^{-1}\,m^{-2}$$

Substrate utilization for maintenance over the active period of 120 days is thus $4{\cdot}1\times10^{-4}\,g\,hr^{-1}\,m^{-2}\times24\,hr\times120=1{\cdot}2\,g\,m^{-2}$, which leaves $70-1{\cdot}2$ or $68{\cdot}8\,g\,m^{-2}$ of substrate for growth.

To estimate the number of generations we can employ equation (19.3):

$$y(S+Nx)=Nx \tag{19.3}$$

where y and x are as equation (19.1),
S represents the amount of available substrate, and
N represents the number of generations.

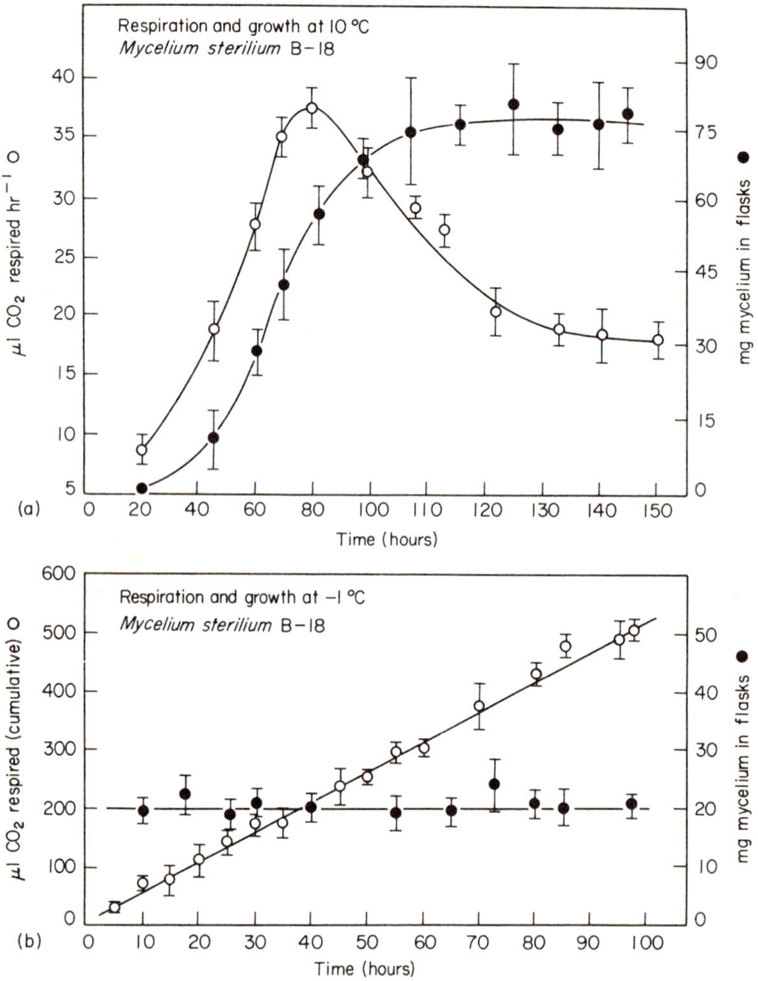

Figure 19.1. Respiration and growth of a common leaf litter fungus on Barrow tundra growing on a mineral salt medium containing a known glucose concentration. Vertical bars represent standard deviations.

Using a value of 68·8 for S, from equation (19.3), we obtain

$$N = \frac{ys}{X(1-y)}$$

$$= \frac{0{\cdot}35\,\mathrm{g\,g^{-1}} \times 68{\cdot}8\,\mathrm{g\,m^{-2}}}{0{\cdot}45\,\mathrm{g\,m^{-2}}\,(1-0{\cdot}35\,\mathrm{g\,g^{-1}})}$$

$$= 82{\cdot}3 \text{ generations}$$

The computed number of generations corresponds to a renewal every 35 hours during the active season or a gross production of 37 g fungal tissue $m^{-2}\,yr^{-1}$.

However 37 g m^{-2} production of fungal tissue would result in 100% loss of standing dead leaves annually. The average annual loss in field leached standing dead is approximately 10 g m^{-2} as measured per unit length of marked dead leaves. Weight losses in litter below, measured in litter bags, is of the same magnitude. On this basis we have assumed that 8·8 g m^{-2} are available annually for microbial growth in standing dead tissue, allowing for 15 generations of the average standing crop m^{-2} of microbes or a production of 4·7 g m^{-2}. These predictions of microbial production are compatible with the availability of substrate and allow us to calculate a yield coefficient (*sensu* Heal & MacLean 1974) of $y' = 4{\cdot}7/8{\cdot}8 = 0{\cdot}54$. Unlike the yield coefficients of equation (19.1), and (19.2), y' allows the recycling of microbial tissue towards its own production. There appears to be no published estimates of fungal production in the vegetation canopy with which we can compare our results. It may be pertinent to note that our estimates of generation time do not fall within the range reported for Parinkina (1974) for bacteria in tundra soils.

To a limited extent we can evaluate the veracity and integrity of our assumptions by independent studies on microbial respiration in standing dead vegetation and weight losses from litter bags placed in the standing dead canopy. During the growing season of the first year following death, standing dead vegetation is diminished by about 10 g m^{-2} or about 14% (10 g m^{-2}/70 g m^{-2}). Flanagan & Veum (1974) predicted weight losses due to microbial respiration of 10·1 – 11·8% and reported measured weight losses from litter bags in the standing dead canopy of 9·9 ± 1·8%. These observations suggest that microbial respiration alone accounts for a significant proportion of the total weight lost, and do not contradict the assumptions made in equations (19.1), (19.2) and (19.3). The high proportion of weight loss due to microbial respiration is similarly compatible with the initial leaching losses (30%), largely complete by the beginning of the active growing season, and the relatively low abundance of invertebrate saprovores feeding aboveground (Bunnell *et al.* 1975c). In summary, we feel that careful examination of individual micro-organisms can lead to understanding of rates of decomposition and organic matter turnover at the 'macro' level of ecosystems. Admittedly the present study was relatively simple because of the single substrate and predominance of a few organisms (Flanagan & Scarborough 1974), but it does suggest that autoecological studies of micro-organisms can contribute to the understanding of rates often viewed at an ecosystem level.

B. MATHEMATICAL SIMULATION MODELS

Flanagan & Veum (1974) utilized empirical relationships to show that weight loss attributable to C loss in respiration was often close to weight losses measured in litter bags, particularly in material 1–3 years of age. They employed generalized moisture and temperature responses of tundra decom-

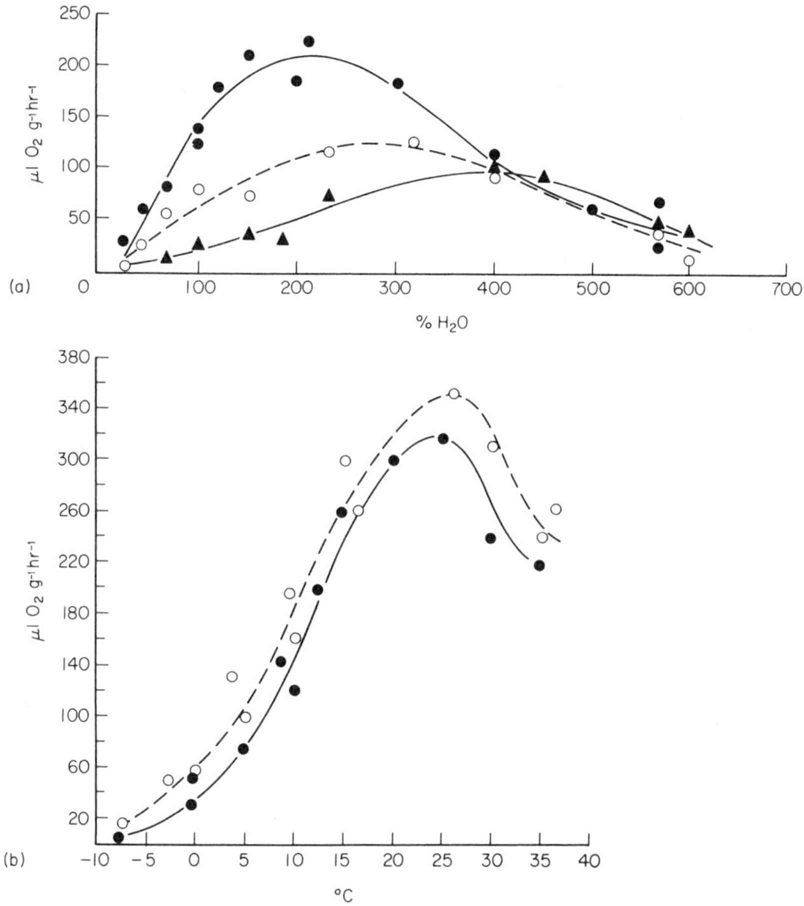

Figure 19.2A. Response of microbial respiration in Barrow litter at 0°C, (▲); 5°C, (○); and 10°C, (●), to changes in moisture content.

Figure 19.2B. Response of Barrow litters (⊙=*Carex aquatilis* and ●=total litter) at 120% H_2O (dry wt) to temperature variation. In A and B lines are plotted by hand and points represent the mean of ten samples each of which was read ten times during the experiments.

poser systems such as those illustrated in Fig. 19.2. Among the tundra mycoflora the Q_{10} value (averaged over 10°C increments from −10°C to 25°C) was 4. Respiration continued down to −5·0°C; was depressed by temperatures above 30°C and was greatest at 25°C. In general cardinal moisture levels were 20%, 200–500% and 1,000% respectively for minimal, optimal and maximal levels. Moisture and temperature responses were interdependent over a wide range of both parameters. Usually, increasing temperature shifted the optimal moisture range towards the Y axis as in Fig. 19.2a. From such data we constructed respiration response surfaces for litter and standing dead vegetation of varying substrate quality (e.g. Fig. 19.3).

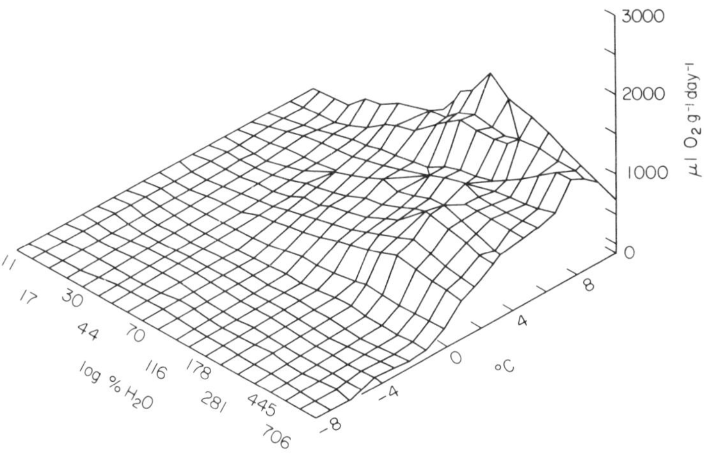

Figure 19.3. Response surface of microbial respiration versus temperature and moisture in total litter at Barrow. Each grid point represents an average of 6 to 30 replicates (Flanagan & Veum 1974).

Having documented general patterns and developed empirical models the next step was to relate specified processes within a biologically sound and mathematically vigorous framework (Bunnell & Tait 1974). The first model formulated was given the mnemonic GRESP because it was based largely on experimentation with Gilson respirometry. This model assumes that the major influences on microbial respiration are moisture, oxygen, temperature, and substrate. It incorporates the influences of these factors on microbial respiration as deduced from laboratory data and integrates the effects over climatic variation within a specific substrate and season to produce independent CO_2 based assessments of weight loss. These predicted losses generated from specific process studies can be related to the integrated weight loss as measured by litter bags.

(i) *The GRESP Model*

The rationale of the model is described in detail by Bunnell & Tait (1974) and Bunnell *et al.* (in press, a). High and low moisture levels adversely influence respiration rates. Equations (19.4) and (19.5) are first approximations of the moisture influence on microbial respiration as incorporated in the model. At moisture levels below optimum, respiration decreases according to equation (19.4), at high moisture levels respiration decreases according to equation (19.5). Thus, at low moisture levels the rate of moisture supply to respiring organisms is determined by coefficient a_1. Coefficient a_2 determines the supply of oxygen to micro-organisms under conditions of high moisture. Both coefficients in equations (19.4) and (19.5) are assumed to follow

Michaelis–Menten relationships with the percentage moisture content of the decomposing substrates.

$$WL=\frac{M}{a_1+M} \tag{19.4}$$

where WL = water limitation reducing respiration,
M = % water content on a dry weight basis, and
a_1 = % water content at which the substrate is half 'saturated' with water.

$$O_a=\frac{M}{a_2+M} \tag{19.5}$$

where O_a = oxygen availability,
M = % water content on a dry weight basis, and
a_2 = water content at which 50% of air spaces are water filled.

The proportion of free channels is therefore

$$1-\left(\frac{M}{a_2+M}\right)=\frac{a_2}{a_2+M}$$

The substrate quality influences on respiration are expressed as

$$a_3\times a_4^{T-10/10} \tag{19.6}$$

where a_3 = the optimal respiration rate at 10°C when neither H_2O nor O_2 are limiting,
a_4 = Q_{10} coefficient, and
T = temperature°C.

The formal statement of the respiration model is:

$$R(TM)=\frac{M}{a_1+M}\times\frac{a_2}{a_2+M}\times a_3\times a_4^{T-10/10} \tag{19.7}$$

where R(TM) = the respiration rate at temperature T and moisture level M.

Estimating model coefficients

Because of the model's nonlinear form, estimation of coefficient values (a_1 through a_4) requires nonlinear optimization techniques (Bunnell *et al.* in press a). Two determinants of the goodness of fit with respect to coefficients a_1 through a_4 were computed and designated r^2 and $r_A{}^2$. The coefficient of determination r^2 is not the square of the correlation coefficient used in linear regression but it can be used in the same way since it determines the proportion of variation in the data accounted for by the model. The statistic $r_A{}^2$ determines the proportion of 'accountable variation' in the data explained in the model. Its derivation is discussed by Bunnell *et al.* (in press, a).

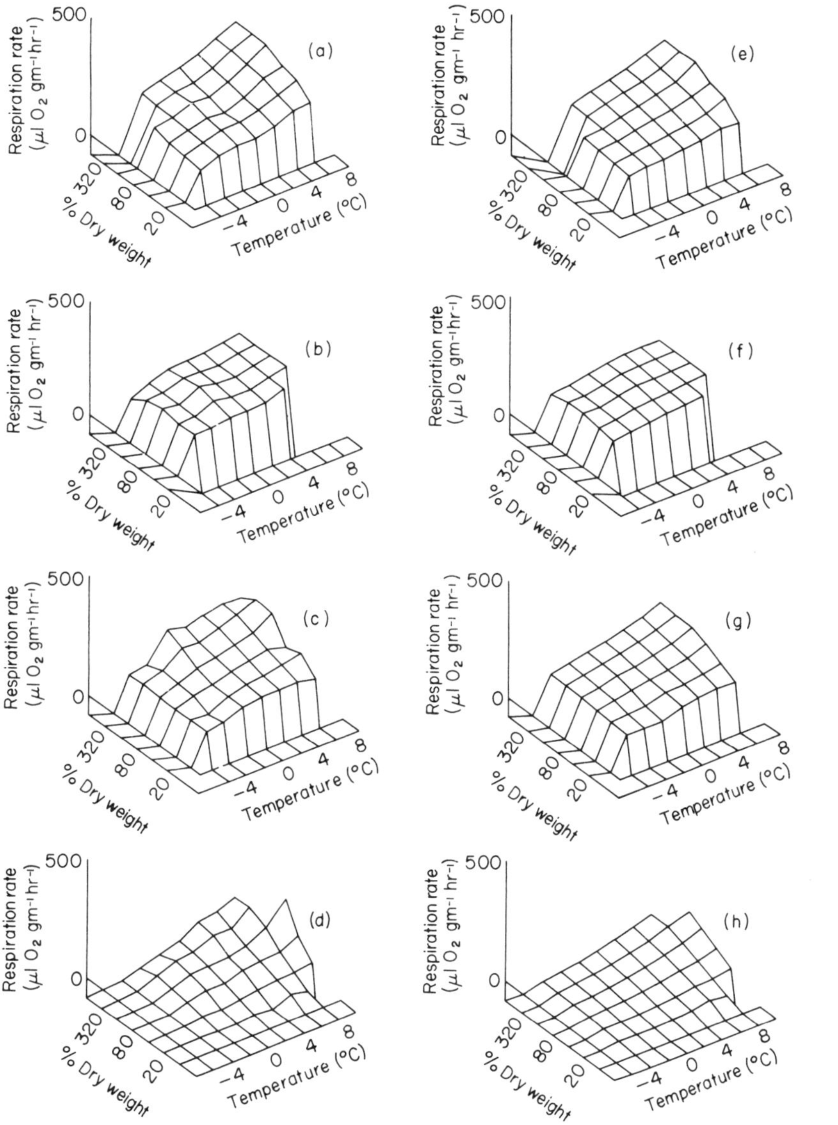

Figure 19.4. Measured (a–d) and simulated (e–h) rates of microbial respiration in standing dead of *Carex aquatilis* (a and e = 1 year old; b and f = 2 year old), and *Eriophorum angustifolium* (c and g) and total litter (d and h) at Barrow, Alaska (from Bunnell *et al.* in press, b).

Table 19.1. 'Goodness of Fit' of the GRESP Function to Data from Different Substrates. (Modified from Bunnell *et al.* in press, a.)

Substrate type	Number of Observations	'best' parameters a_1 (H_2O% dw)	a_2 (H_2O% dw)	a_3 (μlO_2 g^{-1} hr^{-1})	a_4 (Q_{10})	r^2	r_a^2
Eriophorum (new)	54	416	416	777	8·79	0·65	0·96
Eriophorum (1 year old)	79	74	12,709	184	2·56	0·64	0·72
Eriophorum (2 years old)	65	19	3×10^9	107	2·79	0·72	0·77*
Carex (2 years old)	45	213	822	296	2·18	0·63	0·71
Dryas (2 years old)	38	160	160	158	1·75	0·36	0·43†
Carex (1 year old)	96	59	13,300	204	3·33	0·79	0·86
Total Barrow litter	346	116	2,820	232	3·74	0·66	0·71

* An optimum solution had not been found.

† The age of this material was in doubt and the number of treatment replicates at each observation point was only two.

In essence it approximates analysis of variance techniques in a nonlinear system, and permits sampling error to be removed from the fit between model and data. Measures of the goodness of fit of the model to data from different substrates are shown in Table 19.1. Examples of respiration response surfaces generated by the model are compared with measured surfaces in Fig. 19.4.

From Table 19.1, it can be seen: (i) that the model accounts for 71–96% of the accountable variation in respiration data; (ii) the model fits the respiration patterns of a number of diverse substrates equally well; (iii) that the coefficients a_1 through a_4 change between substrates and with age. In particular, it can be seen that as *Eriophorum* material ages, a_1 values decrease while a_2 values increase. As diminution and physical changes occur in the substrate the range over which moisture is not limiting to respiration increases. Bunnell & Tait (1974) predicted this phenomenon stating 'the more

Table 19.2. Results of sensitivity analyses on the GRESP model for total litter.

	Coefficient values that reduce the goodness of fit by 1% of the best value						Maximum departure from best value		
	Analysis number					best	Minimum	Maximum	Max–Min
Coefficient	1	2	3	4	5	fit	as % best	as % best	as % best
a_1 (H_2O% dw)	82	214	79	212	213	116	−32%	+84%	116%
a_2 (H_2O% dw)	4,293	1,085	5,192	1,101	651	2,820	−77%	+84%	161%
a_3 (μl O_2/g/h)	195	345	198	344	389	232	−16%	+68%	84%
a_4 (Q_{10})	3·8	3·6	3·8	3·8	4·1	3·7	−3%	+11%	14%

physically reduced and pitted the substrate the smaller will be the "field capacity" and the larger the "retentive capacity"'; (iv) a_3, a measure of substrate quality, decreases with age of *Eriophorum* substrate. There is a near linear relationship between the best fit value of a_3 and the percent ethanol soluble fraction as determined by Flanagan & Veum (1974) and (v) a_4 is usually higher for younger substrates (8·79 for freshly dead *Eriophorum*) than for older ones (2·56 for one-year-old *Eriophorum*).

Sensitivity analyses employed in evaluating the GRESP model are described by Bunnell *et al.* (in press a). The results of sensitivity analysis (Table 19.2) indicate that microbial respiration in litter is much more sensitive to substrate and temperature variables than to moisture levels. Estimates of the sensitivity of the respiration rate to changes in the coefficients a_1 through a_4 decrease in the order a_4, a_3, a_1 and a_2, i.e. high moisture causing decreased O_2 availability has least influence on predicted respiration rates while temperature has the greatest influence.

These observations are companionable with the theoretical construction of the model and, together with the general goodness of fit of the model to data, indicate that the conceptual framework expressed by equation (19.7) is reasonably correct.

It is important to emphasize that the model is presently limited in several ways, particularly in its applicability to field sites where temperatures are often above 10°C. Further laboratory studies will increase its temperature range. The model does not consider anaerobic decomposition processes, does not account for influences of leaching and microfauna, and in general may underestimate total weight losses occurring during decomposition. The model's use is primarily one of estimating the proportion of weight loss attributable to loss of carbon via microbial respiration. For a specific substrate the closeness of this estimate to field measures of weight loss will reflect to some extent the magnitude of unaccounted microfaunal and leaching influences on decomposition. The model's prediction is substrate specific. Influences of variation in substrate quality on rates of decomposition are addressed by the first extension of GRESP, the model DCOMP.

It is a basic assumption of GRESP that the form of the respiration response is specific for a particular substrate. Acknowledging this assumption, the model may be employed to predict the respiration rate under any set of climatic circumstances. By employing the coefficients for Barrow total litter (Table 19.1), the temperature data for Barrow during 1971 (Weller *et al.* 1971) and moisture data of Flanagan & Veum (1974), a prediction of weight loss in total litter was made for the period 2 June through to 2 September 1971. The model predicted a weight loss of 13·4%. The prediction falls midway between independent measures in litter bag weight loss experiments of 9·9%±1·8% (Flanagan & Veum 1974) and 15% (Benoit *et al.* 1972). On the basis of this performance, it seems that the model can make realistic predictions for decomposition of a specific substrate during a season of known climatic conditions.

In this way interannual differences in weight loss can be related to climatic variation explicitly. Similarly the model allows examination of intraannual weight loss transients and inter- and intrasite variations. The influence of diurnal moisture and temperature patterns can be explored by the model. Bunnell & Tait (1974) have shown that there is a 12% increase in C loss from litter when computed using diurnal temperatures at two-hourly intervals as compared with average daily temperatures. Differences in diurnal climatic patterns may account for some between site variation in weight losses observed by Heal & French (1974). As expressed in equation (19.7) the model allows us to predict weight losses from a specific substrate (e.g. one-year-old litter or cellulose) placed in sites of different climate. Thus, we should be able to predict weight loss in Barrow litter placed, for example, in a tropical grassland. We could not predict weight losses in tropical grass litter unless we constructed respiration temperature/moisture surfaces for that litter.

Where substrate quality transients are steep during decomposition, due to fractions being removed by water flows, microfaunal activities, or to 'soft' substrate decomposition before recalcitrant substrate removal, then the GRESP model cannot be used to predict weight losses.

(ii) *The DCOMP Model*

A second model DCOMP has been designed to assess decomposition rates under the influence of changing substrate quality. The DCOMP model assumes that a decomposing substrate contains a number of basic chemical constituents each having a specific temperature–moisture–chemically determined respiration rate. Basically this concept is an extention of Minderman's (1968) hypothesis, first to include temperature and moisture influences and second, to relate a microbial process, respiration, to weight loss. We assume that the total respiration rate of a substrate containing several chemical groups will be the sum of the respiration of microbes using the different groups. Bunnell *et al.* (in press, b) have discussed several formulations relating substrate chemistry to microbial respiration and weight loss. The assumptions of the version discussed here are as follows: (i) each chemical category will have a temperature–moisture–respiration response surface of specific amplitude; (ii) each chemical response surface will have a similar shape and (iii) any two surfaces will differ in amplitude by a constant chemical-specific factor. The basic response surface used in DCOMP is the best fit of the GRESP model to Barrow total litter. It could just as well be the GRESP fit to cellulose placed in the field at Barrow. It should differ from a specific substrate's temperature–moisture–respiration response surface by a specific and computable constant.

The regression models used in calculating the respiration rate for specific chemical categories over varying temperature and moisture regimes are:

$$RR=\sum_{i=1}^{n} r_i x_i \tag{19.8}$$

where RR = total respiration,

r_i = respiration rate of compound i, and

x_i = amount of compound i present.

To include the influence of temperature (T) and moisture (M) we write

$$RR(T,M)=\sum_{i=1}^{n} r_i(T,M)x_i \tag{19.9}$$

To remove or account for the influence of T and M we write

$$RR(T,M)=\sum_{i=1}^{n} b_i GRESP(T,M)x_i \qquad (19.10)$$

In equation (19.10) the independent variables T and M are expressed as a surface such as those in Fig. 19.4, b=the ratio between the GRESP response surface for total litter and the response surface for compound i (i.e. a chemical specific scaler for the GRESP surface) and GRESP (T,M)= the underlying shape of the substrate specific response surface to T and M. For a substrate containing two different chemical groups, i.e. a 'soft' and a recalcitrant fraction we write:

$$RR(T,M)=b_i GRESP(T,M)x_i+b_j\ GRESP(T,M)x_j \qquad (19.11)$$

where RR(T,M)=total predicted respiration in $\mu lCO_2\ hr^{-1}$,

b_i=the ratio between the amplitude of the GRESP surface for total litter and the amplitude of the respiration response surface for the ethanol soluble compounds,

b_j=the ratio between the amplitude of the GRESP response surface for total litter and the amplitude of the respiration response surface for the recalcitrant compounds,

GRESP(T,M)=the response surface in $\mu lCO_2\ g^{-1}\ hr^{-1}$ of respiration versus T&M for litter from Barrow (see equation (19.7)),

x_i=amount of ethanol soluble compounds i, and

x_j=amount of non-ethanol soluble compounds j

Regression analyses employing field data are summarized in Table 19.3. The regressions are of the form

$$\ln\ (\%\ \text{compound remaining})=b_0+b_1\ (\text{age in years}) \qquad (19.12)$$

The standing dead vegetation used in the analyses are *E. angustifolium* and *C. aquatilis*. Degrees of freedom are small because only one measure of chemical composition was associated with several measures of weight loss (Bunnell *et al.* in press, b). Nevertheless, slopes of all regressions depart significantly from zero ($\alpha = 0{\cdot}05$) which supports Minderman's (1968) hypothesis of exponential decay for specific chemical constituents. The different slopes or decay rates for cellulose placed in different portions of the system reveal the effects of different microclimatic regimes.

Results of analyses employing equation (19.10) are discussed in some detail by Bunnell *et al.* in press, b. Whereas equation (19.12) is simply a least squares fit to field measures, equation (19.10) posits a specific relationship between microbial respiration and abiotic variables, and sums the response of several component substrates. When the two broad chemical groups (equation (19.11)) were grouped by temperature and moisture classes significant relations ($\alpha=0{\cdot}05$) between respiration rate and chemical composition were observed in only one third of the classes.

These findings suggest that there is no simple underlying linear relationship between respiration rate and substrate composition. Broad relationships are, however, apparent. Regression analysis using equation (19.11) with 716 measured respiration rates and nine substrate types yielded coefficients of 3·8 and 0·76 for the ethanol soluble and non-ethanol soluble terms respectively. The associated t values were 5·9 and 15·2. With 7 d.f. we are confident that slopes of the regression differ from zero; $\alpha=0{\cdot}001$.

The model or equation (19.11) estimates the relative respiration rates to be 5:1 for ethanol soluble relative to the more recalcitrant fraction. This ratio approximates that observed for decomposition rates estimated from the standing dead data (0·69/0·12=5·75:1; see Table 19.3).

Table 19.3. Regression coefficients of ln (percent remaining) versus age in years. ln (percent remaining)$=b_0+b_1$ (age). (From Bunnell *et al.* in press, b.)

Substrate	Environment	b_0	b_1	r	Degrees of freedom	α
Cellulose	Standing dead	4·58	−0·015	−0·98	2	0·05
Cellulose	Litter layer	4·58	−0·036	−0·97	2	0·05
Cellulose	Soil	4·60*	−0·014	−0·997	1	0·05
Ethanol soluble	Litter layer	4·12	−0·69	−0·90	8	0·01
Non-ethanol soluble	Litter layer	4·56	−0·12	−0·68	8	0·05

* ln (100%)=4·605.

Assuming that both compound groups contain 45% carbon the respiration rates may be converted to weight losses by separately multiplying the two additive terms of equation (19.11) by

$$\frac{12\,\mu\text{gm C}}{22{\cdot}4\,\mu\text{l}CO_2}\times\frac{1\,\mu\text{g substrate}}{0{\cdot}45\,\mu\text{gC}}$$

and adding the two products.

Using field temperature and moisture data from Barrow the annual decomposition rates of the two fractions were computed to be 0·66 and 0·13 g g^{-1} which is very close to the decomposition rates observed in the field (0·69 and 0·12 g g^{-1}; Table 19.3). Table 19.4 shows measured and simulated values for moisture–temperature–chemical specific rates of weight loss from *E. angustifolium* for the field season of 1973. Because it stimulates respiration and thus decomposition rates for specific chemical components, the DCOMP model also predicts chemical composition (Table 19.4).

Table 19.4. Measured and simulated values for moisture–temperature–chemical specific rates of weight loss from *Eriophorum angustifolium*. Measured values use changes in identified age classes. Simulated values use abiotic data from 1973 to project loss rates (from Bunnell *et al.* in press, b).

	Measured %	Simulated %
% loss per year:		
ethanol soluble	49	48
non-ethanol soluble	11	12
Chemical composition of the substrate after one year:		
ethanol soluble	7	10
non-ethanol soluble	93	90
Total weight loss	27	31

The results summarized in Tables 19.3 and 19.4 document that there is reasonably close agreement between simulated weight losses and substrate composition and the measured values of these. Presently we can assume that the biological principals of decomposition are realistically formulated in the model structure. Also, it appears that chemical composition and, therefore, rate of decomposition loss for specific substrate components can be predicted providing temperature and moisture influences on microbial respiration are incorporated in the model. Consideration of the rates of decomposition losses for specific substrates need not be restricted to carbon compounds as we have done here. There are specific N and P requirements for the decomposition of ethanol soluble and the more recalcitrant substrates. Relationships between C losses and N and P losses formulated in a GRESP or DCOMP model would almost certainly lead to useful insights regarding mineral release from decomposing (respiring) tissues. Consideration of N and P losses accompanying respiration and the possibility of predicting N and P losses and content of substrates from respiration data form a rationale for continuing to develop GRESP and DCOMP as predictive tools for understanding the role and significance of decomposition in ecosystems.

Summary

GRESP and DCOMP have examined the climatic and substrate influences on decomposition rates and represent a more synecological approach than that of the chemostat model discussed earlier. Still, the range and rate of decomposition determinable or predictable using either method are very similar. By two dissimilar routes we have arrived at a similar assessment of decomposition rates for plant leaf tissue at Barrow, Alaska. The simulation models allow exploration of the role of major influencing factors on decomposition while providing a means of examining temporal and spatial variation in the rates of decomposition. Additionally these models allow us to predict overall rates of organic matter decomposition which if assessed in relation to productivity measures provide information on organic accumulation rates. The models ignore, amongst other variables, differences in microbial growth potentials. Rather, they sum all microbial growth potentials and incorporate the universal micro-organisms responding only to temperature, moisture and substrate variables. They thus represent a restatement of Winogradsky (1887) 'Everything is everywhere—the environment selects'.

Conversely, the model of Marr *et al.* (1963) examines microbial growth kinetics in relation to substrate availability ignoring the climatic and substrate quality influences on microbial growth potential. Whether we use one approach or the other we arrive at the same assessment of decomposition rates in the standing dead leaf remains. When we use both approaches we arrive at a paradoxical situation where we are unable to document what are the major determinants of decomposition rates—microbial population decomposition potentials or the influence upon these of environmental factors. It would appear that both constraints apply, particularly at the extremities of decomposition, but that the function decomposition proceeds virtually unregulated in rate across a wide spectrum of micro-organism and climatic variables, at least in tundra.

We may draw support for this statement by considering the breadth of the respiration versus moisture curve particularly at low temperatures (Fig. 19.2a). Also the influences of high and low temperatures are diminished by low and high moisture levels (Fig. 19.2a), and low and high moisture influences are counteracted respectively by high and low temperature levels (Flanagan & Veum 1974). To these observations we may add that micro-organisms in the tundra fall into three basic categories, i.e. psychrophiles, cold tolerant mesophiles and thermo tolerant psychrophiles (Flanagan *et al.* in prep.), and that individual micro-organisms contain different enzymes functioning optimally on specific substrates throughout a temperature range from −8°C to +35°C (Flanagan & Scarborough 1974). Acknowledging these observations it is clear that the range for micro-organism activity is indeed wide and that the influences of climatic and substrate variables upon it are masked because subtle enzymes activities grading to optima within and between micro-organisms occur throughout broad temperature and moisture ranges. The result of such activities and relationships produces a very broad 'optimal' mode for decomposition which lends itself well to prediction by relatively simple modelling techniques.

Acknowledgements

The U.S. Tundra Biome through the National Science Foundation Grant GV 29342 to the University of Alaska supported the research and model development described here. The Naval Arctic Research Laboratory at Pt. Barrow, Alaska, supported the field work. Many individuals, too numerous to mention here, contributed to the research and to development of the model but in particular, we wish to thank Mr. D.E.N. Tait, whose help was crucial particularly in the mathematical formulations of the biological principals.

References

BABIUK L.A. & PAUL E.A. (1970) The use of flourescein isothiocyanate in the determination of the bacterial biomass of grassland soil. *Can. J. Microbiol.* **16**, 57–62.

BENOIT B.E., CAMPBELL W.B. & HARRIS R.W. (1972) Decomposition of organic matter in the wet meadow tundra, Barrow; a revised word model. In: *Proceedings 1972 U.S. Tundra Biome Symposium, Lake Wilderness, Washington*, pp. 111–15. U.S. Tundra Biome.

BUNNELL F.L. (1973) Decomposition: models and the real world. In: *Modern Methods in the Study of Microbial Ecology*, ed. T. Rosswall, pp. 407–15. Bull. from the Ecol. Res. Comm. NFR. No. 17, Stockholm.

BUNNELL F.L. & TAIT D.E.N. (1974) Mathematical simulation models of decomposition processes. In: *Soil Organisms and Decomposition in Tundra*, eds. A.J. Holding, O.W. Heal, S.F. MacLean Jr. & P.W. Flanagan, pp. 207–25. Tundra Biome Steering Committee, Stockholm.

BUNNELL F.L. & SCOULLAR K.A. (1975) ABISKO II—a computer simulation model of carbon flux in tundra ecosystems. In: *Structure and Function of Tundra Ecosystems*, eds. T. Rosswall & O.W. Heal, pp. 425–48. Ecol. Bull. 20. Swedish Natural Sci. Res. Council, Stockholm.

BUNNELL F.L., TAIT D.E.N., FLANAGAN P.W. & VAN CLEVE K. (in press, a) Microbial respiration and substrate weight loss. I. A general model of the influence of abiotic variables. Soil Biol Biochem.

BUNNELL F.L., TAIT D.E.N. & FLANAGAN P.W. (in press, b) Microbial respiration and substrate weight loss. II. A model of the influence of chemical composition. Soil Biol Biochem.

BUNNELL F.L., MACLEAN S.F. JR. & BROWN J. (1975c) Barrow, Alaska. In: *Structure and Function of Tundra Ecosystems*, eds. T. Rosswall & O.W. Heal, pp. 73–124. Ecol. Bull. 20. Swedish Natural Sci. Res. Council, Stockholm.

COLOUMBE H.N. & BROWN J. (1970) The synthesis and modeling of the Barrow, Alaska ecosystem. In: *Proc. of the Conf. on Productivity and Conservation in Northern Circumpolar Lands, Oct. 15–17, 1969*. Morges, Switzerland, I.U.C.N.

FLANAGAN P.W. & SCARBOROUGH A. (1974) Physiological groups of decomposer fungi in tundra plant remains. In: *Soil organisms and decomposition in tundra*, eds. A.J. Holding, O.W. Heal, S.F. MacLean Jr. & P.W. Flanagan, pp. 159–81. Tundra Biome Steering Committee, Stockholm.

FLANAGAN P.W. & VEUM A.K. (1974) Relationships between respiration, weight loss, temperature and moisture in organic residues in tundra. In: *Soil Organisms and Decomposition in Tundra*, eds. A.J. Holding, O.W. Heal, S.F. MacLean Jr. & P.W. Flanagan, pp. 249–77. Tundra Biome Steering Committee, Stockholm.

FLANAGAN P.W., BUNNELL F.L. & BENOIT R.E. (In prep.) Microbial activities and decomposition. Chapt. 17. In: *An Arctic Ecosystem: The Coastal Tundra of Northern Alaska*, eds J. Brown, F.L. Bunnell, P.C. Miller, S.F. MacLean & L.L. Tieszen. Dowden, Hutchinson and Ross, Inc., Stroudsburg, Pa.

GRAY T.R.G. & WILLIAMS S.T. (1971) Microbial productivity in soil. *Symp. Soc. Gen. Microbiol.* **21**, 255–86.

HEAL O.W. & FRENCH D.D. (1974) Decomposition of organic matter in tundra. In: *Soil Organisms and Decomposition in Tundra*, eds. A.J. Holding, O.W. Heal, S.F. MacLean Jr. & P.W. Flanagan, pp. 279–309. Tundra Biome Steering Committee, Stockholm.

HEAL O.W. & MACLEAN JR. (1974) Comparative production in ecosystems-secondary productivity. *Proc. 1st International Cong. of Ecol.*, Sept. 18–24, 1974, p. 37. Centre for Agric. Publishing and Documentation, Wageningen.

JENNY H., GESSEL S.P. & BINGHAM F.T. (1949) Comparative study of decomposition rates of organic matter in temperate and tropical regions. *Soil Science* **2**, 267–72.

JONES H.E. & GORE A.J.P. (In press) A simulation of production and decay in blanket bog. In: *The Ecology of Some British Moors and Montane Grasslands*, eds. O.W. Heal & D.F. Perkins. Springer-Verlag, Berlin.

JONES P.C.T. & MOLLINSON P.E. (1948) A technique for quantitative estimation of soil microorganisms. *J. Gen. Microbiol.* **2**, 54–69.

MARR A.G., NELSON E.H. & CLARK D.J. (1963) The maintenance requirements of *Escherichia coli*. *Ann. N.Y. Acad. Sci.* **102**, 536–48.

MINDERMAN G. (1968) Addition, decomposition and accumulation of organic matter in forests. *J. Ecol.* **56**, 355–62.

OLSON J.S. (1963) Energy storage and the balance of producers and decomposers in ecological systems. *Ecology* **44**, 322–31.

PARINKINA O.M. (1974) Bacterial production in tundra soils. In: *Soil Organisms and Decomposition in Tundra*, eds. A.J. Holding, O.W. Heal, S.F. MacLean Jr. & P.W. Flanagan, pp. 65–77. Tundra Biome Steering Committee, Stockholm.

PITELKA F.A. (1973) Cyclic pattern in lemming populations near Barrow, Alaska. In: *Alaskan Arctic Tundra*, ed. M.E. Britton, pp. 199–215. Arctic Inst. of N. Am., Tech. Paper No. 25.

WELLER G., CUBLEY S., PARKER S., TRABANT D. & BENSON C. (1972) The tundra microclimate during snowmelt at Barrow, Alaska. *Arctic* **25**, 291–300.

WINOGRADSKY S. (1887) Bot. Zeit. **45**, 489. (Original not seen.)

Systematic Index

Subject Index